MOLECULAR WAVE FUNCTIONS
AND PROPERTIES

Molecular Wave Functions and Properties:

TABULATED FROM SCF CALCULATIONS
IN A GAUSSIAN BASIS SET

LAWRENCE C. SNYDER

Bell Telephone Laboratories, Incorporated
Murray Hill, New Jersey

HAROLD BASCH

Scientific Research Staff
Ford Motor Company
Dearborn, Michigan

with extensive help from
ZELDA R. WASSERMAN AND MARY E. BALDACCHINO-DOLAN
Bell Telephone Laboratories, Incorporated
Murray Hill, New Jersey

A WILEY-INTERSCIENCE PUBLICATION

JOHN WILEY & SONS, New York · London · Sydney · Toronto

Library of Congress Cataloging in Publication Data

Snyder, Lawrence C.
    Molecular wave functions and properties.

    Bibliography: p.
    1. Wave function—Tables, etc. 2. Molecules—
Tables, etc.   I. Basch, Harold, joint author.
II. Title
QD462.S68   541'.28   72–5334

ISBN 0–471–81012–6

Printed in the United States of America

10 9 8 7 6 5 4 3 2 1

PREFACE

The purpose of this book is to make a collection of SCF molecular wave functions and derived properties, computed by the authors, available to research scientists, teachers, and students. The presentation of these largely unpublished results together in one book in tabular format, convenient for comparison and interpretation, should provide new insights into the electronic structure and properties of molecules when viewed together by the wider scientific community.

The wave functions of this compendium have been computed in a uniform basis set of contracted Gaussian functions, which we have called the double-zeta basis set. This modest but flexible basis set contains two contracted functions per atomic orbital. Our selection of molecules reflects our recent research interests in thermochemistry, and optical and electron spectroscopy. Included are over fifty molecules built from hydrogen and first-row atoms. The experimental geometries have been assumed when they are known.

The tabular contents of the book include the molecular orbitals, eigenvalues, and several energy expectation values that were generated in the SCF calculations. The first to fourth moments of the charge distribution have been added, together with several properties of the charge distribution at nuclear centers: the potential, charge density, electric field, and electric field gradient. The electronic charge distribution is presented as Mulliken atomic and overlap charges. For many molecules the overlap matrix is given.

We can foresee several ways in which these tables may be used. They will provide a rather large body of data on molecular wave functions and properties within which comparisons can be made and hypotheses tested. In addition, future calculations in this basis will be of enhanced value because of this reference data base. The tabulated wave functions can also serve as a starting point for more extensive calculations. Finally, these wave functions will permit students to make explicit numerical tests of ideas developed in their studies.

In preparing the tables, energy expectation values and eigenvalues were taken from the original SCF computer output. The wave functions which had been saved on cards, were employed with POLYATOM (Version 2) to compute the additional properties presented in the tables. All results for all molecules were collected on a single tape. The tables were prepared from this tape using an Information International FR80 Microfilm Recorder. Thus the tables should be largely free of typographical errors.

The authors wish to thank a number of colleagues at Bell Laboratories who have helped them in the creation of this book.

We are deeply indebted to Mrs. Zelda Wasserman for performing most of the programming and computations required to transfer the data from its original form into the tables presented here. We also acknowledge the contributions of Mrs. Mary Baldacchino-Dolan who adapted graphics programs to draw the molecular pictures.

We thank Dr. Melvin Robin and Norman Kuebler for generously supplying wave functions and expectation values generated in a number of their recent studies; these include the molecules N_2F_2, F_2CO, and N_2H_2.

For reading the manuscript and suggesting a number of improvements, we thank Dr. Robert Ditchfield, Dr. Melvin Robin, Dr. John Tully, and Dr. Edel Wasserman.

We thank Lynne Drake for her skill and patience in typing the text for photographic reproduction.

It is hoped that the generous support of this work by both Bell Laboratories and the Ford Motor Company will be justified by the benefits that will accrue to the scientific community through the general availability of this work.

Lawrence C. Snyder
Harold Basch

CONTENTS

MOLECULAR WAVE FUNCTIONS
AND PROPERTIES

Chapter 1.   INTRODUCTION

This book presents self-consistent field energies, wave functions, and a selected number of one-electron properties that have been computed for atoms and molecules in the course of several research studies.

Over a period of years, the authors have employed the same Gaussian basis set to compute wave functions for over fifty molecules. These wave functions and related properties have been found useful in studies of thermochemistry, electron spectroscopy, optical spectroscopy, NMR, NQR, and general molecular electronic structure correlations. It is clear from these studies that there are many advantages to having electronic wave functions for many molecules in the same basis set. The principal advantage is that errors or inadequacies of the wave functions tend to remain constant from molecule to molecule and thus tend to cancel in predictions of comparative properties. In publications that have reported on these theoretical studies, it has not proved convenient to publish the molecular wave functions and the many derived quantities that have been calculated from them. This compendium of molecular wave functions and properties is being published to make those results available in a complete and readable form to the scientific community.

Tables of atomic Hartree-Fock wave functions[1] and derived properties[2] have been published for several years. Properties of molecules derived from Huckel molecular orbitals have also received wide circulation.[3] A collection of near Hartree-Fock wave functions for diatomic and linear molecules has been published.[4] A compendium[5] has recently been made tabulating energy and other quantities from the available molecular calculations, but except for diatomics and linear molecules, the supply has been meager and in basis sets of variable quality. The tabulation of wave functions and properties presented here should substantially extend this data base of numerical results.

Because we have chosen molecules to serve diverse research needs and interests, the selection of molecules given in this presentation may appear to be arbitrary and haphazard. The molecules included contain only hydrogen and first-row atoms. They range from small linear molecules, such as hydrogen, to the larger nonlinear tetrafluoroethylene and bicyclobutane. They vary in reactivity from the

inert nitrogen molecule to reactive ones such as ozone and difluorocarbene. Our interests have been broad and we assume them to be in
common with those of many other chemists.

The theoretical framework for our calculations is provided by
Roothaan's basis set expansion method for solving the Hartree-Fock self-
consistent field (SCF) equations.[6] Boys emphasized the computational
advantages of Gaussian basis functions in 1950,[7] almost a decade before
the development of computer programs to employ them in a priori calculations implementing Roothaan's formulation. The first of these programs
to become widely available was the POLYATOM system written at MIT.[8]
Early applications by J. W. Moskowitz and others to several large molecules showed that the electronic structure of nonlinear polyatomic molecules could be given a quantitative description by a priori methods.[9]

It has been proven convenient for interpretation and time-saving
in computations to adopt fixed linear combinations of Gaussian functions,
commonly called contracted functions, as a basis set. The contracted
functions were incorporated into the POLYATOM system as part of a
general effort to develop, extend, and improve it.[10] The rewritten
programs, with which the calculations in this book were made, were
largely finished by the beginning of 1967. They are called POLYATOM
(Version 2) and are available through the Quantum Chemistry Program
Exchange.[11]

In early 1967 there were only two high-quality Gaussian basis sets
available in the literature for atoms of the first row of the periodic
table. The first was the exponent-optimized uncontracted set of
Huzinaga.[12] The second was Whitten's set of exponent- and coefficient-
optimized contracted gaussian lobe functions,[13] where p-type atomic
orbitals were simulated by the appropriate placement of two Gaussians
off the nuclear centers. By combining the contracted s-type functions
of Whittens set with the p-type functions of Huzinagas smaller set, the
advantage of having an energy-optimized contracted set was realized
without extensive atom SCF calculations.

This basis set provides two contracted basis functions for each
atomic orbital in the molecule. The contracted functions correspond
to an inner and outer part for each atomic orbital. This division
permits charge distribution at an atom to expand or contract in
response to its Coulombic environment. We shall see that this basis
turns out to be about as good as an exponent-optimized pair of Slater
orbitals for each atomic orbital; for this reason it has been called
the double-zeta (DZ) basis.[14]

A large body of computational experience has shown that, with an
adequate basis set, many properties that depend mainly on the electron
density in the molecular ground state may be fairly accurately computed
from a high-quality SCF wave function.[15] A modest but flexible basis
set is adequate for many chemical uses. Many properties of a molecule
can be characterized by expectation values of its ground state wave
function. In our SCF calculations we have obtained the molecular

orbitals, their eigenvalues, and the total energy. We have also computed expectation values for several parts of the Hamiltonian: the kinetic energy, the potential energy of interaction of the electrons with the nuclei, and the electron-electron repulsion energy. We have computed all moments of the charge distribution up to the fourth relative to the center of mass. These moments exhibit the size and shape of the charge distribution. We have also computed a selection of expectation values for the several nuclear centers of each molecule; these include the potential, charge density, components of the electric field, and the gradient. The atomic and overlap charges defined by Mulliken were computed.[16]

To make this body of numerical data available to ourselves and the scientific community, we have prepared the set of tables of molecular wave functions and properties that are presented at the end of this book. All of the tables have been generated on microfilm by a computer which used other computer output, in the form of tapes and cards from the SCF calculations, as its basic data. The only properties transcribed to the tapes by hand were the eigenvalues and energy expectation values, which were carefully copied from tabulator paper output in some cases. Thus we expect the tables to be almost completely free of typographical errors. We note that the molecular orbital coefficients of Trans-Butadiene and Acrolein were also transcribed from tabulator paper output.

Many applications of these calculated wave functions have been made to elucidate molecular ground state properties. These include thermo-chemistry, nuclear magnetic resonance (NMR), nuclear quadrupole reso-nance (NQR), and molecular structure correlations. Because the occupied molecular orbitals resulting from an SCF calculation often give a quali-tatively correct description of the hole produced by ionization or exci-tation, applications have also been made in photoelectron spectroscopy, ESCA, and optical spectroscopy. Other uses have been made and we hope there will be many more that the authors have not foreseen.

Chapter 2.   BASIC THEORY

Quantum theory provides the basic framework of ideas for understanding the electronic structure and properties of molecules.[17] Solution of the Schroedinger equation would provide molecular electronic wave functions Ψ and characteristic energy values E:

$$H\Psi = E\Psi \tag{1}$$

Here the wave function Ψ is a function of the position and spin coordinates of all nuclei and electrons of a molecule. The position coordinates can be taken to be Cartesian.

For many chemical purposes, it is adequate to employ a nonrelativistic Hamiltonian H which neglects all couplings of electron and nuclear spin to their spin and orbital motions.

We adopt a nonrelativistic Hamiltonian H written in atomic units as

$$H = -\frac{1}{2} \sum_i \nabla^2(i) + \sum_{i>j} \frac{1}{r_{ij}} - \sum_\mu \sum_i \frac{Z_\mu}{r_{\mu i}}$$

$$- \frac{1}{2} \sum_\mu \frac{1}{m_\mu} \nabla^2(\mu) + \sum_{\mu>\nu} \frac{Z_\mu Z_\nu}{r_{\mu\nu}} \tag{2}$$

where the indices i and j run over all electrons, and μ and ν over all nuclei of the molecule.

The atomic units of charge, mass, and action are e, m, and h, respectively, m being the mass of the electron. The atomic unit of length is the Bohr radius $a_0 = 0.529167 \times 10^{-8}$ cm; the unit of energy is the Hartree, $e^2/a_0 = 27.2107$ eV.

Experiment has shown that the only solutions of Eq. (1) corresponding to observed states are antisymmetric with respect to interchange of the spin and spatial coordinates of any pair of identical fermions (particles having half-integral spin), and symmetric with respect to exchange of identical bosons (particles with integral spin).

The energy eigenvalues of Eq. (1) form a continuous set for a free molecule and would correspond to the possible translation, vibration, rotation, and electronic stationary states of the molecule. Transformation of the coordinates of Eq. (1) to center of mass and internal coordinates makes the equation separable for the translational and internal degrees of freedom and gives a simple expression for the eigenvalues and wave functions for translational motion.[17]

Because nuclei are much more massive than the electrons and because the frequency of their motions is much lower than that of the electrons in molecules not having low-lying excited electronic states, it is possible to make the Born-Oppenheimer approximation and assume that the electrons move in a potential provided by fixed nuclei.[18] In this approximation the Schroedinger equation for the electrons is

$$H_t \Psi_e = E_t \Psi_e \tag{3}$$

and the Hamiltonian is

$$H_t = -\tfrac{1}{2} \nabla^2(i) - \sum_\mu \sum_i \frac{Z_\mu}{r_{i\mu}} + \sum_{i>j} \frac{1}{r_{ij}} + \sum_{\mu>\nu} \frac{Z_\mu Z_\nu}{r_{\mu\nu}} \tag{4}$$

This Hamiltonian neglects certain kinetic energy terms (called momentum-coupling terms) that arise in transforming the Hamiltonian from a laboratory frame to an internal coordinate system in order to separate off the translational motion of the center of mass.[19] Now E_t is a function of the nuclear positions for each electronic state and is taken to provide the potential in which the nuclei move. Solution of a second Schroedinger equation for the motion of the nuclei in this potential provides rotational and vibrational energies for the molecule.

It is the solution of Eq. (3), the Schroedinger equation for the electrons moving in the potential of the fixed nuclei of a molecule, which is usually of central concern to chemists. A great deal of theoretical work has been directed toward that goal with the result that fairly accurate eigenvalues E_t and wave functions Ψ_e have been obtained only for systems of up to four electrons; typical examples being H, He, H_2, Be, He_2, and LiH^5.

The principal reason that more progress has not been made toward solutions for the larger molecules of interest to chemists is that the motion of electrons is correlated. In other words, the motion of one electron depends on the location of a second electron. Thus the accurate wave function cannot be represented as a simple product of functions of individual electron coordinates.

Although the motion of the electrons is indeed correlated, computational experience has shown that their motion is mainly determined by their interaction with the nuclei and their mean interaction with each other. This is reflected in the fact that atomic orbitals remain an appropriate language for the discussion of molecular electronic structure.[17]

It can be shown that an approximate wave function for the ground state of a molecule has an expectation value $\bar{E}_t$ for its total energy which has E_t as its lower bound. Here $\bar{E}_t$ is given by

$$\bar{E}_t = \frac{\int \Psi^* H_t \, \Psi \, d\tau}{\int \Psi^* \Psi \, d\tau} \tag{5}$$

This provides a simple means to approximate the true Ψ_e by the adoption of a trial function Ψ depending on some set of parameters; the "best" such trial function being that which minimizes $\bar{E}_t$ with respect to variations of the parameters.

Computational experience has shown that a trial function of determinantal form is often valuable in approximating the wave function of the ground and electronic states of atoms and molecules.[17] It is a simple form of trial function that is properly antisymmetric with respect to interchange of electrons.

$$\Psi = \frac{1}{\sqrt{2n!}} \, |\phi_1(1)\bar{\phi}_1(2)\phi_2(3)\bar{\phi}_2(4)\ldots\phi_n(2n-1)\bar{\phi}_n(2n)| \tag{6}$$

Here we have written a determinantal wave function for a system of 2n electrons that have been assigned in pairs to orbitals ϕ_i, which are one-electron functions of the spatial coordinates of the assigned electron. Each electron has two spin states which are usually denoted by α and β: assignment of an electron to the α state is denoted by an unbarred ϕ, and assignment to the β state by $\bar{\phi}$. An electronic state well described by a trial function that assigns pairs of electrons to each ϕ is called a closed-shell state.

A convenient method to prepare the orbitals ϕ is to expand them in some set of m basis functions χ:

$$\phi_i = \sum_{r=1}^{m} c_{ir}\chi_r \tag{7}$$

For a particular basis set χ, the set of functions ϕ that minimize $\bar{E}_t$ are called SCF orbitals, and the minimum value is E_{SCF}. A complete basis set χ would give the lowest possible value of E_t for a wave function of determinantal form. These would be Hartree-Fock orbitals and the corresponding minimum energy expectation value is E_{HF}. When applied to atoms, the SCF calculation produces a set of ϕ which are called atomic orbitals. When a system contains several nuclei, the functions ϕ are called molecular orbitals.

It has proved convenient to define the difference between E_t obtained by solving the nonrelativistic Schroedinger Eq. (3) and E_{HF} to be the correlation energy[20]:

$$E_t = E_{HF} + E_{corr} \tag{8}$$

This definition and nomenclature is reasonable because, except for the correlation introduced by the Pauli Exclusion Principle, the electrons move independently in Ψ_{HF}.

In determinantal orbital theory the electronic state of an atom or molecule is designated by its electronic configuration--a specific assignment of electrons to one-electron functions. Within the framework of the Pauli Principle, only one electron of each spin state may be assigned to each atomic or molecular orbital.[21]

The orbitals are usually assumed to transform as an irreducible representation of the system's geometric point group as defined by the assumed fixed positions of the nuclei: spherical symmetry for atoms and lower-order space group symmetries for molecules. Thus, the assignment of an insufficient number of electrons to fill all the orbitals of a given degenerate set can give rise to multiplet structure for a given electronic configuration.[17] Here we discuss only closed-shell systems where all members of a space or spin-degenerate set are completely filled with electrons to give a nondegenerate totally symmetrical state. Such an electronic configuration can be written as a normalized single-determinant wave function as, for 2n electrons, was done in Eq. (6).

For atoms, the above assumptions about the orbital transformation properties lead to a factorization of the ϕ's into radial and angular parts in a spherical polar coordinate system. The angular parts are the familiar spherical harmonics. In general, this factorization does not carry over into the molecular case and the complete functional form of the orbital has to be found.[22]

Using the approximate wave function Ψ, the expectation value of the electronic energy $\bar{E}_e$ is given by

$$\bar{E}_e = \bar{E}_t - \sum_{\mu > \nu} \frac{Z_\mu Z_\nu}{R_{\mu\nu}} \tag{9}$$

Evaluated explicitly using Eqs. (4) and (6), $\bar{E}_e$ becomes

$$\bar{E}_e = \sum_{i=1}^{n} \{2H_{ii} + \sum_{j=1}^{n} (2J_{ij} - K_{ij})\} \tag{10}$$

where

$$H_{ii} = <\phi_i(1)| - \tfrac{1}{2} \nabla^2(1) - \sum_\mu \frac{Z_\mu}{r_{1\mu}} |\phi_i(1)> \tag{11}$$

$$J_{ij} = <\phi_i(1)|<\phi_j(2)|\frac{1}{r_{12}} |\phi_j(2)>|\phi_i(1)> \tag{12}$$

$$K_{ij} = <\phi_i(1)|<\phi_j(2)|\frac{1}{r_{12}} |\phi_i(2)>|\phi_j(1)> \tag{13}$$

and the labels i,j now refer to orbitals rather than electrons. J_{ij} and K_{ij} are the familiar coulomb and exchange integrals. A more detailed discussion of the SCF method is given in the next chapter.

Chapter 3.   THE SCF METHOD

The problem of finding self-consistent field orbitals is now one of determining the best possible ϕ_i by the criterion of the variational principle, subject to the orthonormality requirements,

$$<\phi_i|\phi_j> = \delta_{ij} \tag{14}$$

This is accomplished by applying the variational principle to the functional

$$\bar{E}_e - \frac{1}{2} \sum_{i>j} \lambda_{ij} [<\phi_i|\phi_j> + <\phi_j|\phi_i>] \tag{15}$$

calling upon the method of Lagrange,[19] where λ_{ij} are the undetermined Lagrangian multipliers. Thus the $\frac{1}{2}n(n+1)$ constraining equations in (14), which prevent the unrestricted variation of the ϕ_i to achieve an absolute minimum in $\bar{E}_e$, are matched by an equal number of undetermined constants, effectively removing the constraints. The general Hartree-Fock equations for closed-shell electronic configurations are thus obtained.

$$F|\phi_i(1)> = [H + \sum_{j=1}^{n} (2J_j - K_j)]|\phi_i(1)> = \sum_{j=1}^{n} \lambda_{ij}|\phi_j(1)> \tag{16}$$

Here it is convenient to define the operators H, J_j, and K_j as follows:

$$H|\phi_i(1)> = \left[-\frac{1}{2} \nabla^2(1) - \sum_{\mu} \frac{z_\mu}{r_{1\mu}}\right]|\phi_i(1)> \tag{17}$$

$$J_j|\phi_i(1)> = <\phi_j(2)|\frac{1}{r_{12}}|\phi_j(2)>\phi_i(1)> \tag{18}$$

$$K_j |\phi_i(1)> = <\phi_j(2)|\frac{1}{r_{12}} |\phi_i(2)>|\phi_j(1)> \tag{19}$$

The one-electron operator F is called the Fock operator. Fortunately, it is always possible to find a unitary transformation of the orbitals such that the right-hand side of Eq. (16) is brought into diagonal form while simultaneously leaving F and the total wave function Ψ invariant. The canonical Hartree-Fock equations are then obtained as

$$F|\phi_i(1)> = \lambda_{ii}|\phi_i(1)> \tag{20}$$

In practice, the analytic form of (20) is assumed and the ϕ_i thus obtained are the canonical Hartree-Fock molecular orbitals. Notice here that Eq. (10) for the electronic energy $\bar{E}_e$ can be written as

$$\bar{E}_e = \sum_{i=1}^{n} \{2\lambda_{ii} - \sum_{j=1}^{n} (2J_{ij} - K_{ij})\} \tag{21}$$

The Hartree-Fock equations are implicit equations for the ϕ_i since the J_j and K_j operators [Eqs. (18)-(19)] are themselves dependent on all ϕ_j. Equation (20), therefore, has to be solved in a self-consistent manner; given a set of ϕ_j, one constructs Eq. (20) and solves for the ϕ_i. This process is continued until assumed and calculated ϕ_i agree to some predetermined criterion at which point the self-consistent solutions have been found.

The most common method of solving Eq. (20) for molecules involves a basis set expansion for the ϕ_i in terms of a given set of one-electron functions or basis orbitals χ_r as in Eq. (7). Solving for the ϕ_i now reduces to determining the orbital expansion coefficients C_{ir} (for all i and r). The Fock operator [Eq. (20)] becomes,

$$F = H + \sum_{j=1}^{n} \sum_{r=1}^{m} \sum_{s=1}^{m} C_{jr}^* C_{js} [2J_{rs} - K_{rs}] \tag{22}$$

where

$$J_{rs}|\chi_t(1)> = <\chi_r(2)|\frac{1}{r_{12}} |\chi_s(2)>|\chi_t(1)> \tag{23}$$

$$K_{rs}|\chi_t(1)> = <\chi_r(2)|\frac{1}{r_{12}} |\chi_t(2)>|\chi_r(1)> \tag{24}$$

Equation (20) becomes a matrix equation for the set of expansion coefficients

$$\sum_{t=1}^{m} (F_{qt} - \varepsilon_i S_{qt}) C_{it} = 0 \quad q=1,\ldots m$$

$$i=1,\ldots n \qquad (25)$$

with

$$F_{qt} = T_{qt} + V_{qt} + \sum_{j=1}^{n} \sum_{r=1}^{m} \sum_{s=1}^{m} C_{jr}^{*} C_{js} [2J_{qrst} - J_{qrst}] \qquad (26)$$

$$T_{qt} = \langle \chi_q(1) | -\tfrac{1}{2} \nabla^2(1) | \chi_t(1) \rangle \qquad (27)$$

$$V_{qt} = \langle \chi_q(1) | -\sum_{\nu} \frac{Z_\nu}{r_{1\nu}} | \chi_t(1) \rangle \qquad (28)$$

$$J_{qrst} = \langle \chi_q(1) | \langle \chi_r(2) | \frac{1}{r_{12}} | \chi_s(2) \rangle | \chi_t(1) \rangle \qquad (29)$$

$$S_{qt} = \langle \chi_q(1) | \chi_t(1) \rangle \qquad (30)$$

The orbital expansion coefficients on Eq. (7) are obtained in the following manner. (1) The integrals defined by Eqs. (27)-(30) are calculated from an assumed set of basis functions $\{\chi_r\}$. (2) The F matrix, defined in Eq. (26), is constructed using a given set of expansion coefficients $\{C_{ir}\}$. (3) Equations (25) are solved by diagonalization of the F matrix which is first transformed to the basis of the eigenvectors of the S matrix [Eq. (30)]. (4) The resulting set $\{C_{ir}\}$ is compared with the input set of step (2). If they are identical, then the self-consistent solution has been found; if not, a new set $\{C_{ir}\}$ is obtained and the procedure returned to step (2). Method of promoting rapid convergence have been discussed by many investigators.[23]

The closed-shell Hartree-Fock or SCF wave function is invariant to a unitary transformation of the occupied one-electron orbitals among themselves. Even though such a transformation generates a new set of occupied orbitals, the wave function and thus the molecular charge distribution are unchanged. This casts some uncertainty on the meaning of the individual orbitals ϕ_i and the λ_{ii} in Eq. (20). However, the λ_{ii} in Eq. (20) (usually given the symbol ε_i) have a very physical interpretation. If it is assumed[24] that none of the ϕ_j change upon removal of a member electron from some orbital ϕ_m, for example, then the difference in energy between the neutral species ($\bar{E}_e$) and that of the positive ion ($\bar{E}_e{}^+$) is given by

$$\bar{E}_e{}^+ - \bar{E}_e = -<\phi_m(1)|H + \sum_{j=1}^{n} (2J_j - K_j)|\phi_m(1)> = -\varepsilon_m \tag{31}$$

Thus the one-electron orbital energy ε_m is approximately equal to the negative of the ionization potential of an electron out of orbital m. This follows logically from the definition of ε_m as the sum of kinetic and nuclear attraction energies of an electron plus the average energy of interaction of that electron with all the other electrons. It is the fact that ionized states of molecules are often fairly well described as having a hole in an orbital ϕ_m, with the other orbitals ϕ_s unchanged, that gives canonical molecular orbitals some physical meaning.

Chapter 4.   MOLECULAR PROPERTIES

Many experiments can be interpreted in terms of expectation values of
operators over unperturbed stationary states. Other experiments measure
the response of the electronic structure of a molecule to a disturbance.
We provide information on the former by computing several expectation
values of our SCF wave functions, which approximate the molecular ground
states. Our selection of operators and properties is incomplete. We
have chosen a sample that includes both properties readily measured in
experiments and those that reveal basic aspects of the charge distribution.

Each property we have chosen to compute corresponds to the expectation value $\bar{P}$ of some operator P, where P can be written in terms of the
coordinates and momenta of the nuclei and electrons of an atom or
molecule. The expectation value of $\bar{P}$ for a normalized approximate wave
function Ψ is given by

$$\bar{P} = \int \Psi P \Psi \, d\tau \tag{32}$$

The properties of interest to chemists may be broadly classed as
one-electron properties and two-electron properties. The one-electron
properties correspond to operators that can be written as the sum of
operators containing the coordinates of one or less electrons. As we
shall see, examples of one-electron properties are the charge density
and the dipole moment. Two-electron properties correspond to operators
containing the interelectronic distance. Examples are the total energy
and the electron-electron repulsion energy.

Within the basis used, for the Hartree-Fock and SCF wave functions
the expectation value of the energy is stationary with respect to
changes of the one-electron orbitals. The Hartree-Fock wave function
is, of course, an approximation to the true wave function. The missing
electron correlation can, in large part, be introduced by the variational
addition of electron configurations doubly excited from the closed-shell
ground state configuration. The added configurations are coupled by
two-electron operators to the ground configuration, and produce a change
of two-electron ground state properties proportional to the coefficient
of admixture of the doubly excited configurations. However, one-electron
operators do not connect the doubly excited configurations with the

13

ground state, and changes of one-electron expectation values depend on the modification of the one-electron orbitals in response to the addition of the two-electron excitations. This is expected to be of second order and have a smaller effect on one-electron properties.

For the above reason, the charge density and other one-electron properties of molecules are expected to be represented rather well by expectation values of the Hartree-Fock wave function, and two-electron properties are expected to be represented rather more poorly.[25,26]

We define now the several operators P corresponding to expectation values appearing in the tables of this book. Expectation values of the energy are defined in terms of the several parts of the Hamiltonian operator.

The kinetic energy operator H_k:

$$H_k = -\tfrac{1}{2} \sum_i \nabla^2(i) \tag{33}$$

The one-electron potential energy operator H_{ne}:

$$H_{ne} = -\sum_\mu \sum_i \frac{Z_\mu}{r_{i\mu}} \tag{34}$$

The two-electron potential energy operator H_{ee}:

$$H_{ee} = \sum_{i>j} \frac{1}{r_{ij}} \tag{35}$$

The nuclear-repulsion operator H_{nn}:

$$H_{nn} = +\sum_{\mu>\nu} \frac{Z_\mu Z_\nu}{r_{\mu\nu}} \tag{36}$$

In terms of the above operators, we can write the electronic energy operator $H_e(P = H_k + H_{ne} + H_{ee})$, and the total energy operator H_t $(P = H_e + H_{nn})$.

In the tables, we also present expectation values for the electronic potential energy ($P = H_{ne} + H_{ee}$), and the potential energy ($P = H_{ne} + H_{ee} + H_{nn}$).

Five of the energy expectation values are two-electron properties depending on the interelectronic distance. They are: the total energy, the electronic energy, the potential energy, the electronic potential energy, and the two-electron potential energy. The atomic units of all energy expectation values are e^2/a_0, 1 a.u. = 27.2107 eV.

Tabulations of energy expectation values for atoms led Fraga[27] to observe empirically a near constant ratio for the Hartree-Fock total and two-electron potential energies in neutral and negative ion species. Goodisman has given a theoretical discussion of the origin of the constant.[28] The ratio of the potential and kinetic energies, the virial, is a measure of the simultaneous and mutual appropriateness of basis set and geometry when less than Hartree-Fock accuracy is achieved. This point has been thoroughly discussed by McLean[29] for a diatomic molecule and by Shavitt[30] for polyatomics.

The operators corresponding to additional one-electron properties we have computed can be written in the form of a sum of one-particle operators for all nuclei and electrons in the molecule:

$$P = \sum_{\mu} Z_{\mu}\, p'(\tilde{r}_{\mu R}) + \sum_{j} ep(\tilde{r}_{jR}) \tag{37}$$

Here the nuclear charges are given by Z_{μ} and the electronic charge by e (e = -1); and $\tilde{r}_{jR}$ is the vector from the reference position R to the position r_j of the jth particle ($\tilde{r}_{jR} = r_j - R$). The corresponding distance r_{jR} is given by $r_{jR} = |\tilde{r}_j - R|$. The reference position for all moments of the charge distribution is the molecular center of mass. The molecular centers of mass were computed with the following isotopic masses: ^{1}H = 1.008123, ^{11}B = 11.01284, ^{12}C = 12.00382, ^{14}N = 14.00751, ^{16}O = 16.00000, and ^{19}F = 19.00450. For expectation values at atomic centers, the reference position is the atomic center. The operators p' for the nuclei and p for the electrons can be reconstructed from the coordinate products given at the head of each tabulated property. Thus for the first moment, for example, headed in the tables by x, the operators p' and p are given by p = ($X_j - R_x$) = X_{jR} and p' = ($X_{\mu} - R_x$) = $X_{\mu R}$, respectively.

For closed-shell molecules having the determinantal wave functions we have adopted in this study, the expectation value of a one-electron operator can be written in a simple form:

$$\bar{P} = \sum_{\mu} Z_{\mu} p'(\tilde{r}_{\mu R}) + 2\sum_{j} e\langle\phi_j|p(\tilde{r}_R)|\phi_j\rangle \tag{38}$$

Here the first term gives the nuclear contribution and the second, the electronic contribution. For our wave function, the electronic contribution is twice the sum over the occupied molecular orbitals of the expectation value resulting from an electron in each molecular orbital:

$$\langle\phi_j|p(\tilde{r}_R)|\phi_j\rangle = \int\phi_j(1)p(\tilde{r}_{1R})\phi_j(1)\,d\tau \tag{39}$$

The moments of the molecular charge are, like the energy expectation values, properties of the whole molecule. They characterize the size and shape of the charge distribution. The first moment or dipole moment is a vector whose components x, y, and z correspond to the one-electron operators X_{jR}, Y_{jR}, and Z_{jR} for p, respectively. The first moment is origin independent for neutral species. The atomic units of the first moment are ea_0, 1 a.u. = 2.54158 D = 2.54158 esu-cm. The molecular dipole moment holds a prominent place in discussions of polarizabilities and dispersion forces.[31]

The second moments of a charge distribution have components x^2, y^2, z^2, xy, xz, and yz, corresponding to the one-electron operators X_{jR}^2, Y_{jR}^2, Z_{jR}^2, $X_{jR}Y_{jR}$, $X_{jR}Z_{jR}$, and $Y_{jR}Z_{jR}$, respectively. The operator for r^2 is r_{jR}^2. The second moments of the charge distribution for an uncharged molecule with a permanent (nonzero) dipole moment is origin dependent. The atomic units of the second moments are ea_0^2, 1 a.u. = 1.34492×10^{-26} esu-cm^2 (Buckinghams). The two principal methods of obtaining values for second moments, or functions of these moments, are the microwave experiments of Flygare and associates[32] and the optical birefringence method of Buckingham. The former method gives values of $\overline{r^2}$ as measured from the center of mass, whereas by the latter method the molecular quadrupole moment measured is relative to an origin difficult to calculate.[33] Besides being a very good measure of the spatial extent of the wave function, $\overline{r^2}$ also gives the diamagnetic part of the magnetic susceptibility.[34]

The third moments have components x^3, xy^2, xz^2, etc. corresponding to the one-electron operators p = X_{jR}^3, $X_{jR}Y_{jR}^2$, $X_{jR}Z_{jR}^2$, etc. The atomic units of the third moment are ea_0^3, 1 a.u. = 0.711687×10^{-34} esu-cm^3. The fourth moments may be divided into two sets, the even components having only even powers of coordinates and the odd components containing two odd powers of coordinates. The even components x^4, y^4, z^4, x^2y^2, etc. correspond to the operators p = X_{jR}^4, Y_{jR}^4, Z_{jR}^4, $X_{jR}^2Y_{jR}^2$, etc.. The odd components x^3y, xy^3, yxz^2, x^3z, etc. correspond to the one-electron operators p = $X_{jR}^3Y_{jR}$, $X_{jR}Y_{jR}^3$, $X_{jR}Y_{jR}Z_{jR}^2$, $X_{jR}^3Z_{jR}$, etc. The atomic units of the fourth moments are ea_0^4, 1 a.u. = 0.376601×10^{-42} esu-cm^4. Third and higher moments of the charge distribution are quantities that enter into discussions of second- and third-order nonlinear optical effects now routinely observed with laser light sources.[35]

We have evaluated expectation values of several operators at atomic centers. In evaluating the required sums in Eqs. (38) and (39), the nuclear center R at which the property is being evaluated is dropped from the sum giving the nuclear contribution. The potential at nucleus μ due to all other electrons and nuclei is given by the operator p = $r_{j\mu}^{-2}$. The units of this potential are ea_0^{-1}, 1 a.u. = 9.07649×10^{-2} esu-cm^{-1}. This potential at a nucleus is proportional to the diamagnetic part of the nuclear shielding in NMR.[32,34] Recently, the value of the quantum mechanical potential at a nucleus has also been shown to be a significant quantity in determining the chemical shift in x-ray photoelectron spectroscopy.[36,37]

The charge density at nucleus μ is given by the one-electron operator $\delta(r_{j\mu})$. The units of charge density are ea_0^{-3}, 1 a.u. = 32.4140×10^{14} esu-cm^{-3}. The charge density for atoms has been discussed by Fraga and Malli.[2] The charge density is typical of quantities entering the discussion of ESR and Mössbauer spectroscopy.

The electric field at a nucleus μ is given by the gradient of the potential. The operators for the components of the electric field are $p = X_{j\mu}/r_{j\mu}^3$, $Y_{j\mu}/r_{j\mu}^3$, and $Z_{j\mu}/r_{j\mu}^3$. The atomic units of the electric field are ea_0^{-2}, 1 a.u. = 17.1524×10^6 esu-cm^2. The components of the electric field are identified with the Hellmann-Feynman part of the force exerted on an atom in a molecule. In general, values of these operators at any nucleus should vanish only for a Hartree-Fock wave function and optimum geometry. Nonetheless, many investigators have attributed physical significance to values of the force calculated, with approximate MO wave functions.[38] Actually, at any geometry the sum of all forces on all the nuclei should vanish for a Hartree-Fock wave function. Thus, any deviation from this result can be attributed solely to the approximate nature of the wave function.

Values of the electric field have also been found useful in calculating the paramagnetic or high-field part of the nuclear shielding constant for the proton.[39,40] The gradient of the electric field at nucleus μ is a tensor whose components $(3x^2-r^2)r^{-5}$, $(3xy)r^{-5}$, etc. correspond to the operators $(3X_{i\mu}^2-r_{i\mu}^2)\,r_{i\mu}^{-5}$, $(3X_{i\mu}Y_{i\mu})r_i^{-5}$, etc. The atomic units of the field gradient are ea_0^{-3}, 1 a.u. = 32.4140×10^{14} esu-cm^{-3}. The gradient of the electric field is the prominent electronic structure quantity in nuclear quadrupole resonance spectroscopy.[41] Accurate values of electric field gradients can be used to deduce the value of nuclear quadrupole moments[42]; they also give a sensitive test of the electronic charge distribution in the immediate neighborhood of the nucleus. Another interesting aspect of the field gradient is its relationship to vibrational force constants.[43,44]

Chemists have developed several ways to partition the electrons of a molecule among atoms or bonds. The most widely adopted definitions are those of Mulliken.[18]

Let us refer to all basis functions on center μ by the index i, and all basis functions on other centers by j. For our closed-shell determinantal wave functions, the total gross population in basis function i is denoted by $N_{\mu i}$ and defined by

$$N_{\mu i} = 2 \sum_k C_{ki}(C_{ki} + \sum_{i'\neq i} C_{ki'}S_{ii'})$$

$$+ 2\sum_k C_{ki}(\sum_j C_{kj}S_{ij}) \tag{40}$$

where the index k runs over the filled molecular orbitals. The total gross population on atom μ, N_μ, is given by

$$N_\mu = \sum_i N_{\mu i} \tag{41}$$

We refer to the "total gross populations" defined by Mulliken as Mulliken charges in our tables. The subtotal overlap population between basis functions i and j denoted by n(i,j) is defined by

$$n(i,j) = 2 \sum_k 2 C_{ki} C_{kj} S_{ij} \tag{42}$$

The overlap population can be partitioned according to representation of the occupied molecular orbitals. The overlap population between centers μ and ν denoted by $n_{\mu\nu}$ is then given by

$$n_{\mu\nu} = \sum_i \sum_j n(i,j) \tag{43}$$

These definitions of gross atomic and overlap populations suffer from the fact that they are basis set dependent. At present, all definitions of atomic population appear to have arbitrary elements. We present the Mulliken definitions because they are so widely used; however, alternative definitions have been proposed and discussed.[45,46]

The overlap population, defined according to Mulliken, has been taken as a measure of the strength of a given chemical bond and hence correlated with bond energies, vibrational frequencies, and force constants. Recently a relationship between the overlap population and the deuterium quadrupole coupling constant has been suggested.

Attention has recently been focused on breakdown of the total energy according to combinations of basis orbital center contributions, the object being to find the dominant contribution to the binding energy.[47,48] It would have been useful to present such a partitioning of the total energies here, but unfortunately this need was not foreseen, and the two-electron integrals necessary for such computations are, for the most part, no longer available.

The molecular wave function can be used to calculate expectation values of one-electron operators for two very different general purposes.

In the first place, different operators sample different regions of the charge density. A selection of operators and expectation values can give a rather detailed picture of the electron distribution.

In this tabulation we present expectation values of the $1/r^3$, $1/r^2$, $1/r$, $\delta(r)$, r, r^2, r^3, and r^4 operators which, when taken together at several points in the molecular space, characterize the charge distribution almost as much as the wave function itself. These numbers can be used both in comparisons with experiment to evaluate the accuracy of a wave function and on a comparative basis to evaluate specific properties at a particular site in various compounds. Reliable theoretical values

of properties can also be used to test the success of semiempirical
methods.   Second, many model theories make use of such properties as
the electric field gradient, moments of the charge distribution, potentials at a nucleus, overlap integrals, charge densities, and net atomic
populations as intrinsic parts of parameters of these model theories,
which try to explain physical, spectroscopic, or molecular dynamical
observations and processes.

Chapter 5.   THE DOUBLE-ZETA BASIS SET

The basis set we have adopted is a set of contracted functions, which are linear combinations of primitive Gaussian functions[7,10]:

$$\chi_i = \sum_s c_{ij} \eta_j \qquad (44)$$

where a normalized primitive Gaussian η_i centered at A is given by

$$\eta_i(r_A) = N_i^l X_A^l Y_A^m Z_A^n \exp(-\alpha_i r_A^2) \qquad (45)$$

In this work we have used only s and p Gaussians. For the s Gaussians one has $l = m = n = 0$. For the p Gaussians one has a single nonzero exponent with $l=1$ for p_x, $m=1$ for p_y, and $n=1$ for p_z. Here r_A is the distance from the center A and the Cartesian components of the corresponding vector are X_A, Y_A, and Z_A.

For each first-row atom, the s part of our basis is represented by four contractions: S1, S2, S3, and S4; these contain 3, 4, 2, and 1 primitive Gaussian functions, respectively. The s contractions were taken from the atomic SCF calculations of Whitten.[13] Two contractions, S1 and S2, were used for the hydrogen atom. They are taken from Huzinaga's four-term fit to a Slater orbital with exponent 1.2.[12]

The atomic 1s function is almost completely represented by S1 and S2. The 1s population for any first-row atom may be taken to be the sum of the Mulliken populations in S1 and S2 on that atom. The population that may be ascribed to the "2s" atomic orbital is thus the sum of the Mulliken populations of S3 and S4. For each first-row atom, the p parts of our basis are represented by pairs of contractions: for p_x these are X1 and X2, for p_y they are Y1 and Y2, and for p_z they are Z1 and Z2. The inner contraction X1 contains four primitive Gaussians and the outer contraction X2 contains one, as do the corresponding p_y and p_z contractions. The p part of our basis is taken from the atomic SCF calculations of Huzinaga.[12] We have included no d or higher harmonics in our basis set.

The coefficients c_{ij} and exponents α_i of the normalized contracted functions of our double-zeta basis set are listed in Table 1 for the atoms H, B, C, N, O, and F which are contained in molecules of this compendium.

The properties of our double-zeta basis state may be exhibited in several ways. One way is to exhibit expectation values for atoms[1,2] computed in the double-zeta and Hartree-Fock basis sets. We have done that for the atoms of this study in Table 2. Comparison of the total energies computed for the ground states of first-row atoms in this study shows that the double-zeta value is 0.02 a.u. or less above the Hartree-Fock value.[1] The potential and kinetic energies deviate somewhat more. It is interesting to note in Table 2 that while the expectation value of r^{-1} for the DZ and HF[49] descriptions of the atoms differs by about 0.05%, both the charge density at the nucleus δ and the expectation of r^2 are smaller for the DZ basis than for the HF result.[50,51] Thus the DZ wave function appears to be somewhat smaller than the HF value at the nucleus and very far from it. This is not a surprising property of the Gaussian basis.

The term double-zeta[14] originates in the use of Slater functions characterized by an exponent factor zeta in electronic structure calculations. In a molecular context, a basis set containing a single Slater function per atomic orbital, a minimum basis set, having all zeta values chosen to minimize the atomic ground state energy is called a "best-atom single-zeta" basis. If the zetas of that basis have been reoptimized to minimize the total energy of the molecule, it is called a "best-molecule single-zeta" basis set. Corresponding basis sets with two Slater functions per atomic orbital are called "best-atom double-zeta" and "best-molecule double-zeta" basis sets. Table 3 shows the total energy of carbon monoxide computed in several such basis sets of Slater functions, the Hartree-Fock limit, and our "double-zeta" basis.[52,53] The reason for the choice of "double-zeta" as our basis set name is simply that it gives a total energy close to that of the "best-atom double-zeta" basis.

In order to give readers a better idea of how good (or poor) our double-zeta basis set is in applications to molecules, we compare energy and other expectation values for hydrogen-fluoride, carbon monoxide, and formaldehyde computed in our double-zeta (DZ) basis, an extended double-zeta (EDZ) basis employed by Moskowitz and Neumann,[15] and the Hartree-Fock basis of Huo.[52,54] Energy expectation values are presented in Table 4, other properties in Table 5. The EDZ basis[15] referred to in the tables differs from the DZ basis in two ways. First, the S2 and X1 contractions of the DZ basis were split into pairs of smaller contractions employing the same primitive functions. Second, two sets of d primitive Gaussians (x^2, y^2, z^2, xz, yz) were added to the basis for each first-row atom, the exponents α being 0.5 and 1.5, respectively, for the two sets. For hydrogen, p primitives (x, y, z) were added having $\alpha = 1.0$. The main effect of the addition of these functions extending the DZ set is to permit polarization of the atomic charge distribution by the Coulombic environment. It also permits charge to flow with less

Table 1. The Double-Zeta Basis

Hydrogen

	S1		S2	
	C	α	C	α
	0.0328280	19.2406	1.0000000	0.17758
	0.2312081	2.89915		
	0.8172383	0.65341		

Boron

	S1		S2		S3		S4	
	C	α	C	α	C	α	C	α
	0.029411	1726.1683	0.1545436	28.3387	−0.1478468	3.1260	1.0000000	0.0957
	0.1212642	529.3856	0.2442219	11.6417	1.0538343	0.2999		
	0.9049570	108.1104	0.4497834	4.7809				
			0.2726493	1.7114				

	X1		X2	
	C	α	C	α
	0.0179871	11.3413	1.0000000	0.070114
	0.1103417	2.4360		
	0.3831187	0.68358		
	0.6478735	0.21336		

Carbon

	S1		S2		S3		S4	
	c	α	c	α	c	α	c	α
	0.0293140	2548.7260	0.1534800	41.8427	-0.1463020	4.9344	1.0000000	0.1480
	0.1215990	781.6495	0.2433111	17.1893	1.0533749	0.4735		
	0.904751	159.6274	0.4537991	7.0591				
			0.2698321	2.5269				

	X1		X2	
	c	α	c	α
	0.0185330	18.1557000	1.0000000	0.1146000
	0.1154400	3.9864000		
	0.3861999	1.1429300		
	0.6400798	0.3594500		

Nitrogen

	S1		S2		S3		S4	
	c	α	c	α	c	α	c	α
	0.0292226	3489.5253	0.1545241	57.2879	-0.1432535	7.0699	1.0000000	0.20790
	0.1216856	1070.1762	0.2450818	23.5342	1.0524427	0.6784		
	0.9047253	218.5500	0.4569305	9.6648				
			0.2632423	3.4597				

	X1		X2	
	c	α	c	α
	0.0182571	26.7860	1.0000000	0.16537
	0.1164081	5.95635		
	0.3901142	1.70740		
	0.6372268	0.53136		

Table 1. (Continued)

Oxygen

	S1		S2		S3		S4	
	C	α	C	α	C	α	C	α
	0.0292252	4643.4485	0.1527631	76.2320	-0.1403140	9.7044	1.0000000	0.2825
	0.126036	1424.0643	0.2439912	31.3166	1.0515337	0.9311		
	0.9047897	290.7850	0.4582404	12.8607				
			0.2644382	4.6037				

	X1		X2	
	C	α	C	α
	0.0195800	35.1832	1.0000000	0.21373
	0.1241899	7.90403		
	0.3947297	2.30512		
	0.6273796	0.71706		

Fluorine

	S1		S2		S3		S4	
	C	α	C	α	C	α	C	α
	0.0291410	5851.0350	0.1539798	96.0570	-0.1354260	12.9309	1.0000000	0.3574
	0.1213604	1794.4099	0.2460920	39.4609	1.0485966	1.1945		
	0.9049994	355.4519	0.4626529	16.2053				
			0.2557880	5.8010				

	X1		X2	
	C	α		
	0.0208678	44.3555	1.0000000	0.27329
	0.1300922	10.0820		
	0.3962206	2.9958		
	0.6203711	0.93826		

25

Table 2. Comparison of Double-Zeta and Hartree-Fock Expectation Values for Atoms

Energy	Method	Atom State					
		H ^{2}S	B ^{2}P	C ^{3}P	N ^{4}S	O ^{3}P	F ^{2}P
Total	DZ	-0.4993	-24.5243	-37.6812	-54.3897	-74.7931	-99.3863
	HF	-0.5000	-24.5291	-37.6886	-54.4009	-74.8094	-99.4093
Kinetic	DZ	+0.4994	+24.5136	+37.6670	+54.3687	+74.7757	+99.3054
	HF	+0.5000	+24.5294	+37.6885	+54.4016	+74.8098	+99.4086
Potential	DZ	-0.9987	-49.0379	-75.3481	-108.7584	-149.5688	-198.6917
	HF	-1.0000	-49.0584	-75.3772	-108.8024	-149.6192	-198.8179
One-Electron Potential	DZ	-0.9987	-56.8741	-88.1008	-128.3042	-178.0318	-238.5566
	HF	-1.0000	-56.8980	-88.1370	-128.3534	-178.0768	-238.6683
Two-Electron Potential	DZ	0	+7.8362	+12.7527	+19.5458	+28.4630	+39.8649
	HF	0	+7.8396	+12.7598	+19.5510	+28.4576	+39.8504

Other

r^2	DZ	-2.9714	-15.8059	-13.7245	-12.0118	-11.0616	-10.1093
	HF	-3.0000	-15.8472	-13.7950	-12.0722	-11.1647	-10.2358[a]
r^{-1}	DZ	-0.9987	-11.3748	-14.6835	-18.3292	-22.2539	-26.5063
	HF	-1.0000	-11.3796	-14.6895	-18.3362	-22.2596	-26.5187[b]
δ	DZ	0.2695	66.3714	117.7264	190.2740	288.2124	414.2557
	HF	0.3183	69.1135	121.9645	196.4152	296.5419	425.7050

a See Reference 49.
b See Reference 50.

Table 3.   Calculations on the Ground State of CO[a]

Basis	Total Energy
Best-atom single zeta[b]	-112.3261
Best-molecule single zeta[b]	-112.3927
Best-atom double zeta[b]	-112.6755
Best-molecule double zeta[b,c]	-112.7015
Hartree-Fock limit[b,c]	-112.7860
Experimental[c]	-113.377
Our double-zeta gaussian basis[d]	-112.6762

[a] r_e = 2.132 a.u.
[b] Slater-type orbital basis.
[c] See Reference 52.
[d] See Reference 53.

restriction in both pi and sigma bonding regions. In other words, the DZ basis tends to force a more spherical charge distribution about a center than does the EDZ basis.

We can now reach qualitative conclusions on the adequacy of the DZ basis in molecular calculations from the contents of Tables 4 and 5. From Table 4, we see that the DZ total energy is about 0.05 a.u., per first-row atom, above the EDZ and HF results. It is puzzling, and worthy of note, that the DZ potential energy is closer to the HF value than the EDZ result. For all three molecules, the DZ orbital energies are within 0.01 a.u. of the EDZ and HF results, except for the lowest orbital of the 2s band 3SIGMA of CO and 3A1 of formaldehyde, which are about 0.03 a.u. lower in the DZ calculation than the HF result.

Examination of the expectation values compiled in Table 5 shows the first moment z computed with the DZ basis to be within 0.2 a.u. of the EDZ and HF values. This probably reflects the restricted polarization possible with the DZ basis set. The calculated second moments agree within 0.6 a.u. for the DZ and larger basis sets. The DZ value of the potential at each center (r^{-1}) agrees within 0.04 a.u. for all centers with the EDZ and HF results, tending to be less negative in the DZ basis. This probably reflects the fact that a Gaussian function tends to relatively underrepresent the region near the nucleus. The charge density at the nucleus of a first-row atom is larger in the DZ than in the EDZ basis; this coincides with the larger DZ potential energy.

Table 4. Comparison of Energies Computed in the Double-Zeta, Extended Double-Zeta, and Hartree-Fock Basis Sets

Carbon Monoxide

Energy (a.u.)	DZ	EDZ[b]	HF[c]
Total	-112.6763	-112.7622	-112.7860
Kinetic	+112.6526	+112.5067	112.6420
Potential	-225.3290	-225.2689	-225.4280

Orbital	DZ	EDZ[b]	HF[c]
1SIGMA	-20.6724	-20.6614	-20.6612
2SIGMA	-11.4121	-11.3605	-11.3593
3SIGMA	-1.5638	-1.5213	-1.5192
4SIGMA	-0.7984	-0.8031	-0.8024
5SIGMA	-0.5542	-0.5543	-0.5530
1PI	-0.6476	-0.6378	-0.6377

Formaldehyde

	DZ	EDZ[b]
Total	-113.8209	-113.8917
Kinetic	+113.8158	+113.5947
Potential	-227.6367	-227.4864

Orbital	DZ	EDZ[b]
1A1	-20.5844	-20.5738
2A1	-11.3618	-11.3431
3A1	-1.4316	-1.4038
4A1	-0.8694	-0.8646
5A1	-0.6448	-0.6506
1B1	-0.7039	-0.6893
2B1	-0.4426	-0.4402
1B2	-0.5374	-0.5341

Hydrogen Fluoride

Energy (a.u.)	DZ	HF[a]
Total	-100.0150	-100.0705
Kinetic	100.0102	100.0255
Potential	-200.0252	-200.0960

Orbital	DZ	HF[a]
1SIGMA	-26.2813	-26.2950
2SIGMA	-1.6026	-1.6013
3SIGMA	-0.7533	-0.7685
1PI	-0.6432	-0.6505

a See Reference 54.
b See Reference 15.
c See Reference 52.

29

Table 5. Comparison of Expectation Values Computed in the Double-Zeta, Extended Double-Zeta, and Hartree-Fock Basis Sets.

Expectation (a.u.)	Hydrogen Fluoride			Carbon Monoxide				Formaldehyde	
	DZ	HF[a]		DZ	EDZ[b]	HF[c]		DZ	EDZ[b,d]
z	+0.9349	+0.7609		−0.1550	−0.096	−0.1078		−1.2237	−1.110
x^2	−4.1326	−4.2289		−7.4795	−7.5271	−7.516		−8.4828	−8.438
z^2	−2.4791	−2.4992		−9.7528	−9.081	−9.1086		−9.0341	−8.559
r^2	−10.7443	−10.9570		−24.7719	−24.135	−24.141		−26.2896	−25.685
r^{-1} (H) / (C)	−0.8775 (H)	−0.9183 (H)		−14.595 (C)	−14.637 (C)	−14.645 (C)		−14.6096 (C)	−14.616 (C)
r^{-1} (F) / (O)	−26.5906 (F)	−26.5907 (F)		−22.226 (O)	−22.230 (O)	−22.251 (O)		−22.3181 (O)	−22.317 (O)
r^{-1} (H)								−1.0508 (H)	−1.072 (H)
δ (H) / (C)	0.3922 (H)	(H)		117.516 (C)	117.321 (C)			116.5677 (C)	116.363 (C)
δ (F) / (O)	413.4284 (F)	(F)		287.594 (O)	287.218 (O)			287.3473 (O)	286.979 (O)
δ (H)								0.4141 (H)	0.416 (H)

30

Component									
q^e_{zz} (H)	+0.6340	(C)	+0.5226	−0.8607	−1.135	−1.18	(C)	−0.1531	−0.347
q_{xx}	−0.3170		−0.2613	+0.4303	+0.565	+0.59		−0.3386	−0.301
q_{yy}	−0.3170		−0.2613	+0.4303	+0.565	+0.59		+0.4916	+0.648
q_{zz} (F)	+3.1286	(O)	+2.8733	−0.3029	−0.697	−0.679	(O)	+0.6968	+0.4041
q_{xx}	−1.5643		−1.4367	+0.1514	+0.3485	+0.340		−2.6124	−2.2705
q_{yy}	−1.5643		−1.4367	+0.1514	+0.3485	+0.340		+1.9156	+1.866
q_{zz}							(H)	−0.0212	−0.0212
q_{xx}								+0.1632	+0.1501
q_{xz}								−0.1943	−0.1779
q_{yy}								−0.1420	−0.130

a See Reference 50.
b See Reference 15.
c See Reference 48.
d Our coordinate system is used.
e Component of the electric field gradient.

The electric field gradient (q) at first-row atoms is given rather poorly by the DZ basis, it differs by up to 0.3 a.u. from the more accurate EDZ and HF results. The situation for protons is better; in this case the DZ value of the electric field gradient is within 0.1 a.u. of the more accurate EDZ and HF results.

To aid in the interpretation of the Mulliken charges, which are presented for the contracted basis functions in our tables for molecules, we give in Table 6 those charges for atoms in the DZ basis.

Table 6. Mulliken Charges for Atoms in the Double-Zeta Basis.

Atom	State	Contracted Function					
		S1	S2	S3	S4	X1	X2
H	^{2}S	0.53897	0.46103				
B	^{2}P	0.04410	1.95566	1.05041	0.94984	0.25743	0.07590
C	^{3}P	0.04326	1.95639	1.03679	0.96355	0.50947	0.15720
N	^{4}S	0.04356	1.95606	1.04251	0.95788	0.76616	0.23384
O	^{3}P	0.04278	1.95674	1.03956	0.96093	1.00456	0.32877
F	^{2}P	0.04362	1.95579	1.07079	0.92981	1.24714	0.41952

Chapter 6.   THE COMPUTATIONS

The computations of both SCF wave functions and properties have been made with the POLYATOM (Version 2) System of Programs for Quantitative Theoretical Chemistry. The programs and a detailed write-up are available from the Quantum Chemistry Program Exchange (QCPE).[11]

The availability of these programs makes it possible for any person to check any numerical result reported here. It also permits extensions of this work using the same basis and the original programs. We would be grateful if persons carrying out extensions to other molecules communicate their results to the authors for inclusion in any future revisions of these tables.

We have prepared the tables of Chapter 10 on microfilm by means of an Information International FR80 Microfilm Recorder. All numerical results for each molecule were collected on a primary data tape. A program read this primary data tape and prepared a second tape with detailed information for construction of the tables by the microfilm recorder. This method of table production for Chapter 10 should prevent most typographical errors. However, it must be noted that the energy expectation values, and in some cases the eigenvalues, were punched from original paper printout.

The computations at Bell Laboratories were made on a GE 635 computer with a 36 bit word length, and at the Ford Scientific Laboratory on a PHILCO 212 computer with a 48 bit word length. Most numerical quantities reported in the tables were computed in single precision on the GE 635, which places an upper limit of 8 significant decimal figures on them.

A second factor influencing the accuracy of our reported numbers is the degree to which convergence was reached in the SCF iterations. The test of convergence we have employed is on the total energy. In general, the SCF iterations were stopped when the change of total energy between successive iterations was 10^{-5} a.u. or less. Examinations of our original output shows that several molecules reported here were exceptions to this rule. These molecules, together with the difference between the total energy in a.u. at which SCF iterations were halted, include: HCN at 4×10^{-4}, N_2 at 9×10^{-5}, NH_3 at 9×10^{-5}, F_2 at 8×10^{-5}, CH_4N_2 at 3×10^{-5}, CH_2N_2 at 2×10^{-5}, C_2F_4 at 2×10^{-5}, and N_2F_2 at 2×10^{-5}.

Examination of our output has also shown that expectation values of the individual components of the Hamiltonian (for example, the kinetic

energy and the one-electron potential energy) sometimes converge less
rapidly than the total energy. In general, the change of kinetic energy
in atomic units between the last two SCF iterations was 10^{-3} or less.
For a few molecules this final difference was larger. These include
CHOO- at 5×10^{-2}, N_2O_4 at 4×10^{-2}, CF_4 at 4×10^{-2}, CO at 3.5×10^{-2}, C_2F_4 at
8×10^{-2}, and F_2CO at 3×10^{-3}.

The orbital energy eigenvalues also change during the SCF iterations.
In general the largest change of an orbital energy in atomic units
between the last two iterations was 10^{-3} or less. Molecules for which
the largest change of an orbital energy exceeded 10^{-3} atomic units were
the following: N_2O_4 at 10^{-2}, CHOO- at 10^{-2}, CO at 10^{-2}, CF_4 at 3×10^{-3},
FNO at 3×10^{-3}, N_2H_2 at 3×10^{-3}, C_2F_4 at 7×10^{-3}, NH_2CN at 2×10^{-3}, N_2F_2 at
3×10^{-3}, and F_2CO at 5×10^{-3}.

The largest fluctuation of density matrix elements between the last
two SCF iterations was generally smaller than 10^{-3}. Molecules in which
this limit was exceeded include: N_2C_4 at 6×10^{-3}, CH_2N_2 at 4×10^{-3}, CF_4
at 3×10^{-3}, CO at 2×10^{-3}, and NH_2CN at 2×10^{-3}.

The expensive two-electron integrals that are required for most of
these SCF calculations are no longer in our possession. For this reason
we have chosen to outline the deficiencies of convergence insofar as
they can be documented from our original output, rather than restart
these rather expensive computations. It is probable that all quantities
quoted in our tables differ from the true SCF value by less than half
their change between the last two iterations.

In summary, we generally expect the difference from true SCF values
associated with lack of convergence to occur in the fifth figure after
the decimal point for the total energy, the fourth place for orbital
energies, the third place for coefficients in vectors, and the third
place for expectation values of energy components. These rules can be
used to estimate errors in the several tabulated properties, all of
which we expect to be correct in the first two places after the decimal
point. All properties have been accurately computed from the tabulated
vectors from the last SCF iteration to all figures given in the tables.

Chapter 7.   APPLICATIONS

The wave functions and properties reported in this compendium were, for the most part, computed in the course of studies directed toward understanding several physical properties of molecules. These research projects not only produced double-zeta quality SCF wave functions for a large number of polyatomic molecules, but also required the coding of new operators and the testing of various model theories, techniques, and ideas. One rather useful outcome of these studies was the explicit manner in which experiment and theory showed a useful complementarity, the theory providing information that was found to be an indispensable aid in understanding the experiment.

The first studies using the SCF molecular wave functions in a DZ basis were directed toward assignments in the optical absorption spectra of HCOX (X-NH_2,OH,F) compounds,[53,55] ethylene,[56] and small ring compounds.[57,58,59] For the HCOX series the problem centered about the ability to differentiate valence from Rydberg-type electronic transitions. The proper theoretical solution required augmenting the DZ basis with diffuse basis functions on the heavy (first-row) atoms in order to adequately describe the terminating Rydberg states and the use of open-shell SCF techniques to produce excited state wave functions. However, it appears that very extensive configuration interaction will be required to locate many of the excited valence states.[60]

In a study on simple olefins[56] a twisted ethylene was used as a model for the optically active trans-cyclo-octene (TCO) and it was shown that the observed sign of the rotational strength in the $\pi \rightarrow \pi^*$ region of the TCO optical absorption spectrum is as expected for a $\pi \rightarrow \pi^*$ electronic transition. This result was in direct disagreement with the semiempirical results of Yaris et al.[61] Recent improved circular dichroism measurements in the vacuum ultraviolet region confirm the conclusions based on the nonempirical SCF calculations.

A novel feature of the olefin work was the coding of the one-electron linear ($\underline{V}$) and angular ($\underline{r} \times \underline{V}$) momentum operators and their use in a completely nonempirical calculation. Similarly, in the work on small ring compounds, the full one-electron spin-orbit coupling operator $\sum_{i,n} [Z_n(\underline{r_i} \times \underline{V_i}) \cdot S_i] \, r_i^{-3}$ was used in a nonempirical calculation of spin-orbit induced intensity effects in the lowest energy singlet $\leftrightarrow$ triplet electronic transition in diazirine ($CH_2N_2^*$).[57]

35

The first application using almost the complete series of molecules presented here was in a study of heats of reaction.[62] It was found that heats of reaction may be predicted within a few kilocalories of experiment for selected reactions. Moreover, it appears that changes of correlation energy are small for reactions with closed-shell reactants and products.[63,64] The basis is apparently less adequate for unsaturated molecules than for saturated; this is evidently a result of the need for d basis functions which shift charge of a pi bond toward its center.[62]

In the new technique named ESCA (Electron Spectroscopy for Chemical Analysis), developed by Siegbahn and co-workers,[65] a sensitive probe of the valence-shell electron charge distribution in molecules is obtained by the measurement of core electron binding energy. Although the core-shell electron wave function does not change for a given atom in different molecules, the potential that the core electron sees is a function of its chemical environment and gives rise to the small chemical shift effect observed in its ionization potential.

In nonempirical MO theory an electron binding energy can be calculated by neglecting the instantaneous rearrangement of the remaining electrons as a result of the ionization process. In this approximation, for closed-shell electronic configurations, the binding energy is equal to (minus) the appropriate orbital energy, according to Koopmans' theorem.[24] This, of course, also neglects the slower nuclear rearrangements. Although Koopmans' theorem (KT) gives absolute core-shell ionization potentials that are incorrect by tens of electron volts, it has been shown that with flexible basis sets, the binding energy chemical shifts for a given core level in an atom may be computed as differences in orbital energies for different compounds. This approximation reproduces experimental results rather well.[66] The computation of the absolute values of core-ionization potentials requires an account of valence electron relaxation.[67]

Since core-shell binding energies have at the same time a very localized nature and a dependence on molecular environment, it is reasonable to propose models that relate the observed chemical shift effect to the chemical environment in the molecule. A simple and intuitive model is to relate the core electron binding energy chemical shift to the net charge carried by an atom in a molecule. This approach has a firm quantum mechanical basis in that the difference in KT ionization potentials for a given core level can be reproduced by replacing the orbital energies by the quantum mechanical potential at the nucleus.[68]

The net charge is also conveniently computed from the basis orbitals of molecular wave functions as it was defined by Mulliken.[16] There is no need to dwell on the central approximation of the Mulliken population analysis scheme, that is, the equal division of overlap charge distribution between two basis functions. It is sufficient to say that the population analysis of the molecular wave functions has proved to be quite useful. It has been shown that there is a real correlation between net atomic charge and bonding energy chemical shifts, both obtained from the DZ basis molecular wave functions.[66]

One interesting result of the ability to calculate core-electron binding energy chemical shifts by evaluating the quantum mechanical potential at the nucleus is to establish a direct relationship between ESCA and NMR spectroscopies, since in the latter the same operator is involved in calculating the diamagnetic contribution to the nuclear shielding chemical shift effect.[69]

Recent high-resolution microwave Zeeman studies by Flygare and associates have given accurate experimental values for the second moments of the charge distribution in molecules that are used more familiarly as quadrupole moment.[32] Comparison of second moments computed with the DZ basis molecular wave functions with the experimental values shows that in every case where both the experimental and theoretical numbers are available the latter are computed within the experimental uncertainty error limits of the former.[53,55,59,70,71] This is rather surprising since dipole moments have been found usually to be overestimated in the DZ basis by about 0.2 atomic units.

Electron spectroscopy is also a very useful tool for probing the valence-shell electron charge distribution by direct ionization of the valence electrons. Again, using Koopmans' theorem, a test of the accuracy of the DZ basis relative to the limiting Hartree-Fock result can be ascertained by trying to obtain values for the neglected electronic rearrangement energy and the correlation energy error. The electronic rearrangement energy is obtained directly by solving the open-shell SCF equations for the positive ion.[67] The correlation energy error can be estimated from a Mulliken population analysis of the molecular wave function coupled with tables of atom pair correlation energies.[62,64] When both these corrections are made, a recent study on several molecules shows that much better agreement with experimentally determined vertical ionization potentials is obtained.

Fortunately these tedious correction terms usually do not have to be considered, although, for example, in the case of formamide they were important. In most cases a simple scaling of the KT values is sufficient to bring theory in line with experiment, where the latter includes ionizations even out to 30 eV.[59] In addition, theory and experiment show a useful complementarity; the calculations give the number, symmetry type (for possible Jahn-Teller and spin-orbit splittings), and bonding characteristics (for qualitative discussion of Franck-Condon profiles), in a given energy range and spatial region. All this information is found to be a virtually indispensable aid in assigning and understanding the photoelectron spectra.

Chapter 8.   CONTENTS OF THE TABLES

The tables of molecular wave functions and properties are organized in the same basic format for all molecules.

Basic assumptions of the molecular SCF calculation head the first page. These are followed by energy and moment expectation values which are properties of the whole molecule. The second page is headed by a set of expectation values characteristic of the several unique molecular centers. These are followed by overlap populations for the unique pairs of molecular centers. The third page begins the table of molecular orbital energies and vectors. These are followed by the overlap matrix between all unique pairs of functions. All properties are given in atomic units.

The first page for each molecule is headed by the molecule's formula name and its common name. Pages are numbered at the bottom, for each molecule, by the formula name and page number for the molecule. Following the page heading, the assumed molecular symmetry is given and also the molecular geometry as a table of center names, Cartesian coordinates, and nuclear charges for all the atomic centers of the molecule. A computer-drawn picture of the molecule appears to the right with the assumed coordinate system. The Cartesian coordinates of the center of mass, with reference to which all moments are computed, is given below the geometry. Although group theory was not always used to generate symmetry adapted basis functions, as is permitted by POLYATOM (Version 2), we have assigned group theoretical names to the molecular orbitals that resulted from our calculations and have given the number of molecular orbitals belonging to each representation generated by our DZ basis set, and the number of pairs of electrons occupying molecular orbitals of each representation. We have tried to follow the group theory notation and conventions of Herzberg.[72] Insofar as we are able to recall them, the sources of the experimental molecular geometries we have used are given in Table 7 of Chapter 9.

The first molecular properties tabulated are energy expectation values. These include expectation values for the following operators defined in the text: the total energy, the electronic energy, the nuclear repulsion energy, the kinetic energy, the potential energy, the electronic potential energy, the one-electron potential energy, and the two-electron potential energy. The negative of the ratio of the potential energy to the kinetic energy, the virial, is also given. The second

38

set of molecular properties given consists of moments of the charge
distribution relative to the center of mass. These include the first
through fourth moments. Below the symbolic definition of the operator
p corresponding to each moment are listed two numbers: the upper number
is the electronic contribution to the moment, the lower is the nuclear
contribution. We have written our table preparation program to present
only groups of moments such that at least one member moment is greater
than 10^{-4} atomic units. Occasionally this has caused a small moment to
be presented which is zero by symmetry, as the odd fourth moment x^3z of
cyclopropene ($C3H4*$). This has been permitted to reveal any possible
deficiencies of our programs or convergence.

We have tabulated several expectation values for the unique atomic
centers of each molecule. Each expectation value given is the sum of
the electronic and nuclear contributions. The expectation values given
include from left to right in the first row: the potential r^{-1}, the
charge density δ, and the x, y, and z components of the electric field.
In the second row, the six components of the electric field gradient at
each center are given.

Mulliken charges are given for each center. In Mulliken's terminol-
ogy, we give the gross population in each basis function followed at the
extreme right by the gross population on the atom. Mulliken overlap
populations are given for each unique pair of atomic centers. These are
presented as the contribution of filled molecular orbitals of each
representation, with subtotal overlap populations for each representation and the total overlap population for the pair of centers. For
convenience in interpretation of the overlap populations we also give
the distance in atomic units between each pair of centers.

The molecular orbitals and their eigenvalues are given for all
orbitals having an energy less than 0.7 a.u. To the left of each molecular orbital vector is its name, which includes its serial position by
energy and the representation to which it belongs. An asterisk to the
left of this orbital name indicates that the molecular orbital is occupied
by a pair of electrons in the SCF determinantal wave function. The
coefficients of basis functions in each molecular orbital are presented
in columns by basis function type and in rows by the center on which the
basis function is located. The molecular orbital energy, or eigenvalue,
is given directly below the molecular orbital name.

Finally, we give the unique nonzero overlaps between all pairs of
basis functions.

Chapter 9.   MOLECULES TREATED

We list below, in the order of appearance, the molecules for which SCF
wave functions and properties have been tabulated for this book. Molecules are listed in order of increasing number of electrons. Within an
isoelectronic set, they are ordered to bring chemically related molecules into proximity. In Table 7 we give for each molecule, the formula
name, the adopted common name, a reference for the geometry we have
assumed, and the text page at which tables for the molecule begin.

Table 7. Molecules Treated

Formula Name	Common Name	Geometry Reference	Text Page
2-Electron Molecules			
H2	Hydrogen	73	T-2
8-Electron Molecules			
BH3	Borane	74	T-6
10-Electron Molecules			
CH4	Methane	75	T-10
NH3	Ammonia	76	T-14
H20	Water	77	T-18
HF	Hydrogen Fluoride	73	T-22
14-Electron Molecules			
C2H2	Acetylene	76	T-26
HCN	Hydrogen Cyanide	76	T-32
N2	Nitrogen	73	T-36

Table 7. (Continued)

Formula Name	Common Name	Geometry Reference	Text Page
CO	Carbon Monoxide	73	T-40
BF	Boron Fluoride	73	T-44

16-Electron Molecules

B2H6	Diborane	78	T-48
C2H4	Ethylene	76	T-56
N2H2	Diimide	76	T-62
H2CO	Formaldehyde	77	T-68

18-Electron Molecules

C2H6	Ethane	79	T-74
N2H4	Hydrazine	80	T-82
H2O2	Hydrogen Peroxide	81	T-88
F2	Fluorine	73	T-92
CH3OH	Methanol	82	T-96
CH3F	Methyl Fluoride	83	T-104

22-Electron Molecules

CO2	Carbon Dioxide	76	T-110
C3H4	Allene	84	T-116
CH2CO	Ketene	85	T-124
CH2N2	Diazomethane	76	T-132
NNO	Nitrous Oxide	76	T-140
NH2CN	Cyanamide	86	T-146
BH3CO	Borine Carbonyl	87	T-154
CH3CN	Methyl Cyanide	88	T-162
CH3NC	Methyl Isocyanide	89	T-170
C3H4*	Cyclopropene	90	T-178
CH2N2*	Diazirine	91	T-186

24-Electron Molecules

O3	Ozone	92	T-194
CF2	Difluorocarbene	93	T-198
FNO	Nitrozylfluoride	76	T-204
CHONH2	Formamide	94	T-210
CHOOH	Formic Acid	76	T-218
CHOF	Formyl Fluoride	76	T-224
CHOO-	Formate Anion	95	T-230
C3H6	Cyclopropane	96	T-236
C2H5N	Ethylenimine	97	T-246

41

continued

Table 7. (continued)

Formula Name	Common Name	Geometry Reference	Text Page
C2H4O	Ethylene Oxide	98	T-256
CH4N2	Diaziridine	99	T-264

26-Electron Molecules

CH2F2	Difluoromethane	100	T-274

30-Electron Molecules

C4H6	Bicyclobutane	101	T-280
C4H6*	Trans-butadiene	102	T-292
C3H4O	Trans-acrolein	103	T-306

32-Electron Molecules

BF3	Boron Trifluoride	104	T-320
F2CO	Carbonyl Fluoride	76	T-326
N2F2	Difluorodiazine	105	T-332

34-Electron Molecules

C3O2	Carbon Suboxide	106	T-338
CHF3	Trifluoromethane	107	T-346

38-Electron Molecules

CF2N2	Difluorodiazirine	108	T-354

42-Electron Molecules

CF4	Carbon Tetrafluoride	109	T-362

46-Electron Molecules

N2O4	Nitrogen Tetroxide	110	T-368

48-Electron Molecules

C2F4	Tetrafluoroethylene	111	T-378

Chapter 10

TABLES OF WAVE FUNCTIONS AND PROPERTIES

H2   Hydrogen

Molecular Geometry   Symmetry D-h

Center	Coordinates		
	X	Y	Z
H1	0.	0.	0.
H2	0.	0.	1.40165000
Mass	0.	0.	0.70082500

MO Symmetry Types and Occupancy

Representation	SIGM-G	SIGM-U
Number of MOs	2	2
Number Occupied	1	0

Energy Expectation Values

Total	-1.1266	Kinetic	1.1280
Electronic	-1.8400	Potential	-2.2546
Nuclear Repulsion	0.7134	Virial	1.9987

Electronic Potential	-2.9681
One-Electron Potential	-3.6231
Two-Electron Potential	0.6550

Moments of the Charge Distribution

Second

x^2	y^2	z^2	xy	xz	yz	r^2
-1.4636	-1.4636	-2.1245	0.	0.	0.	-5.0517
0.	0.	0.9823	0.	0.	0.	0.9823

Fourth (even)

x^4	y^4	z^4	x^2y^2	x^2z^2	y^2z^2	r^2x^2	r^2y^2	r^2z^2	r^4
-4.4748	-4.4748	-7.4309	-1.4916	-1.9337	-1.9337	-7.9001	-7.9001	-11.2982	-27.0983
0.	0.	0.4825	0.	0.	0.	0.	0.	0.4825	0.4825

Expectation Values at Atomic Centers

Center			Expectation Value		
	r^{-1}	δ	xr^{-3}	yr^{-3}	zr^{-3}
H1	-1.0981	0.3977	0.	0.	-0.0698

	$\dfrac{3x^2-r^2}{r^5}$	$\dfrac{3y^2-r^2}{r^5}$	$\dfrac{3z^2-r^2}{r^5}$	$\dfrac{3xy}{r^5}$	$\dfrac{3xz}{r^5}$	$\dfrac{3yz}{r^5}$
H1	-0.1913	-0.1913	0.3826	0.	0.	0.

Mulliken Charge

Center					Basis Function Type					Total
	S1	S2	S3	S4	X1	X2	Y1	Y2	Z1	Z2	
H1	0.4970	0.5030									1.0000

Overlap Populations

Pair of Atomic Centers

	H2
	H1
1SIGM-G	0.8070
Total SIGM-G	0.8070
Total	0.8070
Distance	1.4016

Molecular Orbitals and Eigenvalues

Center		S1	S2	S3	S4	Basis Function Type X1	X2	Y1	Y2	Z1	Z2
* 1SIGM-G	H1	0.31383	0.28133								
-0.5923	H2	0.31383	0.28133								
1SIGM-U	H1	0.11943	1.63327								
0.2596	H2	-0.11943	-1.63327								

Overlap Matrix

First Function   Second Function and Overlap

H1S1 with
     H1S2   .68301   H2S1   .44839   H2S2   .51604

H1S2 with
     H2S1   .51604   H2S2   .83993

BH3   Borane

Molecular Geometry   Symmetry D₃ₕ

Center	Coordinates		
	X	Y	Z
H1	1.13007049	0.	-1.95733950
H2	1.13007049	0.	1.95733950
H3	-2.26014099	0.	0.
B1	0.	0.	0.
Mass	0.	0.	0.

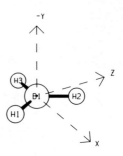

MO Symmetry Types and Occupancy

Representation	A1*	A1**	A2*	A2**	E*	E**
Number of MOs	6	0	0	2	8	0
Number Occupied	2	0	0	0	2	0

Energy Expectation Values

Total	-26.3742	Kinetic	26.3388
Electronic	-33.7773	Potential	-52.7130
Nuclear Repulsion	7.4031	Virial	2.0013

Electronic Potential	-60.1161
One-Electron Potential	-75.2491
Two-Electron Potential	15.1330

Moments of the Charge Distribution

Second

x^2	y^2	z^2	xy	xz	yz	r^2
-14.5590	-5.3116	-14.5590	0.	0.	0.	-34.4296
7.6624	0.	7.6624	0.	0.	0.	15.3247

Third

x^3	xy^2	xz^2	x^2y	y^3	yz^2	x^2z	y^2z	z^3	xyz
8.2676	0.	-8.2676	0.	0.	0.	0.	0.	0.	0.
-8.6590	0.	8.6590	0.	0.	0.	0.	0.	0.	0.

Fourth (even)

x^4	y^4	z^4	x^2y^2	x^2z^2	y^2z^2	r^2x^2	r^2y^2	r^2z^2	r^4
-90.0171	-18.1870	-90.0171	-13.7950	-30.0057	-13.7950	-133.8179	-45.7771	-133.8179	-313.4128
29.3558	0.	29.3558	0.	9.7853	0.	39.1411	0.	39.1411	78.2823

Expectation Values at Atomic Centers

Center			Expectation Value		
	r^{-1}	δ	xr^{-3}	yr^{-3}	zr^{-3}
H1	-1.1364	0.4084	0.0307	0.	-0.0531
B1	-11.4067	65.6327	0.	0.	0.

	$\dfrac{3x^2-r^2}{r^5}$	$\dfrac{3y^2-r^2}{r^5}$	$\dfrac{3z^2-r^2}{r^5}$	$\dfrac{3xy}{r^5}$	$\dfrac{3xz}{r^5}$	$\dfrac{3yz}{r^5}$
H1	-0.0452	-0.0998	0.1450	0.	-0.1647	0.
B1	-0.2842	0.5684	-0.2842	0.	0.	0.

Mulliken Charge

Center				Basis Function Type							Total
	S1	S2	S3	S4	X1	X2	Y1	Y2	Z1	Z2	
H1	0.5116	0.4635									0.9751
B1	0.0436	1.9559	0.7721	0.3075	0.9175	0.0803	0.	0.	0.9175	0.0803	5.0746

Overlap Populations

Pair of Atomic Centers

	H2 H1	B1 H1
1A1*	0.	0.0001
2A1*	0.0288	0.2812
Total A1*	0.0288	0.2813
1E*	-0.0709	0.2935
2E*	0.0204	0.2505
Total E*	-0.0505	0.5441
Total	-0.0218	0.8253
Distance	3.9147	2.2601

Molecular Orbitals and Eigenvalues

Center		Basis Function Type									
		S1	S2	S3	S4	X1	X2	Y1	Y2	Z1	Z2
* 1A1*	H1	0.00109	-0.00018								
-7.6283	H2	0.00109	-0.00018								
	H3	0.00109	-0.00018								
	B1	0.05293	0.97883	0.00432	0.00033	0.	0.	0.	0.	0.	0.
* 2A1*	H1	-0.16088	-0.12459								
-0.7060	H2	-0.16088	-0.12459								
	H3	-0.16088	-0.12459								
	B1	0.00908	0.19295	-0.43973	-0.17061	0.	0.	0.	0.	0.	0.
3A1*	H1	0.09188	1.32942								
0.3345	H2	0.09188	1.32942								
	H3	0.09188	1.32942								
	B1	-0.00022	0.04297	0.34058	-3.38081	0.	0.	0.	0.	0.	0.
4A1*	H1	-0.01346	0.49848								
0.4227	H2	-0.01346	0.49848								
	H3	-0.01346	0.49848								
	B1	0.01221	0.13853	-1.59349	0.47731	0.	0.	0.	0.	0.	0.
1A2**	H1	0.	0.								
0.0570	H2	0.	0.								
	H3	0.	0.								
	B1	0.	0.	0.	0.	0.	0.	0.41857	0.70551	0.	0.
2A2**	H1	0.	0.								
0.3691	H2	0.	0.								
	H3	0.	0.								
	B1	0.	0.	0.	0.	0.	0.	1.12566	-0.97189	0.	0.
* 1E*	H1	0.20217	0.17604								
-0.4939	H2	-0.26284	-0.22888								
	H3	0.06067	0.05283								
	B1	0.	0.	0.	0.	-0.11198	-0.01305	0.	0.	-0.49549	-0.05776
* 2E*	H1	-0.18678	-0.16265								
-0.4939	H2	-0.08169	-0.07114								
	H3	0.26848	0.23378								
	B1	0.	0.	0.	0.	-0.49549	-0.05776	0.	0.	0.11198	0.01305
3E*	H1	-0.01151	-0.16141								
0.2085	H2	-0.04950	-0.69390								
	H3	0.06102	0.85531								
	B1	0.	0.	0.	0.	-0.18300	1.51594	0.	0.	-0.06578	0.54489
4E*	H1	0.06381	0.89443								
0.2085	H2	-0.04188	-0.58700								
	H3	-0.02193	-0.30743								
	B1	0.	0.	0.	0.	0.06578	-0.54489	0.	0.	-0.18300	1.51594
5E*	H1	0.03326	1.33391								
0.4629	H2	-0.02537	-1.01743								
	H3	-0.00789	-0.31648								
	B1	0.	0.	0.	0.	-0.28350	-0.07166	0.	0.	1.21608	0.30738
6E*	H1	0.01009	0.40470								
0.4629	H2	0.02376	0.95285								
	H3	-0.03385	-1.35755								
	B1	0.	0.	0.	0.	-1.21608	-0.30738	0.	0.	-0.28350	-0.07166

Overlap Matrix

First Function	Second Function and Overlap

H1S1 with

H1S2	.68301	H2S1	.00454	H2S2	.07724	B1S1	.00170	B1S2	.03065	B1S3	.29267	B1S4	.32216
B1X1	.22748	B1X2	.16197	B1Z1	-.39401	B1Z2	-.28055						

H1S2 with

H2S1	.07724	H2S2	.25649	B1S1	.00880	B1S2	.11186	B1S3	.54755	B1S4	.67828	B1X1	.24255
B1X2	.28384	B1Z1	-.42010	B1Z2	-.49162								

B1S1 with

B1S2	.35489	B1S3	.00732	B1S4	.01371

B1S2 with

B1S3	.24693	B1S4	.17217

B1S3 with

B1S4	.80606

B1X1 with

B1X2	.55377

B1Y1 with

B1Y2	.55377

B1Z1 with

B1Z2	.55377

CH4   Methane

Molecular Geometry   Symmetry T_d

Center		Coordinates	
	X	Y	Z
H1	1.19309433	1.19309433	1.19309433
H2	-1.19309433	1.19309433	-1.19309433
H3	1.19309433	-1.19309433	-1.19309433
H4	-1.19309433	-1.19309433	1.19309433
C1	0.	0.	0.
Mass	0.	0.	0.

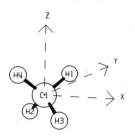

MO Symmetry Types and Occupancy

Representation	A1	A2	E	F1	F2
Number of MOs	6	0	0	0	12
Number Occupied	2	0	0	0	3

Energy Expectation Values

Total	-40.1823	Kinetic	40.1724
Electronic	-53.5741	Potential	-80.3547
Nuclear Repulsion	13.3918	Virial	2.0002
Electronic Potential	-93.7465		
One-Electron Potential	-119.7670		
Two-Electron Potential	26.0205		

Moments of the Charge Distribution

Second

x^2	y^2	z^2	xy	xz	yz	r^2
-11.9447	-11.9447	-11.9447	0.	0.	0.	-35.8342
5.6939	5.6939	5.6939	0.	0.	0.	17.0817

Third

x^3	xy^2	xz^2	x^2y	y^3	yz^2	x^2z	y^2z	z^3	xyz
0.	0.	0.	0.	0.	0.	0.	0.	0.	-5.7756
0.	0.	0.	0.	0.	0.	0.	0.	0.	6.7934

Fourth (even)

x^4	y^4	z^4	x^2y^2	x^2z^2	y^2z^2	r^2x^2	r^2y^2	r^2z^2	r^4
-50.7243	-50.7243	-50.7243	-21.1160	-21.1160	-21.1160	-92.9563	-92.9563	-92.9563	-278.8690
8.1051	8.1051	8.1051	8.1051	8.1051	8.1051	24.3153	24.3153	24.3153	72.9460

Expectation Values at Atomic Centers

Center			Expectation Value		
	r^{-1}	δ	xr^{-3}	yr^{-3}	zr^{-3}
H1	-1.1141	0.4163	0.0372	0.0372	0.0372
C1	-14.7545	116.2910	0.	0.	0.

	$\dfrac{3x^2-r^2}{r^5}$	$\dfrac{3y^2-r^2}{r^5}$	$\dfrac{3z^2-r^2}{r^5}$	$\dfrac{3xy}{r^5}$	$\dfrac{3xz}{r^5}$	$\dfrac{3yz}{r^5}$
H1	0.	0.	0.	0.1681	0.1681	0.1681
C1	0.	0.	0.	0.	0.	0.

Mulliken Charge

Center				Basis Function Type							Total
	S1	S2	S3	S4	X1	X2	Y1	Y2	Z1	Z2	
H1	0.5134	0.3023									0.8157
C1	0.0427	1.9564	0.7562	0.6768	0.8999	0.2019	0.8999	0.2019	0.8999	0.2019	6.7374

Overlap Populations

Pair of Atomic Centers

	H2 H1	C1 H1
1A1	0.	-0.0003
2A1	0.0057	0.1767
Total A1	0.0057	0.1764
1F2	-0.0532	0.1917
2F2	0.0532	0.1917
3F2	-0.0532	0.1917
Total F2	-0.0532	0.5750
Total	-0.0475	0.7514
Distance	3.3746	2.0665

Molecular Orbitals and Eigenvalues

Center		S1	S2	S3	S4	X1	X2	Y1	Y2	Z1	Z2
* 1A1	H1	-0.00056	0.00094								
-11.2080	H2	-0.00056	0.00094								
	H3	-0.00056	0.00094								
	H4	-0.00056	0.00094								
	C1	-0.05190	-0.97895	-0.00446	-0.00343	0.	0.	0.	0.	0.	0.
* 2A1	H1	-0.13215	-0.02511								
-0.9437	H2	-0.13215	-0.02511								
	H3	-0.13215	-0.02511								
	H4	-0.13215	-0.02511								
	C1	0.00936	0.20409	-0.43447	-0.36015	0.	0.	0.	0.	0.	0.
3A1	H1	-0.01812	1.32648								
0.3048	H2	-0.01812	1.32648								
	H3	-0.01812	1.32648								
	H4	-0.01812	1.32648								
	C1	0.00660	0.15339	-0.29876	-3.33841	0.	0.	0.	0.	0.	0.
* 1F2	H1	-0.16800	-0.13546								
-0.5417	H2	0.16800	0.13546								
	H3	-0.16800	-0.13546								
	H4	0.16800	0.13546								
	C1	0.	0.	0.	0.	-0.51837	-0.13122	0.	0.	0.	0.
* 2F2	H1	-0.16800	-0.13546								
-0.5417	H2	-0.16800	-0.13546								
	H3	0.16800	0.13546								
	H4	0.16800	0.13546								
	C1	0.	0.	0.	0.	0.	0.	-0.51837	-0.13122	0.	0.
* 3F2	H1	-0.16800	-0.13546								
-0.5417	H2	0.16800	0.13546								
	H3	0.16800	0.13546								
	H4	-0.16800	-0.13546								
	C1	0.	0.	0.	0.	0.	0.	0.	0.	-0.51837	-0.13122
4F2	H1	-0.07295	-1.12388								
0.3132	H2	0.07295	1.12388								
	H3	-0.07295	-1.12388								
	H4	0.07295	1.12388								
	C1	0.	0.	0.	0.	0.07212	2.01205	0.	0.	0.	0.
5F2	H1	-0.07295	-1.12388								
0.3132	H2	-0.07295	-1.12388								
	H3	0.07295	1.12388								
	H4	0.07295	1.12388								
	C1	0.	0.	0.	0.	0.	0.	0.07212	2.01205	0.	0.
6F2	H1	-0.07295	-1.12388								
0.3132	H2	0.07295	1.12388								
	H3	0.07295	1.12388								
	H4	-0.07295	-1.12388								
	C1	0.	0.	0.	0.	0.	0.	0.	0.	0.07212	2.01205
7F2	H1	0.05967	-0.46951								
0.4894	H2	-0.05967	0.46951								
	H3	0.05967	-0.46951								
	H4	-0.05967	0.46951								
	C1	0.	0.	0.	0.	0.98061	-0.41941	0.	0.	0.	0.

Center		S1	S2	S3	S4	X1	X2	Y1	Y2	Z1	Z2
					Basis Function Type						
8F2	H1	0.05967	-0.46951								
0.4894	H2	0.05967	-0.46951								
	H3	-0.05967	0.46951								
	H4	-0.05967	0.46951								
	C1	0.	0.	0.	0.	0.	0.	0.98061	-0.41941	0.	0.
9F2	H1	0.05967	-0.46951								
0.4894	H2	-0.05967	0.46951								
	H3	-0.05967	0.46951								
	H4	0.05967	-0.46951								
	C1	0.	0.	0.	0.	0.	0.	0.	0.	0.98061	-0.41941

Overlap Matrix

First Function   **Second Function and Overlap**

H1S1 with
    H1S2 .68301   H2S1 .01678   H2S2 .13502   C1S1 .00218   C1S2 .03374   C1S3 .28178   C1S4 .37073
    C1X1 .24724   C1X2 .25032   C1Y1 .24724   C1Y2 .25032   C1Z1 .24724   C1Z2 .25032

H1S2 with
    H2S2 .13502   H2S2 .36382   C1S1 .00762   C1S2 .09734   C1S3 .49455   C1S4 .70403   C1X1 .20808
    C1X2 .35186   C1Y1 .20808   C1Y2 .35186   C1Z1 .20808   C1Z2 .35186

C1S1 with
    C1S2 .35436   C1S3 .00805   C1S4 .01419

C1S2 with
    C1S3 .26226   C1S4 .17745

C1S3 with
    C1S4 .79970

C1X1 with
    C1X2 .53875

C1Y1 with
    C1Y2 .53875

C1Z1 with
    C1Z2 .53875

NH3   Ammonia

Molecular Geometry   Symmetry C$_{3v}$

Center	Coordinates		
	X	Y	Z
H1	1.77599999	0.	0.71996000
H2	-0.88800000	1.53800000	0.71996000
H3	-0.88800000	-1.53800000	0.71996000
N1	0.	0.	0.
Mass	0.	0.	0.12784407

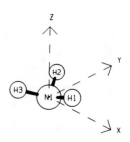

MO Symmetry Types and Occupancy

Representation	A1	A2	E
Number of MOs	8	0	8
Number Occupied	3	0	2

Energy Expectation Values

Total	-56.1714	Kinetic	56.1579
Electronic	-68.1050	Potential	-112.3293
Nuclear Repulsion	11.9336	Virial	2.0002

Electronic Potential	-124.2629
One-Electron Potential	-155.6022
Two-Electron Potential	31.3393

Moments of the Charge Distribution

First

x	y	z	Dipole Moment
0.	0.	0.0356	0.9170
0.	0.	0.8814	

Second

x^2	y^2	z^2	xy	xz	yz	r^2
-9.3223	-9.3221	-7.9353	0.	0.	0.	-26.5798
4.7313	4.7309	1.1662	0.	0.	0.	10.6284

Third

x^3	xy^2	xz^2	x^2y	y^3	yz^2	x^2z	y^2z	z^3	xyz
-3.0968	3.0964	0.	0.	0.	0.	-1.7565	-1.7563	0.7219	0.
4.2014	-4.2010	0.	0.	0.	0.	2.8015	2.8012	0.6082	0.

Fourth (even)

x^4	y^4	z^4	x^2y^2	x^2z^2	y^2z^2	r^2x^2	r^2y^2	r^2z^2	r^4
-36.5166	-36.5145	-29.1302	-12.1719	-10.6696	-10.6694	-59.3581	-59.3557	-50.4692	-169.1830
11.1924	11.1907	0.3706	3.7305	1.6588	1.6587	16.5817	16.5798	3.6881	36.8496

Fourth (odd)

x^3y	xy^3	xyz^2	x^3z	xy^2z	xz^3	x^2yz	y^3z	yz^3
0.	0.	0.	-1.5991	1.5989	-0.0001	0.	0.	0.
0.	0.	0.	2.4877	-2.4875	0.	0.	0.	0.

Expectation Values at Atomic Centers

Center			Expectation Value		
	r^{-1}	δ	xr^{-3}	yr^{-3}	zr^{-3}
H1	-1.0463	0.4213	0.0889	0.	0.0075
N1	-18.3709	188.7119	0.	0.	-0.1711

	$\dfrac{3x^2-r^2}{r^5}$	$\dfrac{3y^2-r^2}{r^5}$	$\dfrac{3z^2-r^2}{r^5}$	$\dfrac{3xy}{r^5}$	$\dfrac{3xz}{r^5}$	$\dfrac{3yz}{r^5}$
H1	0.3779	-0.2004	-0.1775	0.	0.2271	0.
N1	0.5306	0.5306	-1.0612	0.	0.	0.

Mulliken Charge

Center					Basis Function Type						Total
	S1	S2	S3	S4	X1	X2	Y1	Y2	Z1	Z2	
H1	0.5090	0.2023									0.7114
N1	0.0432	1.9560	0.8621	0.8104	0.9202	0.2469	0.9202	0.2469	1.1838	0.6762	7.8659

Overlap Populations

Pair of Atomic Centers

	H2	N1
	H1	H1
1A1	0.	-0.0002
2A1	0.0049	0.1881
3A1	0.0017	0.0112
Total A1	0.0066	0.1990
1E	0.	0.
2E	-0.0781	0.4790
Total E	-0.0782	0.4790
Total	-0.0716	0.6781
Distance	3.0761	1.9164

Molecular Orbitals and Eigenvalues

	Center		Basis Function Type								
		S1	S2	S3	S4	X1	X2	Y1	Y2	Z1	Z2
• 1A1	H1	0.00012	-0.00063								
-15.5210	H2	0.00012	-0.00063								
	H3	0.00012	-0.00063								
	N1	0.05194	0.97899	0.00356	0.00297	-0.00000	-0.00000	-0.00000	0.00000	0.00173	-0.00025
• 2A1	H1	0.13597	0.01169								
-1.1452	H2	0.13598	0.01169								
	H3	0.13598	0.01169								
	N1	-0.00979	-0.21236	0.45734	0.40350	-0.00001	-0.00000	-0.00000	0.00000	0.12101	0.03577
• 3A1	H1	-0.05523	-0.01297								
-0.4125	H2	-0.05522	-0.01296								
	H3	-0.05522	-0.01296								
	N1	-0.00329	-0.06840	0.18393	0.16687	-0.00001	-0.00000	0.00000	0.00000	-0.63926	-0.41709
4A1	H1	0.02622	1.06666								
0.2358	H2	0.02621	1.06663								
	H3	0.02621	1.06659								
	N1	0.00486	0.12265	-0.13414	-2.03223	-0.00002	-0.00002	-0.00000	-0.00001	-0.18269	-0.55982
• 1E	H1	-0.00008	-0.00005								
-0.6236	H2	0.24808	0.16958								
	H3	-0.24801	-0.16953								
	N1	-0.00000	-0.00000	0.00000	-0.00000	-0.00014	-0.00004	0.54087	0.15379	0.00000	-0.00000
• 2E	H1	0.28641	0.19579								
-0.6236	H2	-0.14314	-0.09785								
	H3	-0.14327	-0.09794								
	N1	-0.00000	-0.00000	0.00001	0.00002	0.54086	0.15379	0.00014	0.00004	-0.00001	-0.00000
3E	H1	0.05548	2.01770								
0.3430	H2	-0.03306	-1.20269								
	H3	-0.02241	-0.81520								
	N1	-0.00000	-0.00000	-0.00000	0.00015	-0.39903	-1.43523	0.04424	0.15913	0.00000	0.00004
4E	H1	0.00615	0.22370								
0.3431	H2	0.04497	1.63558								
	H3	-0.05112	-1.85932								
	N1	-0.00000	-0.00000	0.00000	0.00003	-0.04424	-0.15912	-0.39903	-1.43520	0.00000	0.00001
5E	H1	-0.00021	-0.00024								
0.6720	H2	-0.32897	-0.37756								
	H3	0.32918	0.37781								
	N1	0.00000	0.00000	-0.00000	0.00000	-0.00039	0.00089	-0.70731	1.59445	-0.00000	-0.00000
6E	H1	-0.37997	-0.43611								
0.6720	H2	0.19018	0.21832								
	H3	0.18981	0.21790								
	N1	0.00000	0.00000	0.00001	-0.00011	-0.70733	1.59449	0.00040	-0.00089	0.00006	-0.00010

NH3

Overlap Matrix

First Function   Second Function and Overlap

H1S1 with
H1S2	.68301	H2S1	.03204	H2S2	.17749	N1S1	.00254	N1S2	.03621	N1S3	.27096	N1S4	.40406
N1X1	.36157	N1X2	.47351	N1Z1	.14658	N1Z2	.19195						

H1S2 with

H2S1	.17749	H2S2	.43165	N1S1	.00670	N1S2	.08530	N1S3	.44535	N1S4	.70020	N1X1	.24977
N1X2	.54560	N1Z1	.10125	N1Z2	.22118								

N1S1 with

N1S2	.35584	N1S3	.00894	N1S4	.01446

N1S2 with

N1S3	.27246	N1S4	.17925

N1S3 with

N1S4	.79352

N1X1 with

N1X2	.52739

N1Y1 with

N1Y2	.52739

N1Z1 with

N1Z2	.52739

H2O   Water

Molecular Geometry   Symmetry C$_{2v}$

Center		Coordinates	
	X	Y	Z
H1	0.	1.43045600	1.10711800
H2	0.	-1.43045600	1.10711800
O1	0.	0.	0.
Mass	0.	0.	0.12390052

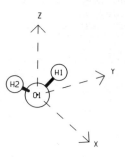

MO Symmetry Types and Occupancy

Representation	A1	A2	B1	B2
Number of MOs	8	0	4	2
Number Occupied	3	0	1	1

Energy Expectation Values

Total	-76.0035	Kinetic	76.0751
Electronic	-85.1985	Potential	-152.0786
Nuclear Repulsion	9.1950	Virial	1.9991

Electronic Potential	-161.2736
One-Electron Potential	-199.0612
Two-Electron Potential	37.7876

Moments of the Charge Distribution

First

x	y	z	Dipole Moment
0.	0.	0.0795	1.0548
0.	0.	0.9752	

Second

x^2	y^2	z^2	xy	xz	yz	r^2
-5.4885	-7.2251	-6.5062	0.	0.	0.	-19.2198
0.	4.0924	2.0562	0.	0.	0.	6.1487

Third

x^3	xy^2	xz^2	x^2y	y^3	yz^2	x^2z	y^2z	z^3	xyz
0.	0.	0.	0.	0.	0.	0.1470	-2.4318	-0.7962	0.
0.	0.	0.	0.	0.	0.	0.	4.0237	1.8858	0.

Fourth (even)

x^4	y^4	z^4	x^2y^2	x^2z^2	y^2z^2	r^2x^2	r^2y^2	r^2z^2	r^4
-15.3089	-23.4559	-18.6936	-5.6945	-5.5165	-8.2256	-26.5199	-37.3759	-32.4357	-96.3315
0.	8.3739	1.8710	0.	0.	3.9562	0.	12.3301	5.8272	18.1573

Expectation Values at Atomic Centers

Center			Expectation Value		
	r^{-1}	δ	xr^{-3}	yr^{-3}	zr^{-3}
H1	-0.9721	0.4155	0.	0.1010	0.0372
O1	-22.3409	286.8897	0.	0.	-0.2482

	$\dfrac{3x^2-r^2}{r^5}$	$\dfrac{3y^2-r^2}{r^5}$	$\dfrac{3z^2-r^2}{r^5}$	$\dfrac{3xy}{r^5}$	$\dfrac{3xz}{r^5}$	$\dfrac{3yz}{r^5}$
H1	-0.3132	0.2950	0.0182	0.	0.	0.3806
O1	-2.0487	1.8086	0.2400	0.	0.	0.

Mulliken Charge

Center				Basis Function Type							Total
	S1	S2	S3	S4	X1	X2	Y1	Y2	Z1	Z2	
H1	0.4930	0.1211									0.6142
O1	0.0426	1.9567	0.9325	0.9225	1.3640	0.6360	0.9740	0.2849	1.1378	0.5206	8.7717

Overlap Populations

Pair of Atomic Centers

	H2	O1
	H1	H1
1A1	0.	-0.0001
2A1	0.0055	0.2022
3A1	0.0253	0.0339
Total A1	0.0307	0.2359
1B1	-0.1032	0.3310
Total B1	-0.1032	0.3310
1B2	0.	0.
Total B2	0.	0.
Total	-0.0725	0.5669
Distance	2.8609	1.8088

Molecular Orbitals and Eigenvalues

Center		S1	S2	S3	S4	X1	X2	Y1	Y2	Z1	Z2
· 1A1	H1	-0.00010	-0.00035								
-20.5544	H2	-0.00010	-0.00035								
	O1	0.05112	0.97934	0.00373	0.00220	0.	0.	0.	0.	0.00166	-0.00025
· 2A1	H1	0.13814	0.00678								
-1.3624	H2	0.13814	0.00678								
	O1	-0.01032	-0.22752	0.48846	0.44416	0.	0.	0.	0.	0.12892	0.03249
· 3A1	H1	0.13999	0.06026								
-0.5669	H2	0.13999	0.06026								
	O1	0.00372	0.08113	-0.19087	-0.26781	0.	0.	0.	0.	0.62613	0.32605
4A1	H1	-0.05233	-1.05961								
0.2183	H2	-0.05233	-1.05961								
	O1	-0.00377	-0.09577	0.10504	1.25806	0.	0.	0.	0.	0.26811	0.52605
· 1B1	H1	-0.25564	-0.13108								
-0.7174	H2	0.25564	0.13108								
	O1	0.	0.	0.	0.	0.	0.	-0.57445	-0.17668	0.	0.
2B1	H1	-0.03877	-1.57215								
0.3109	H2	0.03877	1.57215								
	O1	0.	0.	0.	0.	0.	0.	0.40183	1.02747	0.	0.
· 1B2	H1	0.	0.								
-0.5064	H2	0.	0.								
	O1	0.	0.	0.	0.	-0.72883	-0.40888	0.	0.	0.	0.

Overlap Matrix

H1S1 with
    H1S2 .68301   H2S1 .04941   H2S2 .21283   O1S1 .00266   O1S2 .03666   O1S3 .25351   O1S4 .42136
    O1Y1 .27590   O1Y2 .44025   O1Z1 .21354   O1Z2 .34074

H1S2 with
    H2S1 .21283   H2S2 .48350   O1S1 .00581   O1S2 .07448   O1S3 .39468   O1S4 .67246   O1Y1 .16340
    O1Y2 .43420   O1Z1 .12647   O1Z2 .33605

O1S1 with
    O1S2 .35461   O1S3 .00974   O1S4 .01469

O1S2 with
    O1S3 .28286   O1S4 .18230

O1S3 with
    O1S4 .79029

O1X1 with
    O1X2 .50607

O1Y1 with
    O1Y2 .50607

O1Z1 with
    O1Z2 .50607

H2O   T-21

HF   Hydrogen Fluoride

Molecular Geometry   Symmetry C-ᵥ

Center   Coordinates
                X   Y   Z
H1   0.   0.   1.73300000
F1   0.   0.   0.

Mass   0.   0.   0.08729876

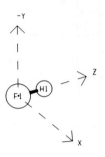

MO Symmetry Types and Occupancy

Representation   SIGMA PI-X PI-Y
Number of MOs   8   2   2
Number Occupied   3   1   1

Energy Expectation Values

Total   -100.0150   Kinetic   100.0102
Electronic   -105.2083   Potential   -200.0252
Nuclear Repulsion   5.1933   Virial   2.0000

Electronic Potential   -205.2185
One-Electron Potential -250.5796
Two-Electron Potential   45.3612

Moments of the Charge Distribution

First

     x   y   z   Dipole Moment
    0.   0.   0.0749   0.9349
    0.   0.   0.8600

Second

    x^2 y^2 z^2 xy   xz   yz r^2
  -4.1326   -4.1326   -5.2560   0.   0.   0.   -13.5212
    0.   0.   2.7769   0.   0.   0.   2.7769

Third

    x^3 xy^2 xz^2 x^2y y^3 yz^2 x^2z y^2z z^3 xyz
    0.   0.   0.   0.   0.   0.   0.0178   0.0178   -2.2409   0.
    0.   0.   0.   0.   0.   0.   0.   0.   4.4511   0.

Fourth (even)

    x^4 y^4 z^4 x^2y^2 x^2z^2 y^2z^2 r^2x^2 r^2y^2 r^2z^2 r^4
  -8.7354   -8.7354   -14.3964   -2.9118   -3.3297   -3.3297   -14.9769   -14.9769   -21.0557   -51.0095
    0.   0.   7.3356   0.   0.   0.   0.   0.   7.3356   7.3356

Expectation Values at Atomic Centers

Center			Expectation Value		
	r^{-1}	δ	xr^{-3}	yr^{-3}	zr^{-3}
H1	-0.8775	0.3922	0.	0.	0.1168
F1	-26.5906	413.4284	0.	0.	-0.2435

	$\dfrac{3x^2-r^2}{r^5}$	$\dfrac{3y^2-r^2}{r^5}$	$\dfrac{3z^2-r^2}{r^5}$	$\dfrac{3xy}{r^5}$	$\dfrac{3xz}{r^5}$	$\dfrac{3yz}{r^5}$
H1	-0.3170	-0.3170	0.6340	0.	0.	0.
F1	-1.5643	-1.5643	3.1286	0.	0.	0.

Mulliken Charge

Center				Basis Function Type							Total
	S1	S2	S3	S4	X1	X2	Y1	Y2	Z1	Z2	
H1	0.4598	0.0548									0.5146
F1	0.0435	1.9558	1.0168	0.9384	1.4253	0.5747	1.4253	0.5747	1.1140	0.4169	9.4854

Overlap Populations

Pair of Atomic Centers

	F1 H1
1SIGMA	0.
2SIGMA	0.1948
3SIGMA	0.2662
Total SIGMA	0.4609
1PI-X	0.
Total PI-X	0.
1PI-Y	0.
Total PI-Y	0.
Total	0.4609
Distance	1.7330

HF

Molecular Orbitals and Eigenvalues

Center		S1	S2	S3	S4	X1	X2	Y1	Y2	Z1	Z2
• 1SIGMA	H1	0.00009	0.00008								
-26.2813	F1	-0.05174	-0.97892	-0.00449	-0.00109	0.	0.	0.	0.	-0.00114	0.00028
• 2SIGMA	H1	-0.13402	0.00141								
-1.6026	F1	0.01130	0.24298	-0.54200	-0.47764	0.	0.	0.	0.	-0.10044	-0.02384
• 3SIGMA	H1	-0.28768	-0.08672								
-0.7533	F1	-0.00302	-0.06484	0.15100	0.21752	0.	0.	0.	0.	-0.62980	-0.26094
4SIGMA	H1	-0.09726	-1.47634								
0.2123	F1	-0.00281	-0.06854	0.08596	0.80044	0.	0.	0.	0.	0.30878	0.51505
• 1PI-X	H1	0.	0.								
-0.6432	F1	0.	0.	0.	0.	-0.75524	-0.38094	0.	0.	0.	0.
• 1PI-Y	H1	0.	0.								
-0.6432	F1	0.	0.	0.	0.	0.	0.	-0.75524	-0.38094	0.	0.

Basis Function Type

Overlap Matrix

First Function   Second Function and Overlap

H1S1 with
      H1S2 .68302   F1S1 .00267   F1S2 .03566   F1S3 .23749   F1S4 .42731   F1Z1 .30394   F1Z2 .58878

H1S2 with
      F1S1 .00512   F1S2 .06523   F1S3 .35366   F1S4 .64005   F1Z1 .15953   F1Z2 .49897

F1S1 with
      F1S2 .35647   F1S3 .01003   F1S4 .01474

F1S2 with
      F1S3 .28653   F1S4 .18096

F1S3 with
      F1S4 .78531

F1X1 with
      F1X2 .49444

F1Y1 with
      F1Y2 .49444

F1Z1 with
      F1Z2 .49444

C2H2   Acetylene

Molecular Geometry   Symmetry D-_h

Center	X	Coordinates Y	Z
H1	0.	0.	3.13970000
H2	0.	0.	-3.13970000
C1	0.	0.	1.13670000
C2	0.	0.	-1.13670000
Mass	0.	0.	0.

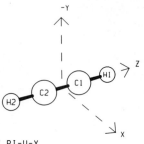

MO Symmetry Types and Occupancy

Representation	SIGM-G	SIGM-U	PI-G-X	PI-U-X	PI-G-Y	PI-U-Y
Number of MOs	8	8	2	2	2	2
Number Occupied	3	2	0	1	0	1

Energy Expectation Values

Total	-76.7919	Kinetic	76.7035
Electronic	-101.5835	Potential	-153.4953
Nuclear Repulsion	24.7917	Virial	2.0012

Electronic Potential	-178.2870
One-Electron Potential	-228.1976
Two-Electron Potential	49.9106

Moments of the Charge Distribution

Second

x^2	y^2	z^2	xy	xz	yz	r^2
-10.3603	-10.3603	-40.2700	0.	0.	0.	-60.9906
0.	0.	35.2205	0.	0.	0.	35.2205

Fourth (even)

x^4	y^4	z^4	x^2y^2	x^2z^2	y^2z^2	r^2x^2	r^2y^2	r^2z^2	r^4
-43.7637	-43.7637	-307.3413	-14.5879	-32.5953	-32.5954	-90.9470	-90.9470	-372.5321	-554.4260
0.	0.	214.3830	0.	0.	0.	0.	0.	214.3830	214.3830

Expectation Values at Atomic Centers

Center			Expectation Value		
	r^{-1}	8	xr^{-3}	yr^{-3}	zr^{-3}
H1	-1.0158	0.3983	0.	0.	0.0668
C1	-14.7152	116.6025	0.	0.	0.0225

	$\dfrac{3x^2-r^2}{r^5}$	$\dfrac{3y^2-r^2}{r^5}$	$\dfrac{3z^2-r^2}{r^5}$	$\dfrac{3xy}{r^5}$	$\dfrac{3xz}{r^5}$	$\dfrac{3yz}{r^5}$
H1	-0.1927	-0.1927	0.3854	0.	0.	0.
C1	0.0625	0.0625	-0.1250	0.	0.	0.

Mulliken Charge

Center					Basis Function Type						Total
	S1	S2	S3	S4	X1	X2	Y1	Y2	Z1	Z2	
H1	0.4967	0.2450									0.7417
C1	0.0428	1.9564	0.7802	0.4278	0.7495	0.2505	0.7495	0.2505	0.9988	0.0522	6.2583

Overlap Populations

		Pair of Atomic Centers		
	H2	C1	C1	C2
	H1	H1	H2	C1
1SIGM-G	0.	-0.0003	0.	0.0016
2SIGM-G	0.	0.0406	0.0052	0.7130
3SIGM-G	0.0023	0.3419	-0.0390	0.1131
Total SIGM-G	0.0023	0.3822	-0.0338	0.8277
1SIGM-U	0.	0.0001	0.	-0.0034
2SIGM-U	-0.0018	0.3702	0.0048	0.0361
Total SIGM-U	-0.0018	0.3703	0.0048	0.0327
1PI-U-X	0.	0.	0.	0.5617
Total PI-U-X	0.	0.	0.	0.5617
1PI-U-Y	0.	0.	0.	0.5617
Total PI-U-Y	0.	0.	0.	0.5617
Total	0.0005	0.7525	-0.0291	1.9838
Distance	6.2794	2.0030	4.2764	2.2734

Molecular Orbitals and Eigenvalues

Center		S1	S2	S3	S4	X1	X2	Y1	Y2	Z1	Z2
* 1SIGM-G	H1	0.00083	-0.00125								
-11.2595	H2	0.00083	-0.00125								
	C1	0.03669	0.69188	0.00373	0.00134	0.	0.	0.	0.	-0.00112	0.00122
	C2	0.03669	0.69188	0.00373	0.00134	0.	0.	0.	0.	0.00112	-0.00122
* 2SIGM-G	H1	-0.08081	-0.00554								
-1.0411	H2	-0.08081	-0.00554								
	C1	0.00816	0.17814	-0.37151	-0.18008	0.	0.	0.	0.	0.18226	-0.03485
	C2	0.00816	0.17814	-0.37151	-0.18008	0.	0.	0.	0.	-0.18226	0.03485
* 3SIGM-G	H1	0.21013	0.12136								
-0.6821	H2	0.21013	0.12136								
	C1	-0.00044	-0.00595	0.04710	0.07563	0.	0.	0.	0.	0.43817	0.04027
	C2	-0.00044	-0.00595	0.04710	0.07563	0.	0.	0.	0.	-0.43817	-0.04027
4SIGM-G	H1	0.05781	0.76002								
0.3456	H2	0.05781	0.76002								
	C1	0.00066	0.02935	0.06153	-0.55491	0.	0.	0.	0.	-0.51093	0.33425
	C2	0.00066	0.02935	0.06153	-0.55491	0.	0.	0.	0.	0.51093	-0.33425
5SIGM-G	H1	0.23295	2.31631								
0.5362	H2	0.23295	2.31631								
	C1	-0.00461	-0.02591	0.75936	-3.02691	0.	0.	0.	0.	0.08381	-2.66314
	C2	-0.00461	-0.02591	0.75936	-3.02691	0.	0.	0.	0.	-0.08381	2.66314
6SIGM-G	H1	-0.09621	3.13353								
0.6664	H2	-0.09621	3.13353								
	C1	0.00557	0.06368	-0.70523	-2.31007	0.	0.	0.	0.	-0.10474	-3.22138
	C2	0.00557	0.06368	-0.70523	-2.31007	0.	0.	0.	0.	0.10474	3.22138
* 1SIGM-U	H1	0.00031	0.00023								
-11.2558	H2	-0.00031	-0.00023								
	C1	0.03672	0.69263	0.00196	0.00382	0.	0.	0.	0.	0.00064	-0.00160
	C2	-0.03672	-0.69263	-0.00196	-0.00382	0.	0.	0.	0.	0.00064	-0.00160
* 2SIGM-U	H1	0.21539	0.10631								
-0.7647	H2	-0.21539	-0.10631								
	C1	-0.00527	-0.11573	0.24417	0.14885	0.	0.	0.	0.	0.26134	0.04281
	C2	0.00527	0.11573	-0.24417	-0.14885	0.	0.	0.	0.	0.26134	0.04281
3SIGM-U	H1	-0.04785	0.30535								
0.2361	H2	0.04785	-0.30535								
	C1	-0.00357	-0.10203	0.02824	3.72628	0.	0.	0.	0.	0.10835	-2.28387
	C2	0.00357	0.10203	-0.02824	-3.72628	0.	0.	0.	0.	0.10835	-2.28387
4SIGM-U	H1	-0.14901	-2.14272								
0.3619	H2	0.14901	2.14272								
	C1	-0.00295	-0.05494	0.22862	-0.76006	0.	0.	0.	0.	-0.04206	2.29301
	C2	0.00295	0.05494	-0.22862	0.76006	0.	0.	0.	0.	-0.04206	2.29301
1PI-G-X	H1	0.	0.								
0.1760	H2	0.	0.								
	C1	0.	0.	0.	0.	0.37759	0.97902	0.	0.	0.	0.
	C2	0.	0.	0.	0.	-0.37759	-0.97902	0.	0.	0.	0.
2PI-G-X	H1	0.	0.								
0.6354	H2	0.	0.								
	C1	0.	0.	0.	0.	-0.84350	1.23133	0.	0.	0.	0.
	C2	0.	0.	0.	0.	0.84350	-1.23133	0.	0.	0.	0.

Center		Basis Function Type									
		S1	S2	S3	S4	X1	X2	Y1	Y2	Z1	Z2
* 1PI-U-X	H1	0.	0.								
-0.4136	H2	0.	0.								
	C1	0.	0.	0.	0.	0.48931	0.17188	0.	0.	0.	0.
	C2	0.	0.	0.	0.	0.48931	0.17188	0.	0.	0.	0.
2PI-U-X	H1	0.	0.								
0.4636	H2	0.	0.								
	C1	0.	0.	0.	0.	0.61070	-0.64189	0.	0.	0.	0.
	C2	0.	0.	0.	0.	0.61070	-0.64189	0.	0.	0.	0.
1PI-G-Y	H1	0.	0.								
0.1760	H2	0.	0.								
	C1	0.	0.	0.	0.	0.	0.	0.37759	0.97902	0.	0.
	C2	0.	0.	0.	0.	0.	0.	-0.37759	-0.97902	0.	0.
2PI-G-Y	H1	0.	0.								
0.6354	H2	0.	0.								
	C1	0.	0.	0.	0.	0.	0.	-0.84350	1.23133	0.	0.
	C2	0.	0.	0.	0.	0.	0.	0.84350	-1.23133	0.	0.
* 1PI-U-Y	H1	0.	0.								
-0.4136	H2	0.	0.								
	C1	0.	0.	0.	0.	0.	0.	0.48931	0.17188	0.	0.
	C2	0.	0.	0.	0.	0.	0.	0.48931	0.17188	0.	0.
2PI-U-Y	H1	0.	0.								
0.4636	H2	0.	0.								
	C1	0.	0.	0.	0.	0.	0.	0.61070	-0.64189	0.	0.
	C2	0.	0.	0.	0.	0.	0.	0.61070	-0.64189	0.	0.

Overlap Matrix

First Function		Second Function and Overlap				

H1S1 with
H1S2 .68302   H2S2 .00257   C1S1 .00258   C1S2 .03907   C1S3 .30301   C1S4 .38268   C1Z1 .44764
C1Z2 .43112

H1S2 with
H2S1 .00257   H2S2 .03017   C1S1 .00798   C1S2 .10171   C1S3 .51116   C1S4 .71887   C1Z1 .36090
C1Z2 .60144

H2S1 with
C1S2 .00002   C1S3 .00565   C1S4 .06662   C1Z1 -.02481   C1Z2 -.22460

H2S2 with
C1S1 .00063   C1S2 .00898   C1S3 .08206   C1S4 .22708   C1Z1 -.12977   C1Z2 -.47508

C1S1 with
C1S2 .35436   C1S3 .00805   C1S4 .01419   C2S3 .00311   C2S4 .00661   C2Z1 .00786   C2Z2 .00997

C1S2 with
C1S3 .26226   C1S4 .17745   C2S2 .00012   C2S3 .04458   C2S4 .08483   C2Z1 .10188   C2Z2 .12472

C1S3 with
C1S4 .79970   C2S1 .00311   C2S2 .04458   C2S3 .31236   C2S4 .44887   C2Z1 .43765   C2Z2 .55128

C1S4 with
C2S1 .00661   C2S2 .08483   C2S3 .44887   C2S4 .68218   C2Z1 .31634   C2Z2 .61373

C1X1 with
C1X2 .53875   C2X1 .25731   C2X2 .33798

C1X2 with
C2X1 .33798   C2X2 .74368

C1Y1 with
C1Y2 .53875   C2Y1 .25731   C2Y2 .33798

C1Y2 with
C2Y1 .33798   C2Y2 .74368

C1Z1 with
C1Z2 .53875   C2S1 -.00786   C2S2 -.10188   C2S3 -.43765   C2S4 -.31634   C2Z1 -.34453   C2Z2 .02324

C1Z2 with
C2S1 -.00997   C2S2 -.12472   C2S3 -.55128   C2S4 -.61373   C2Z1 .02324   C2Z2 .30320

HCN   Hydrogen Cyanide

Molecular Geometry   Symmetry C-v

Center		Coordinates	
	X	Y	Z
H1	0.	0.	-2.00913000
C1	0.	0.	0.
N1	0.	0.	2.18331999
Mass	0.	0.	1.05692096

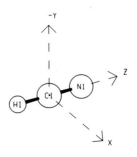

MO Symmetry Types and Occupancy

Representation	SIGMA	PI-X	PI-Y
Number of MOs	14	4	4
Number Occupied	5	1	1

Energy Expectation Values

Total	-92.8286	Kinetic	92.6959
Electronic	-116.7214	Potential	-185.5245
Nuclear Repulsion	23.8928	Virial	2.0014

Electronic Potential	-209.4173
One-Electron Potential	-264.9227
Two-Electron Potential	55.5054

Moments of the Charge Distribution

First

x	y	z	Dipole Moment
0.	0.	0.2153	1.3075
0.	0.	-1.5228	

Second

x^2	y^2	z^2	xy	xz	yz	r^2
-8.8310	-8.8327	-32.2789	0.	0.	0.	-49.9426
0.	0.	24.9846	0.	0.	0.	24.9846

Third

x^3	xy^2	xz^2	x^2y	y^3	yz^2	x^2z	y^2z	z^3	xyz
0.	0.	0.	0.	0.	0.	0.6218	0.6247	16.9158	0.
0.	0.	0.	0.	0.	0.	0.	0.	-25.9029	0.

Fourth (even)

x^4	y^4	z^4	x^2y^2	x^2z^2	y^2z^2	r^2x^2	r^2y^2	r^2z^2	r^4
-31.2358	-31.2565	-195.0185	-10.4154	-24.0545	-24.0592	-65.7057	-65.7310	-243.1322	-374.5690
0.	0.	107.1283	0.	0.	0.	0.	0.	107.1283	107.1283

Expectation Values at Atomic Centers

Center			Expectation Value		
	r^{-1}	δ	xr^{-3}	yr^{-3}	zr^{-3}
H1	-0.9490	0.3857	0.	0.	-0.0665
C1	-14.6686	116.7476	0.	0.	-0.0452
N1	-18.3110	189.6934	0.	0.	0.2089

Center	$\dfrac{3x^2-r^2}{r^5}$	$\dfrac{3y^2-r^2}{r^5}$	$\dfrac{3z^2-r^2}{r^5}$	$\dfrac{3xy}{r^5}$	$\dfrac{3xz}{r^5}$	$\dfrac{3yz}{r^5}$
H1	-0.1843	-0.1844	0.3687	0.	0.	0.
C1	0.1044	0.1033	-0.2077	0.	0.	0.
N1	0.4020	0.4068	-0.8088	0.	0.	0.

Mulliken Charge

Center	Basis Function Type										Total
	S1	S2	S3	S4	X1	X2	Y1	Y2	Z1	Z2	
H1	0.4829	0.2339									0.7168
C1	0.0429	1.9565	0.8376	0.4024	0.7411	0.2018	0.7416	0.2024	1.0103	0.0936	6.2302
N1	0.0434	1.9560	0.9710	0.7018	0.7779	0.2791	0.7769	0.2791	1.1195	0.1482	7.0530

Overlap Populations

	Pair of Atomic Centers		
	C1 H1	N1 H1	N1 C1
1SIGMA	0.	-0.0001	-0.0017
2SIGMA	-0.0002	0.	0.
3SIGMA	0.0071	0.0002	0.7595
4SIGMA	0.6866	-0.0007	0.0525
5SIGMA	-0.0135	-0.0048	-0.0416
Total SIGMA	0.6800	-0.0054	0.7687
1PI-X	0.	0.	0.5147
Total PI-X	0.	0.	0.5147
1PI-Y	0.	0.	0.5149
Total PI-Y	0.	0.	0.5149
Total	0.6800	-0.0054	1.7983
Distance	2.0091	4.1924	2.1833

Molecular Orbitals and Eigenvalues

	Center	S1	S2	S3	S4	X1	X2	Y1	Y2	Z1	Z2
* 1SIGMA	H1	0.00023	-0.00215								
-15.6294	C1	0.00001	0.00004	0.00011	-0.00124	0.	0.	0.	0.	-0.00012	-0.00320
	N1	0.05196	0.97917	0.00225	0.00548	0.	0.	0.	0.	-0.00251	0.00010
* 2SIGMA	H1	0.00091	-0.00092								
-11.3139	C1	0.05189	0.97895	0.00484	0.00155	0.	0.	0.	0.	0.00063	-0.00042
	N1	-0.00005	-0.00096	0.00116	-0.00063	0.	0.	0.	0.	-0.00145	0.00064
* 3SIGMA	H1	0.04625	-0.00209								
-1.2693	C1	-0.00715	-0.15406	0.32888	0.14623	0.	0.	0.	0.	0.24513	-0.01600
	N1	-0.00893	-0.19115	0.43250	0.17422	0.	0.	0.	0.	-0.22990	0.04244
* 4SIGMA	H1	-0.29432	-0.13002								
-0.8113	C1	0.00631	0.13535	-0.31655	-0.19368	0.	0.	0.	0.	0.40687	0.06666
	N1	-0.00338	-0.07294	0.16346	0.07930	0.	0.	0.	0.	0.01641	0.03840
* 5SIGMA	H1	0.07178	0.12179								
-0.5756	C1	0.00306	0.06266	-0.18215	-0.06262	0.	0.	0.	0.	-0.31313	0.05437
	N1	-0.00544	-0.11702	0.27661	0.44072	0.	0.	0.	0.	0.56739	0.13519
6SIGMA	H1	0.14075	0.83539								
0.2302	C1	0.00364	0.08944	-0.13300	-2.13913	0.	0.	0.	0.	0.25511	-0.95476
	N1	-0.00306	-0.07851	0.07597	1.57434	0.	0.	0.	0.	-0.16366	-0.52137
7SIGMA	H1	0.15871	3.19808								
0.4241	C1	0.00089	0.04103	0.09461	-1.04406	0.	0.	0.	0.	0.11015	3.47824
	N1	0.00121	0.05179	0.11184	-2.20160	0.	0.	0.	0.	-0.28080	0.29497
8SIGMA	H1	-0.32189	0.18140								
0.5284	C1	0.00590	0.05264	-0.84652	2.37795	0.	0.	0.	0.	0.63733	0.03542
	N1	0.00364	0.07348	-0.23550	-1.16185	0.	0.	0.	0.	-0.08683	1.04459
9SIGMA	H1	-0.11510	1.27982								
0.6552	C1	0.00674	0.05170	-1.02613	-0.95279	0.	0.	0.	0.	-0.20858	1.09724
	N1	-0.00040	-0.02893	-0.11170	0.38595	0.	0.	0.	0.	0.34653	-1.66714
* 1PI-X	H1	0.	0.								
-0.5043	C1	0.	0.	0.	0.	0.49070	0.15026	0.	0.	0.	0.
	N1	0.	0.	0.	0.	0.51423	0.18591	0.	0.	0.	0.
2PI-X	H1	0.	0.								
0.1603	C1	0.	0.	0.	0.	0.38347	0.88867	0.	0.	0.	0.
	N1	0.	0.	0.	0.	-0.45917	-0.71815	0.	0.	0.	0.
3PI-X	H1	0.	0.								
0.4682	C1	0.	0.	0.	0.	0.98940	-1.02205	0.	0.	0.	0.
	N1	0.	0.	0.	0.	0.00938	-0.09047	0.	0.	0.	0.
* 1PI-Y	H1	0.	0.								
-0.5041	C1	0.	0.	0.	0.	0.	0.	0.49086	0.15060	0.	0.
	N1	0.	0.	0.	0.	0.	0.	0.51381	0.18586	0.	0.
2PI-Y	H1	0.	0.								
0.1604	C1	0.	0.	0.	0.	0.	0.	0.38327	0.88864	0.	0.
	N1	0.	0.	0.	0.	0.	0.	-0.45926	-0.71847	0.	0.
3PI-Y	H1	0.	0.								
0.4682	C1	0.	0.	0.	0.	0.	0.	0.98941	-1.02210	0.	0.
	N1	0.	0.	0.	0.	0.	0.	0.00945	-0.09041	0.	0.

Overlap Matrix

First Function   Second Function and Overlap

H1S1 with
    H1S2 .68301   C1S1 .00254   C1S2 .03853   C1S3 .30092   C1S4 .38152   C1Z1 -.44578   C1Z2 -.43139
    N1S2 .00001   N1S3 .00249   N1S4 .04324   N1Z1 -.01047   N1Z2 -.17228

H1S2 with
    C1S1 .00795   C1S2 .10129   C1S3 .50956   C1S4 .71745   C1Z1 -.36088   C1Z2 -.60224   N1S1 .00057
    N1S2 .00780   N1S3 .06360   N1S4 .18489   N1Z1 -.08716   N1Z2 -.39159

C1S1 with
    C1S2 .35436   C1S3 .00805   C1S4 .01419   N1S3 .00186   N1S4 .00680   N1Z1 -.00603   N1Z2 -.01244

C1S2 with
    C1S3 .26226   C1S4 .17745   N1S2 .00007   N1S3 .03055   N1S4 .08731   N1Z1 -.08091   N1Z2 -.15331

C1S3 with
    C1S4 .79970   N1S1 .00296   N1S2 .04081   N1S3 .27555   N1S4 .45357   N1Z1 -.36127   N1Z2 -.60899

C1S4 with
    N1S1 .00554   N1S2 .07096   N1S3 .38522   N1S4 .64813   N1Z1 -.23053   N1Z2 -.57662

C1X1 with
    C1X2 .53875   N1X1 .21755   N1X2 .38556

C1X2 with
    N1X1 .24856   N1X2 .69456

C1Y1 with
    C1Y2 .53875   N1Y1 .21755   N1Y2 .38556

C1Y2 with
    N1Y1 .24856   N1Y2 .69456

C1Z1 with
    C1Z2 .53875   N1S1 .00700   N1S2 .09011   N1S3 .42960   N1S4 .38929   N1Z1 -.33912   N1Z2 -.05386

C1Z2 with
    N1S1 .00792   N1S2 .09963   N1S3 .47657   N1S4 .62755   N1Z1 .01990   N1Z2 .24633

N1S1 with
    N1S2 .35584   N1S3 .00894   N1S4 .01446

N1S2 with
    N1S3 .27246   N1S4 .17925

N1S3 with
    N1S4 .79352

N1X1 with
    N1X2 .52739

N1Y1 with
    N1Y2 .52739

N1Z1 with
    N1Z2 .52739

N2   Nitrogen

Molecular Geometry   Symmetry D-$_h$

Center		Coordinates	
	X	Y	Z
N1	0.	0.	-1.03704999
N2	0.	0.	1.03704999
Mass	0.	0.	0.

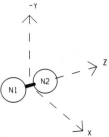

MO Symmetry Types and Occupancy

Representation	SIGM-G	SIGM-U	PI-G-X	PI-U-X	PI-G-Y	PI-U-Y
Number of MOs	6	6	2	2	2	2
Number Occupied	3	2	0	1	0	1

Energy Expectation Values

Total	-108.8695	Kinetic	108.7360
Electronic	-132.4942	Potential	-217.6055
Nuclear Repulsion	23.6247	Virial	2.0012

Electronic Potential	-241.2302
One-Electron Potential	-302.5057
Two-Electron Potential	61.2755

Moments of the Charge Distribution

Second

x^2	y^2	z^2	xy	xz	yz	r^2
-7.5772	-7.5765	-24.4415	0.	0.	0.	-39.5952
0.	0.	15.0566	0.	0.	0.	15.0566

Fourth (even)

x^4	y^4	z^4	x^2y^2	x^2z^2	y^2z^2	r^2x^2	r^2y^2	r^2z^2	r^4
-22.2518	-22.2450	-98.7529	-7.4161	-16.4293	-16.4277	-46.0972	-46.0888	-131.6099	-223.7959
0.	0.	16.1930	0.	0.	0.	0.	0.	16.1930	16.1930

Expectation Values at Atomic Centers

Center			Expectation Value		
	r^{-1}	δ	xr^{-3}	yr^{-3}	zr^{-3}
N1	-18.2326	189.9486	0.	0.	-0.2489

	$\dfrac{3x^2-r^2}{r^5}$	$\dfrac{3y^2-r^2}{r^5}$	$\dfrac{3z^2-r^2}{r^5}$	$\dfrac{3xy}{r^5}$	$\dfrac{3xz}{r^5}$	$\dfrac{3yz}{r^5}$
N1	0.4386	0.4377	-0.8763	0.	0.	0.

Mulliken Charge

Center				Basis Function Type							Total
	S1	S2	S3	S4	X1	X2	Y1	Y2	Z1	Z2	
N1	0.0435	1.9560	0.9924	0.8358	0.7684	0.2316	0.7685	0.2315	1.1173	0.0550	7.0000

Overlap Populations

Pair of Atomic Centers

	N2 N1
1SIGM-G	0.0012
2SIGM-G	0.7595
3SIGM-G	0.0278
Total SIGM-G	0.7885
1SIGM-U	-0.0050
2SIGM-U	-0.5727
Total SIGM-U	-0.5777
1PI-U-X	0.4844
Total PI-U-X	0.4844
1PI-U-Y	0.4843
Total PI-U-Y	0.4843
Total	1.1794
Distance	2.0741

Molecular Orbitals and Eigenvalues

Center		S1	S2	S3	S4	X1	X2	Y1	Y2	Z1	Z2
* 1SIGM-G	N1	0.03671	0.69182	0.00266	0.00094	0.	0.	0.	0.	0.00222	-0.00104
-15.7190	N2	0.03674	0.69240	0.00266	0.00095	0.	0.	0.	0.	-0.00222	0.00103
* 2SIGM-G	N1	-0.00800	-0.17058	0.38408	0.15415	0.	0.	0.	0.	0.25750	-0.02583
-1.5262	N2	-0.00800	-0.17058	0.38408	0.15416	0.	0.	0.	0.	-0.25750	0.02582
* 3SIGM-G	N1	-0.00256	-0.05869	0.10355	0.31945	0.	0.	0.	0.	-0.52271	-0.09455
-0.6247	N2	-0.00256	-0.05869	0.10354	0.31945	0.	0.	0.	0.	0.52271	0.09454
4SIGM-G	N1	-0.00077	0.00724	0.20437	0.15632	0.	0.	0.	0.	0.39717	-1.11315
0.5799	N2	-0.00077	0.00725	0.20435	0.15624	0.	0.	0.	0.	-0.39717	1.11321
* 1SIGM-U	N1	-0.03676	-0.69289	-0.00137	-0.00607	0.	0.	0.	0.	-0.00175	-0.00172
-15.7156	N2	0.03673	0.69232	0.00137	0.00607	0.	0.	0.	0.	-0.00175	-0.00172
* 2SIGM-U	N1	0.00739	0.15764	-0.38308	-0.44252	0.	0.	0.	0.	0.22763	0.04267
-0.7738	N2	-0.00739	-0.15764	0.38308	0.44253	0.	0.	0.	0.	0.22764	0.04268
3SIGM-U	N1	-0.00343	-0.09090	0.07465	3.11001	0.	0.	0.	0.	-0.00154	1.97311
0.4121	N2	0.00343	0.09090	-0.07466	-3.11001	0.	0.	0.	0.	-0.00152	1.97306
1PI-G-X	N1	0.	0.	0.	0.	0.51129	0.66986	0.	0.	0.	0.
0.1434	N2	0.	0.	0.	0.	-0.51129	-0.66986	0.	0.	0.	0.
* 1PI-U-X	N1	0.	0.	0.	0.	0.51274	0.16447	0.	0.	0.	0.
-0.6251	N2	0.	0.	0.	0.	0.51274	0.16447	0.	0.	0.	0.
2PI-U-X	N1	0.	0.	0.	0.	-0.60642	0.64527	0.	0.	0.	0.
0.5990	N2	0.	0.	0.	0.	-0.60642	0.64527	0.	0.	0.	0.
1PI-G-Y	N1	0.	0.	0.	0.	0.	0.	0.51133	0.66979	0.	0.
0.1433	N2	0.	0.	0.	0.	0.	0.	-0.51133	-0.66979	0.	0.
* 1PI-U-Y	N1	0.	0.	0.	0.	0.	0.	0.51284	0.16437	0.	0.
-0.6252	N2	0.	0.	0.	0.	0.	0.	0.51284	0.16437	0.	0.
2PI-U-Y	N1	0.	0.	0.	0.	0.	0.	-0.60634	0.64530	0.	0.
0.5990	N2	0.	0.	0.	0.	0.	0.	-0.60634	0.64530	0.	0.

Basis Function Type

Overlap Matrix

First Function   Second Function and Overlap

N1S1 with
    N1S2 .35584   N1S3 .00894   N1S4 .01446   N2S3 .00201   N2S4 .00592   N2Z1 -.00583   N2Z2 -.01009

N1S2 with
    N1S3 .27246   N1S4 .17925   N2S2 .00004   N2S3 .03012   N2S4 .07567   N2Z1 -.07640   N2Z2 -.12531

N1S3 with
    N1S4 .79352   N2S1 .00201   N2S2 .03012   N2S3 .24850   N2S4 .40251   N2Z1 -.37211   N2Z2 -.55066

N1S4 with
    N2S1 .00592   N2S2 .07567   N2S3 .40251   N2S4 .63943   N2Z1 -.29741   N2Z2 -.62601

N1X1 with
    N1X2 .52739   N2X1 .19579   N2X2 .30044

N1X2 with
    N2X1 .30044   N2X2 .70068

N1Y1 with
    N1Y2 .52739   N2Y1 .19579   N2Y2 .30044

N1Y2 with
    N2Y1 .30044   N2Y2 .70068

N1Z1 with
    N1Z2 .52739   N2S1 .00583   N2S2 .07640   N2S3 .37211   N2S4 .29741   N2Z1 -.35081   N2Z2 -.03713

N1Z2 with
    N2S1 .01009   N2S2 .12531   N2S3 .55066   N2S4 .62601   N2Z1 -.03713   N2Z2 .20221

CO   Carbon Monoxide

Molecular Geometry   Symmetry C-ᵥ

Center		Coordinates	
	X	Y	Z
C1	0.	0.	0.
O1	0.	0.	2.13200000
Mass	0.	0.	1.21811953

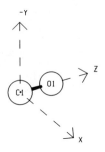

MO Symmetry Types and Occupancy

Representation	SIGMA	PI-X	PI-Y
Number of MOs	12	4	4
Number Occupied	5	1	1

Energy Expectation Values

Total	-112.6763	Kinetic	112.6526
Electronic	-135.1904	Potential	-225.3290
Nuclear Repulsion	22.5141	Virial	2.0002

Electronic Potential	-247.8430
One-Electron Potential	-310.4046
Two-Electron Potential	62.5616

Moments of the Charge Distribution

First

x	y	z	Dipole Moment
0.	0.	-0.1573	0.1550
0.	0.	0.0023	

Second

x^2	y^2	z^2	xy	xz	yz	r^2
-7.4795	-7.4795	-25.3371	0.	0.	0.	-40.2962
0.	0.	15.5843	0.	0.	0.	15.5843

Third

x^3	xy^2	xz^2	x^2y	y^3	yz^2	x^2z	y^2z	z^3	xyz
0.	0.	0.	0.	0.	0.	1.2307	1.2307	13.1420	0.
0.	0.	0.	0.	0.	0.	0.	0.	-4.7388	0.

Fourth (even)

x^4	y^4	z^4	x^2y^2	x^2z^2	y^2z^2	r^2x^2	r^2y^2	r^2z^2	r^4
-22.0565	-22.0565	-115.0623	-7.3522	-18.0390	-18.0390	-47.4477	-47.4477	-151.1402	-246.0355
0.	0.	18.7904	0.	0.	0.	0.	0.	18.7904	18.7904

Expectation Values at Atomic Centers

Center			Expectation Value		
	r^{-1}	δ	xr^{-3}	yr^{-3}	zr^{-3}
C1	-14.5945	117.5158	0.	0.	-0.1871
O1	-22.2262	287.5940	0.	0.	0.2907

Center	$\dfrac{3x^2-r^2}{r^5}$	$\dfrac{3y^2-r^2}{r^5}$	$\dfrac{3z^2-r^2}{r^5}$	$\dfrac{3xy}{r^5}$	$\dfrac{3xz}{r^5}$	$\dfrac{3yz}{r^5}$
C1	0.4303	0.4303	-0.8607	0.	0.	0.
O1	0.1514	0.1514	-0.3029	0.	0.	0.

Mulliken Charge

Center				Basis Function Type							Total
	S1	S2	S3	S4	X1	X2	Y1	Y2	Z1	Z2	
C1	0.0432	1.9564	0.9885	0.8504	0.4288	0.0421	0.4288	0.0421	0.9793	0.0397	5.7991
O1	0.0427	1.9567	0.9984	0.7828	1.0870	0.4422	1.0870	0.4422	1.2177	0.1443	8.2009

Overlap Populations

Pair of Atomic Centers

	O1 C1
1SIGMA	-0.0006
2SIGMA	-0.0016
3SIGMA	0.6441
4SIGMA	-0.0678
5SIGMA	-0.1461
Total SIGMA	0.4281
1PI-X	0.3726
Total PI-X	0.3726
1PI-Y	0.3726
Total PI-Y	0.3726
Total	1.1733
Distance	2.1320

Molecular Orbitals and Eigenvalues

Center		S1	S2	S3	S4	X1	X2	Y1	Y2	Z1	Z2
* 1SIGMA	C1	0.00000	-0.00004	-0.00010	-0.00102	0.	0.	0.	0.	-0.00027	-0.00071
-20.6724	O1	0.05112	0.97934	0.00356	0.00295	0.	0.	0.	0.	-0.00198	0.00012
* 2SIGMA	C1	-0.05192	-0.97933	-0.00389	-0.00260	0.	0.	0.	0.	-0.00376	0.00066
-11.4121	O1	0.00003	0.00071	-0.00162	0.00261	0.	0.	0.	0.	0.00136	-0.00169
* 3SIGMA	C1	-0.00568	-0.12025	0.26460	0.05678	0.	0.	0.	0.	0.23904	-0.03755
-1.5638	O1	-0.00981	-0.21377	0.47906	0.28575	0.	0.	0.	0.	-0.21695	0.02907
* 4SIGMA	C1	0.00656	0.13454	-0.37341	-0.13503	0.	0.	0.	0.	-0.13975	0.05510
-0.7984	O1	-0.00560	-0.12329	0.27622	0.41805	0.	0.	0.	0.	0.54318	0.13104
* 5SIGMA	C1	-0.00650	-0.14307	0.31890	0.54043	0.	0.	0.	0.	-0.52053	-0.04981
-0.5542	O1	0.00066	0.01279	-0.04550	0.00680	0.	0.	0.	0.	0.31745	0.07629
6SIGMA	C1	0.00089	0.03143	0.02849	-1.18486	0.	0.	0.	0.	-0.03737	-1.51003
0.3053	O1	-0.00334	-0.08567	0.09179	1.41008	0.	0.	0.	0.	-0.04864	-0.29097
7SIGMA	C1	-0.00829	-0.09899	1.03275	-1.90191	0.	0.	0.	0.	-0.72036	0.24696
0.5906	O1	-0.00196	-0.04510	0.09199	0.72767	0.	0.	0.	0.	0.13165	-0.69271
* 1PI-X	C1	0.	0.	0.	0.	-0.35437	-0.03962	0.	0.	0.	0.
-0.6476	O1	0.	0.	0.	0.	-0.63741	-0.28505	0.	0.	0.	0.
2PI-X	C1	0.	0.	0.	0.	0.52071	0.75413	0.	0.	0.	0.
0.1228	O1	0.	0.	0.	0.	-0.39998	-0.48253	0.	0.	0.	0.
3PI-X	C1	0.	0.	0.	0.	1.01388	-1.05947	0.	0.	0.	0.
0.4610	O1	0.	0.	0.	0.	-0.05725	-0.00287	0.	0.	0.	0.
* 1PI-Y	C1	0.	0.	0.	0.	0.	0.	-0.35437	-0.03962	0.	0.
-0.6476	O1	0.	0.	0.	0.	0.	0.	-0.63741	-0.28505	0.	0.
2PI-Y	C1	0.	0.	0.	0.	0.	0.	0.52071	0.75413	0.	0.
0.1228	O1	0.	0.	0.	0.	0.	0.	-0.39998	-0.48253	0.	0.
3PI-Y	C1	0.	0.	0.	0.	0.	0.	1.01388	-1.05947	0.	0.
0.4610	O1	0.	0.	0.	0.	0.	0.	-0.05725	-0.00287	0.	0.

Basis Function Type

Overlap Matrix

First Function   Second Function and Overlap

C1S1 with
    C1S2 .35436   C1S3 .00805   C1S4 .01419   O1S3 .00087   O1S4 .00639   O1Z1 -.00406   O1Z2 -.01394

C1S2 with
    C1S3 .26226   C1S4 .17745   O1S2 .00005   O1S3 .01835   O1S4 .08270   O1Z1 -.05904   O1Z2 -.16994

C1S3 with
    C1S4 .79970   O1S1 .00266   O1S2 .03604   O1S3 .23656   O1S4 .43533   O1Z1 -.29601   O1Z2 -.62513

C1S4 with
    O1S1 .00462   O1S2 .05960   O1S3 .32716   O1S4 .59542   O1Z1 -.17399   O1Z2 -.52853

C1X1 with
    C1X2 .53875   O1X1 .17932   O1X2 .40340

C1X2 with
    O1X1 .18834   O1X2 .63219

C1Y1 with
    C1Y2 .53875   O1Y1 .17932   O1Y2 .40340

C1Y2 with
    O1Y1 .18834   O1Y2 .63219

C1Z1 with
    C1Z2 .53875   O1S1 .00604   O1S2 .07815   O1S3 .39736   O1S4 .44283   O1Z1 -.31226   O1Z2 -.12422

C1Z2 with
    O1S1 .00641   O1S2 .08150   O1S3 .40781   O1S4 .61156   O1Z1 .01623   O1Z2 .20345

O1S1 with
    O1S2 .35461   O1S3 .00974   O1S4 .01469

O1S2 with
    O1S3 .28286   O1S4 .18230

O1S3 with
    O1S4 .79029

O1X1 with
    O1X2 .50607

O1Y1 with
    O1Y2 .50607

O1Z1 with
    O1Z2 .50607

BF   Boron Fluoride

Molecular Geometry   Symmetry C-ᵥ

Center		Coordinates	
	X	Y	Z
B1	0.	0.	-1.19526701
F1	0.	0.	1.19526701
Mass	0.	0.	0.31822165

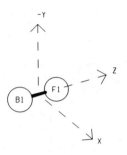

MO Symmetry Types and Occupancy

Representation	SIGMA	PI-X	PI-Y
Number of MOs	12	4	4
Number Occupied	5	1	1

Energy Expectation Values

Total	-124.0719	Kinetic	124.1673
Electronic	-142.8962	Potential	-248.2393
Nuclear Repulsion	18.8242	Virial	1.9992
Electronic Potential	-267.0635		
One-Electron Potential	-332.7367		
Two-Electron Potential	65.6731		

Moments of the Charge Distribution

First

x	y	z	Dipole Moment
0.	0.	-0.1164	0.2095
0.	0.	0.3260	

Second

x^2	y^2	z^2	xy	xz	yz	r^2
-7.6548	-7.6548	-29.8455	0.	0.	0.	-45.1551
0.	0.	18.3761	0.	0.	0.	18.3761

Third

x^3	xy^2	xz^2	x^2y	y^3	yz^2	x^2z	y^2z	z^3	xyz
0.	0.	0.	0.	0.	0.	4.3873	4.3873	35.1582	0.
0.	0.	0.	0.	0.	0.	0.	0.	-11.2627	0.

Fourth (even)

x^4	y^4	z^4	x^2y^2	x^2z^2	y^2z^2	r^2x^2	r^2y^2	r^2z^2	r^4
-28.3715	-28.3715	-194.6597	-9.4572	-27.7495	-27.7495	-65.5782	-65.5782	-250.1587	-381.3150
0.	0.	31.5605	0.	0.	0.	0.	0.	31.5605	31.5605

Expectation Values at Atomic Centers

Center	r^{-1}	δ	xr^{-3}	yr^{-3}	zr^{-3}
			Expectation Value		
B1	-11.2935	66.2026	0.	0.	-0.1306
F1	-26.5134	413.6515	0.	0.	0.2662

Center	$\dfrac{3x^2-r^2}{r^5}$	$\dfrac{3y^2-r^2}{r^5}$	$\dfrac{3z^2-r^2}{r^5}$	$\dfrac{3xy}{r^5}$	$\dfrac{3xz}{r^5}$	$\dfrac{3yz}{r^5}$
B1	0.1819	0.1819	-0.3638	0.	0.	0.
F1	-0.3017	-0.3017	0.6034	0.	0.	0.

Mulliken Charge

Center					Basis Function Type						Total
	S1	S2	S3	S4	X1	X2	Y1	Y2	Z1	Z2	
B1	0.0440	1.9558	0.9574	0.8719	0.1209	-0.0045	0.1209	-0.0045	0.6365	0.0209	4.7194
F1	0.0436	1.9557	1.0365	0.8643	1.3446	0.5390	1.3446	0.5390	1.3146	0.2987	9.2806

Overlap Populations

Pair of Atomic Centers

	F1 B1
1SIGMA	-0.0001
2SIGMA	-0.0016
3SIGMA	0.2856
4SIGMA	0.2014
5SIGMA	-0.2456
Total SIGMA	0.2396
1PI-X	0.1415
Total PI-X	0.1415
1PI-Y	0.1415
Total PI-Y	0.1415
Total	0.5227
Distance	2.3905

Molecular Orbitals and Eigenvalues

		S1	S2	S3	S4	X1	X2	Y1	Y2	Z1	Z2
• 1SIGMA	B1	0.00000	0.00001	0.00009	0.00018	0.	0.	0.	0.	0.00022	0.00013
-26.3659	F1	-0.05174	-0.97888	-0.00456	-0.00133	0.	0.	0.	0.	0.00109	-0.00019
• 2SIGMA	B1	0.05295	0.97925	0.00370	0.00125	0.	0.	0.	0.	0.00505	-0.00180
-7.7673	F1	-0.00004	-0.00070	0.00208	-0.00210	0.	0.	0.	0.	-0.00227	0.00178
• 3SIGMA	B1	-0.00306	-0.06580	0.11085	-0.00512	0.	0.	0.	0.	0.12003	-0.03128
-1.7170	F1	-0.01133	-0.24282	0.54992	0.43698	0.	0.	0.	0.	-0.12578	-0.00858
• 4SIGMA	B1	0.00541	0.11020	-0.28699	-0.00149	0.	0.	0.	0.	-0.23185	0.08417
-0.8487	F1	-0.00305	-0.06590	0.14763	0.25137	0.	0.	0.	0.	0.67695	0.21095
• 5SIGMA	B1	0.00824	0.17403	-0.44981	-0.55775	0.	0.	0.	0.	0.38614	0.02998
-0.4139	F1	-0.00112	-0.02352	0.06048	0.09799	0.	0.	0.	0.	-0.19752	-0.08128
6SIGMA	B1	0.00091	0.03761	0.06799	-0.55568	0.	0.	0.	0.	-0.05821	-1.13128
0.1858	F1	-0.00267	-0.06198	0.10395	0.61348	0.	0.	0.	0.	-0.04857	0.01105
7SIGMA	B1	-0.00878	-0.08575	1.24232	-1.54938	0.	0.	0.	0.	-0.70844	0.39388
0.4049	F1	-0.00105	-0.02593	0.03103	0.37385	0.	0.	0.	0.	0.01465	-0.14709
• 1PI-X	B1	0.	0.	0.	0.	0.15513	-0.00760	0.	0.	0.	0.
-0.7494	F1	0.	0.	0.	0.	0.73003	0.34880	0.	0.	0.	0.
2PI-X	B1	0.	0.	0.	0.	-0.48221	-0.72080	0.	0.	0.	0.
0.0777	F1	0.	0.	0.	0.	0.23303	0.23981	0.	0.	0.	0.
3PI-X	B1	0.	0.	0.	0.	1.11215	-0.98638	0.	0.	0.	0.
0.3214	F1	0.	0.	0.	0.	-0.11409	-0.04925	0.	0.	0.	0.
• 1PI-Y	B1	0.	0.	0.	0.	0.	0.	0.15513	-0.00760	0.	0.
-0.7494	F1	0.	0.	0.	0.	0.	0.	0.73003	0.34880	0.	0.
2PI-Y	B1	0.	0.	0.	0.	0.	0.	-0.48221	-0.72080	0.	0.
0.0777	F1	0.	0.	0.	0.	0.	0.	0.23303	0.23981	0.	0.
3PI-Y	B1	0.	0.	0.	0.	0.	0.	1.11215	-0.98638	0.	0.
0.3214	F1	0.	0.	0.	0.	0.	0.	-0.11409	-0.04925	0.	0.

Overlap Matrix

First Function   Second Function and Overlap

B1S1 with
    B1S2 .35489   B1S3 .00732   B1S4 .01371   F1S3 .00011   F1S4 .00479   F1Z1 -.00105   F1Z2 -.01577

B1S2 with
    B1S3 .24693   B1S4 .17217   F1S2 .00004   F1S3 .00673   F1S4 .06632   F1Z1 -.02549   F1Z2 -.18951

B1S3 with
    B1S4 .80606   F1S1 .00246   F1S2 .03203   F1S3 .19605   F1S4 .40145   F1Z1 -.18574   F1Z2 -.57532

B1S4 with
    F1S1 .00318   F1S2 .04082   F1S3 .23424   F1S4 .47913   F1Z1 -.08456   F1Z2 -.35481

B1X1 with
    B1X2 .55377   F1X1 .12351   F1X2 .42571

B1X2 with
    F1X1 .09296   F1X2 .42424

B1Y1 with
    B1Y2 .55377   F1Y1 .12351   F1Y2 .42571

B1Y2 with
    F1Y1 .09296   F1Y2 .42424

B1Z1 with
    B1Z2 .55377   F1S1 .00497   F1S2 .06371   F1S3 .34825   F1S4 .53321   F1Z1 -.20994   F1Z2 -.25253

B1Z2 with
    F1S1 .00369   F1S2 .04709   F1S3 .25950   F1S4 .48250   F1Z1 .02311   F1Z2 .15369

F1S1 with
    F1S2 .35647   F1S3 .01003   F1S4 .01474

F1S2 with
    F1S3 .28653   F1S4 .18096

F1S3 with
    F1S4 .78531

F1X1 with
    F1X2 .49444

F1Y1 with
    F1Y2 .49444

F1Z1 with
    F1Z2 .49444

B2H6   Diborane

Molecular Geometry   Symmetry D₂ₕ

Center		Coordinates	
	X	Y	Z
H1	0.	-1.94740340	-2.82426137
H2	0.	1.94740340	-2.82426137
H3	0.	-1.94740340	2.82426137
H4	0.	1.94740340	2.82426137
H5	-1.89547500	0.	0.
H6	1.89547500	0.	0.
B1	0.	0.	-1.67715313
B2	0.	0.	1.67715313
Mass	0.	0.	0.

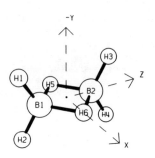

MO Symmetry Types and Occupancy

Representation	AG	AU	B1G	B1U	B2G	B2U	B3G	B3U
Number of MOs	10	0	0	8	2	4	4	4
Number Occupied	3	0	0	2	0	1	1	1

Energy Expectation Values

Total	-52.7675	Kinetic	52.7123
Electronic	-84.5136	Potential	-105.4798
Nuclear Repulsion	31.7461	Virial	2.0010

Electronic Potential	-137.2259
One-Electron Potential	-183.9906
Two-Electron Potential	46.7647

Moments of the Charge Distribution

Second

x^2	y^2	z^2	xy	xz	yz	r^2
-18.5861	-28.5461	-74.2260	0.	0.	0.	-121.3581
7.1857	15.1695	60.0342	0.	0.	0.	82.3894

Fourth (even)

x^4	y^4	z^4	x^2y^2	x^2z^2	y^2z^2	r^2x^2	r^2y^2	r^2z^2	r^4
-93.5330	-176.8780	-678.2579	-33.2990	-66.2909	-195.6863	-193.1229	-405.8634	-940.2351	-1539.2214
25.8168	57.5286	333.6160	0.	0.	120.9990	25.8168	178.5275	454.6149	658.9593

Expectation Values at Atomic Centers

Center	r^{-1}	δ	xr^{-3}	yr^{-3}	zr^{-3}
H1	-1.1357	0.4043	0.	-0.0441	-0.0276
H5	-1.0292	0.4091	-0.0869	0.	0.
B1	-11.3897	65.4980	0.	0.	-0.0494

Center	$\dfrac{3x^2-r^2}{r^5}$	$\dfrac{3y^2-r^2}{r^5}$	$\dfrac{3z^2-r^2}{r^5}$	$\dfrac{3xy}{r^5}$	$\dfrac{3xz}{r^5}$	$\dfrac{3yz}{r^5}$
H1	-0.1108	0.1394	-0.0287	0.	0.	0.1595
H5	0.0985	-0.1647	0.0662	0.	0.	0.
B1	0.2834	-0.3074	0.0240	0.	0.	0.

Mulliken Charge

Center	S1	S2	S3	S4	X1	X2	Y1	Y2	Z1	Z2	Total
H1	0.5072	0.4090									0.9162
H5	0.5198	0.4094									0.9292
B1	0.0435	1.9557	0.7272	0.3310	0.4106	0.0008	0.9654	0.1189	0.7039	-0.0186	5.2384

Overlap Populations

Pair of Atomic Centers

	H2 / H1	H3 / H1	H3 / H2	H5 / H1	H6 / H5	B1 / H1	B1 / H3	B1 / H5	B2 / B1
1AG	0.	0.	0.	0.	0.	0.	0.	-0.0002	0.0010
2AG	0.0027	0.0004	0.0001	0.0057	0.0139	0.0465	0.0088	0.1408	0.2875
3AG	0.0245	0.0041	0.0009	-0.0244	0.0278	0.1716	-0.0082	0.0244	0.0767
Total AG	0.0272	0.0045	0.0010	-0.0188	0.0416	0.2180	0.0006	0.1650	0.3652
1B1U	0.	0.	0.	0.	0.	0.	0.	0.	-0.0021
2B1U	0.0233	-0.0038	-0.0009	0.	0.	0.2258	-0.0185	0.	-0.0996
Total B1U	0.0233	-0.0038	-0.0009	0.	0.	0.2258	-0.0185	0.	-0.1016
1B2U	-0.0241	0.0039	-0.0009	0.	0.	0.1765	0.0186	0.	0.1491
Total B2U	-0.0241	0.0039	-0.0009	0.	0.	0.1765	0.0186	0.	0.1491
1B3G	-0.0355	-0.0059	0.0013	0.	0.	0.2431	-0.0328	0.	-0.2358
Total B3G	-0.0355	-0.0059	0.0013	0.	0.	0.2431	-0.0328	0.	-0.2358
1B3U	0.	0.	0.	0.	-0.1238	0.	0.	0.1993	0.0669
Total B3U	0.	0.	0.	0.	-0.1238	0.	0.	0.1993	0.0669
Total	-0.0091	-0.0013	0.0006	-0.0188	-0.0822	0.8634	-0.0322	0.3643	0.2437
Distance	3.8948	5.6485	6.8611	3.9194	3.7910	2.2601	4.9046	2.5309	3.3543

Molecular Orbitals and Eigenvalues

Center		S1	S2	S3	S4	X1	X2	Y1	Y2	Z1	Z2
• 1AG	H1	0.00087	-0.00040								
-7.6389	H2	0.00087	-0.00040								
	H3	0.00087	-0.00040								
	H4	0.00087	-0.00040								
	H5	0.00053	-0.00078								
	H6	0.00053	-0.00078								
	B1	0.03742	0.69192	0.00342	0.00114	0.	0.	0.	0.	0.00132	-0.00043
	B2	0.03742	0.69192	0.00342	0.00114	0.	0.	0.	0.	-0.00132	0.00043
• 2AG	H1	-0.06241	-0.03452								
-0.8893	H2	-0.06241	-0.03452								
	H3	-0.06241	-0.03452								
	H4	-0.06241	-0.03452								
	H5	-0.16367	-0.06865								
	H6	-0.16367	-0.06865								
	B1	0.00640	0.13636	-0.29980	-0.09140	0.	0.	0.	0.	-0.13377	0.01727
	B2	0.00640	0.13636	-0.29980	-0.09140	0.	0.	0.	0.	0.13377	-0.01727
• 3AG	H1	0.14101	0.11536								
-0.5155	H2	0.14101	0.11536								
	H3	0.14101	0.11536								
	H4	0.14101	0.11536								
	H5	-0.11838	-0.12360								
	H6	-0.11838	-0.12360								
	B1	-0.00151	-0.03280	0.06860	0.11715	0.	0.	0.	0.	-0.36506	0.00799
	B2	-0.00151	-0.03280	0.06860	0.11715	0.	0.	0.	0.	0.36506	-0.00799
4AG	H1	0.04033	0.40930								
0.2027	H2	0.04033	0.40930								
	H3	0.04033	0.40930								
	H4	0.04033	0.40930								
	H5	-0.06510	-0.72040								
	H6	-0.06510	-0.72040								
	B1	-0.00102	-0.01619	0.10066	-0.48332	0.	0.	0.	0.	0.09311	1.17766
	B2	-0.00102	-0.01619	0.10066	-0.48332	0.	0.	0.	0.	-0.09311	-1.17766
5AG	H1	0.02310	0.13785								
0.2890	H2	0.02310	0.13785								
	H3	0.02310	0.13785								
	H4	0.02310	0.13785								
	H5	-0.07330	-0.70329								
	H6	-0.07330	-0.70329								
	B1	-0.00317	-0.02928	0.46241	0.07776	0.	0.	0.	0.	0.61359	-0.47535
	B2	-0.00317	-0.02928	0.46241	0.07776	0.	0.	0.	0.	-0.61359	0.47535
6AG	H1	-0.03464	-0.08474								
0.4049	H2	-0.03464	-0.08474								
	H3	-0.03464	-0.08474								
	H4	-0.03464	-0.08474								
	H5	-0.04504	-0.77877								
	H6	-0.04504	-0.77877								
	B1	0.00683	0.05004	-1.07600	1.58582	0.	0.	0.	0.	0.33578	0.04703
	B2	0.00683	0.05004	-1.07600	1.58582	0.	0.	0.	0.	-0.33578	-0.04703

Center		S1	S2	S3	S4	X1	X2	Y1	Y2	Z1	Z2
7AG	H1	0.04580	1.48215								
0.5294	H2	0.04580	1.48215								
	H3	0.04580	1.48215								
	H4	0.04580	1.48215								
	H5	0.11631	1.93319								
	H6	0.11631	1.93319								
	B1	0.00352	0.08698	-0.14889	-3.21918	0.	0.	0.	0.	0.19066	-0.10798
	B2	0.00352	0.08698	-0.14889	-3.21918	0.	0.	0.	0.	-0.19066	0.10798
* 1B1U	H1	0.00082	-0.00028								
-7.6384	H2	0.00082	-0.00028								
	H3	-0.00082	0.00028								
	H4	-0.00082	0.00028								
	H5	0.	0.								
	H6	0.	0.								
	B1	0.03743	0.69212	0.00310	0.00259	0.	0.	0.	0.	0.00130	0.00056
	B2	-0.03743	-0.69212	-0.00310	-0.00259	0.	0.	0.	0.	0.00130	0.00056
* 2B1U	H1	0.14675	0.11036								
-0.6413	H2	0.14675	0.11036								
	H3	-0.14675	-0.11036								
	H4	-0.14675	-0.11036								
	H5	0.	0.								
	H6	0.	0.								
	B1	-0.00581	-0.12391	0.28542	0.16071	0.	0.	0.	0.	-0.15473	0.00057
	B2	0.00581	0.12391	-0.28542	-0.16071	0.	0.	0.	0.	-0.15473	0.00057
3B1U	H1	-0.00747	-0.04872								
0.1657	H2	-0.00747	-0.04872								
	H3	0.00747	0.04872								
	H4	0.00747	0.04872								
	H5	0.	0.								
	H6	0.	0.								
	B1	0.00286	0.07942	-0.03633	-2.26039	0.	0.	0.	0.	-0.02795	-1.69779
	B2	-0.00286	-0.07942	0.03633	2.26039	0.	0.	0.	0.	-0.02795	-1.69779
4B1U	H1	0.07750	1.20067								
0.3496	H2	0.07750	1.20067								
	H3	-0.07750	-1.20067								
	H4	-0.07750	-1.20067								
	H5	0.	0.								
	H6	0.	0.								
	B1	0.00392	0.09705	-0.15004	-3.00205	0.	0.	0.	0.	-0.10844	-0.00525
	B2	-0.00392	-0.09705	0.15004	3.00205	0.	0.	0.	0.	-0.10844	-0.00525
5B1U	H1	0.02232	-0.39830								
0.5334	H2	0.02232	-0.39830								
	H3	-0.02232	0.39830								
	H4	-0.02232	0.39830								
	H5	0.	0.								
	H6	0.	0.								
	B1	0.00148	0.03925	-0.03247	-0.19727	0.	0.	0.	0.	-1.09800	-0.05951
	B2	-0.00148	-0.03925	0.03247	0.19727	0.	0.	0.	0.	-1.09800	-0.05951

Basis Function Type

Center		S1	S2	S3	S4	X1	X2	Y1	Y2	Z1	Z2
						Basis Function Type					
6B1U	H1	-0.01173	-0.29756								
0.6985	H2	-0.01173	-0.29756								
	H3	0.01173	0.29756								
	H4	0.01173	0.29756								
	H5	0.	0.								
	H6	0.	0.								
	B1	0.00949	0.04793	-1.65563	5.07302	0.	0.	0.	0.	-0.02643	1.59171
	B2	-0.00949	-0.04793	1.65563	-5.07302	0.	0.	0.	0.	-0.02643	1.59171
1B2G	H1	0.	0.								
0.0925	H2	0.	0.								
	H3	0.	0.								
	H4	0.	0.								
	H5	0.	0.								
	H6	0.	0.								
	B1	0.	0.	0.	0.	-0.40078	-0.79631	0.	0.	0.	0.
	B2	0.	0.	0.	0.	0.40078	0.79631	0.	0.	0.	0.
2B2G	H1	0.	0.								
0.3779	H2	0.	0.								
	H3	0.	0.								
	H4	0.	0.								
	H5	0.	0.								
	H6	0.	0.								
	B1	0.	0.	0.	0.	-0.80827	1.18210	0.	0.	0.	0.
	B2	0.	0.	0.	0.	0.80827	-1.18210	0.	0.	0.	0.
* 1B2U	H1	-0.15362	-0.11106								
-0.5366	H2	0.15362	0.11106								
	H3	-0.15362	-0.11106								
	H4	0.15362	0.11106								
	H5	0.	0.								
	H6	0.	0.								
	B1	0.	0.	0.	0.	0.	0.	0.37308	0.04471	0.	0.
	B2	0.	0.	0.	0.	0.	0.	0.37308	0.04471	0.	0.
2B2U	H1	0.05684	0.54715								
0.2223	H2	-0.05684	-0.54715								
	H3	0.05684	0.54715								
	H4	-0.05684	-0.54715								
	H5	0.	0.								
	H6	0.	0.								
	B1	0.	0.	0.	0.	0.	0.	-0.19666	0.93009	0.	0.
	B2	0.	0.	0.	0.	0.	0.	-0.19666	0.93009	0.	0.
3B2U	H1	-0.07374	-0.80355								
0.3906	H2	0.07374	0.80355								
	H3	-0.07374	-0.80355								
	H4	0.07374	0.80355								
	H5	0.	0.								
	H6	0.	0.								
	B1	0.	0.	0.	0.	0.	0.	-0.80113	-0.17135	0.	0.
	B2	0.	0.	0.	0.	0.	0.	-0.80113	-0.17135	0.	0.

		S1	S2	S3	S4	X1	X2	Y1	Y2	Z1	Z2
* 1B3G	H1	0.17775	0.13694								
-0.4653	H2	-0.17775	-0.13694								
	H3	-0.17775	-0.13694								
	H4	0.17775	0.13694								
	H5	0.	0.								
	H6	0.	0.								
	B1	0.	0.	0.	0.	0.	0.	-0.38117	-0.10636	0.	0.
	B2	0.	0.	0.	0.	0.	0.	0.38117	0.10636	0.	0.
2B3G	H1	-0.00574	-0.69644								
0.2471	H2	0.00574	0.69644								
	H3	0.00574	0.69644								
	H4	-0.00574	-0.69644								
	H5	0.	0.								
	H6	0.	0.								
	B1	0.	0.	0.	0.	0.	0.	0.24445	-2.32981	0.	0.
	B2	0.	0.	0.	0.	0.	0.	-0.24445	2.32981	0.	0.
3B3G	H1	-0.03342	-1.37511								
0.6443	H2	0.03342	1.37511								
	H3	0.03342	1.37511								
	H4	-0.03342	-1.37511								
	H5	0.	0.								
	H6	0.	0.								
	B1	0.	0.	0.	0.	0.	0.	-1.02778	-1.62794	0.	0.
	B2	0.	0.	0.	0.	0.	0.	1.02778	1.62794	0.	0.
* 1B3U	H1	0.	0.								
-0.5491	H2	0.	0.								
	H3	0.	0.								
	H4	0.	0.								
	H5	-0.25913	-0.25868								
	H6	0.25913	0.25868								
	B1	0.	0.	0.	0.	0.29853	0.00062	0.	0.	0.	0.
	B2	0.	0.	0.	0.	0.29853	0.00062	0.	0.	0.	0.
2B3U	H1	0.	0.								
0.1805	H2	0.	0.								
	H3	0.	0.								
	H4	0.	0.								
	H5	-0.06686	-0.78500								
	H6	0.06686	0.78500								
	B1	0.	0.	0.	0.	0.10316	-0.88262	0.	0.	0.	0.
	B2	0.	0.	0.	0.	0.10316	-0.88262	0.	0.	0.	0.
3B3U	H1	0.	0.								
0.5310	H2	0.	0.								
	H3	0.	0.								
	H4	0.	0.								
	H5	0.09612	1.23718								
	H6	-0.09612	-1.23718								
	B1	0.	0.	0.	0.	0.95871	0.13036	0.	0.	0.	0.
	B2	0.	0.	0.	0.	0.95871	0.13036	0.	0.	0.	0.

Overlap Matrix

H1S1 with
    H1S2 .68302   H2S1 .00478   H2S2 .07896   H3S1 .00002   H3S2 .00741   H5S1 .00449   H5S2 .07684
    B1S1 .00170   B1S2 .03065   B1S3 .29267   B1S4 .32216   B1Y1 -.39201   B1Y2 -.27912   B1Z1 -.23091
    B1Z2 -.16441

H1S2 with
    H2S1 .07896   H2S2 .26004   H3S1 .00741   H3S2 .05884   H5S1 .07684   H5S2 .25565   B1S1 .00880
    B1S2 .11186   B1S3 .54754   B1S4 .67827   B1Y1 -.41798   B1Y2 -.48913   B1Z1 -.24621   B1Z2 -.28812

H2S1 with
    H3S2 .00088

H2S2 with
    H3S1 .00088   H3S2 .01530

H3S1 with
    B1S3 .00565   B1S4 .06522   B1Y1 -.01293   B1Y2 -.08329   B1Z1 .02988   B1Z2 .19253

H3S2 with
    B1S1 .00031   B1S2 .00477   B1S3 .06767   B1S4 .20878   B1Y1 -.05767   B1Y2 -.18871   B1Z1 .13330
    B1Z2 .43621

H5S1 with
    H5S2 .68302   H6S1 .00623   H6S2 .08842   B1S1 .00073   B1S2 .01551   B1S3 .22374   B1S4 .28876
    B1X1 -.28916   B1X2 -.25008   B1Z1 .25585   B1Z2 .22128

H5S2 with
    H6S1 .08842   H6S2 .27914   B1S1 .00699   B1S2 .09008   B1S3 .47494   B1S4 .62569   B1X1 -.35394
    B1X2 -.44603   B1Z1 .31317   B1Z2 .39465

B1S1 with
    B1S2 .35489   B1S3 .00732   B1S4 .01371   B2S3 .00117   B2S4 .00467   B2Z1 -.00462   B2Z2 -.00876

B1S2 with
    B1S3 .24693   B1S4 .17217   B2S2 .00001   B2S3 .01840   B2S4 .06083   B2Z1 -.06140   B2Z2 -.11091

B1S3 with
    B1S4 .80606   B2S1 .00117   B2S2 .01840   B2S3 .19941   B2S4 .35898   B2Z1 -.34831   B2Z2 -.53667

B1S4 with
    B2S1 .00467   B2S2 .06083   B2S3 .35898   B2S4 .58369   B2Z1 -.32057   B2Z2 -.63862

B1X1 with
    B1X2 .55377   B2X1 .18649   B2X2 .29896

B1X2 with
    B2X1 .29896   B2X2 .67406

B1Y1 with
    B1Y2 .55377   B2Y1 .18649   B2Y2 .29896

B1Y2 with
    B2Y1 .29896   B2Y2 .67406

B1Z1 with
    B1Z2 .55377   B2S1 .00462   B2S2 .06140   B2S3 .34831   B2S4 .32057   B2Z1 -.35326   B2Z2 -.06893

B1Z2 with
    B2S1 .00876   B2S2 .11091   B2S3 .53667   B2S4 .63862   B2Z1 -.06893   B2Z2 .14231

C2H4   Ethylene

Molecular Geometry   Symmetry D$_{2h}$

Center	Coordinates		
	X	Y	Z
H1	1.75550000	0.	-2.32780001
H2	-1.75550000	0.	-2.32780001
H3	-1.75550000	0.	2.32780001
H4	1.75550000	0.	2.32780001
C1	0.	0.	-1.26520000
C2	0.	0.	1.26520000
Mass	0.	0.	0.

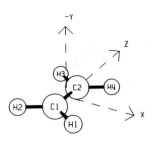

MO Symmetry Types and Occupancy

Representation	AG	B1G	B1U	B2G	B2U	B3U
Number of MOs	8	4	2	2	4	8
Number Occupied	3	1	1	0	1	2

Energy Expectation Values

Total	-78.0054	Kinetic	77.9429
Electronic	-111.2719	Potential	-155.9483
Nuclear Repulsion	33.2665	Virial	2.0008

Electronic Potential	-189.2149
One-Electron Potential	-247.6380
Two-Electron Potential	58.4232

Moments of the Charge Distribution

Second

x^2	y^2	z^2	xy	xz	yz	r^2
-21.4298	-12.0538	-49.9634	0.	0.	0.	-83.4470
12.3271	0.	40.8834	0.	0.	0.	53.2105

Fourth (even)

x^4	y^4	z^4	x^2y^2	x^2z^2	y^2z^2	r^2x^2	r^2y^2	r^2z^2	r^4
-106.8933	-53.3503	-325.2155	-22.1145	-100.7009	-41.7032	-229.7087	-117.1680	-467.6196	-814.4963
37.9895	0.	148.1953	0.	66.7964	0.	104.7859	0.	214.9917	319.7775

Expectation Values at Atomic Centers

Center			Expectation Value		
	r^{-1}	δ	xr^{-3}	yr^{-3}	zr^{-3}
H1	-1.0950	0.4189	0.0560	0.	-0.0348
C1	-14.7271	116.4638	0.	0.	-0.0097

	$\dfrac{3x^2-r^2}{r^5}$	$\dfrac{3y^2-r^2}{r^5}$	$\dfrac{3z^2-r^2}{r^5}$	$\dfrac{3xy}{r^5}$	$\dfrac{3xz}{r^5}$	$\dfrac{3yz}{r^5}$
H1	0.2056	-0.1790	-0.0267	0.	-0.2269	0.
C1	-0.1355	0.2723	-0.1367	0.	0.	0.

Mulliken Charge

Center				Basis Function Type						Total
	S1	S2	S3	S4	X1	X2	Y1	Y2	Z1	Z2	
H1	0.5174	0.3077									0.8251
C1	0.0428	1.9564	0.7939	0.4760	0.9312	0.1659	0.6780	0.3220	0.9354	0.0481	6.3498

Overlap Populations

	Pair of Atomic Centers					
	H2	H3	H3	C1	C1	C2
	H1	H1	H2	H1	H3	C1
1AG	0.	0.	0.	-0.0001	0.	0.0010
2AG	0.0025	0.0002	0.0007	0.0686	0.0150	0.5327
3AG	0.0242	0.0023	0.0084	0.1094	-0.0497	0.3138
Total AG	0.0267	0.0025	0.0091	0.1779	-0.0348	0.8475
1B1G	-0.0591	0.0057	-0.0205	0.2504	-0.0496	-0.2680
Total B1G	-0.0591	0.0057	-0.0205	0.2504	-0.0496	-0.2680
1B1U	0.	0.	0.	0.	0.	0.5341
Total B1U	0.	0.	0.	0.	0.	0.5341
1B2U	-0.0299	-0.0027	0.0101	0.1598	0.0249	0.1777
Total B2U	-0.0299	-0.0027	0.0101	0.1598	0.0249	0.1777
1B3U	0.	0.	0.	-0.0001	0.	-0.0021
2B3U	0.0182	-0.0015	-0.0058	0.2123	-0.0213	-0.1211
Total B3U	0.0182	-0.0015	-0.0058	0.2123	-0.0213	-0.1232
Total	-0.0440	0.0039	-0.0071	0.8004	-0.0807	1.1681
Distance	3.5110	5.8311	4.6556	2.0520	3.9989	2.5304

Molecular Orbitals and Eigenvalues

Center		S1	S2	S3	S4	X1	X2	Y1	Y2	Z1	Z2
* 1AG	H1	-0.00051	0.00053								
-11.2419	H2	-0.00051	0.00053								
	H3	-0.00051	0.00053								
	H4	-0.00051	0.00053								
	C1	-0.03669	-0.69206	-0.00354	-0.00111	0.	0.	0.	0.	-0.00064	0.00045
	C2	-0.03669	-0.69206	-0.00354	-0.00111	0.	0.	0.	0.	0.00064	-0.00045
* 2AG	H1	-0.08398	-0.02031								
-1.0386	H2	-0.08398	-0.02031								
	H3	-0.08398	-0.02031								
	H4	-0.08398	-0.02031								
	C1	0.00759	0.16420	-0.35638	-0.17976	0.	0.	0.	0.	-0.12314	0.01348
	C2	0.00759	0.16420	-0.35638	-0.17976	0.	0.	0.	0.	0.12314	-0.01348
* 3AG	H1	0.12121	0.09647								
-0.5851	H2	0.12121	0.09647								
	H3	0.12121	0.09647								
	H4	0.12121	0.09647								
	C1	0.00046	0.01011	-0.02314	0.03466	0.	0.	0.	0.	-0.47669	-0.06352
	C2	0.00046	0.01011	-0.02314	0.03466	0.	0.	0.	0.	0.47669	0.06352
4AG	H1	-0.02685	-1.12898								
0.2904	H2	-0.02685	-1.12898								
	H3	-0.02685	-1.12898								
	H4	-0.02685	-1.12898								
	C1	-0.00287	-0.06689	0.12829	1.70861	0.	0.	0.	0.	-0.12360	-1.07936
	C2	-0.00287	-0.06689	0.12829	1.70861	0.	0.	0.	0.	0.12360	1.07936
5AG	H1	-0.04350	0.23838								
0.4140	H2	-0.04350	0.23838								
	H3	-0.04350	0.23838								
	H4	-0.04350	0.23838								
	C1	0.00165	0.03666	-0.08704	0.04025	0.	0.	0.	0.	0.51068	-0.90565
	C2	0.00165	0.03666	-0.08704	0.04025	0.	0.	0.	0.	-0.51068	0.90565
* 1B1G	H1	-0.19158	-0.15003								
-0.5047	H2	0.19158	0.15003								
	H3	-0.19158	-0.15003								
	H4	0.19158	0.15003								
	C1	0.	0.	0.	0.	-0.37279	-0.12703	0.	0.	0.	0.
	C2	0.	0.	0.	0.	0.37279	0.12703	0.	0.	0.	0.
2B1G	H1	-0.06586	-1.53920								
0.4713	H2	0.06586	1.53920								
	H3	-0.06586	-1.53920								
	H4	0.06586	1.53920								
	C1	0.	0.	0.	0.	-0.50992	4.15735	0.	0.	0.	0.
	C2	0.	0.	0.	0.	0.50992	-4.15735	0.	0.	0.	0.
3B1G	H1	0.01054	-3.48862								
0.6880	H2	-0.01054	3.48862								
	H3	0.01054	-3.48862								
	H4	-0.01054	3.48862								
	C1	0.	0.	0.	0.	0.56986	6.01314	0.	0.	0.	0.
	C2	0.	0.	0.	0.	-0.56986	-6.01314	0.	0.	0.	0.

Center		S1	S2	S3	S4	X1	X2	Y1	Y2	Z1	Z2
* 1B1U	H1	0.	0.								
-0.3752	H2	0.	0.								
	H3	0.	0.								
	H4	0.	0.								
	C1	0.	0.	0.	0.	0.	0.	-0.46243	-0.21422	0.	0.
	C2	0.	0.	0.	0.	0.	0.	-0.46243	-0.21422	0.	0.
2B1U	H1	0.	0.								
0.4895	H2	0.	0.								
	H3	0.	0.								
	H4	0.	0.								
	C1	0.	0.	0.	0.	0.	0.	-0.65512	0.63892	0.	0.
	C2	0.	0.	0.	0.	0.	0.	-0.65512	0.63892	0.	0.
1B2G	H1	0.	0.								
0.1459	H2	0.	0.								
	H3	0.	0.								
	H4	0.	0.								
	C1	0.	0.	0.	0.	0.	0.	-0.39794	-0.83702	0.	0.
	C2	0.	0.	0.	0.	0.	0.	0.39794	0.83702	0.	0.
2B2G	H1	0.	0.								
0.6143	H2	0.	0.								
	H3	0.	0.								
	H4	0.	0.								
	C1	0.	0.	0.	0.	0.	0.	-0.80294	1.18450	0.	0.
	C2	0.	0.	0.	0.	0.	0.	0.80294	-1.18450	0.	0.
* 1B2U	H1	0.15238	0.10222								
-0.6467	H2	-0.15238	-0.10222								
	H3	-0.15238	-0.10222								
	H4	0.15238	0.10222								
	C1	0.	0.	0.	0.	0.37958	0.05998	0.	0.	0.	0.
	C2	0.	0.	0.	0.	0.37958	0.05998	0.	0.	0.	0.
2B2U	H1	0.07940	1.31697								
0.3683	H2	-0.07940	-1.31697								
	H3	-0.07940	-1.31697								
	H4	0.07940	1.31697								
	C1	0.	0.	0.	0.	-0.37271	-1.02549	0.	0.	0.	0.
	C2	0.	0.	0.	0.	-0.37271	-1.02549	0.	0.	0.	0.
3B2U	H1	-0.08322	-0.34650								
0.4409	H2	0.08322	0.34650								
	H3	0.08322	0.34650								
	H4	-0.08322	-0.34650								
	C1	0.	0.	0.	0.	-0.54128	0.95648	0.	0.	0.	0.
	C2	0.	0.	0.	0.	-0.54128	0.95648	0.	0.	0.	0.
* 1B3U	H1	-0.00040	0.00032								
-11.2403	H2	-0.00040	0.00032								
	H3	0.00040	-0.00032								
	H4	0.00040	-0.00032								
	C1	-0.03671	-0.69247	-0.00249	-0.00307	0.	0.	0.	0.	0.00041	-0.00086
	C2	0.03671	0.69247	0.00249	0.00307	0.	0.	0.	0.	0.00041	-0.00086

C2H4

		S1	S2	S3	S4	X1	X2	Y1	Y2	Z1	Z2
• 2B3U	H1	0.14630	0.07280								
-0.7917	H2	0.14630	0.07280								
	H3	-0.14630	-0.07280								
	H4	-0.14630	-0.07280								
	C1	-0.00584	-0.12666	0.28342	0.21224	0.	0.	0.	0.	-0.19436	-0.01677
	C2	0.00584	0.12666	-0.28342	-0.21224	0.	0.	0.	0.	-0.19436	-0.01677
3B3U	H1	0.00070	0.06772								
0.2604	H2	0.00070	0.06772								
	H3	-0.00070	-0.06772								
	H4	-0.00070	-0.06772								
	C1	0.00351	0.09675	-0.05478	-2.96239	0.	0.	0.	0.	0.05071	-1.83029
	C2	-0.00351	-0.09675	0.05478	2.96239	0.	0.	0.	0.	0.05071	-1.83029
4B3U	H1	-0.07250	-1.48685								
0.3997	H2	-0.07250	-1.48685								
	H3	0.07250	1.48685								
	H4	0.07250	1.48685								
	C1	-0.00240	-0.05530	0.11356	0.89199	0.	0.	0.	0.	-0.14792	-1.50738
	C2	0.00240	0.05530	-0.11356	-0.89199	0.	0.	0.	0.	-0.14792	-1.50738

Overlap Matrix

H1S1 with
      H1S2 .68301   H2S1 .01227   H2S2 .11817   H3S1 .00001   H3S2 .00552   C1S1 .00227   C1S2 .03490
      C1S3 .28654   C1S4 .37345   C1X1 .37014   C1X2 .37049   C1Z1 -.22405   C1Z2 -.22426

H1S2 with
      H2S1 .11817   H2S2 .33471   H3S1 .00552   H3S2 .04885   C1S1 .00770   C1S2 .09833   C1S3 .49833
      C1S4 .70742   C1X1 .30847   C1X2 .51987   C1Z1 -.18672   C1Z2 -.31467

H2S1 with
      H3S1 .00056   H3S2 .03143

H2S2 with
      H3S1 .03143   H3S2 .14596

H3S1 with
      C1S2 .00005   C1S3 .01067   C1S4 .08821   C1X1 -.01769   C1X2 -.11566   C1Z1 .03620   C1Z2 .23672

H3S2 with
      C1S1 .00095   C1S2 .01327   C1S3 .11019   C1S4 .27333   C1X1 -.07078   C1X2 -.22885   C1Z1 .14487
      C1Z2 .46838

C1S1 with
      C1S2 .35436   C1S3 .00805   C1S4 .01419   C2S3 .00173   C2S4 .00550   C2Z1 -.00547   C2Z2 -.00963

C1S2 with
      C1S3 .26226   C1S4 .17745   C2S2 .00002   C2S3 .02645   C2S4 .07112   C2Z1 -.07227   C2Z2 -.12097

C1S3 with
      C1S4 .79970   C2S1 .00173   C2S2 .02645   C2S3 .23544   C2S4 .39098   C2Z1 -.36829   C2Z2 -.54802

C1S4 with
      C2S1 .00550   C2S2 .07112   C2S3 .39098   C2S4 .62262   C2Z1 -.30744   C2Z2 -.63076

C1X1 with
      C1X2 .53875   C2X1 .19552   C2X2 .30241

C1X2 with
      C2X1 .30241   C2X2 .69289

C1Y1 with
      C1Y2 .53875   C2Y1 .19552   C2Y2 .30241

C1Y2 with
      C2Y1 .30241   C2Y2 .69289

C1Z1 with
      C1Z2 .53875   C2S1 .00547   C2S2 .07227   C2S3 .36829   C2S4 .30744   C2Z1 -.35275   C2Z2 -.04626

C1Z2 with
      C2S1 .00963   C2S2 .12097   C2S3 .54802   C2S4 .63076   C2Z1 -.04626   C2Z2 .18446

N2H2   Diimide

Molecular Geometry   Symmetry C_{2h}

Center	Coordinates		
	X	Y	Z
H1	-1.90560000	1.78060000	0.
H2	1.90560000	-1.78060000	0.
N1	-1.16260000	0.	0.
N2	1.16260000	0.	0.
Mass	0.	0.	0.

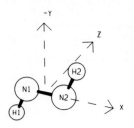

MO Symmetry Types and Occupancy

Representation	AG	AU	BG	BU
Number of MOs	10	2	2	10
Number Occupied	4	1	0	3

Energy Expectation Values

Total	-109.9418	Kinetic	110.0095
Electronic	-142.4096	Potential	-219.9513
Nuclear Repulsion	-109.9418	Virial	1.9994

Electronic Potential	-252.4191
One-Electron Potential	-323.0972
Two-Electron Potential	70.6781

Moments of the Charge Distribution

Second

x^2	y^2	z^2	xy	xz	yz	r^2
-35.7902	-14.2639	-9.0947	3.4601	0.	0.	-59.1488
26.1856	6.3411	0.	-6.7862	0.	0.	32.5266

Fourth (even)

x^4	y^4	z^4	x^2y^2	x^2z^2	y^2z^2	r^2x^2	r^2y^2	r^2z^2	r^4
-167.5416	-58.5418	-28.8329	-45.5878	-24.1426	-12.3678	-237.2720	-116.4975	-65.3432	-419.1128
51.9498	20.1046	0.	23.0264	0.	0.	74.9762	43.1310	0.	118.1072

Fourth (odd)

x^3y	xy^3	xyz^2	x^3z	xy^2z	xz^3	x^2yz	y^3z	yz^3
16.5571	18.9232	1.2430	0.	0.	0.	0.	0.	0.
-24.6429	-21.5160	0.	0.	0.	0.	0.	0.	0.

Expectation Values at Atomic Centers

Center			Expectation Value		
	r^{-1}	δ	xr^{-3}	yr^{-3}	zr^{-3}
H1	-1.0078	0.4288	-0.0427	0.0802	0.
N1	-18.2961	189.2738	-0.1611	-0.1395	0.

	$\dfrac{3x^2-r^2}{r^5}$	$\dfrac{3y^2-r^2}{r^5}$	$\dfrac{3z^2-r^2}{r^5}$	$\dfrac{3xy}{r^5}$	$\dfrac{3xz}{r^5}$	$\dfrac{3yz}{r^5}$
H1	-0.1026	0.3095	-0.2069	-0.2456	0.	0.
N1	-0.0519	-1.0210	1.0728	-0.7903	0.	0.

Mulliken Charge

Center				Basis Function Type							Total
	S1	S2	S3	S4	X1	X2	Y1	Y2	Z1	Z2	
H1	0.5174	0.1876									0.7051
N1	0.0433	1.9560	0.9315	0.7914	0.9943	0.0813	1.1518	0.3453	0.6845	0.3155	7.2949

Overlap Populations

		Pair of Atomic Centers		
	H2	N1	N1	N2
	H1	H1	H2	N1
1AG	0.	-0.0001	0.	0.0006
2AG	0.	0.0351	0.0019	0.6332
3AG	0.0094	0.2534	-0.0680	0.1328
4AG	0.0077	0.0112	-0.0705	-0.5291
Total AG	0.0170	0.2995	-0.1367	0.2374
1AU	0.	0.	0.	0.4604
Total AU	0.	0.	0.	0.4604
1BU	0.	0.	0.	-0.0030
2BU	-0.0028	0.3374	-0.0444	-0.3732
3BU	-0.0008	0.0054	0.0214	0.2771
Total BU	-0.0036	0.3428	-0.0230	-0.0991
Total	0.0134	0.6423	-0.1596	0.5988
Distance	5.2161	1.9294	3.5474	2.3252

Molecular Orbitals and Eigenvalues

Center		S1	S2	S3	S4	X1	X2	Y1	Y2	Z1	Z2
• 1AG	H1	0.00021	-0.00052								
-15.6384	H2	0.00021	-0.00052								
	N1	0.03673	0.69221	0.00259	0.00132	0.00140	-0.00072	0.00108	-0.00031	0.	0.
	N2	0.03673	0.69221	0.00259	0.00132	-0.00140	0.00072	-0.00108	0.00031	0.	0.
• 2AG	H1	-0.07091	0.01148								
-1.3897	H2	-0.07091	0.01148								
	N1	0.00765	0.16359	-0.36777	-0.20748	-0.19000	0.00924	-0.04294	-0.01527	0.	0.
	N2	0.00765	0.16359	-0.36777	-0.20748	0.19000	-0.00924	0.04294	0.01527	0.	0.
• 3AG	H1	0.19665	0.13334								
-0.6959	H2	0.19665	0.13334								
	N1	-0.00074	-0.01742	0.02649	0.08196	-0.44205	-0.07074	0.17100	0.03067	0.	0.
	N2	-0.00074	-0.01742	0.02649	0.08196	0.44205	0.07074	-0.17100	-0.03067	0.	0.
• 4AG	H1	-0.15760	-0.12315								
-0.3928	H2	-0.15760	-0.12315								
	N1	-0.00215	-0.04874	0.09285	0.28920	-0.25981	-0.12415	-0.42713	-0.23853	0.	0.
	N2	-0.00215	-0.04874	0.09285	0.28920	0.25981	0.12415	0.42713	0.23853	0.	0.
5AG	H1	-0.00199	1.50128								
0.2690	H2	-0.00199	1.50128								
	N1	0.00241	0.05778	-0.08768	-0.97595	0.11696	0.34070	-0.22797	-0.97673	0.	0.
	N2	0.00241	0.05778	-0.08768	-0.97595	-0.11696	-0.34070	0.22797	0.97673	0.	0.
6AG	H1	-0.10383	-0.10009								
0.6079	H2	-0.10383	-0.10009								
	N1	-0.00033	0.00626	0.11177	0.31438	0.46168	-1.04182	-0.10608	0.42247	0.	0.
	N2	-0.00033	0.00626	0.11177	0.31438	-0.46168	1.04182	0.10608	-0.42247	0.	0.
• 1AU	H1	0.	0.								
-0.5224	H2	0.	0.								
	N1	0.	0.	0.	0.	0.	0.	0.	0.	-0.47886	-0.21581
	N2	0.	0.	0.	0.	0.	0.	0.	0.	-0.47886	-0.21581
2AU	H1	0.	0.								
0.6557	H2	0.	0.								
	N1	0.	0.	0.	0.	0.	0.	0.	0.	0.65506	-0.64054
	N2	0.	0.	0.	0.	0.	0.	0.	0.	0.65506	-0.64054
1BG	H1	0.	0.								
0.1171	H2	0.	0.								
	N1	0.	0.	0.	0.	0.	0.	0.	0.	0.48582	0.61556
	N2	0.	0.	0.	0.	0.	0.	0.	0.	-0.48582	-0.61556
• 1BU	H1	0.00000	-0.00006								
-15.6369	H2	-0.00000	0.00006								
	N1	0.03674	0.69243	0.00179	0.00450	0.00115	0.00130	0.00100	-0.00030	0.	0.
	N2	-0.03674	-0.69243	-0.00179	-0.00450	0.00115	0.00130	0.00100	-0.00030	0.	0.
• 2BU	H1	-0.18735	-0.06370								
-0.9565	H2	0.18735	0.06370								
	N1	0.00600	0.12897	-0.29682	-0.32632	0.13838	-0.01198	-0.19638	-0.02753	0.	0.
	N2	-0.00600	-0.12897	0.29682	0.32632	0.13838	-0.01198	-0.19638	-0.02753	0.	0.
• 3BU	H1	0.07231	0.03688								
-0.6378	H2	-0.07231	-0.03688								
	N1	0.00394	0.08324	-0.20914	-0.18230	0.12557	0.07712	0.40624	0.14144	0.	0.
	N2	-0.00394	-0.08324	0.20914	0.18230	0.12557	0.07712	0.40624	0.14144	0.	0.

Center		Basis Function Type									
		S1	S2	S3	S4	X1	X2	Y1	Y2	Z1	Z2
4BU	H1	-0.06550	-1.42251								
0.2630	H2	0.06550	1.42251								
	N1	-0.00308	-0.07596	0.09997	1.38392	-0.01887	0.14304	0.23902	0.49095	0.	0.
	N2	0.00308	0.07596	-0.09997	-1.38392	-0.01887	0.14304	0.23902	0.49095	0.	0.
5BU	H1	0.04876	0.84826								
0.4338	H2	-0.04876	-0.84826								
	N1	-0.00266	-0.07145	0.05081	2.09135	0.14712	2.02897	-0.20038	-0.24850	0.	0.
	N2	0.00266	0.07145	-0.05081	-2.09135	0.14712	2.02897	-0.20038	-0.24850	0.	0.

Overlap Matrix

First Function			Second Function and Overlap			

H1S1 with

H1S2	.68301	H2S1	.00009	H2S2	.01438	N1S1	.00246	N1S2	.03516	N1S3	.26630	N1S4	.40081
N1X1	-.14860	N1X2	-.19676	N1Y1	.35612	N1Y2	.47154						

H1S2 with

H2S1	.01438	H2S2	.08931	N1S1	.00664	N1S2	.08457	N1S3	.44225	N1S4	.69686	N1X1	-.10377
N1X2	-.22728	N1Y1	.24869	N1Y2	.54467								

H2S1 with

N1S1	.00001	N1S2	.00021	N1S3	.01319	N1S4	.09625	N1X1	.03473	N1X2	.24644	N1Y1	-.02016
N1Y2	-.14302												

H2S2 with

N1S1	.00138	N1S2	.01841	N1S3	.12799	N1S4	.29823	N1X1	.12642	N1X2	.43944	N1Y1	-.07337
N1Y2	-.25503												

N1S1 with

N1S2	.35584	N1S3	.00894	N1S4	.01446	N2S3	.00095	N2S4	.00470	N2X1	-.00360	N2X2	-.00942

N1S2 with

N1S3	.27246	N1S4	.17925	N2S2	.00001	N2S3	.01548	N2S4	.06064	N2X1	-.04871	N2X2	-.11763

N1S3 with

N1S4	.79352	N2S1	.00095	N2S2	.01548	N2S3	.17248	N2S4	.33807	N2X1	-.29098	N2X2	-.53358

N1S4 with

N2S1	.00470	N2S2	.06064	N2S3	.33807	N2S4	.57006	N2X1	-.28061	N2X2	-.63391

N1X1 with

N1X2	.52739	N2S1	.00360	N2S2	.04871	N2S3	.29098	N2S4	.28061	N2X1	-.32974	N2X2	-.10690

N1X2 with

N2S1	.00942	N2S2	.11763	N2S3	.53358	N2S4	.63391	N2X1	-.10690	N2X2	.06774

N1Y1 with

N1Y2	.52739	N2Y1	.13765	N2Y2	.26010

N1Y2 with

N2Y1	.26010	N2Y2	.63952

N1Z1 with

N1Z2	.52739	N2Z1	.13765	N2Z2	.26010

N1Z2 with

N2Z1	.26010	N2Z2	.63952

H2CO   Formaldehyde

Molecular Geometry   Symmetry C$_{2v}$

Center	Coordinates		
	X	Y	Z
H1	1.79338799	0.	-1.10977690
H2	-1.79338799	0.	-1.10977690
C1	0.	0.	0.
O1	0.	0.	2.28250000
Mass	0.	0.	1.14198339

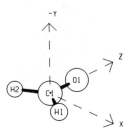

MO Symmetry Types and Occupancy

Representation	A1	A2	B1	B2
Number of MOs	14	0	6	4
Number Occupied	5	0	2	1

Energy Expectation Values

Total	-113.8209	Kinetic	113.8158
Electronic	-144.9890	Potential	-227.6367
Nuclear Repulsion	31.1681	Virial	2.0000
Electronic Potential	-258.8048		
One-Electron Potential	-330.6402		
Two-Electron Potential	71.8354		

Moments of the Charge Distribution

First

x	y	z	Dipole Moment
0.	0.	1.0076	1.2237
0.	0.	-2.2313	

Second

x^2	y^2	z^2	xy	xz	yz	r^2
-14.9153	-8.7727	-37.4059	0.	0.	0.	-61.0939
6.4325	0.	28.3718	0.	0.	0.	34.8043

Third

x^3	xy^2	xz^2	x^2y	y^3	yz^2	x^2z	y^2z	z^3	xyz
0.	0.	0.	0.	0.	0.	14.4540	1.9871	22.8062	0.
0.	0.	0.	0.	0.	0.	-14.4844	0.	-19.9020	0.

Fourth (even)

x^4	y^4	z^4	x^2y^2	x^2z^2	y^2z^2	r^2x^2	r^2y^2	r^2z^2	r^4
-66.1803	-27.0823	-193.9331	-12.8786	-58.1365	-24.8236	-137.1954	-64.7846	-276.8933	-478.8732
20.6884	0.	75.1591	0.	32.6154	0.	53.3038	0.	107.7745	161.0783

Expectation Values at Atomic Centers

Center			Expectation Value		
	r^{-1}	δ	xr^{-3}	yr^{-3}	zr^{-3}
H1	-1.0508	0.4141	0.0557	0.	-0.0302
C1	-14.6096	116.5677	0.	0.	-0.0580
O1	-22.3181	287.3473	0.	0.	0.2626

	$\dfrac{3x^2-r^2}{r^5}$	$\dfrac{3y^2-r^2}{r^5}$	$\dfrac{3z^2-r^2}{r^5}$	$\dfrac{3xy}{r^5}$	$\dfrac{3xz}{r^5}$	$\dfrac{3yz}{r^5}$
H1	0.1632	-0.1420	-0.0212	0.	-0.1943	0.
C1	-0.3386	0.4917	-0.1531	0.	0.	0.
O1	-2.6124	1.9156	0.6968	0.	0.	0.

Mulliken Charge

Center				Basis Function Type							Total
	S1	S2	S3	S4	X1	X2	Y1	Y2	Z1	Z2	
H1	0.5133	0.3412									0.8545
C1	0.0429	1.9565	0.8521	0.4366	0.9688	0.1214	0.5570	0.1503	0.8819	0.0173	5.9848
O1	0.0426	1.9567	0.9720	0.8515	1.4012	0.4764	0.8686	0.4240	1.0989	0.2142	8.3062

Overlap Populations

	Pair of Atomic Centers			
	H2	C1	O1	O1
	H1	H1	H1	C1
1A1	0.	0.	0.	-0.0005
2A1	0.	-0.0002	0.	-0.0002
3A1	-0.0001	0.0032	-0.0008	0.5882
4A1	0.0189	0.2950	-0.0105	-0.1002
5A1	0.0146	0.0605	-0.0151	0.0966
Total A1	0.0335	0.3585	-0.0264	0.5839
1B1	-0.0281	0.2164	0.0236	0.2587
2B1	-0.1318	0.1368	-0.1146	-0.2050
Total B1	-0.1599	0.3532	-0.0910	0.0537
1B2	0.	0.	0.	0.4326
Total B2	0.	0.	0.	0.4326
Total	-0.1265	0.7116	-0.1174	1.0702
Distance	3.5868	2.1090	3.8372	2.2825

Molecular Orbitals and Eigenvalues

Center		S1	S2	S3	S4	X1	X2	Y1	Y2	Z1	Z2
* 1A1	H1	0.00004	0.00000								
-20.5844	H2	0.00004	0.00000								
	C1	-0.00000	-0.00002	-0.00006	-0.00103	0.	0.	0.	0.	-0.00031	-0.00065
	O1	0.05112	0.97938	0.00340	0.00291	0.	0.	0.	0.	-0.00184	0.00006
* 2A1	H1	-0.00074	0.00077								
-11.3618	H2	-0.00074	0.00077								
	C1	-0.05191	-0.97907	-0.00447	-0.00190	0.	0.	0.	0.	-0.00102	0.00055
	O1	0.00002	0.00045	-0.00075	0.00050	0.	0.	0.	0.	0.00109	-0.00044
* 3A1	H1	-0.02845	0.00592								
-1.4316	H2	-0.02845	0.00592								
	C1	0.00551	0.11719	-0.25943	-0.08215	0.	0.	0.	0.	-0.20468	0.03152
	O1	0.00964	0.21101	-0.46565	-0.32844	0.	0.	0.	0.	0.18361	-0.01066
* 4A1	H1	-0.17609	-0.07174								
-0.8694	H2	-0.17609	-0.07174								
	C1	0.00803	0.16989	-0.41919	-0.24203	0.	0.	0.	0.	0.21400	0.04464
	O1	-0.00440	-0.09654	0.21734	0.24921	0.	0.	0.	0.	0.14861	0.05350
* 5A1	H1	0.08828	0.07989								
-0.6448	H2	0.08828	0.07989								
	C1	0.00115	0.01893	-0.09583	0.07171	0.	0.	0.	0.	-0.44952	0.01843
	O1	-0.00356	-0.07896	0.17210	0.31334	0.	0.	0.	0.	0.57069	0.16915
6A1	H1	0.04221	0.85894								
0.2751	H2	0.04221	0.85894								
	C1	0.00486	0.10901	-0.25306	-1.94687	0.	0.	0.	0.	0.20966	-0.75613
	O1	-0.00249	-0.06661	0.04763	1.07844	0.	0.	0.	0.	-0.13860	-0.33972
7A1	H1	0.05114	1.43999								
0.2851	H2	0.05114	1.43999								
	C1	0.00282	0.06209	-0.14888	-1.37741	0.	0.	0.	0.	0.22606	1.68182
	O1	0.00184	0.04851	-0.03947	-0.76757	0.	0.	0.	0.	0.01671	0.00642
8A1	H1	-0.26339	-0.72991								
0.6091	H2	-0.26339	-0.72991								
	C1	0.00571	0.02499	-0.99630	3.23625	0.	0.	0.	0.	0.66474	-0.62105
	O1	0.00291	0.06435	-0.15433	-0.81386	0.	0.	0.	0.	0.00273	0.81084
* 1B1	H1	-0.16867	-0.09858								
-0.7039	H2	0.16867	0.09858								
	C1	0.	0.	0.	0.	-0.49395	-0.08608	0.	0.	0.	0.
	O1	0.	0.	0.	0.	-0.39741	-0.12784	0.	0.	0.	0.
* 2B1	H1	-0.18938	-0.26111								
-0.4426	H2	0.18938	0.26111								
	C1	0.	0.	0.	0.	-0.24508	0.00855	0.	0.	0.	0.
	O1	0.	0.	0.	0.	0.63491	0.34601	0.	0.	0.	0.
3B1	H1	0.07442	1.20018								
0.3807	H2	-0.07442	-1.20018								
	C1	0.	0.	0.	0.	0.53098	-2.54396	0.	0.	0.	0.
	O1	0.	0.	0.	0.	0.11122	0.52333	0.	0.	0.	0.
4B1	H1	-0.00107	-1.85428								
0.4361	H2	0.00107	1.85428								
	C1	0.	0.	0.	0.	0.84084	1.63038	0.	0.	0.	0.
	O1	0.	0.	0.	0.	-0.27228	-0.39830	0.	0.	0.	0.

Center		S1	S2	S3	S4	X1	X2	Y1	Y2	Z1	Z2
• 1B2	H1	0.	0.								
-0.5374	H2	0.	0.								
	C1	0.	0.	0.	0.	0.	0.	-0.41813	-0.12319	0.	0.
	O1	0.	0.	0.	0.	0.	0.	-0.55790	-0.27140	0.	0.
2B2	H1	0.	0.								
0.1080	H2	0.	0.								
	C1	0.	0.	0.	0.	0.	0.	-0.46625	-0.72728	0.	0.
	O1	0.	0.	0.	0.	0.	0.	0.43558	0.53145	0.	0.
3B2	H1	0.	0.								
0.4622	H2	0.	0.								
	C1	0.	0.	0.	0.	0.	0.	1.01195	-1.08798	0.	0.
	O1	0.	0.	0.	0.	0.	0.	-0.06860	0.05509	0.	0.

Overlap Matrix

First Function	Second Function and Overlap

H1S1 with

H1S2	.68301	H2S1	.01026	H2S2	.10949	C1S1	.00194	C1S2	.03052	C1S3	.26808	C1S4	.36275
C1X1	.35301	C1X2	.36972	C1Z1	-.21845	C1Z2	-.22879	O1S2	.00004	O1S3	.00294	O1S4	.04171
O1X1	.00505	O1X2	.08337	O1Z1	-.00955	O1Z2	-.15769						

H1S2 with

H2S1	.10949	H2S2	.31910	C1S1	.00739	C1S2	.09444	C1S3	.48346	C1S4	.69402	C1X1	.30585
C1X2	.52239	C1Z1	-.18927	C1Z2	-.32327	O1S1	.00076	O1S2	.01023	O1S3	.07203	O1S4	.19292
O1X1	.03851	O1X2	.17927	O1Z1	-.07285	O1Z2	-.33910						

C1S1 with

C1S2	.35436	C1S3	.00805	C1S4	.01419	O1S3	.00047	O1S4	.00530	O1Z1	-.00270	O1Z2	-.01295

C1S2 with

C1S3	.26226	C1S4	.17745	O1S2	.00002	O1S3	.01120	O1S4	.06939	O1Z1	-.04169	O1Z2	-.15896

C1S3 with

C1S4	.79970	O1S1	.00194	O1S2	.02682	O1S3	.19299	O1S4	.38774	O1Z1	-.26053	O1Z2	-.60788

C1S4 with

O1S1	.00419	O1S2	.05412	O1S3	.30064	O1S4	.55822	O1Z1	-.17137	O1Z2	-.53390

C1X1 with

C1X2	.53875	O1X1	.14715	O1X2	.36665

C1X2 with

O1X1	.17618	O1X2	.60162

C1Y1 with

C1Y2	.53875	O1Y1	.14715	O1Y2	.36665

C1Y2 with

O1Y1	.17618	O1Y2	.60162

C1Z1 with

C1Z2	.53875	O1S1	.00495	O1S2	.06441	O1S3	.34665	O1S4	.42254	O1Z1	-.30341	O1Z2	-.18204

C1Z2 with

O1S1	.00636	O1S2	.08095	O1S3	.40807	O1S4	.62021	O1Z1	-.00833	O1Z2	.13398

O1S1 with

O1S2	.35461	O1S3	.00974	O1S4	.01469

O1S2 with

O1S3	.28286	O1S4	.18230

O1S3 with

O1S4	.79029

O1X1 with

O1X2	.50607

O1Y1 with

O1Y2	.50607

O1Z1 with

O1Z2	.50607

C2H6   Ethane

Molecular Geometry   Symmetry D_{3d}

Center	Coordinates		
	X	Y	Z
H1	1.69855762	-0.98066292	-0.69996772
H2	-1.69855762	-0.98066292	-0.69996772
H3	0.	1.96132584	-0.69996772
H4	1.69855762	0.98066292	3.61586168
H5	0.	-1.96132579	3.61586168
H6	-1.69855762	0.98066292	3.61586168
C1	0.	0.	0.
C2	0.	0.	2.91589999
Mass	0.	0.	1.45794939

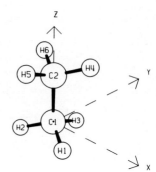

MO Symmetry Types and Occupancy

Representation	A1G	A1U	A2G	A2U	EG	EU
Number of MOs	8	0	0	8	8	8
Number Occupied	3	0	0	2	2	2

Energy Expectation Values

Total	-79.1981	Kinetic	79.1524
Electronic	-121.1291	Potential	-158.3505
Nuclear Repulsion	41.9310	Virial	2.0006

Electronic Potential	-200.2815
One-Electron Potential	-267.5387
Two-Electron Potential	67.2572

Moments of the Charge Distribution

Second

x^2	y^2	z^2	xy	xz	yz	r^2
-22.6239	-22.6239	-65.3088	0.	0.	0.	-110.5566
11.5404	11.5404	53.4470	0.	0.	0.	76.5278

Third

x^3	xy^2	xz^2	x^2y	y^3	yz^2	x^2z	y^2z	z^3	xyz
0.	0.	0.	0.	0.	0.	0.0001	0.0001	0.0005	0.
0.	0.	0.	0.	0.	0.	0.	0.	-0.0002	0.

Fourth (even)

x^4	y^4	z^4	x^2y^2	x^2z^2	y^2z^2	r^2x^2	r^2y^2	r^2z^2	r^4
-110.8770	-110.8775	-437.3185	-36.9590	-105.7261	-105.7264	-253.5621	-253.5630	-648.7710	-1155.8961
33.2952	33.2952	184.3223	11.0984	53.7389	53.7390	98.1325	98.1325	291.8002	488.0653

Fourth (odd)

x^3y	xy^3	xyz^2	x^3z	xy^2z	xz^3	x^2yz	y^3z	yz^3
0.0001	0.0001	0.0001	0.0001	0.	0.	-20.9316	20.9319	0.0001
0.	0.	0.	0.	0.	0.	24.4216	-24.4216	0.

Expectation Values at Atomic Centers

Center			Expectation Value		
	r^{-1}	δ	xr^{-3}	yr^{-3}	zr^{-3}
H1	-1.1223	0.4183	0.0484	-0.0279	-0.0251
C1	-14.7453	116.2805	0.	0.	-0.0003

	$\dfrac{3x^2-r^2}{r^5}$	$\dfrac{3y^2-r^2}{r^5}$	$\dfrac{3z^2-r^2}{r^5}$	$\dfrac{3xy}{r^5}$	$\dfrac{3xz}{r^5}$	$\dfrac{3yz}{r^5}$
H1	0.1579	-0.0556	-0.1023	-0.1849	-0.1317	0.0760
C1	-0.0225	-0.0225	0.0450	0.	0.	0.

Mulliken Charge

Center	Basis Function Type										Total
	S1	S2	S3	S4	X1	X2	Y1	Y2	Z1	Z2	
H1	0.5156	0.3236									0.8392
C1	0.0427	1.9564	0.7497	0.6295	0.8963	0.1881	0.8963	0.1881	0.8492	0.0860	6.4824

Overlap Populations

	Pair of Atomic Centers					
	H2	H4	H4	C1	C1	C2
	H1	H1	H2	H1	H4	C1
1A1G	0.	0.	0.	-0.0002	0.	0.0006
2A1G	0.0015	0.0003	0.0001	0.0654	0.0106	0.3898
3A1G	0.0195	0.0060	0.0019	0.0305	-0.0598	0.4916
Total A1G	0.0209	0.0063	0.0020	0.0958	-0.0492	0.8821
1A2U	0.	0.	0.	-0.0001	0.	-0.0019
2A2U	0.0086	-0.0021	-0.0006	0.1511	-0.0238	-0.2679
Total A2U	0.0086	-0.0021	-0.0006	0.1510	-0.0238	-0.2697
1EG	0.0260	0.0078	0.0025	0.0726	-0.0148	-0.1725
2EG	-0.0782	-0.0233	0.0071	0.2356	-0.0413	-0.1725
Total EG	-0.0521	-0.0156	0.0096	0.3082	-0.0561	-0.3450
1EU	0.0124	-0.0035	-0.0010	0.0574	0.0080	0.1427
2EU	-0.0373	0.0106	-0.0032	0.1582	0.0262	0.1427
Total EU	-0.0249	0.0071	-0.0042	0.2156	0.0342	0.2854
Total	-0.0475	-0.0044	0.0068	0.7705	-0.0949	0.5527
Distance	3.3971	4.7406	5.8321	2.0825	4.1135	2.9159

Molecular Orbitals and Eigenvalues

Center		S1	S2	S3	S4	X1	X2	Y1	Y2	Z1	Z2
* 1A1G	H1	0.00042	-0.00071								
-11.2161	H2	0.00042	-0.00071								
	H3	0.00042	-0.00071								
	H4	0.00042	-0.00071								
	H5	0.00042	-0.00071								
	H6	0.00042	-0.00071								
	C1	0.03652	0.68878	0.00315	0.00201	-0.00000	0.00000	-0.00000	0.00000	0.00023	-0.00036
	C2	0.03688	0.69552	0.00318	0.00205	0.00000	-0.00000	0.00000	-0.00000	-0.00023	0.00035
* 2A1G	H1	0.08422	0.01004								
-1.0161	H2	0.08422	0.01004								
	H3	0.08422	0.01003								
	H4	0.08422	0.01004								
	H5	0.08422	0.01004								
	H6	0.08422	0.01004								
	C1	-0.00699	-0.15228	0.32270	0.22519	-0.00000	0.00000	-0.00000	0.00000	0.06470	-0.00783
	C2	-0.00699	-0.15228	0.32270	0.22523	0.00000	-0.00000	0.00000	-0.00000	-0.06469	0.00780
* 3A1G	H1	-0.07795	-0.09007								
-0.4947	H2	-0.07795	-0.09007								
	H3	-0.07793	-0.09007								
	H4	-0.07795	-0.09006								
	H5	-0.07793	-0.09006								
	H6	-0.07795	-0.09006								
	C1	-0.00075	-0.01612	0.03904	0.04397	-0.00000	-0.00000	0.00002	0.00002	0.47982	0.11680
	C2	-0.00075	-0.01612	0.03905	0.04390	0.00000	-0.00000	-0.00002	-0.00002	-0.47981	-0.11678
4A1G	H1	-0.00138	0.91712								
0.2729	H2	-0.00138	0.91709								
	H3	-0.00137	0.91772								
	H4	-0.00141	0.91655								
	H5	-0.00140	0.91736								
	H6	-0.00140	0.91678								
	C1	0.00375	0.08727	-0.16709	-1.86516	-0.00000	-0.00006	-0.00002	-0.00063	0.08724	0.78530
	C2	0.00375	0.08734	-0.16726	-1.86630	0.00001	0.00015	0.00003	0.00067	-0.08712	-0.78337
5A1G	H1	0.07503	-0.18064								
0.4546	H2	0.07504	-0.18062								
	H3	0.07504	-0.18051								
	H4	0.07504	-0.18058								
	H5	0.07505	-0.18044								
	H6	0.07505	-0.18054								
	C1	-0.00220	-0.03773	0.18866	-0.14971	-0.00000	0.00001	-0.00000	-0.00013	-0.54135	0.85742
	C2	-0.00220	-0.03773	0.18869	-0.14976	-0.00001	0.00004	-0.00000	0.00015	0.54132	-0.85756
* 1A2U	H1	0.00033	-0.00063								
-11.2155	H2	0.00033	-0.00063								
	H3	0.00033	-0.00063								
	H4	-0.00032	0.00062								
	H5	-0.00032	0.00062								
	H6	-0.00032	0.00062								
	C1	0.03689	0.69567	0.00266	0.00406	0.00000	0.00000	-0.00000	0.00000	-0.00010	0.00072
	C2	-0.03653	-0.68892	-0.00263	-0.00404	-0.00000	0.00000	-0.00000	0.00000	-0.00010	0.00072

C2H6

Center		S1	S2	S3	S4	X1	X2	Y1	Y2	Z1	Z2
* 2A2U	H1	-0.11474	-0.04258								
-0.8353	H2	-0.11474	-0.04258								
	H3	-0.11474	-0.04258								
	H4	0.11474	0.04259								
	H5	0.11474	0.04258								
	H6	0.11474	0.04258								
	C1	0.00614	0.13427	-0.28630	-0.28845	-0.00000	0.00000	0.00000	0.00000	0.14182	-0.00202
	C2	-0.00614	-0.13427	0.28630	0.28843	0.00000	-0.00000	0.00000	-0.00000	0.14182	-0.00202
3A2U	H1	-0.00876	0.15834								
0.2513	H2	-0.00876	0.15837								
	H3	-0.00877	0.15793								
	H4	0.00875	-0.15728								
	H5	0.00876	-0.15712								
	H6	0.00875	-0.15733								
	C1	0.00358	0.09956	-0.04988	-2.46968	0.00000	0.00003	0.00003	0.00027	-0.05865	-1.60831
	C2	-0.00358	-0.09947	0.04979	2.46767	-0.00000	-0.00004	0.00002	0.00005	-0.05879	-1.60904
4A2U	H1	0.03685	1.23227								
0.3897	H2	0.03688	1.23254								
	H3	0.03655	1.22879								
	H4	-0.03689	-1.23305								
	H5	-0.03653	-1.22887								
	H6	-0.03685	-1.23255								
	C1	0.00201	0.05815	-0.01776	-1.94981	0.00002	0.00007	0.00030	0.00192	0.28242	1.06446
	C2	-0.00201	-0.05819	0.01795	1.95029	0.00002	0.00031	0.00031	0.00226	0.28248	1.06468
* 1EG	H1	-0.10237	-0.09677								
-0.4830	H2	-0.10845	-0.10252								
	H3	0.21083	0.19930								
	H4	-0.10845	-0.10252								
	H5	0.21082	0.19930								
	H6	-0.10236	-0.09676								
	C1	0.00000	0.00000	-0.00000	0.00000	0.00615	0.00175	0.36887	0.10498	-0.00002	-0.00001
	C2	0.00000	0.00000	-0.00000	-0.00001	-0.00615	-0.00175	-0.36887	-0.10497	0.00002	0.00001
* 2EG	H1	0.18433	0.17425								
-0.4830	H2	-0.18082	-0.17093								
	H3	-0.00351	-0.00332								
	H4	-0.18082	-0.17093								
	H5	-0.00351	-0.00332								
	H6	0.18433	0.17425								
	C1	-0.00000	-0.00000	0.00000	0.00000	0.36888	0.10498	-0.00615	-0.00175	0.00000	0.00000
	C2	-0.00000	0.00000	0.00000	-0.00001	-0.36887	-0.10498	0.00615	0.00175	-0.00000	-0.00000
3EG	H1	-0.02828	-1.32072								
0.3169	H2	0.03018	1.40961								
	H3	-0.00189	-0.08854								
	H4	0.03014	1.40872								
	H5	-0.00190	-0.08865								
	H6	-0.02825	-1.32013								
	C1	0.00000	0.00001	-0.00000	-0.00020	0.10676	2.37940	0.00601	0.13382	0.00001	0.00013
	C2	0.00000	0.00000	-0.00003	0.00000	-0.10666	-2.37871	-0.00600	-0.13380	0.00001	0.00002

C2H6

Center		S1	S2	S3	S4	X1	X2	Y1	Y2	Z1	Z2
4EG	H1	-0.01854	-0.86532								
0.3169	H2	-0.01525	-0.71176								
	H3	0.03376	1.57619								
	H4	-0.01522	-0.71106								
	H5	0.03371	1.57523								
	H6	-0.01850	-0.86453								
	C1	-0.00000	-0.00002	0.00006	0.00050	0.00601	0.13383	-0.10677	-2.37952	-0.00000	-0.00039
	C2	-0.00000	-0.00002	0.00009	0.00034	-0.00600	-0.13379	0.10666	2.37858	-0.00003	0.00011
5EG	H1	0.07496	-0.31290								
0.5204	H2	-0.07831	0.32684								
	H3	0.00334	-0.01396								
	H4	-0.07831	0.32686								
	H5	0.00334	-0.01395								
	H6	0.07497	-0.31288								
	C1	0.00000	0.00000	-0.00000	0.00001	0.69862	-0.74973	0.02640	-0.02833	-0.00000	-0.00001
	C2	0.00000	0.00000	-0.00001	-0.00001	-0.69861	0.74972	-0.02640	0.02833	-0.00001	-0.00001
6EG	H1	-0.04715	0.19677								
0.5204	H2	-0.04136	0.17259								
	H3	0.08850	-0.36937								
	H4	-0.04136	0.17259								
	H5	0.08850	-0.36935								
	H6	-0.04715	0.19678								
	C1	-0.00000	-0.00000	-0.00000	0.00001	-0.02640	0.02832	0.69861	-0.74971	-0.00000	-0.00001
	C2	0.00000	0.00000	-0.00001	0.00001	0.02640	-0.02832	-0.69861	0.74972	-0.00001	0.00001
* 1EU	H1	-0.09211	-0.06607								
-0.5960	H2	-0.08644	-0.06200								
	H3	0.17855	0.12808								
	H4	0.08644	0.06200								
	H5	-0.17855	-0.12807								
	H6	0.09211	0.06607								
	C1	-0.00000	-0.00000	-0.00000	0.00002	-0.00671	-0.00149	0.36588	0.08136	0.00000	0.00001
	C2	0.00000	0.00000	0.00000	-0.00002	-0.00671	-0.00149	0.36589	0.08137	0.00000	0.00001
* 2EU	H1	0.15299	0.10974								
-0.5960	H2	-0.15626	-0.11209								
	H3	0.00328	0.00235								
	H4	0.15627	0.11208								
	H5	-0.00328	-0.00235								
	H6	-0.15299	-0.10974								
	C1	0.00000	0.00000	-0.00000	-0.00000	0.36588	0.08136	0.00671	0.00149	-0.00000	0.00000
	C2	0.00000	0.00000	-0.00000	0.00001	0.36589	0.08137	0.00671	0.00149	0.00000	0.00000
3EU	H1	0.09623	1.26555								
0.3678	H2	-0.09970	-1.31077								
	H3	0.00349	0.04608								
	H4	0.09972	1.31157								
	H5	-0.00349	-0.04606								
	H6	-0.09625	-1.26622								
	C1	0.00000	0.00002	-0.00001	-0.00047	-0.16563	-1.23143	-0.00510	-0.03790	0.00004	0.00026
	C2	-0.00000	-0.00001	-0.00001	0.00038	-0.16569	-1.23272	-0.00511	-0.03795	0.00006	0.00018

Center   Basis Function Type

Center		S1	S2	S3	S4	X1	X2	Y1	Y2	Z1	Z2
4EU	H1	-0.05953	-0.78107								
0.3679	H2	-0.05350	-0.70179								
	H3	0.11319	1.48945								
	H4	0.05352	0.70234								
	H5	-0.11322	-1.49057								
	H6	0.05955	0.78172								
	C1	0.00000	0.00012	-0.00003	-0.00378	0.00510	0.03789	-0.16567	-1.23114	0.00043	0.00168
	C2	-0.00000	-0.00011	0.00002	0.00373	0.00510	0.03797	-0.16576	-1.23288	0.00043	0.00165
5EU	H1	-0.06164	0.15933								
0.4663	H2	0.06313	-0.16321								
	H3	-0.00150	0.00386								
	H4	-0.06313	0.16324								
	H5	0.00150	-0.00386								
	H6	0.06163	-0.15935								
	C1	0.00000	0.00000	-0.00000	0.00001	-0.66585	0.52378	-0.01383	0.01088	0.00000	-0.00002
	C2	0.00000	0.00000	-0.00000	-0.00001	-0.66586	0.52375	-0.01383	0.01088	-0.00001	0.00000
6EU	H1	0.03732	-0.09642								
0.4663	H2	0.03473	-0.08971								
	H3	-0.07205	0.18619								
	H4	-0.03473	0.08970								
	H5	0.07205	-0.18617								
	H6	-0.03732	0.09642								
	C1	0.00000	0.00000	-0.00001	-0.00004	0.01383	-0.01087	-0.66584	0.52382	0.00001	0.00000
	C2	-0.00000	-0.00000	0.00000	0.00003	0.01383	-0.01087	-0.66584	0.52384	0.00000	0.00001

Overlap Matrix

H1S1 with

H1S2	.68301	H2S1	.01595	H2S2	.13213	H4S1	.00043	H4S2	.02808	C1S1	.00209	C1S2	.03250		
C1S3	.27658	C1S4	.36773	C1X1	.34527	C1X2	.35404	C1Y1	-.19934	C1Y2	-.20440	C1Z1	-.14228		
C1Z2	-.14590														

H1S2 with

H2S1	.13213	H2S2	.35892	H4S1	.02808	H4S2	.13597	C1S1	.00753	C1S2	.09624	C1S3	.49038		
C1S4	.70027	C1X1	.29377	C1X2	.49862	C1Y1	-.16961	C1Y2	-.28788	C1Z1	-.12106	C1Z2	-.20548		

H2S1 with

H4S1	.00001	H4S2	.00551

H2S2 with

H4S1	.00551	H4S2	.04880

H4S1 with

C1S2	.00003	C1S3	.00825	C1S4	.07873	C1X1	.01368	C1X2	.10209	C1Y1	.00790	C1Y2	.05894		
C1Z1	.02912	C1Z2	.21733												

H4S2 with

C1S1	.00081	C1S2	.01133	C1S3	.09779	C1S4	.25356	C1X1	.06104	C1X2	.20754	C1Y1	.03524		
C1Y2	.11982	C1Z1	.12993	C1Z2	.44180										

C1S1 with

C1S2	.35436	C1S3	.00805	C1S4	.01419	C2S3	.00064	C2S4	.00404	C2Z1	-.00292	C2Z2	-.00873

C1S2 with

C1S3	.26226	C1S4	.17745	C2S3	.01090	C2S4	.05270	C2Z1	-.04035	C2Z2	-.11032

C1S3 with

C1S4	.79970	C2S1	.00064	C2S2	.01090	C2S3	.14491	C2S4	.30913	C2Z1	-.26562	C2Z2	-.52104

C1S4 with

C2S1	.00404	C2S2	.05270	C2S3	.30913	C2S4	.53303	C2Z1	-.28141	C2Z2	-.63468

C1X1 with

C1X2	.53875	C2X1	.12465	C2X2	.25035

C1X2 with

C2X1	.25035	C2X2	.61435

C1Y1 with

C1Y2	.53875	C2Y1	.12465	C2Y2	.25035

C1Y2 with

C2Y1	.25035	C2Y2	.61435

C1Z1 with

C1Z2	.53875	C2S1	.00292	C2S2	.04035	C2S3	.26562	C2S4	.28141	C2Z1	-.32147	C2Z2	-.13253

C1Z2 with

C2S1	.00873	C2S2	.11032	C2S3	.52104	C2S4	.63468	C2Z1	-.13253	C2Z2	.01574

N2H4   Hydrazine

Molecular Geometry   Symmetry C$_2$

Center	Coordinates		
	X	Y	Z
H1	1.85177501	1.54242601	1.07766700
H2	1.85177501	-1.54242601	1.07766700
H3	-1.85177501	-0.96744770	-1.61384299
H4	-1.85177501	-1.18263599	1.46349400
N1	1.41636901	0.	0.
N2	-1.41636901	0.	0.
Mass	0.	-0.06763548	0.06307109

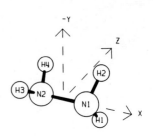

MO Symmetry Types and Occupancy

Representation	A	B
Number of MOs	14	14
Number Occupied	5	4

Energy Expectation Values

Total	-111.1261	Kinetic	111.1354
Electronic	-151.8872	Potential	-222.2615
Nuclear Repulsion	40.7611	Virial	1.9999
Electronic Potential	-263.0226		
One-Electron Potential	-342.4822		
Two-Electron Potential	79.4597		

Moments of the Charge Distribution

First

x	y	z	Dipole Moment
-0.0003	0.1102	-0.1029	1.1245
0.	-0.9326	0.8697	

Second

x^2	y^2	z^2	xy	xz	yz	r^2
-52.8151	-16.1646	-16.1521	-2.0911	-2.2423	0.0885	-85.1318
41.8017	6.8842	6.8877	3.9815	4.2696	0.0250	55.5737

Third

x^3	xy^2	xz^2	x^2y	y^3	yz^2	x^2z	y^2z	z^3	xyz
-0.0037	-2.4539	2.4532	1.0859	0.5622	3.2870	-1.0128	-3.7684	0.1775	-0.1717
0.	5.0265	-5.0265	-4.5455	-1.1443	-4.5740	4.2388	5.2162	0.1163	0.3515

Fourth (even)

x^4	y^4	z^4	x^2y^2	x^2z^2	y^2z^2	r^2x^2	r^2y^2	r^2z^2	r^4
-283.1726	-61.5206	-61.5343	-59.0602	-58.9778	-23.8511	-401.2106	-144.4319	-144.3632	-690.0056
103.3763	13.6521	13.8734	23.5154	23.5393	9.6227	150.4309	46.7903	47.0354	244.2566

Fourth (odd)

x^3y	xy^3	xyz^2	x^3z	xy^2z	xz^3	x^2yz	y^3z	yz^3
-9.0112	-5.7489	-7.2419	-9.6618	-6.7175	-7.2128	0.5887	0.3006	-0.4035
13.6527	5.7050	8.9927	14.6408	8.2470	7.5144	0.1706	0.2603	1.3218

Expectation Values at Atomic Centers

Center			Expectation Value		
	r^{-1}	δ	xr^{-3}	yr^{-3}	zr^{-3}
H1	-1.0428	0.4218	0.0284	0.0750	0.0211
H2	-1.0353	0.4328	0.0462	-0.0729	0.0273
N1	-18.3431	188.8921	0.0800	0.0105	-0.1624

	$\dfrac{3x^2-r^2}{r^5}$	$\dfrac{3y^2-r^2}{r^5}$	$\dfrac{3z^2-r^2}{r^5}$	$\dfrac{3xy}{r^5}$	$\dfrac{3xz}{r^5}$	$\dfrac{3yz}{r^5}$
H1	-0.1630	0.2132	-0.0502	0.1188	0.0987	0.2763
H2	-0.1523	0.2084	-0.0561	-0.1148	0.0980	-0.2815
N1	0.7923	0.2981	-1.0903	-0.0201	0.6243	-0.0426

Mulliken Charge

Center				Basis Function Type							Total
	S1	S2	S3	S4	X1	X2	Y1	Y2	Z1	Z2	
H1	0.5084	0.1837									0.6921
H2	0.5201	0.2190									0.7391
N1	0.0432	1.9560	0.8782	0.8228	0.8098	0.1950	0.9434	0.2037	1.1784	0.5383	7.5688

Overlap Populations

	Pair of Atomic Centers								
	H2	H3	H3	H4	N1	N1	N1	N1	N2
	H1	H1	H2	H2	H1	H2	H3	H4	N1
1A	0.	0.	0.	0.	-0.0001	-0.0001	0.	0.	0.0002
2A	0.0003	0.	-0.0001	0.	0.0555	0.0674	0.0045	0.0082	0.4030
3A	-0.0460	0.0036	-0.0115	0.0448	0.0951	0.1754	0.0153	0.0102	0.1300
4A	0.0261	0.0089	0.0069	0.0062	0.1446	0.0023	-0.1103	-0.0340	0.3060
5A	-0.0147	0.0022	-0.0040	0.0074	0.0392	0.0239	-0.0198	-0.0288	-0.2038
Total A	-0.0343	0.0146	-0.0086	0.0583	0.3343	0.2690	-0.1104	-0.0444	0.6355
1B	0.	0.	0.	0.	-0.0001	-0.0001	0.	0.	-0.0014
2B	0.0074	-0.0006	-0.0012	-0.0032	0.1747	0.1649	-0.0221	-0.0271	-0.3449
3B	-0.0436	-0.0047	0.0105	-0.0275	0.1582	0.1404	0.0278	-0.0069	0.1024
4B	-0.0189	-0.0010	0.0053	-0.0340	0.0105	0.0913	-0.0117	-0.0712	-0.0624
Total B	-0.0550	-0.0064	0.0146	-0.0647	0.3433	0.3964	-0.0060	-0.1052	-0.3063
Total	-0.0893	0.0082	0.0059	-0.0064	0.6776	0.6654	-0.1164	-0.1496	0.3291
Distance	3.0849	5.2211	4.6142	3.7409	1.9313	1.9313	3.7711	3.7711	2.8327

Molecular Orbitals and Eigenvalues

Center		S1	S2	S3	S4	X1	X2	Y1	Y2	Z1	Z2
• 1A	H1	0.00016	-0.00052								
-15.5790	H2	0.00012	-0.00037								
	H3	0.00016	-0.00053								
	H4	0.00012	-0.00037								
	N1	0.03647	0.68728	0.00239	0.00162	-0.00067	0.00039	-0.00005	0.00001	0.00117	-0.00023
	N2	0.03700	0.69727	0.00242	0.00169	0.00067	-0.00037	-0.00118	0.00023	-0.00003	0.00001
• 2A	H1	-0.08057	0.00659								
-1.2640	H2	-0.08771	0.00429								
	H3	-0.08056	0.00656								
	H4	-0.08771	0.00426								
	N1	0.00709	0.15359	-0.32999	-0.26630	0.08893	0.00718	0.01150	0.00312	-0.07235	-0.02779
	N2	0.00709	0.15360	-0.32998	-0.26642	-0.08893	-0.00729	0.07299	0.02792	-0.00643	-0.00118
• 3A	H1	-0.11695	-0.08284								
-0.6769	H2	0.19301	0.14201								
	H3	-0.11705	-0.08285								
	H4	0.19311	0.14209								
	N1	0.00096	0.02016	-0.05188	-0.05871	0.08139	0.01004	-0.34968	-0.07155	0.15206	0.05710
	N2	0.00096	0.02014	-0.05184	-0.05874	-0.08140	-0.01004	-0.17603	-0.06191	0.33845	0.06748
• 4A	H1	0.16776	0.13356								
-0.5788	H2	0.05162	0.05787								
	H3	0.16774	0.13360								
	H4	0.05160	0.05785								
	N1	0.00019	0.00493	-0.00369	-0.04989	0.41642	0.13677	0.12859	0.05069	0.15852	0.04391
	N2	0.00019	0.00493	-0.00373	-0.04972	-0.41642	-0.13665	-0.14915	-0.04026	-0.13932	-0.05359
• 5A	H1	-0.10616	-0.06319								
-0.4123	H2	0.02808	0.07111								
	H3	-0.10618	-0.06314								
	H4	0.02811	0.07126								
	N1	-0.00208	-0.04384	0.11305	0.14823	0.23986	0.12427	-0.11444	-0.03895	-0.40059	-0.25765
	N2	-0.00208	-0.04382	0.11298	0.14832	-0.23982	-0.12403	0.39152	0.25430	0.14212	0.05680
6A	H1	0.01720	0.64916								
0.2142	H2	0.02258	1.06597								
	H3	0.01729	0.65072								
	H4	0.02263	1.06646								
	N1	0.00292	0.07228	-0.09049	-1.13155	-0.10554	-0.24418	0.08091	0.09299	-0.16200	-0.46493
	N2	0.00293	0.07244	-0.09060	-1.13426	0.10529	0.24310	0.16736	0.47071	-0.06934	-0.05985
7A	H1	0.03264	1.32770								
0.3005	H2	-0.03992	-0.86636								
	H3	0.03269	1.32808								
	H4	-0.03989	-0.86700								
	N1	0.00074	0.01762	-0.02730	-0.28696	-0.00925	-0.06143	-0.28074	-0.96993	-0.03119	-0.29169
	N2	0.00073	0.01750	-0.02709	-0.28548	0.00966	0.06280	0.01149	0.22329	0.28235	0.98829
8A	H1	-0.17529	-0.13129								
0.6845	H2	-0.12804	-0.02522								
	H3	-0.17564	-0.13183								
	H4	-0.12802	-0.02494								
	N1	0.00038	0.00148	-0.06398	0.51246	-0.49696	0.81387	-0.10915	0.35834	-0.23935	0.59119
	N2	0.00038	0.00148	-0.06443	0.51438	0.49589	-0.81229	0.23225	-0.56614	0.12590	-0.39940

Center		S1	S2	S3	S4	X1	X2	Y1	Y2	Z1	Z2

Basis Function Type

Center		S1	S2	S3	S4	X1	X2	Y1	Y2	Z1	Z2
9A	H1	0.13636	0.15020								
0.6989	H2	-0.23074	-0.38840								
	H3	0.13667	0.15082								
	H4	-0.23082	-0.38873								
	N1	-0.00074	-0.01697	0.03457	0.14311	0.09446	-0.06337	0.43153	-0.94289	-0.35824	0.31822
	N2	-0.00074	-0.01695	0.03479	0.14171	-0.09380	0.06224	0.38648	-0.38204	-0.40580	0.91920
* 1B	H1	0.00008	0.00035								
-15.5787	H2	0.00009	0.00052								
	H3	-0.00008	-0.00035								
	H4	-0.00010	-0.00051								
	N1	-0.03700	-0.69729	-0.00178	-0.00407	0.00065	0.00089	0.00009	0.00003	-0.00122	0.00009
	N2	0.03647	0.68730	0.00175	0.00405	0.00064	0.00089	-0.00121	0.00008	0.00001	0.00004
* 2B	H1	-0.12789	-0.02650								
-1.0254	H2	-0.12375	-0.02313								
	H3	0.12789	0.02649								
	H4	0.12374	0.02311								
	N1	0.00672	0.14576	-0.31842	-0.34151	-0.09617	0.00816	-0.00957	0.00632	-0.12033	-0.03199
	N2	-0.00672	-0.14576	0.31840	0.34154	-0.09618	0.00813	-0.11936	-0.03236	-0.01795	0.00408
* 3B	H1	0.16871	0.09088								
-0.6613	H2	-0.15438	-0.11057								
	H3	-0.16864	-0.09079								
	H4	0.15427	0.11053								
	N1	0.00110	0.02258	-0.06192	-0.02656	-0.02576	-0.01507	0.36847	0.09565	0.13565	0.06635
	N2	-0.00110	-0.02259	0.06197	0.02651	-0.02569	-0.01505	0.10975	0.05960	0.37683	0.10002
* 4B	H1	-0.02189	-0.05071								
-0.4376	H2	0.14396	0.12995								
	H3	0.02185	0.05070								
	H4	-0.14395	-0.12987								
	N1	0.00267	0.05471	-0.15379	-0.09818	-0.08567	-0.10111	-0.14134	-0.03617	0.42200	0.24449
	N2	-0.00267	-0.05472	0.15383	0.09817	-0.08572	-0.10113	0.43095	0.24650	-0.11151	-0.01902
5B	H1	0.02729	0.77967								
0.2552	H2	0.01467	0.63835								
	H3	-0.02733	-0.78004								
	H4	-0.01461	-0.63589								
	N1	0.00417	0.10546	-0.11527	-1.68232	0.18087	0.63954	-0.01139	-0.04169	-0.11833	-0.33065
	N2	-0.00416	-0.10537	0.11510	1.68099	0.18109	0.63976	-0.11700	-0.32653	-0.01998	-0.06579
6B	H1	0.05975	1.44676								
0.3489	H2	0.02499	-0.17693								
	H3	-0.05975	-1.44563								
	H4	-0.02502	0.17643								
	N1	-0.00080	-0.01623	0.04929	0.01183	-0.29463	-0.91529	-0.21593	-0.58271	-0.11827	-0.29760
	N2	0.00080	0.01624	-0.04934	-0.01221	-0.29472	-0.91510	-0.10286	-0.25598	-0.22342	-0.60132
7B	H1	0.02173	-0.38826								
0.3835	H2	0.03739	1.73175								
	H3	-0.02173	0.38821								
	H4	-0.03737	-1.73201								
	N1	-0.00039	-0.00470	0.04513	-0.23376	-0.24900	-0.70725	0.27414	0.88919	-0.11390	-0.51838
	N2	0.00039	0.00468	-0.04505	0.23388	-0.24912	-0.70718	-0.13277	-0.57918	0.26560	0.85090

Center		S1	S2	S3	S4	X1	X2	Y1	Y2	Z1	Z2
8B	H1	0.14914	0.29128								
0.6955	H2	-0.03056	-0.13545								
	H3	-0.14851	-0.29065								
	H4	0.03043	0.13479								
	N1	-0.00030	0.00830	0.11529	-0.75750	-0.37988	0.63294	0.11087	-0.30253	0.53883	-0.67291
	N2	0.00030	-0.00833	-0.11514	0.75659	-0.38119	0.63493	0.53005	-0.64968	0.14718	-0.34617

Overlap Matrix

First Function   Second Function and Overlap

H1S1 with
      H1S2 .68301   H2S1 .03146   H2S2 .17614   H3S1 .00009   H3S2 .01428   N1S1 .00245   N1S2 .03500
      N1S3 .26562   N1S4 .40033   N1X1 .08685   N1X2 .11519   N1Y1 .30768   N1Y2 .40806   N1Z1 .21497
      N1Z2 .28510

H1S2 with
      H2S1 .17614   H2S2 .42959   H3S1 .01428   H3S2 .08889   N1S1 .00663   N1S2 .08446   N1S3 .44179
      N1S4 .69636   N1X1 .06075   N1X2 .13310   N1Y1 .21520   N1Y2 .47152   N1Z1 .15036   N1Z2 .32944

H2S1 with
      H2S2 .68301   H3S1 .00064   H3S2 .03319   H4S1 .00705   H4S2 .09327   N1S1 .00245   N1S2 .03500
      N1S3 .26562   N1S4 .40033   N1X1 .08685   N1X2 .11519   N1Y1 -.30768   N1Y2 -.40806   N1Z1 .21497
      N1Z2 .28510

H2S2 with
      H3S1 .03319   H3S2 .15101   H4S1 .09327   H4S2 .28865   N1S1 .00663   N1S2 .08446   N1S3 .44179
      N1S4 .69636   N1X1 .06075   N1X2 .13310   N1Y1 -.21520   N1Y2 -.47152   N1Z1 .15036   N1Z2 .32944

H3S1 with
      N1S2 .00008   N1S3 .00762   N1S4 .07402   N1X1 -.02244   N1X2 -.21068   N1Y1 -.00664   N1Y2 -.06237
      N1Z1 -.01108   N1Z2 -.10404

H3S2 with
      N1S1 .00103   N1S2 .01389   N1S3 .10177   N1S4 .25496   N1X1 -.10757   N1X2 -.40687   N1Y1 -.03184
      N1Y2 -.12044   N1Z1 -.05312   N1Z2 -.20091

H4S1 with
      N1S2 .00008   N1S3 .00762   N1S4 .07402   N1X1 -.02244   N1X2 -.21068   N1Y1 -.00812   N1Y2 -.07624
      N1Z1 .01005   N1Z2 .09435

H4S2 with
      N1S1 .00103   N1S2 .01389   N1S3 .10177   N1S4 .25496   N1X1 -.10757   N1X2 -.40687   N1Y1 -.03893
      N1Y2 -.14723   N1Z1 .04817   N1Z2 .18220

N1S1 with
      N1S2 .35584   N1S3 .00894   N1S4 .01446   N2S3 .00016   N2S4 .00273   N2X1 .00109   N2X2 .00745

N1S2 with
      N1S3 .27246   N1S4 .17925   N2S3 .00322   N2S4 .03589   N2X1 .01638   N2X2 .09411

N1S3 with
      N1S4 .79352   N2S1 .00016   N2S2 .00322   N2S3 .07194   N2S4 .22354   N2X1 .15385   N2X2 .46009

N1S4 with
      N2S1 .00273   N2S2 .03589   N2S3 .22354   N2S4 .43425   N2X1 .22741   N2X2 .60681

N1X1 with
      N1X2 .52739   N2S1 -.00109   N2S2 -.01638   N2S3 -.15385   N2S4 -.22741   N2X1 -.23461   N2X2 -.20160

N1X2 with
      N2S1 -.00745   N2S2 -.09411   N2S3 -.46009   N2S4 -.60681   N2X1 -.20160   N2X2 -.16842

N1Y1 with
      N1Y2 .52739   N2Y1 .06185   N2Y2 .18489

N1Y2 with
      N2Y1 .18489   N2Y2 .51505

N1Z1 with
      N1Z2 .52739   N2Z1 .06185   N2Z2 .18489

N1Z2 with
      N2Z1 .18489   N2Z2 .51505

H2O2 Hydrogen Peroxide

Molecular Geometry   Symmetry C_2

Center	Coordinates		
	X	Y	Z
H1	1.66448300	0.65565740	-1.54391301
H2	0.	-1.78896400	1.54391301
O1	0.	0.	-1.39369000
O2	0.	0.	1.39369000
Mass	0.04932948	-0.03358726	0.

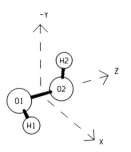

MO Symmetry Types and Occupancy

Representation	A	B
Number of MOs	12	12
Number Occupied	5	4

Energy Expectation Values

Total	-150.7373	Kinetic	150.8889
Electronic	-187.4960	Potential	-301.6262
Nuclear Repulsion	36.7588	Virial	1.9990
Electronic Potential	-338.3850		
One-Electron Potential	-431.7720		
Two-Electron Potential	93.3870		

Moments of the Charge Distribution

First

x	y	z	Dipole Moment
0.0397	-0.0274	-0.0001	0.9878
0.7766	-0.5287	0.	

Second

x^2	y^2	z^2	xy	xz	yz	r^2
-10.6888	-11.0862	-45.1452	-0.5055	1.2347	1.8133	-66.9203
2.6501	3.5745	35.8453	1.1733	-2.5698	-3.7743	42.0698

Third

x^3	xy^2	xz^2	x^2y	y^3	yz^2	x^2z	y^2z	z^3	xyz
-2.2178	-0.2692	0.1545	-1.1369	2.8271	-0.1064	1.8393	-1.8393	-0.0014	0.7256
4.2114	0.6144	2.1993	1.7951	-5.0809	-1.4975	-4.0239	4.0239	0.	-1.5850

Fourth (even)

x^4	y^4	z^4	x^2y^2	x^2z^2	y^2z^2	r^2x^2	r^2y^2	r^2z^2	r^4
-31.5843	-33.9948	-195.7675	-10.1449	-31.2896	-32.4452	-73.0187	-76.5848	-259.5022	-409.1058
6.8055	9.7204	71.7287	1.2468	6.2997	8.5123	14.3521	19.4796	86.5408	120.3725

Fourth (odd)

x^3y	xy^3	xyz^2	x^3z	xy^2z	xz^3	x^2yz	y^3z	yz^3
-2.0829	-0.9842	-1.4696	5.5082	1.2930	4.1896	2.4057	7.5839	6.1534
2.9043	0.7956	2.8085	-6.5054	-1.4193	-6.1256	-2.7826	-8.8564	-8.9966

Expectation Values at Atomic Centers

Center			Expectation Value		
	r^{-1}	δ	xr^{-3}	yr^{-3}	zr^{-3}
H1	-0.9361	0.4261	0.1093	0.0525	-0.0368
O1	-22.2740	287.3831	-0.1832	-0.0535	-0.1436

	$\dfrac{3x^2-r^2}{r^5}$	$\dfrac{3y^2-r^2}{r^5}$	$\dfrac{3z^2-r^2}{r^5}$	$\dfrac{3xy}{r^5}$	$\dfrac{3xz}{r^5}$	$\dfrac{3yz}{r^5}$
H1	0.4389	-0.2070	-0.2319	0.3061	-0.0962	-0.0401
O1	-0.2643	-2.5802	2.8445	1.1099	-0.6398	-0.1691

Mulliken Charge

Center				Basis Function Type							Total
	S1	S2	S3	S4	X1	X2	Y1	Y2	Z1	Z2	
H1	0.4986	0.0853									0.5839
O1	0.0427	1.9567	0.9655	0.9362	1.1690	0.3855	1.3984	0.5346	0.7665	0.2610	8.4161

Overlap Populations

	Pair of Atomic Centers			
	H2	O1	O1	O2
	H1	H1	H2	O1
1A	0.	-0.0001	0.	0.
2A	-0.0003	0.0711	0.0051	0.3541
3A	0.0129	0.1335	-0.0372	0.2179
4A	0.0051	0.0303	-0.0484	0.1759
5A	0.0008	0.0304	-0.0102	-0.2064
Total A	0.0185	0.2651	-0.0908	0.5415
1B	0.	0.	0.	-0.0007
2B	-0.0010	0.1867	-0.0213	-0.2886
3B	-0.0089	0.0647	0.0385	0.1378
4B	-0.0042	0.0232	-0.0300	-0.2775
Total B	-0.0141	0.2745	-0.0128	-0.4289
Total	0.0043	0.5396	-0.1036	0.1125
Distance	4.2757	1.7953	3.4395	2.7874

Molecular Orbitals and Eigenvalues

Center		S1	S2	S3	S4	X1	X2	Y1	Y2	Z1	Z2
• 1A	H1	0.00002	-0.00030								
-20.6285	H2	0.00003	-0.00030								
	O1	0.03690	0.70682	0.00266	0.00125	0.00088	-0.00018	0.00030	-0.00012	0.00078	-0.00033
	O2	0.03539	0.67797	0.00258	0.00114	0.00004	0.00004	-0.00089	0.00020	-0.00075	0.00036
• 2A	H1	-0.09091	0.01198								
-1.4986	H2	-0.09093	0.01199								
	O1	0.00733	0.16140	-0.34569	-0.29920	-0.06483	-0.02122	-0.01451	-0.00157	-0.09040	-0.01778
	O2	0.00733	0.16140	-0.34567	-0.29922	-0.01024	-0.00627	0.06571	0.02034	0.09042	0.01780
• 3A	H1	0.19149	0.08679								
-0.7112	H2	0.19064	0.08641								
	O1	0.00119	0.02674	-0.05528	-0.10568	0.41556	0.14707	-0.05426	-0.03312	-0.17445	-0.06011
	O2	0.00118	0.02640	-0.05447	-0.10466	0.20314	0.08500	-0.36463	-0.12392	0.17442	0.06028
• 4A	H1	0.05598	0.06670								
-0.6029	H2	0.05613	0.06682								
	O1	-0.00146	-0.03038	0.08444	0.07258	-0.12790	-0.08881	0.28260	0.13324	-0.39450	-0.17871
	O2	-0.00145	-0.03032	0.08431	0.07237	-0.30974	-0.15658	0.01514	0.03365	0.39449	0.17866
• 5A	H1	0.10190	0.01380								
-0.5284	H2	0.10198	0.01364								
	O1	0.00172	0.03772	-0.08642	-0.14574	0.11040	0.05954	0.44817	0.22917	0.24751	0.11685
	O2	0.00172	0.03774	-0.08647	-0.14570	-0.37396	-0.19004	-0.26755	-0.13954	-0.24785	-0.11701
6A	H1	0.04875	1.16579								
0.2192	H2	0.04869	1.16391								
	O1	0.00238	0.05864	-0.07939	-0.73986	-0.20130	-0.44024	-0.05893	-0.18863	0.05249	0.05715
	O2	0.00237	0.05834	-0.07888	-0.73694	-0.01913	0.01394	0.20862	0.47809	-0.05415	-0.05962
• 1B	H1	0.00025	0.00005								
-20.6283	H2	-0.00024	-0.00006								
	O1	-0.03539	-0.67797	-0.00203	-0.00260	-0.00090	0.00017	-0.00027	-0.00000	-0.03068	-0.00069
	O2	0.03690	0.70681	0.00214	0.00265	0.00008	-0.00006	-0.00097	0.00016	-0.00071	-0.00068
• 2B	H1	-0.13876	-0.01276								
-1.2376	H2	0.13878	0.01276								
	O1	0.00753	0.16520	-0.36480	-0.36369	-0.10639	-0.02196	-0.04471	-0.00622	0.06449	0.00154
	O2	-0.00753	-0.16519	0.36477	0.36367	-0.00267	0.00220	-0.11541	-0.02275	0.06444	0.00154
• 3B	H1	0.15764	0.07210								
-0.7178	H2	-0.15861	-0.07237								
	O1	0.00270	0.05858	-0.13933	-0.16850	0.29872	0.10320	0.30660	0.12708	0.05258	0.04957
	O2	-0.00271	-0.05879	0.13975	0.16923	0.17500	0.08000	0.39256	0.14334	0.05233	0.04945
• 4B	H1	0.07985	0.05498								
-0.5003	H2	-0.07957	-0.05521								
	O1	0.00098	0.02092	-0.05336	-0.06019	0.35304	0.17254	-0.41847	-0.23051	0.03626	0.03936
	O2	-0.00097	-0.02079	0.05308	0.05974	-0.52060	-0.27884	0.17406	0.07547	0.03514	0.03880
5B	H1	0.01361	0.51920								
0.1945	H2	-0.01395	-0.52362								
	O1	0.00334	0.07905	-0.13434	-0.74857	-0.06714	-0.15014	-0.03490	-0.08499	-0.44563	-0.59248
	O2	-0.00335	-0.07929	0.13467	0.75150	-0.00770	-0.02403	-0.07594	-0.17236	-0.44537	-0.59218
6B	H1	0.07440	1.27624								
0.2878	H2	-0.07436	-1.27617								
	O1	0.00000	0.00520	0.03498	-0.33782	-0.22885	-0.49746	-0.10010	-0.17806	0.27962	0.48762
	O2	-0.00000	-0.00520	-0.03498	0.33782	-0.00921	0.01676	-0.24959	-0.52810	0.27969	0.48771

H2O2

Overlap Matrix

First Function   Second Function and Overlap

H1S1 with

H1S2 .68301	H2S1 .00172	H2S2 .05081	O1S1 .00275	O1S2 .03777	O1S3 .25859	O1S4 .42558
O1X1 .32757	O1X2 .51646	O1Y1 .12903	O1Y2 .20344	O1Z1 -.02956	O1Z2 -.04661	

H1S2 with

H2S1 .05081	H2S2 .19727	O1S1 .00586	O1S2 .07511	O1S3 .39756	O1S4 .67606	O1X1 .19150
O1X2 .50764	O1Y1 .07544	O1Y2 .19997	O1Z1 -.01728	O1Z2 -.04582		

H2S1 with

O1S1 .00001	O1S2 .00022	O1S3 .00896	O1S4 .07459	O1Y1 -.01397	O1Y2 -.13386	O1Z1 .02294
O1Z2 .21980						

H2S2 with

O1S1 .00127	O1S2 .01689	O1S3 .11073	O1S4 .26449	O1Y1 -.05856	O1Y2 -.23677	O1Z1 .09615
O1Z2 .38880						

O1S1 with

O1S2 .35461	O1S3 .00974	O1S4 .01469	O2S3 .00003	O2S4 .00164	O2Z1 -.00034	O2Z2 -.00584

O1S2 with

O1S3 .28286	O1S4 .18230	O2S3 .00070	O2S4 .02201	O2Z1 -.00569	O2Z2 -.07473

O1S3 with

O1S4 .79029	O2S1 .00003	O2S2 .00070	O2S3 .02953	O2S4 .14851	O2Z1 -.07905	O2Z2 -.38199

O1S4 with

O2S1 .00164	O2S2 .02201	O2S3 .14851	O2S4 .33372	O2Z1 -.17540	O2Z2 -.56185	

O1X1 with

O1X2 .50607	O2X1 .02924	O2X2 .13478

O1X2 with

O2X1 .13478	O2X2 .43592

O1Y1 with

O1Y2 .50607	O2Y1 .02924	O2Y2 .13478

O1Y2 with

O2Y1 .13478	O2Y2 .43592

O1Z1 with

O1Z2 .50607	O2S1 .00034	O2S2 .00569	O2S3 .07905	O2S4 .17540	O2Z1 -.14875	O2Z2 -.22068

O1Z2 with

O2S1 .00584	O2S2 .07473	O2S3 .38199	O2S4 .56185	O2Z1 -.22068	O2Z2 -.28796

F2   Fluorine

Molecular Geometry   Symmetry D-h

Center   Coordinates
              X   Y   Z
F1   0.   0.   -1.34000000
F2   0.   0.   1.34000000

Mass   0.   0.   0.

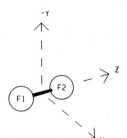

MO Symmetry Types and Occupancy

Representation	SIGM-G	SIGM-U	PI-G-X	PI-U-X	PI-G-Y	PI-U-Y
Number of MOs	6	6	2	2	2	2
Number Occupied	3	2	1	1	1	1

Energy Expectation Values

Total	-198.6932	Kinetic	198.5430
Electronic	-228.9171	Potential	-397.2362
Nuclear Repulsion	30.2239	Virial	2.0008

Electronic Potential	-427.4601
One-Electron Potential	-536.6403
Two-Electron Potential	109.1802

Moments of the Charge Distribution

Second

x^2	y^2	z^2	xy	xz	yz	r^2
-7.0485	-7.0485	-39.1860	0.	0.	0.	-53.2831
0.	0.	32.3208	0.	0.	0.	32.3208

Fourth (even)

x^4	y^4	z^4	x^2y^2	x^2z^2	y^2z^2	r^2x^2	r^2y^2	r^2z^2	r^4
-14.1149	-14.1149	-138.0664	-4.7050	-17.3604	-17.3604	-36.1802	-36.1803	-172.7872	-245.1477
0.	0.	58.0352	0.	0.	0.	0.	0.	58.0352	58.0352

Expectation Values at Atomic Centers

Center			Expectation Value		
	r^{-1}	δ	xr^{-3}	yr^{-3}	zr^{-3}
F1	-26.4551	414.3493	0.	0.	-0.1744

	$\dfrac{3x^2-r^2}{r^5}$	$\dfrac{3y^2-r^2}{r^5}$	$\dfrac{3z^2-r^2}{r^5}$	$\dfrac{3xy}{r^5}$	$\dfrac{3xz}{r^5}$	$\dfrac{3yz}{r^5}$
F1	-3.3735	-3.3735	6.7469	0.	0.	0.

Mulliken Charge

Center					Basis Function Type						Total
	S1	S2	S3	S4	X1	X2	Y1	Y2	Z1	Z2	
F1	0.0436	1.9558	1.0718	0.9138	1.5150	0.4850	1.5150	0.4850	0.7447	0.2703	9.0000

Overlap Populations

Pair of Atomic Centers

F2
F1

1SIGM-G	0.
2SIGM-G	0.3114
3SIGM-G	0.1264
Total SIGM-G	0.4377
1SIGM-U	-0.0002
2SIGM-U	-0.2794
Total SIGM-U	-0.2797
1PI-G-X	-0.2521
Total PI-G-X	-0.2521
1PI-U-X	0.1888
Total PI-U-X	0.1888
1PI-G-Y	-0.2521
Total PI-G-Y	-0.2521
1PI-U-Y	0.1888
Total PI-U-Y	0.1888
Total	0.0316
Distance	2.6800

Molecular Orbitals and Eigenvalues

Center		S1	S2	S3	S4	X1	X2	Y1	Y2	Z1	Z2
* 1SIGM-G	F1	0.03668	0.69405	0.00307	0.00062	0.	0.	0.	0.	0.00070	-0.00029
-26.4289	F2	0.03649	0.69047	0.00306	0.00062	0.	0.	0.	0.	-0.00070	0.00029
* 2SIGM-G	F1	0.00803	0.17188	-0.38952	-0.31229	0.	0.	0.	0.	-0.08358	-0.02121
-1.7768	F2	0.00803	0.17188	-0.38952	-0.31229	0.	0.	0.	0.	0.08358	0.02122
* 3SIGM-G	F1	-0.00194	-0.04051	0.10427	0.13534	0.	0.	0.	0.	-0.50794	-0.22303
-0.7411	F2	-0.00194	-0.04051	0.10427	0.13534	0.	0.	0.	0.	0.50794	0.22303
* 1SIGM-U	F1	0.03649	0.69047	0.00271	0.00129	0.	0.	0.	0.	0.00055	0.00025
-26.4287	F2	-0.03668	-0.69406	-0.00273	-0.00129	0.	0.	0.	0.	0.00056	0.00025
* 2SIGM-U	F1	0.00868	0.18516	-0.42952	-0.39233	0.	0.	0.	0.	0.05540	0.00425
-1.5055	F2	-0.00868	-0.18516	0.42952	0.39233	0.	0.	0.	0.	0.05540	0.00425
3SIGM-U	F1	-0.00247	-0.05359	0.12230	0.31727	0.	0.	0.	0.	0.56703	0.52633
0.0635	F2	0.00247	0.05359	-0.12230	-0.31727	0.	0.	0.	0.	0.56703	0.52633
* 1PI-G-X	F1	0.	0.	0.	0.	0.58383	0.26404	0.	0.	0.	0.
-0.6770	F2	0.	0.	0.	0.	-0.58384	-0.26404	0.	0.	0.	0.
* 1PI-U-X	F1	0.	0.	0.	0.	0.53498	0.22185	0.	0.	0.	0.
-0.8167	F2	0.	0.	0.	0.	0.53498	0.22185	0.	0.	0.	0.
* 1PI-G-Y	F1	0.	0.	0.	0.	0.	0.	0.58383	0.26404	0.	0.
-0.6770	F2	0.	0.	0.	0.	0.	0.	-0.58384	-0.26404	0.	0.
* 1PI-U-Y	F1	0.	0.	0.	0.	0.	0.	0.53498	0.22185	0.	0.
-0.8167	F2	0.	0.	0.	0.	0.	0.	0.53498	0.22185	0.	0.

Overlap Matrix

First Function   Second Function and Overlap

F1S1 with
F1S2	.35647	F1S3	.01003	F1S4	.01474	F2S3	.00001	F2S4	.00113	F2Z1	-.00012	F2Z2	-.00475

F1S2 with
F1S3	.28653	F1S4	.18096	F2S3	.00022	F2S4	.01526	F2Z1	-.00221	F2Z2	-.06057

F1S3 with
F1S4	.78531	F2S1	.00001	F2S2	.00022	F2S3	.01503	F2S4	.11029	F2Z1	-.04478	F2Z2	-.32422

F1S4 with
F2S1	.00113	F2S2	.01526	F2S3	.11029	F2S4	.27707	F2Z1	-.13762	F2Z2	-.51510

F1X1 with
F1X2	.49444	F2X1	.01527	F2X2	.10277

F1X2 with
F2X1	.10277	F2X2	.37477

F1Y1 with
F1Y2	.49444	F2Y1	.01527	F2Y2	.10277

F1Y2 with
F2Y1	.10277	F2Y2	.37477

F1Z1 with
F1Z2	.49444	F2S1	.00012	F2S2	.00221	F2S3	.04478	F2S4	.13762	F2Z1	-.09502	F2Z2	-.21891

F1Z2 with
F2S1	.00475	F2S2	.06057	F2S3	.32422	F2S4	.51510	F2Z1	-.21891	F2Z2	-.36086

CH3OH   Methanol

Molecular Geometry   Symmetry C_{1h}

Center	Coordinates		
	X	Y	Z
H1	0.	1.94687420	-0.70621413
H2	1.68604252	-0.97343710	-0.70621413
H3	-1.68604252	-0.97343710	-0.70621413
H4	0.	-1.70987320	3.28114152
C1	0.	0.	0.
O1	0.	0.	2.69700000
Mass	0.	-0.05380652	1.38355321

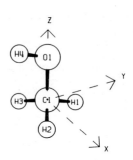

MO Symmetry Types and Occupancy

Representation	A*	A**
Number of MOs	22	6
Number Occupied	7	2

Energy Expectation Values

Total	-115.0059	Kinetic	115.0920
Electronic	-155.1952	Potential	-230.0979
Nuclear Repulsion	40.1893	Virial	1.9993

Electronic Potential	-270.2873
One-Electron Potential	-351.5595
Two-Electron Potential	81.2722

Moments of the Charge Distribution

First

x	y	z	Dipole Moment
0.	-0.0004	1.5613	0.9567
0.	-0.7414	-2.1655	

Second

x^2	y^2	z^2	xy	xz	yz	r^2
-15.8637	-17.1108	-51.9194	0.	0.	1.4979	-84.8939
5.6855	8.4773	41.9887	0.	0.	-3.3612	56.1514

Third

x^3	xy^2	xz^2	x^2y	y^3	yz^2	x^2z	y^2z	z^3	xyz
0.	0.	0.	4.4847	-2.7572	0.9773	13.2983	10.3978	25.6372	0.
0.	0.	0.	-5.2285	1.9130	-3.8977	-11.8813	-6.6888	-18.3094	0.

Fourth (even)

x^4	y^4	z^4	x^2y^2	x^2z^2	y^2z^2	r^2x^2	r^2y^2	r^2z^2	r^4
-66.2365	-73.9686	-283.0214	-22.1068	-61.8609	-67.1162	-150.2042	-163.1916	-411.9985	-725.3943
16.1623	24.9740	115.9758	4.8083	24.8292	34.8159	45.7999	64.5982	175.6209	286.0190

Fourth (odd)

x^3y	xy^3	xyz^2	x^3z	xy^2z	xz^3	x^2yz	y^3z	yz^3
0.	0.	0.	0.	0.	0.	-8.8658	18.1789	7.9856
0.	0.	0.	0.	0.	0.	10.9264	-22.1028	-12.6686

Expectation Values at Atomic Centers

Center	r^{-1}	δ	Expectation Value xr^{-3}	yr^{-3}	zr^{-3}
H1	-1.0992	0.4217	0.	0.0623	-0.0143
H2	-1.1030	0.4290	0.0596	-0.0291	-0.0255
H4	-0.9832	0.4195	0.	-0.0867	0.0584
C1	-14.6800	116.2961	0.	0.0076	-0.0294
O1	-22.3515	286.8545	0.	0.1930	0.1342

	$\dfrac{3x^2-r^2}{r^5}$	$\dfrac{3y^2-r^2}{r^5}$	$\dfrac{3z^2-r^2}{r^5}$	$\dfrac{3xy}{r^5}$	$\dfrac{3xz}{r^5}$	$\dfrac{3yz}{r^5}$
H1	-0.1683	0.2652	-0.0970	0.	0.	-0.1530
H2	0.1562	-0.0619	-0.0944	-0.1904	-0.1357	0.0743
H4	-0.3107	0.4547	-0.1441	0.	0.	-0.2700
C1	-0.1777	-0.2252	0.4028	0.	0.	0.0110
O1	-2.1552	0.6796	1.4756	0.	0.	-0.8668

Mulliken Charge

Center	S1	S2	S3	S4	X1	X2	Y1	Y2	Z1	Z2	Total
H1	0.5192	0.2897									0.8089
H2	0.5257	0.3142									0.8399
H4	0.4967	0.1206									0.6173
C1	0.0427	1.9564	0.7755	0.5813	0.9321	0.1360	0.9541	0.1936	0.6780	0.0232	6.2728
O1	0.0426	1.9567	0.9247	0.9130	1.3773	0.5914	1.0945	0.3991	0.9984	0.3236	8.6212

Overlap Populations

Pair of Atomic Centers

	H2 / H1	H3 / H2	H4 / H1	H4 / H2	C1 / H1	C1 / H2	C1 / H4	O1 / H1	O1 / H2	O1 / H4
1A*	0.	0.	0.	0.	0.	0.	0.	0.	0.	0.
2A*	0.	0.	0.	0.	-0.0004	-0.0004	0.	0.	0.	0.
3A*	0.	0.	0.	-0.0001	0.0051	0.0035	0.0077	0.0030	0.0004	0.1771
4A*	0.0065	0.0052	-0.0010	-0.0026	0.1920	0.1727	-0.0482	-0.0003	-0.0022	0.1274
5A*	-0.0034	0.0146	-0.0006	0.0101	0.0128	0.0918	0.0047	0.0096	-0.0152	0.1717
6A*	-0.0045	0.0003	0.0039	-0.0006	0.3909	0.0023	-0.0427	-0.0586	0.0048	0.0413
7A*	-0.0589	0.0401	0.0056	-0.0111	0.1556	0.1074	-0.0158	-0.0190	-0.0459	0.0457
Total A*	-0.0603	0.0602	0.0079	-0.0044	0.7560	0.3773	-0.0943	-0.0652	-0.0580	0.5631
1A**	0.	-0.0610	0.	0.	0.	0.2364	0.	0.	0.0323	0.
2A**	0.	-0.0822	0.	0.	0.	0.1526	0.	0.	-0.0776	0.
Total A**	0.	-0.1431	0.	0.	0.	0.3890	0.	0.	-0.0453	0.
Total	-0.0603	-0.0830	0.0079	-0.0044	0.7560	0.7663	-0.0943	-0.0652	-0.1034	0.5631
Distance	3.3721	3.3721	5.4103	4.3914	2.0710	2.0710	3.6999	3.9207	3.9207	1.8069

Overlap Populations (continued)

	O1 C1
1A'	-0.0005
2A'	-0.0004
3A'	0.2681
4A'	-0.0772
5A'	0.2220
6A'	0.1713
7A'	-0.1381
Total A'	0.4452
1A''	0.1851
2A''	-0.2291
Total A''	-0.0440
Total	0.4012
Distance	2.6970

Molecular Orbitals and Eigenvalues

	Center				Basis Function Type						
		S1	S2	S3	S4	X1	X2	Y1	Y2	Z1	Z2
* 1A*	H1	0.00005	0.00058								
-20.5425	H2	0.00009	0.00027								
	H3	0.00009	0.00027								
	H4	-0.00012	-0.00003								
	C1	-0.00000	0.00003	0.00029	-0.00202	-0.00000	0.00000	0.00002	-0.00032	-0.00022	-0.00040
	O1	0.05112	0.97934	0.00353	0.00282	0.00000	-0.00000	-0.00137	0.00050	-0.00087	-0.00026
* 2A*	H1	-0.00067	0.00115								
-11.2808	H2	-0.00064	0.00124								
	H3	-0.00064	0.00124								
	H4	-0.00005	0.00036								
	C1	-0.05191	-0.97901	-0.00412	-0.00418	0.00000	-0.00000	0.00010	0.00004	-0.00081	0.00037
	O1	0.00001	0.00013	-0.00039	0.00087	-0.00000	0.00000	-0.00002	0.00015	0.00041	-0.00063
* 3A*	H1	0.02121	0.00375								
-1.3623	H2	0.02292	-0.00307								
	H3	0.02292	-0.00307								
	H4	0.13294	0.00178								
	C1	-0.00339	-0.07360	0.15184	0.06071	-0.00000	0.00000	-0.00945	-0.00597	0.10324	-0.01307
	O1	-0.00997	-0.22001	0.46972	0.42798	0.00000	-0.00000	-0.09637	-0.02485	-0.07005	-0.01328
* 4A*	H1	0.13642	0.03436								
-0.9333	H2	0.13498	0.02229								
	H3	0.13498	0.02229								
	H4	-0.11611	-0.05088								
	C1	-0.00869	-0.18736	0.42267	0.32805	-0.00000	0.00000	0.00616	-0.01251	-0.07011	-0.01936
	O1	0.00289	0.06418	-0.13712	-0.18018	-0.00000	-0.00000	0.10167	0.01532	-0.19083	-0.03835
* 5A*	H1	0.04528	0.00962								
-0.6848	H2	-0.10975	-0.06474								
	H3	-0.10975	-0.06474								
	H4	-0.22717	-0.11105								
	C1	0.00087	0.01996	-0.03478	-0.04417	0.00000	0.00000	0.24795	0.06384	0.23857	0.02231
	O1	-0.00211	-0.04667	0.10363	0.15461	0.00000	0.00000	0.42986	0.15058	-0.22371	-0.05463
* 6A*	H1	0.23053	0.14919								
-0.5932	H2	-0.02350	-0.00659								
	H3	-0.02350	-0.00659								
	H4	0.07072	0.06288								
	C1	-0.00010	-0.00309	-0.00005	0.05068	-0.00000	0.00000	0.35577	0.10293	-0.31800	-0.03506
	O1	-0.00115	-0.02419	0.06518	0.05532	-0.00000	-0.00000	0.07797	0.04326	0.41993	0.18448
* 7A*	H1	0.17356	0.18478								
-0.4935	H2	-0.11925	-0.12546								
	H3	-0.11925	-0.12546								
	H4	0.13265	0.07286								
	C1	-0.00052	-0.00993	0.03464	-0.05407	-0.00000	0.00000	0.33260	0.04614	0.15656	-0.00249
	O1	0.00252	0.05462	-0.13271	-0.17637	-0.00000	-0.00000	-0.42387	-0.23160	-0.28233	-0.17146
8A*	H1	-0.01289	-0.71733								
0.2185	H2	-0.00005	-0.36336								
	H3	-0.00005	-0.36336								
	H4	-0.05666	-0.96576								
	C1	-0.00005	-0.00530	-0.02672	0.69825	0.00000	0.00000	-0.05317	0.39407	-0.20272	-0.90145
	O1	-0.00343	-0.08649	0.10021	1.06842	-0.00000	-0.00000	-0.21583	-0.46257	-0.07319	-0.03067

CH3OH

Center		S1	S2	S3	S4	X1	X2	Y1	Y2	Z1	Z2
9A*	H1	-0.00366	0.94162								
0.2933	H2	-0.05460	0.04322								
	H3	-0.05460	0.04322								
	H4	0.00713	1.18568								
	C1	0.00406	0.09044	-0.21325	-1.47436	-0.00000	-0.00000	0.06516	-0.87558	-0.00261	-1.15509
	O1	-0.00116	-0.03091	0.02322	0.41650	0.00000	0.00000	0.15215	0.66589	-0.22287	-0.69520
10A*	H1	0.04640	-0.32737								
0.3203	H2	-0.02236	-1.80268								
	H3	-0.02236	-1.80268								
	H4	0.02731	0.30418								
	C1	-0.00521	-0.12057	0.24553	2.67000	0.00000	0.00000	-0.02885	-1.18178	-0.19036	-0.50567
	O1	0.00050	0.01749	0.01831	-0.41231	-0.00000	-0.00000	0.12438	0.34070	0.02825	0.17977
11A*	H1	-0.11192	-2.37671								
0.3561	H2	0.02559	0.29824								
	H3	0.02559	0.29824								
	H4	0.06543	1.28434								
	C1	-0.00149	-0.03291	0.08172	0.83311	0.00000	0.00000	0.23263	1.84441	-0.04047	-0.80637
	O1	0.00135	0.03261	-0.04963	-0.40012	-0.00000	-0.00000	0.08882	0.15850	-0.14222	-0.22931
12A*	H1	-0.09945	0.39189								
0.4606	H2	0.02261	-0.34010								
	H3	0.02261	-0.34010								
	H4	-0.04274	0.57846								
	C1	0.00015	0.00171	-0.01863	0.11205	-0.00000	0.00000	-0.93692	0.73058	0.06120	-0.31101
	O1	0.00071	0.01485	-0.04213	-0.13762	0.00000	-0.00000	0.10907	0.01973	-0.05772	-0.05214
13A*	H1	0.01548	0.27845								
0.6314	H2	0.04822	0.25861								
	H3	0.04822	0.25861								
	H4	0.06173	0.07633								
	C1	0.00359	0.06966	-0.25992	-0.93460	0.00000	0.00000	-0.05111	0.05681	-1.14995	0.79428
	O1	-0.00295	-0.05631	0.21547	0.01480	-0.00000	-0.00000	0.03420	0.06531	-0.39448	-0.59651
* 1A**	H1	-0.00000	-0.00000								
-0.6238	H2	0.18285	0.14394								
	H3	-0.18285	-0.14394								
	H4	0.00000	-0.00000								
	C1	-0.00000	-0.00000	-0.00000	0.00000	0.45807	0.07833	-0.00000	-0.00000	-0.00000	-0.00000
	O1	0.00000	0.00000	-0.00000	-0.00000	0.37599	0.16917	-0.00000	-0.00000	0.00000	0.00000
* 2A**	H1	-0.00000	-0.00000								
-0.4473	H2	-0.16394	-0.18171								
	H3	0.16394	0.18171								
	H4	-0.00000	0.00000								
	C1	0.00000	0.00000	-0.00000	0.00000	-0.28102	-0.04827	0.00000	0.00000	-0.00000	-0.00000
	O1	-0.00000	-0.00000	-0.00000	0.00000	0.63173	0.37797	0.00000	0.00000	0.00000	0.00000
3A**	H1	0.00000	0.00000								
0.3471	H2	-0.08686	-1.77624								
	H3	0.08686	1.77624								
	H4	-0.00000	-0.00000								
	C1	0.00000	0.00000	-0.00000	-0.00000	0.12385	2.37288	-0.00000	-0.00000	0.00000	0.00000
	O1	-0.00000	-0.00000	0.00000	0.00000	-0.14022	-0.43832	-0.00000	-0.00000	0.00000	0.00000

CH3OH

Center		Basis Function Type									
		S1	S2	S3	S4	X1	X2	Y1	Y2	Z1	Z2
4A**	H1	0.00000	-0.00000								
0.4521	H2	-0.06746	0.58460								
	H3	0.06746	-0.58460								
	H4	0.00000	-0.00000								
	C1	-0.00000	-0.00000	0.00000	0.00000	-0.97250	0.53839	0.00000	-0.00000	-0.00000	0.00000
	O1	-0.00000	-0.00000	0.00000	0.00000	0.02940	-0.09451	-0.00000	0.00000	0.00000	0.00000

Overlap Matrix

First Function   Second Function and Overlap

H1S1 with

H1S2	.68301	H2S1	.01688	H2S2	.13535	H4S1	.00005	H4S2	.01075	C1S1	.00216	C1S2	.03339

H1S2 .68301 H2S1 .01688 H2S2 .13535 H4S1 .00005 H4S2 .01075 C1S1 .00216 C1S2 .03339
C1S3 .28031 C1S4 .36989 C1Y1 .40126 C1Y2 .40771 C1Z1 -.14555 C1Z2 -.14789 O1S2 .00003
O1S3 .00230 O1S4 .03662 O1Y1 .00437 O1Y2 .08136 O1Z1 -.00764 O1Z2 -.14222

H1S2 with

H2S1 .13535 H2S2 .36436 H4S1 .01075 H4S2 .07436 C1S1 .00760 C1S2 .09703 C1S3 .49338
C1S4 .70297 C1Y1 .33874 C1Y2 .57341 C1Z1 -.12288 C1Z2 -.20800 O1S1 .00068 O1S2 .00914
O1S3 .06541 O1S4 .17975 O1Y1 .03804 O1Y2 .18275 O1Z1 -.06650 O1Z2 -.31945

H2S1 with

H2S2 .68301 H3S1 .01688 H3S2 .13535 H4S1 .00124 H4S2 .04409 C1S1 .00216 C1S2 .03339
C1S3 .28031 C1S4 .36989 C1X1 .34750 C1X2 .35309 C1Y1 -.20063 C1Y2 -.20386 C1Z1 -.14555
C1Z2 -.14789 O1S2 .00003 O1S3 .00230 O1S4 .03662 O1X1 .00378 O1X2 .07046 O1Y1 -.00219
O1Y2 -.04068 O1Z1 -.00764 O1Z2 -.14222

H2S2 with

H3S1 .13535 H3S2 .36436 H4S1 .04409 H4S2 .18047 C1S1 .00760 C1S2 .09703 C1S3 .49338
C1S4 .70297 C1X1 .29336 C1X2 .49659 C1Y1 -.16937 C1Y2 -.28671 C1Z1 -.12288 C1Z2 -.20800
O1S1 .00068 O1S2 .00914 O1S3 .06541 O1S4 .17975 O1X1 .03295 O1X2 .15827 O1Y1 -.01902
O1Y2 -.09137 O1Z1 -.06650 O1Z2 -.31945

H4S1 with

H4S2 .68301 C1S2 .00018 C1S3 .02021 C1S4 .11690 C1Y1 -.03018 C1Y2 -.14140 C1Z1 .05791
C1Z2 .27134 O1S1 .00268 O1S2 .03681 O1S3 .25424 O1S4 .42196 O1Y1 -.33075 O1Y2 -.52686
O1Z1 .11299 O1Z2 .17999

H4S2 with

C1S1 .00143 C1S2 .01962 C1S3 .14805 C1S4 .32914 C1Y1 -.09174 C1Y2 -.26166 C1Z1 .17604
C1Z2 .50211 O1S1 .00581 O1S2 .07457 O1S3 .39509 O1S4 .67297 O1Y1 -.19552 O1Y2 -.51937
O1Z1 .06680 O1Z2 .17743

C1S1 with

C1S2 .35436 C1S3 .00805 C1S4 .01419 O1S3 .00007 O1S4 .00296 O1Z1 -.00073 O1Z2 -.00985

C1S2 with

C1S3 .26226 C1S4 .17745 O1S3 .00245 O1S4 .04024 O1Z1 -.01391 O1Z2 -.12349

C1S3 with

C1S4 .79970 O1S1 .00073 O1S2 .01072 O1S3 .10199 O1S4 .27042 O1Z1 -.16710 O1Z2 -.53252

C1S4 with

O1S1 .00309 O1S2 .04012 O1S3 .23120 O1S4 .45682 O1Z1 -.15631 O1Z2 -.52668

C1X1 with

C1X2 .53875 O1X1 .08176 O1X2 .27280

C1X2 with

O1X1 .14317 O1X2 .51576

C1Y1 with

C1Y2 .53875 O1Y1 .08176 O1Y2 .27280

C1Y2 with

O1Y1 .14317 O1Y2 .51576

C1Z1 with

C1Z2 .53875 O1S1 .00269 O1S2 .03573 O1S3 .22265 O1S4 .34998 O1Z1 -.24767 O1Z2 -.29423

C1Z2 with

O1S1 .00593 O1S2 .07578 O1S3 .39086 O1S4 .61934 O1Z1 -.06611 O1Z2 -.04397

O1S1 with

O1S2 .35461 O1S3 .00974 O1S4 .01469

O1S2 with

O1S3 .28286 O1S4 .18230

O1S3 with

O1S4 .79029

Overlap Matrix   (continued)

First Function   Second Function and Overlap

01X1 with
    01X2   .50607

01Y1 with
    01Y2   .50607

01Z1 with
    01Z2   .50607

CH3F   Methyl Fluoride

Molecular Geometry   Symmetry C$_{3v}$

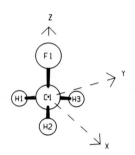

Center	Coordinates		
	X	Y	Z
H1	-1.69302851	-0.97747046	-0.68972185
H2	1.69302851	-0.97747046	-0.68972185
H3	0.	1.95494090	-0.68972185
C1	0.	0.	0.
F1	0.	0.	2.61464089
Mass	0.	0.	1.39877190

MO Symmetry Types and Occupancy

Representation	A1	A2	E
Number of MOs	14	0	12
Number Occupied	5	0	4

Energy Expectation Values

Total	-139.0203	Kinetic	139.0360
Electronic	-176.2745	Potential	-278.0562
Nuclear Repulsion	37.2542	Virial	1.9999
Electronic Potential	-315.3105		
One-Electron Potential	-405.1112		
Two-Electron Potential	89.8007		

Moments of the Charge Distribution

First

x	y	z	Dipole Moment
0.	0.	2.6936	1.0217
0.	0.	-3.7153	

Second

x^2	y^2	z^2	xy	xz	yz	r^2
-14.4352	-14.4352	-47.2359	0.	0.	0.	-76.1063
5.7327	5.7327	38.1298	0.	0.	0.	49.5952

Third

x^3	xy^2	xz^2	x^2y	y^3	yz^2	x^2z	y^2z	z^3	xyz
0.	0.	0.	4.7579	-4.7579	0.	15.0874	15.0874	37.4382	0.
0.	0.	0.	-5.6035	5.6035	0.	-11.9727	-11.9727	-27.5723	0.

Fourth (even)

x^4	y^4	z^4	x^2y^2	x^2z^2	y^2z^2	r^2x^2	r^2y^2	r^2z^2	r^4
-58.7699	-58.7699	-240.9087	-19.5900	-54.7355	-54.7355	-133.0954	-133.0954	-350.3798	-616.5705
16.4319	16.4319	99.7142	5.4773	25.0049	25.0049	46.9141	46.9140	149.7240	243.5521

Fourth (odd)

x^3y	xy^3	xyz^2	x^3z	xy^2z	xz^3	x^2yz	y^3z	yz^3
0.	0.	0.	0.	0.	0.	-9.7469	9.7469	0.
0.	0.	0.	0.	0.	0.	11.7030	-11.7029	0.

Expectation Values at Atomic Centers

Center			Expectation Value		
	r^{-1}	δ	xr^{-3}	yr^{-3}	zr^{-3}
H1	-1.0799	0.4245	-0.0565	-0.0326	-0.0165
C1	-14.6436	116.3220	0.	0.	-0.0449
F1	-26.6113	413.4674	0.	0.	0.2248

	$\dfrac{3x^2-r^2}{r^5}$	$\dfrac{3y^2-r^2}{r^5}$	$\dfrac{3z^2-r^2}{r^5}$	$\dfrac{3xy}{r^5}$	$\dfrac{3xz}{r^5}$	$\dfrac{3yz}{r^5}$
H1	0.1557	-0.0617	-0.0939	0.1882	0.1281	0.0740
C1	-0.2942	-0.2942	0.5883	0.	0.	0.
F1	-1.6334	-1.6334	3.2668	0.	0.	0.

Mulliken Charge

Center				Basis Function Type							Total
	S1	S2	S3	S4	X1	X2	Y1	Y2	Z1	Z2	
H1	0.5218	0.3040									0.8258
C1	0.0428	1.9564	0.7943	0.5348	0.9662	0.1414	0.9662	0.1414	0.5842	0.0118	6.1395
F1	0.0435	1.9558	1.0141	0.9372	1.4308	0.5375	1.4308	0.5375	1.0937	0.4023	9.3831

Overlap Populations

	Pair of Atomic Centers			
	H2	C1	F1	F1
	H1	H1	H1	C1
1A1	0.	0.	0.	-0.0002
2A1	0.	-0.0004	0.	-0.0003
3A1	0.	0.0017	0.0012	0.2120
4A1	0.0067	0.1815	0.0009	-0.0303
5A1	0.0083	0.0686	-0.0265	0.1705
Total A1	0.0150	0.2515	-0.0244	0.3516
1E	-0.0189	0.1166	0.0229	0.2061
2E	0.0063	0.0389	0.0076	0.2061
3E	-0.1066	0.2643	-0.0559	-0.2278
4E	0.0355	0.0881	-0.0186	-0.2278
Total E	-0.0837	0.5079	-0.0441	-0.0436
Total	-0.0686	0.7594	-0.0685	0.3080
Distance	3.3861	2.0730	3.8393	2.6146

Molecular Orbitals and Eigenvalues

Center		S1	S2	S3	S4	X1	X2	Y1	Y2	Z1	Z2
• 1A1	H1	0.00003	0.00019								
-26.2618	H2	0.00003	0.00019								
	H3	0.00003	0.00019								
	C1	-0.00000	0.00002	0.00017	-0.00087	0.	0.	0.	0.	-0.00010	-0.00014
	F1	0.05174	0.97892	0.00437	0.00144	0.	0.	0.	0.	-0.00104	0.00016
• 2A1	H1	0.00071	-0.00121								
-11.3175	H2	0.00071	-0.00121								
	H3	0.00071	-0.00121								
	C1	0.05191	0.97904	0.00415	0.00379	0.	0.	0.	0.	0.00115	-0.00067
	F1	-0.00001	-0.00012	0.00041	-0.00087	0.	0.	0.	0.	-0.00046	0.00048
• 3A1	H1	-0.01311	-0.00073								
-1.5866	H2	-0.01311	-0.00073								
	H3	-0.01311	-0.00073								
	C1	0.00268	0.05806	-0.11935	-0.02713	0.	0.	0.	0.	-0.09619	0.01390
	F1	0.01113	0.23954	-0.53127	-0.47590	0.	0.	0.	0.	0.09038	0.02316
• 4A1	H1	-0.13808	-0.02888								
-0.9602	H2	-0.13808	-0.02888								
	H3	-0.13808	-0.02888								
	C1	0.00902	0.19307	-0.44872	-0.29045	0.	0.	0.	0.	0.06144	0.03753
	F1	-0.00318	-0.06875	0.15484	0.20558	0.	0.	0.	0.	0.22043	0.07526
• 5A1	H1	0.08376	0.04896								
-0.6617	H2	0.08376	0.04896								
	H3	0.08376	0.04896								
	C1	-0.00073	-0.01811	0.01884	0.14194	0.	0.	0.	0.	-0.39524	-0.00512
	F1	-0.00175	-0.03630	0.09691	0.09347	0.	0.	0.	0.	0.58826	0.27448
6A1	H1	0.02116	0.56719								
0.2287	H2	0.02116	0.56719								
	H3	0.02116	0.56719								
	C1	-0.00101	-0.01526	0.10041	-0.68834	0.	0.	0.	0.	0.27711	1.35561
	F1	0.00283	0.06804	-0.09294	-0.72070	0.	0.	0.	0.	0.21562	0.24653
7A1	H1	0.01433	-1.38237								
0.3069	H2	0.01433	-1.38237								
	H3	0.01433	-1.38237								
	C1	-0.00709	-0.15484	0.39411	2.84024	0.	0.	0.	0.	-0.16374	-0.16460
	F1	0.00102	0.02803	-0.00887	-0.44306	0.	0.	0.	0.	0.12828	0.26938
8A1	H1	0.04857	0.32510								
0.6130	H2	0.04857	0.32510								
	H3	0.04857	0.32510								
	C1	0.00329	0.06947	-0.19421	-0.94795	0.	0.	0.	0.	-1.15191	0.95145
	F1	-0.00243	-0.04478	0.18019	-0.00538	0.	0.	0.	0.	-0.43229	-0.38188
• 1E	H1	-0.12533	-0.07409								
-0.6943	H2	0.12533	0.07409								
	H3	0.	0.								
	C1	0.	0.	0.	0.	0.35495	0.07287	0.	0.	0.	0.
	F1	0.	0.	0.	0.	0.57205	0.25822	0.	0.	0.	0.
• 2E	H1	-0.07236	-0.04278								
-0.6943	H2	-0.07236	-0.04278								
	H3	0.14471	0.08555								
	C1	0.	0.	0.	0.	0.	0.	0.35495	0.07287	0.	0.
	F1	0.	0.	0.	0.	0.	0.	0.57205	0.25822	0.	0.

Center		S1	S2	S3	S4	X1	X2	Y1	Y2	Z1	Z2
• 3E	H1	-0.20833	-0.20181								
-0.5302	H2	0.20833	0.20181								
	H3	0.	0.								
	C1	0.	0.	0.	0.	0.42373	0.05844	0.	0.	0.	0.
	F1	0.	0.	0.	0.	-0.49805	-0.28165	0.	0.	0.	0.
• 4E	H1	0.12028	0.11651								
-0.5302	H2	0.12028	0.11651								
	H3	-0.24056	-0.23303								
	C1	0.	0.	0.	0.	0.	0.	-0.42373	-0.05844	0.	0.
	F1	0.	0.	0.	0.	0.	0.	0.49805	0.28165	0.	0.
5E	H1	-0.09182	-1.72002								
0.3281	H2	0.09182	1.72002								
	H3	0.	0.								
	C1	0.	0.	0.	0.	-0.11175	-2.26806	0.	0.	0.	0.
	F1	0.	0.	0.	0.	0.13757	0.33066	0.	0.	0.	0.
6E	H1	0.05301	0.99305								
0.3281	H2	0.05301	0.99305								
	H3	-0.10602	-1.98610								
	C1	0.	0.	0.	0.	0.	0.	0.11175	2.26806	0.	0.
	F1	0.	0.	0.	0.	0.	0.	-0.13757	-0.33066	0.	0.
7E	H1	-0.05463	0.59761								
0.4327	H2	0.05463	-0.59761								
	H3	0.	0.								
	C1	0.	0.	0.	0.	0.96955	-0.50734	0.	0.	0.	0.
	F1	0.	0.	0.	0.	-0.04038	0.08067	0.	0.	0.	0.
8E	H1	-0.03154	0.34503								
0.4327	H2	-0.03154	0.34503								
	H3	0.06308	-0.69006								
	C1	0.	0.	0.	0.	0.	0.	0.96955	-0.50734	0.	0.
	F1	0.	0.	0.	0.	0.	0.	-0.04038	0.08067	0.	0.

Overlap Matrix

First Function			Second Function and Overlap				

H1S1 with

H1S2 .68301	H2S1 .01635	H2S2 .13354	C1S1 .00214	C1S2 .03323	C1S3 .27965	C1S4 .36950
C1X1 -.34808	C1X2 -.35426	C1Y1 -.20097	C1Y2 -.20453	C1Z1 -.14181	C1Z2 -.14432	F1S2 .00003
F1S3 .00158	F1S4 .02643	F1X1 -.00237	F1X2 -.05596	F1Y1 -.00137	F1Y2 -.03231	F1Z1 -.00463
F1Z2 -.10923						

H1S2 with

H2S1 .13354	H2S2 .36132	C1S1 .00759	C1S2 .09689	C1S3 .49284	C1S4 .70249	C1X1 -.29426
C1X2 -.49835	C1Y1 -.16989	C1Y2 -.28772	C1Z1 -.11988	C1Z2 -.20302	F1S1 .00064	F1S2 .00844
F1S3 .05793	F1S4 .15903	F1X1 -.02610	F1X2 -.13780	F1Y1 -.01507	F1Y2 -.07956	F1Z1 -.05095
F1Z2 -.26896						

C1S1 with

C1S2 .35436	C1S3 .00805	C1S4 .01419	F1S3 .00002	F1S4 .00240	F1Z1 -.00030	F1Z2 -.00949

C1S2 with

C1S3 .26226	C1S4 .17745	F1S3 .00131	F1S4 .03381	F1Z1 -.00775	F1Z2 -.11955

C1S3 with

C1S4 .79970	F1S1 .00076	F1S2 .01061	F1S3 .09157	F1S4 .25258	F1Z1 -.13658	F1Z2 -.50886

C1S4 with

F1S1 .00277	F1S2 .03563	F1S3 .20645	F1S4 .42454	F1Z1 -.12067	F1Z2 -.46476

C1X1 with

C1X2 .53875	F1X1 .06769	F1X2 .26712

C1X2 with

F1X1 .11200	F1X2 .45798

C1Y1 with

C1Y2 .53875	F1Y1 .06769	F1Y2 .26712

C1Y2 with

F1Y1 .11200	F1Y2 .45798

C1Z1 with

C1Z2 .53875	F1S1 .00258	F1S2 .03362	F1S3 .20666	F1S4 .36114	F1Z1 -.20990	F1Z2 -.34692

C1Z2 with

F1S1 .00508	F1S2 .06457	F1S3 .34336	F1S4 .58816	F1Z1 -.04641	F1Z2 -.04761

F1S1 with

F1S2 .35647	F1S3 .01003	F1S4 .01474

F1S2 with

F1S3 .28653	F1S4 .18096

F1S3 with

F1S4 .78531

F1X1 with

F1X2 .49444

F1Y1 with

F1Y2 .49444

F1Z1 with

F1Z2 .49444

CO2   Carbon Dioxide

Molecular Geometry   Symmetry D-$_h$

Center		Coordinates	
	X	Y	Z
C1	0.	0.	0.
O1	0.	0.	2.19440001
O2	0.	0.	-2.19440001
Mass	0.	0.	0.

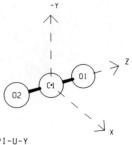

MO Symmetry Types and Occupancy

Representation	SIGM-G	SIGM-U	PI-G-X	PI-G-Y	PI-U-X	PI-U-Y
Number of MOs	10	8	2	2	4	4
Number Occupied	4	3	1	1	1	1

Energy Expectation Values

Total	-187.5377	Kinetic	187.5403
Electronic	-245.8680	Potential	-375.0780
Nuclear Repulsion	58.3303	Virial	2.0000

Electronic Potential	-433.4083
One-Electron Potential	-559.2007
Two-Electron Potential	125.7924

Moments of the Charge Distribution

Second

x^2	y^2	z^2	xy	xz	yz	r^2
-11.0184	-11.0184	-93.5293	0.	0.	0.	-115.5661
0.	0.	77.0463	0.	0.	0.	77.0463

Fourth (even)

x^4	y^4	z^4	x^2y^2	x^2z^2	y^2z^2	r^2x^2	r^2y^2	r^2z^2	r^4
-29.6716	-29.6716	-661.7365	-9.8905	-51.1350	-51.1350	-90.6971	-90.6972	-764.0065	-945.4007
0.	0.	371.0079	0.	0.	0.	0.	0.	371.0079	371.0079

Expectation Values at Atomic Centers

Center			Expectation Value		
	r^{-1}	δ	xr^{-3}	yr^{-3}	zr^{-3}
C1	-14.4675	116.7739	0.	0.	0.
O1	-22.2335	287.3835	0.	0.	0.3022

	$\dfrac{3x^2-r^2}{r^5}$	$\dfrac{3y^2-r^2}{r^5}$	$\dfrac{3z^2-r^2}{r^5}$	$\dfrac{3xy}{r^5}$	$\dfrac{3xz}{r^5}$	$\dfrac{3yz}{r^5}$
C1	0.2076	0.2076	-0.4153	0.	0.	0.
O1	-0.6275	-0.6275	1.2549	0.	0.	0.

Mulliken Charge

Center					Basis Function Type						Total
	S1	S2	S3	S4	X1	X2	Y1	Y2	Z1	Z2	
C1	0.0429	1.9566	0.8673	0.0277	0.6708	0.0760	0.6708	0.0760	0.9897	0.0355	5.4133
O1	0.0427	1.9566	0.9784	0.8266	1.1730	0.4536	1.1730	0.4536	1.0406	0.1952	8.2933

Overlap Populations

Pair of Atomic Centers

	O1	O2
	C1	O1
1SIGM-G	-0.0003	0.
2SIGM-G	0.0006	0.
3SIGM-G	0.2852	0.0441
4SIGM-G	-0.0386	-0.0100
Total SIGM-G	0.2469	0.0341
1SIGM-U	-0.0006	0.
2SIGM-U	0.3575	0.0027
3SIGM-U	-0.0104	0.0053
Total SIGM-U	0.3464	0.0081
1PI-G-X	0.	-0.0656
Total PI-G-X	0.	-0.0656
1PI-G-Y	0.	-0.0656
Total PI-G-Y	0.	-0.0656
1PI-U-X	0.2614	0.0190
Total PI-U-X	0.2614	0.0190
1PI-U-Y	0.2614	0.0190
Total PI-U-Y	0.2614	0.0190
Total	1.1161	-0.0510
Distance	2.1944	4.3888

Molecular Orbitals and Eigenvalues

Center		S1	S2	S3	S4	X1	X2	Y1	Y2	Z1	Z2
• 1SIGM-G	C1	0.00000	-0.00007	-0.00039	-0.00147	0.	0.	0.	0.	0.	0.
-20.6702	O1	0.03615	0.69252	0.00231	0.00220	0.	0.	0.	0.	-0.00151	-0.00011
	O2	0.03615	0.69252	0.00231	0.00220	0.	0.	0.	0.	0.00151	0.00011
• 2SIGM-G	C1	0.05190	0.97902	0.00443	-0.00032	0.	0.	0.	0.	0.	0.
-11.5152	O1	-0.00002	-0.00058	0.00075	0.00055	0.	0.	0.	0.	-0.00144	-0.00016
	O2	-0.00002	-0.00058	0.00075	0.00055	0.	0.	0.	0.	0.00144	0.00016
• 3SIGM-G	C1	0.00749	0.15934	-0.34633	-0.00658	0.	0.	0.	0.	0.	0.
-1.5713	O1	0.00656	0.14368	-0.31448	-0.24362	0.	0.	0.	0.	0.14500	0.01700
	O2	0.00656	0.14368	-0.31448	-0.24362	0.	0.	0.	0.	-0.14500	-0.01700
• 4SIGM-G	C1	0.00686	0.13817	-0.41102	-0.06661	0.	0.	0.	0.	0.	0.
-0.7972	O1	-0.00447	-0.09823	0.22330	0.30416	0.	0.	0.	0.	0.36014	0.11278
	O2	-0.00447	-0.09823	0.22330	0.30416	0.	0.	0.	0.	-0.36014	-0.11278
5SIGM-G	C1	-0.00182	-0.06482	-0.06078	2.37061	0.	0.	0.	0.	0.	0.
0.2013	O1	0.00330	0.07807	-0.13644	-0.92428	0.	0.	0.	0.	0.26108	0.49923
	O2	0.00330	0.07807	-0.13644	-0.92428	0.	0.	0.	0.	-0.26108	-0.49923
• 1SIGM-U	C1	0.	0.	0.	0.	0.	0.	0.	0.	-0.00073	-0.00188
-20.6703	O1	0.03615	0.69251	0.00224	0.00301	0.	0.	0.	0.	-0.00172	0.00035
	O2	-0.03615	-0.69251	-0.00224	-0.00301	0.	0.	0.	0.	-0.00172	0.00035
• 2SIGM-U	C1	0.	0.	0.	0.	0.	0.	0.	0.	0.40312	-0.01222
-1.5188	O1	-0.00713	-0.15572	0.34557	0.17229	0.	0.	0.	0.	-0.13264	0.03823
	O2	0.00713	0.15572	-0.34557	-0.17229	0.	0.	0.	0.	-0.13264	0.03823
• 3SIGM-U	C1	0.	0.	0.	0.	0.	0.	0.	0.	-0.46175	0.07360
-0.7393	O1	-0.00340	-0.07475	0.17091	0.30389	0.	0.	0.	0.	0.42453	0.07659
	O2	0.00340	0.07475	-0.17091	-0.30389	0.	0.	0.	0.	0.42453	0.07659
4SIGM-U	C1	0.	0.	0.	0.	0.	0.	0.	0.	0.06307	2.17946
0.3726	O1	0.00250	0.06848	-0.03693	-1.33938	0.	0.	0.	0.	-0.08614	0.13165
	O2	-0.00250	-0.06848	0.03693	1.33938	0.	0.	0.	0.	-0.08614	0.13165
• 1PI-G-X	C1	0.	0.	0.	0.	0.	0.	0.	0.	0.	0.
-0.5405	O1	0.	0.	0.	0.	0.53044	0.28571	0.	0.	0.	0.
	O2	0.	0.	0.	0.	-0.53044	-0.28571	0.	0.	0.	0.
• 1PI-G-Y	C1	0.	0.	0.	0.	0.	0.	0.	0.	0.	0.
-0.5405	O1	0.	0.	0.	0.	0.	0.	0.53044	0.28571	0.	0.
	O2	0.	0.	0.	0.	0.	0.	-0.53044	-0.28571	0.	0.
• 1PI-U-X	C1	0.	0.	0.	0.	0.45750	0.06039	0.	0.	0.	0.
-0.7389	O1	0.	0.	0.	0.	0.40766	0.13950	0.	0.	0.	0.
	O2	0.	0.	0.	0.	0.40766	0.13950	0.	0.	0.	0.
2PI-U-X	C1	0.	0.	0.	0.	-0.64076	-0.80862	0.	0.	0.	0.
0.1712	O1	0.	0.	0.	0.	0.38641	0.46014	0.	0.	0.	0.
	O2	0.	0.	0.	0.	0.38641	0.46014	0.	0.	0.	0.
3PI-U-X	C1	0.	0.	0.	0.	0.90654	-1.27092	0.	0.	0.	0.
0.3834	O1	0.	0.	0.	0.	-0.02383	0.13820	0.	0.	0.	0.
	O2	0.	0.	0.	0.	-0.02383	0.13820	0.	0.	0.	0.
• 1PI-U-Y	C1	0.	0.	0.	0.	0.	0.	0.45750	0.06039	0.	0.
-0.7389	O1	0.	0.	0.	0.	0.	0.	0.40766	0.13950	0.	0.
	O2	0.	0.	0.	0.	0.	0.	0.40766	0.13950	0.	0.

Center		S1	S2	S3	S4	Basis Function Type X1	X2	Y1	Y2	Z1	Z2
2PI-U-Y	C1	0.	0.	0.	0.	0.	0.	-0.64076	-0.80861	0.	0.
0.1712	01	0.	0.	0.	0.	0.	0.	0.38641	0.46014	0.	0.
	02	0.	0.	0.	0.	0.	0.	0.38641	0.46014	0.	0.
3PI-U-Y	C1	0.	0.	0.	0.	0.	0.	0.90654	-1.27092	0.	0.
0.3834	01	0.	0.	0.	0.	0.	0.	-0.02383	0.13820	0.	0.
	02	0.	0.	0.	0.	0.	0.	-0.02383	0.13820	0.	0.

Overlap Matrix

First Function   Second Function and Overlap

C1S1 with
     C1S2 .35436   C1S3 .00805   C1S4 .01419   O1S3 .00068   O1S4 .00592   O1Z1 -.00344   O1Z2 -.01354

C1S2 with
     C1S3 .26226   C1S4 .17745   O1S2 .00003   O1S3 .01501   O1S4 .07701   O1Z1 -.05127   O1Z2 -.16557

C1S3 with
     C1S4 .79970   O1S1 .00234   O1S2 .03196   O1S3 .21781   O1S4 .41533   O1Z1 -.28139   O1Z2 -.61877

C1S4 with
     O1S1 .00444   O1S2 .05731   O1S3 .31611   O1S4 .58002   O1Z1 -.17312   O1Z2 -.53130

C1X1 with
     C1X2 .53875   O1X1 .16537   O1X2 .38803

C1X2 with
     O1X1 .18330   O1X2 .61958

C1Y1 with
     C1Y2 .53875   O1Y1 .16537   O1Y2 .38803

C1Y2 with
     O1Y1 .18330   O1Y2 .61958

C1Z1 with
     C1Z2 .53875   O1S1 .00557   O1S2 .07221   O1S3 .37613   O1S4 .43494   O1Z1 -.30965   O1Z2 -.14925

C1Z2 with
     O1S1 .00639   O1S2 .08137   O1S3 .40838   O1S4 .61575   O1Z1 .00585   O1Z2 .17444

O1S1 with
     O1S2 .35461   O1S3 .00974   O1S4 .01469   O2S4 .00006   O2Z2 .00079

O1S2 with
     O1S3 .28286   O1S4 .18230   O2S4 .00097   O2Z1 .00001   O2Z2 .01082

O1S3 with
     O1S4 .79029   O2S3 .00014   O2S4 .01243   O2Z1 .00112   O2Z2 .08237

O1S4 with
     O2S1 .00006   O2S2 .00097   O2S3 .01243   O2S4 .06583   O2Z1 .02506   O2Z2 .21852

O1X1 with
     O1X2 .50607   O2X1 .00040   O2X2 .01933

O1X2 with
     O2X1 .01933   O2X2 .12766

O1Y1 with
     O1Y2 .50607   O2Y1 .00040   O2Y2 .01933

O1Y2 with
     O2Y1 .01933   O2Y2 .12766

O1Z1 with
     O1Z2 .50607   O2S2 -.00001   O2S3 -.00112   O2S4 -.02506   O2Z1 -.00523   O2Z2 -.10600

O1Z2 with
     O2S1 -.00079   O2S2 -.01082   O2S3 -.08237   O2S4 -.21852   O2Z1 -.10600   O2Z2 -.39789

C3H4   Allene

Molecular Geometry   Symmetry D_{2d}

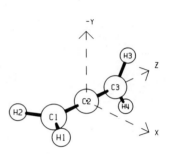

Center		Coordinates	
	X	Y	Z
H1	1.75567000	0.	-3.52519000
H2	-1.75567000	0.	-3.52519000
H3	0.	-1.75567000	3.52519000
H4	0.	1.75567000	3.52519000
C1	0.	0.	-2.47860000
C2	0.	0.	0.
C3	0.	0.	2.47860000
Mass	0.	0.	0.

MO Symmetry Types and Occupancy

Representation	A1	A2	B1	B2	E
Number of MOs	12	0	0	10	16
Number Occupied	4	0	0	3	4

Energy Expectation Values

Total	-115.8203	Kinetic	115.7466
Electronic	-174.9088	Potential	-231.5669
Nuclear Repulsion	59.0885	Virial	2.0006
Electronic Potential	-290.6554		
One-Electron Potential	-387.2674		
Two-Electron Potential	96.6120		

Moments of the Charge Distribution

Second

x^2	y^2	z^2	xy	xz	yz	r^2
-21.0005	-21.0005	-135.0104	0.	0.	0.	-177.0113
6.1648	6.1648	123.4294	0.	0.	0.	135.7589

Third

x^3	xy^2	xz^2	x^2y	y^3	yz^2	x^2z	y^2z	z^3	xyz
0.	0.	0.	0.	0.	0.	15.9799	-15.9799	-0.0001	0.
0.	0.	0.	0.	0.	0.	-21.7319	21.7319	0.	0.

Fourth (even)

x^4	y^4	z^4	x^2y^2	x^2z^2	y^2z^2	r^2x^2	r^2y^2	r^2z^2	r^4
-97.8259	-97.8259	-1530.0516	-28.2207	-168.5164	-168.5164	-294.5630	-294.5630	-1867.0844	-2456.2105
19.0021	19.0021	1070.6227	0.	76.6092	76.6092	95.6113	95.6113	1223.8411	1415.0636

Expectation Values at Atomic Centers

Center	r^{-1}	δ	Expectation Value xr^{-3}	yr^{-3}	zr^{-3}
H1	-1.0767	0.4121	0.0569	0.	-0.0343
C1	-14.7178	116.3766	0.	0.	-0.0248
C2	-14.6950	116.7316	0.	0.	0.

Center	$\dfrac{3x^2-r^2}{r^5}$	$\dfrac{3y^2-r^2}{r^5}$	$\dfrac{3z^2-r^2}{r^5}$	$\dfrac{3xy}{r^5}$	$\dfrac{3xz}{r^5}$	$\dfrac{3yz}{r^5}$
H1	0.2114	-0.1855	-0.0259	0.	-0.2294	0.
C1	-0.1183	0.0796	0.0387	0.	0.	0.
C2	0.1667	0.1667	-0.3334	0.	0.	0.

Mulliken Charge

Center	S1	S2	S3	S4	X1	X2	Y1	Y2	Z1	Z2	Total
H1	0.5095	0.2841									0.7935
C1	0.0428	1.9564	0.7669	0.5207	0.9462	0.1620	0.7663	0.3114	0.9021	0.0808	6.4556
C2	0.0429	1.9565	0.8304	0.2352	0.6917	0.2696	0.6917	0.2696	1.0143	-0.0871	5.9148

Overlap Populations

	Pair of Atomic Centers						
	H2 H1	H3 H1	C1 H1	C1 H3	C2 H1	C2 C1	C3 C1
1A1	0.	0.	0.	0.	0.	-0.0003	0.
2A1	0.	0.	-0.0001	0.	0.	-0.0007	0.0002
3A1	0.0007	0.	0.0312	0.0005	0.0057	0.3374	0.0512
4A1	0.0198	0.0002	0.1655	-0.0018	-0.0334	0.0391	0.0012
Total A1	0.0205	0.0002	0.1966	-0.0013	-0.0277	0.3756	0.0526
1B2	0.	0.	0.	0.	0.	-0.0014	-0.0003
2B2	0.0052	0.	0.0986	-0.0020	0.0029	0.1906	-0.0669
3B2	0.0145	-0.0001	0.0832	0.0054	-0.0094	0.2201	-0.0644
Total B2	0.0197	-0.0002	0.1818	0.0033	-0.0065	0.4093	-0.1317
1E	0.	0.	0.	0.0012	0.	0.0241	0.0093
2E	-0.0605	0.	0.3248	0.	0.0172	0.1263	0.0093
3E	-0.0152	0.	0.0776	0.	-0.0398	-0.2482	-0.0521
4E	0.	0.	0.	-0.0045	0.	0.5526	-0.0521
Total E	-0.0757	0.	0.4024	-0.0033	-0.0226	0.4548	-0.0857
Total	-0.0354	0.0000	0.7808	-0.0012	-0.0569	1.2397	-0.1649
Distance	3.5113	7.4748	2.0439	6.2552	3.9382	2.4786	4.9572

Molecular Orbitals and Eigenvalues

Center		S1	S2	S3	S4	X1	X2	Y1	Y2	Z1	Z2
* 1A1	H1	0.00010	-0.00023								
-11.2837	H2	0.00010	-0.00023								
	H3	0.00010	-0.00023								
	H4	0.00010	-0.00023								
	C1	0.00132	0.02481	0.00094	-0.00103	0.	0.	0.	0.	0.00092	-0.00096
	C2	0.05187	0.97839	0.00389	0.00328	0.	0.	0.	0.	0.	0.
	C3	0.00132	0.02481	0.00094	-0.00103	0.	0.	0.	0.	-0.00092	0.00096
* 2A1	H1	-0.00045	0.00052								
-11.2478	H2	-0.00045	0.00052								
	H3	-0.00045	0.00052								
	H4	-0.00045	0.00052								
	C1	-0.03668	-0.69179	-0.00312	-0.00259	0.	0.	0.	0.	-0.00047	-0.00035
	C2	0.00190	0.03599	-0.00094	0.00312	0.	0.	0.	0.	0.	0.
	C3	-0.03668	-0.69179	-0.00312	-0.00259	0.	0.	0.	0.	0.00047	0.00035
* 3A1	H1	-0.05761	-0.00821								
-1.0812	H2	-0.05761	-0.00821								
	H3	-0.05761	-0.00821								
	H4	-0.05761	-0.00821								
	C1	0.00568	0.12274	-0.26795	-0.15495	0.	0.	0.	0.	-0.11497	-0.00060
	C2	0.00811	0.17485	-0.38333	-0.10367	0.	0.	0.	0.	0.	0.
	C3	0.00568	0.12274	-0.26795	-0.15495	0.	0.	0.	0.	0.11497	0.00060
* 4A1	H1	0.13528	0.08041								
-0.7157	H2	0.13528	0.08041								
	H3	0.13528	0.08041								
	H4	0.13528	0.08041								
	C1	-0.00316	-0.06825	0.15674	0.12730	0.	0.	0.	0.	-0.31203	-0.03731
	C2	0.00570	0.12370	-0.28066	-0.15163	0.	0.	0.	0.	0.	0.
	C3	-0.00316	-0.06825	0.15674	0.12730	0.	0.	0.	0.	0.31203	0.03731
5A1	H1	-0.01390	-0.09964								
0.2944	H2	-0.01390	-0.09964								
	H3	-0.01390	-0.09964								
	H4	-0.01390	-0.09964								
	C1	0.00202	0.05609	-0.02991	-1.51009	0.	0.	0.	0.	0.08651	-1.71981
	C2	-0.00329	-0.09305	0.03773	4.06591	0.	0.	0.	0.	0.	0.
	C3	0.00202	0.05609	-0.02991	-1.51009	0.	0.	0.	0.	-0.08651	1.71981
6A1	H1	0.02458	1.17000								
0.3113	H2	0.02458	1.17000								
	H3	0.02458	1.17000								
	H4	0.02458	1.17000								
	C1	0.00439	0.09373	-0.25606	-1.67126	0.	0.	0.	0.	0.06101	0.68011
	C2	0.00005	-0.01354	-0.10016	0.33206	0.	0.	0.	0.	0.	0.
	C3	0.00439	0.09373	-0.25606	-1.67126	0.	0.	0.	0.	-0.06101	-0.68011
7A1	H1	-0.07599	-0.54011								
0.5490	H2	-0.07599	-0.54011								
	H3	-0.07599	-0.54011								
	H4	-0.07599	-0.54011								
	C1	0.00444	0.04614	-0.59267	0.48922	0.	0.	0.	0.	-0.60885	-0.44555
	C2	0.00015	-0.06325	-0.45206	2.13741	0.	0.	0.	0.	0.	0.
	C3	0.00444	0.04614	-0.59267	0.48922	0.	0.	0.	0.	0.60885	0.44555

C3H4

Center		S1	S2	S3	S4	X1	X2	Y1	Y2	Z1	Z2
* 1B2	H1	-0.00039	0.00023								
-11.2479	H2	-0.00039	0.00023								
	H3	0.00039	-0.00023								
	H4	0.00039	-0.00023								
	C1	-0.03670	-0.69221	-0.00267	-0.00504	0.	0.	0.	0.	-0.00038	-0.00086
	C2	0.	0.	0.	0.	0.	0.	0.	0.	0.00108	-0.00449
	C3	0.03670	0.69221	0.00267	0.00504	0.	0.	0.	0.	-0.00038	-0.00086
* 2B2	H1	0.09331	0.03502								
-0.9649	H2	0.09331	0.03502								
	H3	-0.09331	-0.03502								
	H4	-0.09331	-0.03502								
	C1	-0.00686	-0.15061	0.30977	0.21904	0.	0.	0.	0.	0.04559	-0.00042
	C2	0.	0.	0.	0.	0.	0.	0.	0.	-0.34124	0.05991
	C3	0.00686	0.15061	-0.30977	-0.21904	0.	0.	0.	0.	0.04559	-0.00042
* 3B2	H1	-0.11518	-0.06903								
-0.6283	H2	-0.11518	-0.06903								
	H3	0.11518	0.06903								
	H4	0.11518	0.06903								
	C1	0.00055	0.00980	-0.04350	-0.00318	0.	0.	0.	0.	0.38181	0.06397
	C2	0.	0.	0.	0.	0.	0.	0.	0.	-0.44295	0.07083
	C3	-0.00055	-0.00980	0.04350	0.00318	0.	0.	0.	0.	0.38181	0.06397
4B2	H1	-0.00017	0.04184								
0.2601	H2	-0.00017	0.04184								
	H3	0.00017	-0.04184								
	H4	0.00017	-0.04184								
	C1	0.00261	0.08674	0.06041	-4.21871	0.	0.	0.	0.	0.18602	-1.76101
	C2	0.	0.	0.	0.	0.	0.	0.	0.	-0.13429	-4.49646
	C3	-0.00261	-0.08674	-0.06041	4.21871	0.	0.	0.	0.	0.18602	-1.76101
5B2	H1	0.05684	1.30459								
0.3382	H2	0.05684	1.30459								
	H3	-0.05684	-1.30459								
	H4	-0.05684	-1.30459								
	C1	0.00172	0.04147	-0.06967	-1.03199	0.	0.	0.	0.	0.08951	1.55751
	C2	0.	0.	0.	0.	0.	0.	0.	0.	0.06739	0.44181
	C3	-0.00172	-0.04147	0.06967	1.03199	0.	0.	0.	0.	0.08951	1.55751
6B2	H1	0.03468	0.54951								
0.5279	H2	0.03468	0.54951								
	H3	-0.03468	-0.54951								
	H4	-0.03468	-0.54951								
	C1	0.00026	0.00935	0.01001	2.27890	0.	0.	0.	0.	0.22202	0.40727
	C2	0.	0.	0.	0.	0.	0.	0.	0.	-0.57647	4.48697
	C3	-0.00026	-0.00935	-0.01001	-2.27890	0.	0.	0.	0.	0.22202	0.40727
* 1E	H1	0.	0.								
-0.6128	H2	0.	0.								
	H3	-0.21693	-0.14553								
	H4	0.21693	0.14553								
	C1	0.	0.	0.	0.	0.	0.	0.08528	0.02774	0.	0.
	C2	0.	0.	0.	0.	0.	0.	0.18812	0.02360	0.	0.
	C3	0.	0.	0.	0.	0.	0.	0.51445	0.10407	0.	0.

C3H4

Center		Basis Function Type									
		S1	S2	S3	S4	X1	X2	Y1	Y2	Z1	Z2
• 2E	H1	0.21693	0.14552								
-0.6128	H2	-0.21693	-0.14552								
	H3	0.	0.								
	H4	0.	0.								
	C1	0.	0.	0.	0.	0.51445	0.10408	0.	0.	0.	0.
	C2	0.	0.	0.	0.	0.18812	0.02359	0.	0.	0.	0.
	C3	0.	0.	0.	0.	0.08528	0.02774	0.	0.	0.	0.
• 3E	H1	-0.10305	-0.07440								
-0.3768	H2	0.10305	0.07440								
	H3	0.	0.								
	H4	0.	0.								
	C1	0.	0.	0.	0.	-0.14687	-0.13445	0.	0.	0.	0.
	C2	0.	0.	0.	0.	0.43092	0.22142	0.	0.	0.	0.
	C3	0.	0.	0.	0.	0.49132	0.20976	0.	0.	0.	0.
• 4E	H1	0.	0.								
-0.3768	H2	0.	0.								
	H3	0.10305	0.07440								
	H4	-0.10305	-0.07440								
	C1	0.	0.	0.	0.	0.	0.	0.49132	0.20976	0.	0.
	C2	0.	0.	0.	0.	0.	0.	0.43092	0.22142	0.	0.
	C3	0.	0.	0.	0.	0.	0.	-0.14687	-0.13445	0.	0.
5E	H1	0.09962	0.30331								
0.1593	H2	-0.09962	-0.30331								
	H3	0.	0.								
	H4	0.	0.								
	C1	0.	0.	0.	0.	-0.07066	0.09792	0.	0.	0.	0.
	C2	0.	0.	0.	0.	-0.49273	-0.86092	0.	0.	0.	0.
	C3	0.	0.	0.	0.	0.37065	0.86153	0.	0.	0.	0.
6E	H1	0.	0.								
0.1593	H2	0.	0.								
	H3	-0.09962	-0.30331								
	H4	0.09962	0.30331								
	C1	0.	0.	0.	0.	0.	0.	0.37065	0.86153	0.	0.
	C2	0.	0.	0.	0.	0.	0.	-0.49273	-0.86092	0.	0.
	C3	0.	0.	0.	0.	0.	0.	-0.07066	0.09792	0.	0.
7E	H1	-0.07718	-1.25543								
0.4192	H2	0.07718	1.25543								
	H3	0.	0.								
	H4	0.	0.								
	C1	0.	0.	0.	0.	-0.46549	2.68677	0.	0.	0.	0.
	C2	0.	0.	0.	0.	-0.31792	-0.86335	0.	0.	0.	0.
	C3	0.	0.	0.	0.	-0.43154	0.79607	0.	0.	0.	0.
8E	H1	0.	0.								
0.4192	H2	0.	0.								
	H3	0.07718	1.25537								
	H4	-0.07718	-1.25537								
	C1	0.	0.	0.	0.	0.	0.	-0.43155	0.79605	0.	0.
	C2	0.	0.	0.	0.	0.	0.	-0.31792	-0.86330	0.	0.
	C3	0.	0.	0.	0.	0.	0.	-0.46550	2.68668	0.	0.

Center		S1	S2	S3	S4	X1	X2	Y1	Y2	Z1	Z2
9E	H1	-0.04832	1.66501								
0.4934	H2	0.04832	-1.66501								
	H3	0.	0.								
	H4	0.	0.								
	C1	0.	0.	0.	0.	-0.82867	-1.51444	0.	0.	0.	0.
	C2	0.	0.	0.	0.	0.16761	1.02048	0.	0.	0.	0.
	C3	0.	0.	0.	0.	0.40535	-0.99698	0.	0.	0.	0.
10E	H1	0.	0.								
0.4934	H2	0.	0.								
	H3	-0.04832	1.66498								
	H4	0.04832	-1.66498								
	C1	0.	0.	0.	0.	0.	0.	-0.40536	0.99699	0.	0.
	C2	0.	0.	0.	0.	0.	0.	-0.16762	-1.02044	0.	0.
	C3	0.	0.	0.	0.	0.	0.	0.82867	1.51439	0.	0.
11E	H1	0.06893	2.52021								
0.5951	H2	-0.06893	-2.52021								
	H3	0.	0.								
	H4	0.	0.								
	C1	0.	0.	0.	0.	-0.15488	-4.23043	0.	0.	0.	0.
	C2	0.	0.	0.	0.	-0.05687	2.35344	0.	0.	0.	0.
	C3	0.	0.	0.	0.	-0.69304	-0.13177	0.	0.	0.	0.
12E	H1	0.	0.								
0.5951	H2	0.	0.								
	H3	0.06893	2.52023								
	H4	-0.06893	-2.52023								
	C1	0.	0.	0.	0.	0.	0.	0.69304	0.13176	0.	0.
	C2	0.	0.	0.	0.	0.	0.	0.05684	-2.35340	0.	0.
	C3	0.	0.	0.	0.	0.	0.	0.15490	4.23043	0.	0.
13E	H1	0.07213	0.44168								
0.6544	H2	-0.07213	-0.44168								
	H3	0.	0.								
	H4	0.	0.								
	C1	0.	0.	0.	0.	0.18468	-1.74083	0.	0.	0.	0.
	C2	0.	0.	0.	0.	-0.95836	2.54092	0.	0.	0.	0.
	C3	0.	0.	0.	0.	0.48682	-1.39056	0.	0.	0.	0.
14E	H1	0.	0.								
0.6544	H2	0.	0.								
	H3	0.07213	0.44175								
	H4	-0.07213	-0.44175								
	C1	0.	0.	0.	0.	0.	0.	-0.48681	1.39057	0.	0.
	C2	0.	0.	0.	0.	0.	0.	0.95836	-2.54098	0.	0.
	C3	0.	0.	0.	0.	0.	0.	-0.18467	1.74094	0.	0.

Basis Function Type

C3H4

Overlap Matrix

First Function   Second Function and Overlap

H1S1 with
  H1S2 .68302   H2S1 .01226   H2S2 .11813   H3S2 .00025   C1S1 .00232   C1S2 .03556   C1S3 .28922
  C1S4 .37497   C1X1 .37378   C1X2 .37174   C1Z1 -.22282 C1Z2 -.22160 C2S2 .00007   C2S3 .01220
  C2S4 .09357   C2X1 .01988   C2X2 .12131   C2Z1 -.03992 C2Z2 -.24357

H1S2 with
  H2S1 .11813   H2S2 .33463   H3S1 .00025   H3S2 .00701   C1S1 .00775   C1S2 .09888   C1S3 .50045
  C1S4 .70931   C1X1 .30980   C1X2 .52112   C1Z1 -.18468 C1Z2 -.31065 C2S1 .00104   C2S2 .01440
  C2S3 .11722   C2S4 .28417   C2X1 .07515   C2X2 .23668   C2Z1 -.15089 C2Z2 -.47523

H3S1 with
  C1S3 .00002   C1S4 .00525   C1Y1 -.00008 C1Y2 -.01182 C1Z1 .00026   C1Z2 .04042

H3S2 with
  C1S1 .00002   C1S2 .00026   C1S3 .00562   C1S4 .04223   C1Y1 -.00417 C1Y2 -.04568 C1Z1 .01425
  C1Z2 .15621

C1S1 with
  C1S2 .35436   C1S3 .00805   C1S4 .01419   C2S3 .00196   C2S4 .00572   C2Z1 -.00590 C2Z2 -.00972
  C3S4 .00037   C3Z1 -.00002 C3Z2 -.00236

C1S2 with
  C1S3 .26226   C1S4 .17745   C2S2 .00003   C2S3 .02951   C2S4 .07380   C2Z1 -.07765 C2Z2 -.12197
  C3S3 .00001   C3S4 .00532   C3Z1 -.00035 C3Z2 -.03129

C1S3 with
  C1S4 .79970   C2S1 .00196   C2S2 .02951   C2S3 .24990   C2S4 .40249   C2Z1 -.38250 C2Z2 -.54971
  C3S2 .00001   C3S3 .00330   C3S4 .05102   C3Z1 -.01506 C3Z2 -.20292

C1S4 with
  C2S1 .00572   C2S2 .07380   C2S3 .40249   C2S4 .63469   C2Z1 -.30985 C2Z2 -.62829 C3S1 .00037
  C3S2 .00532   C3S3 .05102   C3S4 .16227   C3Z1 -.08326 C3Z2 -.38213

C1X1 with
  C1X2 .53875   C2X1 .20699   C2X2 .30956   C3X1 .00536   C3X2 .05932

C1X2 with
  C2X1 .30956   C2X2 .70327   C3X1 .05932   C3X2 .24461

C1Y1 with
  C1Y2 .53875   C2Y1 .20699   C2Y2 .30956   C3Y1 .00536   C3Y2 .05932

C1Y2 with
  C2Y1 .30956   C2Y2 .70327   C3Y1 .05932   C3Y2 .24461

C1Z1 with
  C1Z2 .53875   C2S1 .00590   C2S2 .07765   C2S3 .38250   C2S4 .30985   C2Z1 -.35336 C2Z2 -.03293
  C3S1 .00002   C3S2 .00035   C3S3 .01506   C3S4 .08326   C3Z1 -.04391 C3Z2 -.20096

C1Z2 with
  C2S1 .00972   C2S2 .12197   C2S3 .54971   C2S4 .62829   C2Z1 -.03293 C2Z2 .20814   C3S1 .00236
  C3S2 .03129   C3S3 .20292   C3S4 .38213   C3Z1 -.20096 C3Z2 -.44426

C2S1 with
  C2S2 .35436   C2S3 .00805   C2S4 .01419

C2S2 with
  C2S3 .26226   C2S4 .17745

C2S3 with
  C2S4 .79970

C2X1 with
  C2X2 .53875

C2Y1 with
  C2Y2 .53875

C2Z1 with
  C2Z2 .53875

CH2CO   Ketene

Molecular Geometry   Symmetry C$_{2v}$

Center		Coordinates	
	X	Y	Z
H1	1.79502000	0.	-3.46621999
H2	-1.79502000	0.	-3.46621999
C1	0.	0.	-2.48313001
C2	0.	0.	0.
O1	0.	0.	2.19400001
Mass	0.	0.	-0.04025800

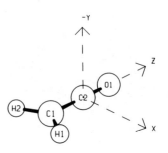

MO Symmetry Types and Occupancy

Representation	A1	A2	B1	B2
Number of MOs	20	0	8	6
Number Occupied	7	0	2	2

Energy Expectation Values

Total	-151.6595	Kinetic	151.6665
Electronic	-210.2085	Potential	-303.3259
Nuclear Repulsion	58.5490	Virial	2.0000
Electronic Potential	-361.8750		
One-Electron Potential	-472.9271		
Two-Electron Potential	111.0521		

Moments of the Charge Distribution

First

x	y	z	Dipole Moment
0.	0.	2.5962	0.7973
0.	0.	-3.3935	

Second

x^2	y^2	z^2	xy	xz	yz	r^2
-17.8561	-14.4832	-113.7846	0.	0.	0.	-146.1238
6.4442	0.	99.2252	0.	0.	0.	105.6694

Third

x^3	xy^2	xz^2	x^2y	y^3	yz^2	x^2z	y^2z	z^3	xyz
0.	0.	0.	0.	0.	0.	22.9775	8.3333	81.4916	0.
0.	0.	0.	0.	0.	0.	-22.0776	0.	-78.6653	0.

Fourth (even)

x^4	y^4	z^4	x^2y^2	x^2z^2	y^2z^2	r^2x^2	r^2y^2	r^2z^2	r^4
-74.4185	-56.8738	-1064.8125	-19.6388	-135.2636	-84.1236	-229.3210	-160.6362	-1284.1997	-1674.1569
20.7638	0.	688.5529	0.	75.6369	0.	96.4007	0.	764.1898	860.5905

Expectation Values at Atomic Centers

Center			Expectation Value		
	r^{-1}	δ	xr^{-3}	yr^{-3}	zr^{-3}
H1	-1.0407	0.3972	0.0471	0.	-0.0355
C1	-14.7045	116.2653	0.	0.	-0.0501
C2	-14.5676	116.8565	0.	0.	-0.0615
O1	-22.2644	287.3654	0.	0.	0.2854

Center	$\dfrac{3x^2-r^2}{r^5}$	$\dfrac{3y^2-r^2}{r^5}$	$\dfrac{3z^2-r^2}{r^5}$	$\dfrac{3xy}{r^5}$	$\dfrac{3xz}{r^5}$	$\dfrac{3yz}{r^5}$
H1	0.2233	-0.1954	-0.0279	0.	-0.2155	0.
C1	-0.0568	-0.2628	0.3196	0.	0.	0.
C2	0.5023	-0.0588	-0.4435	0.	0.	0.
O1	0.7631	-1.5503	0.7872	0.	0.	0.

Mulliken Charge

Center	Basis Function Type										Total
	S1	S2	S3	S4	X1	X2	Y1	Y2	Z1	Z2	
H1	0.4939	0.2594									0.7533
C1	0.0427	1.9563	0.7388	0.5315	0.9529	0.2141	0.9140	0.4147	0.8203	0.0642	6.6496
C2	0.0430	1.9565	0.8762	0.1280	0.5398	0.0999	0.7616	0.1636	1.0087	0.0166	5.5940
O1	0.0427	1.9566	0.9796	0.8160	1.0121	0.4126	1.2815	0.4645	1.0983	0.1859	8.2498

Overlap Populations

	Pair of Atomic Centers						
	H2	C1	C2	C2	O1	O1	O1
	H1	H1	H1	C1	H1	C1	C2
1A1	0.	0.	0.	0.	0.	-0.0001	-0.0008
2A1	0.	0.	0.	-0.0005	0.	0.	-0.0003
3A1	0.	-0.0002	0.	-0.0017	0.	0.	0.
4A1	0.	-0.0007	-0.0002	0.0134	-0.0002	0.0175	0.6133
5A1	0.0031	0.0885	0.0154	0.5984	0.0004	0.0118	0.0349
6A1	0.0125	0.1337	-0.0460	-0.2021	-0.0015	0.0019	0.0041
7A1	0.0236	0.1728	-0.0190	0.1997	0.0009	-0.0001	0.0411
Total A1	0.0392	0.3941	-0.0499	0.6072	-0.0004	0.0311	0.6923
1B1	-0.0291	0.1794	0.0254	0.1702	0.0019	0.0177	0.1771
2B1	-0.0199	0.1970	-0.0250	-0.2108	-0.0025	-0.0519	0.2578
Total B1	-0.0490	0.3764	0.0004	-0.0406	-0.0006	-0.0342	0.4349
1B2	0.	0.	0.	0.0780	0.	0.0128	0.3695
2B2	0.	0.	0.	0.3986	0.	-0.0896	-0.3022
Total B2	0.	0.	0.	0.4766	0.	-0.0768	0.0673
Total	-0.0098	0.7705	-0.0495	1.0432	-0.0010	-0.0799	1.1945
Distance	3.5900	2.0466	3.9034	2.4831	5.9380	4.6771	2.1940

Molecular Orbitals and Eigenvalues

Center		Basis Function Type									
		S1	S2	S3	S4	X1	X2	Y1	Y2	Z1	Z2
* 1A1	H1	-0.00005	0.00011								
-20.6390	H2	-0.00005	0.00011								
	C1	0.00000	-0.00001	-0.00002	0.00177	0.	0.	0.	0.	0.00017	0.00036
	C2	-0.00000	0.00005	0.00039	-0.00044	0.	0.	0.	0.	0.00044	0.00214
	O1	-0.05113	-0.97936	-0.00324	-0.00368	0.	0.	0.	0.	0.00217	-0.00037
* 2A1	H1	-0.00011	0.00039								
-11.4164	H2	-0.00011	0.00039								
	C1	-0.00028	-0.00516	-0.00120	0.00150	0.	0.	0.	0.	-0.00123	0.00115
	C2	-0.05190	-0.97905	-0.00419	-0.00268	0.	0.	0.	0.	-0.00150	0.00113
	O1	0.00003	0.00058	-0.00091	0.00050	0.	0.	0.	0.	0.00132	-0.00072
* 3A1	H1	-0.00062	0.00068								
-11.2579	H2	-0.00062	0.00068								
	C1	-0.05190	-0.97889	-0.00439	-0.00491	0.	0.	0.	0.	-0.00116	-0.00061
	C2	0.00030	0.00579	-0.00060	0.00349	0.	0.	0.	0.	0.00065	-0.00178
	O1	-0.00000	-0.00001	0.00003	-0.00027	0.	0.	0.	0.	-0.00002	0.00067
* 4A1	H1	0.00099	-0.00572								
-1.5077	H2	0.00099	-0.00572								
	C1	-0.00051	-0.00883	0.04013	0.03140	0.	0.	0.	0.	0.03619	0.00918
	C2	-0.00569	-0.12044	0.26757	0.02688	0.	0.	0.	0.	0.23847	-0.01868
	O1	-0.00969	-0.21173	0.46846	0.30396	0.	0.	0.	0.	-0.19697	0.01476
* 5A1	H1	0.09307	0.02414								
-1.0681	H2	0.09307	0.02414								
	C1	-0.00839	-0.18406	0.37495	0.22882	0.	0.	0.	0.	0.13655	0.00298
	C2	-0.00583	-0.12480	0.28669	0.04751	0.	0.	0.	0.	-0.30511	0.01084
	O1	0.00211	0.04570	-0.10692	-0.04525	0.	0.	0.	0.	-0.01634	-0.04556
* 6A1	H1	0.10531	0.06750								
-0.7539	H2	0.10531	0.06750								
	C1	-0.00324	-0.07222	0.14365	0.21455	0.	0.	0.	0.	-0.19733	0.00973
	C2	0.00580	0.11997	-0.32545	-0.17404	0.	0.	0.	0.	-0.15021	0.09557
	O1	-0.00483	-0.10591	0.24248	0.34184	0.	0.	0.	0.	0.46669	0.11745
* 7A1	H1	-0.15415	-0.09064								
-0.6801	H2	-0.15415	-0.09064								
	C1	0.00243	0.04946	-0.14329	-0.06539	0.	0.	0.	0.	0.40406	0.06405
	C2	-0.00180	-0.04142	0.07665	0.02698	0.	0.	0.	0.	-0.42193	0.07261
	O1	-0.00192	-0.04272	0.09272	0.20310	0.	0.	0.	0.	0.34743	0.05350
8A1	H1	0.02782	1.19582								
0.2679	H2	0.02782	1.19582								
	C1	0.00524	0.10346	-0.36171	-1.07541	0.	0.	0.	0.	-0.08558	0.88749
	C2	-0.00040	-0.02828	-0.10510	0.78988	0.	0.	0.	0.	0.08426	0.84353
	O1	0.00288	0.07082	-0.09943	-0.97991	0.	0.	0.	0.	0.13513	0.28948
9A1	H1	0.01563	-0.16819								
0.2795	H2	0.01563	-0.16819								
	C1	-0.00382	-0.10707	0.04911	3.38987	0.	0.	0.	0.	-0.18760	2.23178
	C2	0.00230	0.06548	-0.02441	-3.38870	0.	0.	0.	0.	0.14390	1.76692
	O1	0.00040	0.01210	0.00297	-0.24709	0.	0.	0.	0.	-0.07601	-0.31758
10A1	H1	0.06282	1.14484								
0.3281	H2	0.06282	1.14484								
	C1	0.00023	0.02209	0.10271	-1.83678	0.	0.	0.	0.	0.30576	1.26076
	C2	0.00168	0.04821	-0.01954	-1.52816	0.	0.	0.	0.	-0.01534	-1.26071
	O1	-0.00292	-0.07477	0.07999	1.21627	0.	0.	0.	0.	-0.08726	-0.28935

CH2CO

| | | Basis Function Type | | | | | | | | | |
		S1	S2	S3	S4	X1	X2	Y1	Y2	Z1	Z2
11A1	H1	0.08207	0.64625								
0.5366	H2	0.08207	0.64625								
	C1	-0.00307	-0.02598	0.45101	0.96616	0.	0.	0.	0.	0.56051	0.48843
	C2	-0.00105	0.03865	0.47157	-2.14811	0.	0.	0.	0.	-0.34121	2.29755
	O1	-0.00014	0.00841	0.08832	-0.57941	0.	0.	0.	0.	-0.33213	-0.23518
* 1B1	H1	-0.16046	-0.10317								
-0.6692	H2	0.16046	0.10317								
	C1	0.	0.	0.	0.	-0.40082	-0.06281	0.	0.	0.	0.
	C2	0.	0.	0.	0.	-0.32621	-0.05526	0.	0.	0.	0.
	O1	0.	0.	0.	0.	-0.34600	-0.13737	0.	0.	0.	0.
* 2B1	H1	-0.16678	-0.07716								
-0.5626	H2	0.16678	0.07716								
	C1	0.	0.	0.	0.	-0.35637	-0.17152	0.	0.	0.	0.
	C2	0.	0.	0.	0.	0.24028	0.09684	0.	0.	0.	0.
	O1	0.	0.	0.	0.	0.50431	0.23118	0.	0.	0.	0.
3B1	H1	0.11460	0.53903								
0.1196	H2	-0.11460	-0.53903								
	C1	0.	0.	0.	0.	-0.02652	-0.32414	0.	0.	0.	0.
	C2	0.	0.	0.	0.	-0.59016	-0.51034	0.	0.	0.	0.
	O1	0.	0.	0.	0.	0.41727	0.44917	0.	0.	0.	0.
4B1	H1	-0.05939	0.05876								
0.4075	H2	0.05939	-0.05876								
	C1	0.	0.	0.	0.	-0.68194	0.67137	0.	0.	0.	0.
	C2	0.	0.	0.	0.	-0.54053	0.60229	0.	0.	0.	0.
	O1	0.	0.	0.	0.	-0.03751	-0.03183	0.	0.	0.	0.
5B1	H1	0.08923	1.54095								
0.4894	H2	-0.08923	-1.54095								
	C1	0.	0.	0.	0.	0.40954	-3.61697	0.	0.	0.	0.
	C2	0.	0.	0.	0.	-0.54949	2.46274	0.	0.	0.	0.
	O1	0.	0.	0.	0.	-0.13055	-0.39781	0.	0.	0.	0.
6B1	H1	-0.00332	-2.84750								
0.5481	H2	0.00332	2.84750								
	C1	0.	0.	0.	0.	0.57937	3.70922	0.	0.	0.	0.
	C2	0.	0.	0.	0.	-0.58613	-1.23825	0.	0.	0.	0.
	O1	0.	0.	0.	0.	0.21680	0.33263	0.	0.	0.	0.
* 1B2	H1	0.	0.								
-0.6569	H2	0.	0.								
	C1	0.	0.	0.	0.	0.	0.	0.14064	0.03079	0.	0.
	C2	0.	0.	0.	0.	0.	0.	0.40120	0.06237	0.	0.
	O1	0.	0.	0.	0.	0.	0.	0.59469	0.23715	0.	0.
* 2B2	H1	0.	0.								
-0.3663	H2	0.	0.								
	C1	0.	0.	0.	0.	0.	0.	-0.53873	-0.28685	0.	0.
	C2	0.	0.	0.	0.	0.	0.	-0.29561	-0.11054	0.	0.
	O1	0.	0.	0.	0.	0.	0.	0.38386	0.26739	0.	0.
3B2	H1	0.	0.								
0.1926	H2	0.	0.								
	C1	0.	0.	0.	0.	0.	0.	-0.32557	-0.87344	0.	0.
	C2	0.	0.	0.	0.	0.	0.	0.60377	0.95287	0.	0.
	O1	0.	0.	0.	0.	0.	0.	-0.32315	-0.45154	0.	0.

CH2CO

Center		S1	S2	S3	S4	X1	X2	Y1	Y2	Z1	Z2
4B2	H1	0.	0.								
0.4353	H2	0.	0.								
	C1	0.	0.	0.	0.	0.	0.	0.49031	-0.24078	0.	0.
	C2	0.	0.	0.	0.	0.	0.	0.69314	-1.07720	0.	0.
	O1	0.	0.	0.	C.	0.	0.	0.06497	0.11207	0.	0.
5B2	H1	0.	0.								
0.5601	H2	0.	0.								
	C1	0.	0.	0.	0.	0.	0.	0.89028	-1.43495	0.	0.
	C2	0.	0.	0.	0.	0.	0.	-0.59683	1.22860	0.	0.
	O1	0.	0.	0.	0.	0.	0.	-0.01318	-0.11885	0.	0.

Basis Function Type

Overlap Matrix

First Function   Second Function and Overlap

H1S1 with
    H1S2 .68301   H2S1 .01018   H2S2 .10912   C1S1 .00230   C1S2 .03534   C1S3 .28834   C1S4 .37448
    C1X1 .38095   C1X2 .37967   C1Z1 -.20864   C1Z2 -.20793   C2S2 .00008   C2S3 .01316   C2S4 .09674
    C2X1 .02172   C2X2 .12741   C2Z1 -.04193   C2Z2 -.24603   O1S4 .00070   O1X2 .00291   O1Z1 -.00001
    O1Z2 -.00919

H1S2 with
    H2S1 .10912   H2S2 .31844   C1S1 .00774   C1S2 .09870   C1S3 .49975   C1S4 .70870   C1X1 .31630
    C1X2 .53240   C1Z1 -.17323   C1Z2 -.29158   C2S1 .00109   C2S2 .01508   C2S3 .12139   C2S4 .29050
    C2X1 .07947   C2X2 .24663   C2Z1 -.15346   C2Z2 -.47624   O1S1 .00002   O1S2 .00029   O1S3 .00340
    O1S4 .02055   O1X1 .00196   O1X2 .02448   O1Z1 -.00619   O1Z2 -.07720

C1S1 with
    C1S2 .35436   C1S3 .00805   C1S4 .01419   C2S3 .00194   C2S4 .00570   C2Z1 -.00586   C2Z2 -.00971
    O1S4 .00005   O1Z2 -.00076

C1S2 with
    C1S3 .26226   C1S4 .17745   C2S2 .00003   C2S3 .02923   C2S4 .07357   C2Z1 -.07717   C2Z2 -.12189
    O1S4 .00086   O1Z2 -.01107

C1S3 with
    C1S4 .79970   C2S1 .00194   C2S2 .02923   C2S3 .24862   C2S4 .40148   C2Z1 -.38126   C2Z2 -.54958
    O1S2 .00002   O1S3 .00106   O1S4 .02077   O1Z1 -.00406   O1Z2 -.10983

C1S4 with
    C2S1 .00570   C2S2 .07357   C2S3 .40148   C2S4 .63364   C2Z1 -.30965   C2Z2 -.62852   O1S1 .00036
    O1S2 .00482   O1S3 .03603   O1S4 .11063   O1Z1 -.04362   O1Z2 -.25474

C1X1 with
    C1X2 .53875   C2X1 .20596   C2X2 .30893   O1X1 .00197   O1X2 .03496

C1X2 with
    C2X1 .30893   C2X2 .70236   O1X1 .03306   O1X2 .17353

C1Y1 with
    C1Y2 .53875   C2Y1 .20596   C2Y2 .30893   O1Y1 .00197   O1Y2 .03496

C1Y2 with
    C2Y1 .30893   C2Y2 .70236   O1Y1 .03306   O1Y2 .17353

C1Z1 with
    C1Z2 .53875   C2S1 .00586   C2S2 .07717   C2S3 .38126   C2S4 .30965   C2Z1 -.35335   C2Z2 -.03411
    O1S1 .00002   O1S2 .00041   O1S3 .00791   O1S4 .05310   O1Z1 -.01917   O1Z2 -.17746

C1Z2 with
    C2S1 .00971   C2S2 .12189   C2S3 .54958   C2S4 .62852   C2Z1 -.03411   C2Z2 .20606   O1S1 .00193
    O1S2 .02530   O1S3 .15342   O1S4 .32661   O1Z1 -.11199   O1Z2 -.39285

C2S1 with
    C2S2 .35436   C2S3 .00805   C2S4 .01419   O1S3 .00068   O1S4 .00592   O1Z1 -.00345   O1Z2 -.01354

C2S2 with
    C2S3 .26226   C2S4 .17745   O1S2 .00003   O1S3 .01503   O1S4 .07704   O1Z1 -.05132   O1Z2 -.16560

C2S3 with
    C2S4 .79970   O1S1 .00234   O1S2 .03199   O1S3 .21793   O1S4 .41546   O1Z1 -.28149   O1Z2 -.61881

C2S4 with
    O1S1 .00444   O1S2 .05732   O1S3 .31618   O1S4 .58011   O1Z1 -.17312   O1Z2 -.53129

C2X1 with
    C2X2 .53875   O1X1 .16546   O1X2 .38813

C2X2 with
    O1X1 .18333   O1X2 .61967

C2Y1 with
    C2Y2 .53875   O1Y1 .16546   O1Y2 .38813

C2Y2 with
    O1Y1 .18333   O1Y2 .61967

Overlap Matrix (continued)

First Function   Second Function and Overlap

C2Z1 with
     C2Z2 .53875   O1S1 .00557   O1S2 .07224   O1S3 .37626   O1S4 .43499   O1Z1 -.30967   O1Z2 -.14909

C2Z2 with
     O1S1 .00639   O1S2 .08137   O1S3 .40837   O1S4 .61573   O1Z1 .00592   O1Z2 .17462

O1S1 with
     O1S2 .35461   O1S3 .00974   O1S4 .01469

O1S2 with
     O1S3 .28286   O1S4 .18230

O1S3 with
     O1S4 .79029

O1X1 with
     O1X2 .50607

O1Y1 with
     O1Y2 .50607

O1Z1 with
     O1Z2 .50607

CH2N2   Diazomethane

Molecular Geometry   Symmetry C$_{2v}$

Center		Coordinates		
	X	Y	Z
H1	1.82649100	0.	-3.10513601
H2	-1.82649100	0.	-3.10513601
C1	0.	0.	-2.19447199
N1	0.	0.	0.30000000
N2	0.	0.	2.41652200
Mass	0.	0.	0.12962848

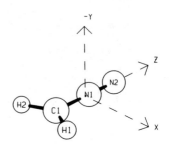

MO Symmetry Types and Occupancy

Representation	A1	A2	B1	B2
Number of MOs	20	0	8	6
Number Occupied	7	0	2	2

Energy Expectation Values

Total	-147.7702	Kinetic	147.7536
Electronic	-209.0510	Potential	-295.5237
Nuclear Repulsion	61.2808	Virial	2.0001
Electronic Potential	-356.8046		
One-Electron Potential	-467.8158		
Two-Electron Potential	111.0112		

Moments of the Charge Distribution

First

x	y	z	Dipole Moment
-0.0001	0.	2.6532	0.5601
0.	0.	-3.2133	

Second

x^2	y^2	z^2	xy	xz	yz	r^2
-18.2745	-14.7868	-105.0074	0.	0.0002	0.	-138.0688
6.6721	0.	90.1484	0.	0.	0.	96.8206

Third

x^3	xy^2	xz^2	x^2y	y^3	yz^2	x^2z	y^2z	z^3	xyz
-0.0008	-0.0001	-0.0005	0.	0.	0.	22.0285	6.1903	57.3457	0.
0.	0.	0.	0.	0.	0.	-21.5828	0.	-59.2603	0.

Fourth (even)

x^4	y^4	z^4	x^2y^2	x^2z^2	y^2z^2	r^2x^2	r^2y^2	r^2z^2	r^4
-76.8156	-57.5455	-972.7607	-19.9351	-128.6953	-84.7455	-225.4461	-162.2261	-1186.2015	-1573.8737
22.2587	0.	585.4992	0.	69.8153	0.	92.0740	0.	655.3144	747.3884

Fourth (odd)

x^3y	xy^3	xyz^2	x^3z	xy^2z	xz^3	x^2yz	y^3z	yz^3
0.	0.	0.	0.0022	0.0003	0.0016	0.	0.	0.
0.	0.	0.	0.	0.	0.	0.	0.	0.

Expectation Values at Atomic Centers

Center			Expectation Value		
	r^{-1}	δ	xr^{-3}	yr^{-3}	zr^{-3}
H1	-1.0560	0.4063	0.0516	0.	-0.0349
C1	-14.6796	116.1966	0.	0.	-0.0651
N1	-18.1810	189.1605	0.	0.	-0.0859
N2	-18.2567	189.7313	0.	0.	0.2521

	$\dfrac{3x^2-r^2}{r^5}$	$\dfrac{3y^2-r^2}{r^5}$	$\dfrac{3z^2-r^2}{r^5}$	$\dfrac{3xy}{r^5}$	$\dfrac{3xz}{r^5}$	$\dfrac{3yz}{r^5}$
H1	0.2435	-0.2051	-0.0385	0.	-0.2051	0.
C1	-0.1361	-0.4437	0.5798	0.	0.	0.
N1	-0.1941	0.2823	-0.0882	0.	0.	0.
N2	1.3466	-1.1589	-0.1877	0.	0.	0.

Mulliken Charge

Center	Basis Function Type										Total
	S1	S2	S3	S4	X1	X2	Y1	Y2	Z1	Z2	
H1	0.5042	0.2650									0.7692
C1	0.0427	1.9564	0.7549	0.4656	0.9666	0.1777	0.9612	0.4155	0.6902	0.0318	6.4625
N1	0.0433	1.9561	0.9370	0.4580	0.9815	0.3458	0.8896	0.3297	1.0944	-0.0675	6.9679
N2	0.0434	1.9559	0.9602	0.8337	0.5759	0.1335	1.0811	0.3229	1.0053	0.1191	7.0311

Overlap Populations

	Pair of Atomic Centers						
	H2	C1	N1	N1	N2	N2	N2
	H1	H1	H1	C1	H1	C1	N1
1A1	0.	0.	0.	-0.0007	0.	0.	-0.0002
2A1	0.	0.	0.	0.	0.	-0.0004	-0.0064
3A1	0.	-0.0003	0.	-0.0019	0.	-0.0001	0.
4A1	0.	0.	0.0002	0.0036	-0.0001	-0.0186	0.7094
5A1	0.0008	0.0499	0.0048	0.5987	-0.0009	-0.0927	-0.1285
6A1	0.0288	0.2955	-0.0406	-0.0380	0.0053	0.0077	-0.3364
7A1	0.0057	0.0094	0.0138	0.0196	0.0030	-0.0130	0.2070
Total A1	0.0353	0.3544	-0.0218	0.5812	0.0074	-0.1171	0.4449
1B1	-0.0218	0.1263	0.0356	0.1948	0.0015	0.0077	0.2000
2B1	-0.0299	0.2759	-0.0467	-0.3559	-0.0028	-0.0390	0.2689
Total B1	-0.0517	0.4021	-0.0111	-0.1611	-0.0013	-0.0313	0.4689
1B2	0.	0.	0.	0.1270	0.	0.0098	0.3970
2B2	0.	0.	0.	0.2161	0.	-0.1548	-0.2081
Total B2	0.	0.	0.	0.3431	0.	-0.1451	0.1889
Total	-0.0163	0.7565	-0.0328	0.7633	0.0061	-0.2935	1.1026
Distance	3.6530	2.0409	3.8641	2.4945	5.8159	4.6110	2.1165

Molecular Orbitals and Eigenvalues

Center		Basis Function Type									
		S1	S2	S3	S4	X1	X2	Y1	Y2	Z1	Z2
* 1A1	H1	0.00006	-0.00004								
-15.7495	H2	0.00006	-0.00004								
	C1	0.00000	-0.00003	0.00040	-0.00176	0.00000	-0.00000	0.	0.	0.00037	-0.00097
	N1	0.05192	0.97866	0.00355	0.00347	-0.00000	0.00000	0.	0.	0.00098	-0.00109
	N2	0.00128	0.02412	0.00060	-0.00041	-0.00000	-0.00000	0.	0.	-0.00073	0.00067
* 2A1	H1	0.00003	0.00005								
-15.6941	H2	0.00003	0.00005								
	C1	0.00000	-0.00002	-0.00049	-0.00486	0.00000	-0.00000	0.	0.	-0.00050	-0.00136
	N1	-0.00131	-0.02474	0.00044	-0.00529	-0.00000	0.00000	0.	0.	0.00027	-0.00889
	N2	0.05194	0.97883	0.00168	0.01270	0.00000	-0.00000	0.	0.	-0.00333	-0.00263
* 3A1	H1	0.00075	-0.00103								
-11.2740	H2	0.00075	-0.00103								
	C1	0.05191	0.97897	0.00473	0.00477	0.00000	0.00000	0.	0.	0.00186	0.00016
	N1	-0.00002	-0.00036	0.00073	-0.00147	0.00000	-0.00000	0.	0.	-0.00052	0.00282
	N2	-0.00000	0.00002	0.00029	-0.00199	-0.00000	0.00000	0.	0.	0.00000	0.00036
* 4A1	H1	0.00803	-0.00177								
-1.5093	H2	0.00803	-0.00177								
	C1	-0.00149	-0.03183	0.07145	-0.03636	-0.00000	0.00000	0.	0.	0.06027	-0.02757
	N1	-0.00870	-0.18551	0.42208	0.17120	-0.00000	-0.00000	0.	0.	0.20433	-0.07861
	N2	-0.00730	-0.15682	0.34164	0.20843	-0.00000	0.00000	0.	0.	-0.23453	-0.00100
* 5A1	H1	-0.06964	-0.00910								
-1.1257	H2	-0.06964	-0.00912								
	C1	0.00721	0.15573	-0.33289	-0.22052	0.00000	-0.00002	0.	0.	-0.15870	-0.00327
	N1	0.00424	0.09015	-0.21519	-0.18984	-0.00000	0.00001	0.	0.	0.42946	-0.07481
	N2	-0.00422	-0.09260	0.19143	0.23350	-0.00000	-0.00000	0.	0.	-0.07931	-0.02291
* 6A1	H1	-0.18193	-0.10138								
-0.7512	H2	-0.18192	-0.10149								
	C1	0.00559	0.12080	-0.27740	-0.20319	-0.00001	-0.00007	0.	0.	0.33092	0.02406
	N1	-0.00434	-0.09204	0.23010	0.26960	-0.00001	0.00003	0.	0.	-0.14961	0.05149
	N2	0.00385	0.08351	-0.18927	-0.30778	-0.00001	-0.00001	0.	0.	-0.11736	-0.00251
* 7A1	H1	-0.06897	-0.04801								
-0.6528	H2	-0.06897	-0.04806								
	C1	0.00078	0.01756	-0.03749	0.08946	0.00000	-0.00003	0.	0.	0.20504	0.07022
	N1	0.00156	0.03150	-0.09496	-0.02592	0.00000	0.00001	0.	0.	-0.35528	0.14562
	N2	-0.00593	-0.12619	0.31032	0.30215	0.00000	-0.00000	0.	0.	0.49978	0.14908
8A1	H1	-0.01561	-0.56955								
0.2625	H2	-0.01561	-0.57065								
	C1	-0.00061	-0.00322	0.09851	-1.37905	0.00002	-0.00085	0.	0.	-0.03642	-2.15936
	N1	-0.00232	-0.05656	0.08019	0.94383	0.00004	0.00029	0.	0.	-0.09386	1.82347
	N2	-0.00140	-0.03812	0.02060	1.54266	0.00000	-0.00013	0.	0.	0.01773	-0.45926
9A1	H1	0.05482	1.53037								
0.3012	H2	0.05487	1.53466								
	C1	0.00594	0.13538	-0.29009	-3.09116	-0.00015	0.00344	0.	0.	0.15956	-0.08312
	N1	-0.00157	-0.04254	0.02577	0.76509	-0.00017	-0.00110	0.	0.	-0.08804	-0.85951
	N2	0.00051	0.00367	-0.07816	0.53306	-0.00007	0.00049	0.	0.	0.09629	-0.11527
10A1	H1	0.03655	0.27610								
0.4031	H2	0.03652	0.27587								
	C1	-0.00343	-0.06042	0.28846	-2.21094	0.00011	-0.00033	0.	0.	0.30647	-0.20798
	N1	0.00343	0.09451	-0.05200	-3.01686	0.00001	0.00008	0.	0.	0.06497	-3.30048
	N2	-0.00516	-0.12379	0.20025	4.01890	0.00003	-0.00004	0.	0.	-0.15204	-2.12962

CH2N2

Center		S1	S2	S3	S4	X1	X2	Y1	Y2	Z1	Z2
11A1	H1	-0.10568	-0.38465								
0.5399	H2	-0.10569	-0.38347								
	C1	0.00518	0.05869	-0.65946	1.27001	0.00023	0.00053	0.	0.	-0.31648	-0.24944
	N1	0.00052	-0.00808	-0.16099	-0.90150	-0.00016	-0.00003	0.	0.	-0.31330	0.14466
	N2	0.00111	-0.01007	-0.29308	0.92083	-0.00007	0.00014	0.	0.	0.45588	-1.44995
12A1	H1	0.26927	0.41301								
0.6782	H2	0.26928	0.41341								
	C1	-0.00344	0.00987	0.76869	-3.31204	0.00000	0.00038	0.	0.	-0.88853	0.10873
	N1	-0.00517	-0.09182	0.42354	0.33797	0.00012	-0.00028	0.	0.	-0.13542	-1.63424
	N2	-0.00002	-0.00300	-0.01629	1.13240	0.00026	-0.00023	0.	0.	0.33665	-0.53798
* 1B1	H1	0.13312	0.09434								
-0.7081	H2	-0.13313	-0.09427								
	C1	0.00000	0.00000	0.00000	-0.00008	0.36076	0.01800	0.	0.	0.00000	0.00001
	N1	-0.00000	-0.00000	0.00001	0.00001	0.46725	0.17041	0.	0.	-0.00000	-0.00002
	N2	0.00000	0.00000	-0.00001	0.00000	0.26893	0.05230	0.	0.	-0.00001	-0.00000
* 2B1	H1	0.19826	0.10017								
-0.5511	H2	-0.19826	-0.10012								
	C1	0.00000	0.00000	0.00000	-0.00006	0.40781	0.18481	0.	0.	0.00000	0.00001
	N1	0.00000	-0.00000	-0.00000	0.00000	-0.36871	-0.20897	0.	0.	-0.00000	-0.00001
	N2	0.00000	-0.00000	-0.00000	0.00001	-0.33820	-0.09274	0.	0.	0.00000	-0.00000
3B1	H1	0.08475	0.28534								
0.1241	H2	-0.08475	-0.28554								
	C1	-0.00000	-0.00001	-0.00000	0.00017	-0.04586	-0.05712	0.	0.	-0.00001	-0.00005
	N1	0.00000	0.00000	-0.00000	-0.00002	-0.48159	-0.56522	0.	0.	0.00000	0.00001
	N2	-0.00000	-0.00000	0.00000	0.00000	0.53611	0.67877	0.	0.	-0.00000	-0.00000
4B1	H1	-0.06547	-1.78853								
0.3576	H2	0.06540	1.78604								
	C1	-0.00001	-0.00013	0.00045	0.00164	-0.41181	3.30106	0.	0.	-0.00012	-0.00032
	N1	0.00000	0.00003	0.00000	-0.00073	-0.14850	-0.92796	0.	0.	0.00007	-0.00013
	N2	-0.00000	-0.00003	0.00012	0.00034	-0.16391	0.44559	0.	0.	-0.00012	-0.00019
5B1	H1	0.04984	-1.84981								
0.4994	H2	-0.04979	1.84943								
	C1	-0.00000	-0.00005	0.00039	-0.00007	0.85850	1.45816	0.	0.	-0.00003	-0.00002
	N1	-0.00000	-0.00000	0.00007	0.00044	-0.37999	-0.26088	0.	0.	0.00008	0.00013
	N2	-0.00000	0.00001	0.00010	-0.00049	-0.05440	0.32717	0.	0.	-0.00013	0.00053
* 1B2	H1	0.	0.								
-0.6803	H2	0.	0.								
	C1	0.	0.	0.	0.	0.	0.	-0.18331	-0.02312	0.	0.
	N1	0.	0.	0.	0.	0.	0.	-0.54131	-0.19365	0.	0.
	N2	0.	0.	0.	0.	0.	0.	-0.44181	-0.10228	0.	0.
* 2B2	H1	0.	0.								
-0.3220	H2	0.	0.								
	C1	0.	0.	0.	0.	0.	0.	-0.55181	-0.32240	0.	0.
	N1	0.	0.	0.	0.	0.	0.	-0.12205	-0.09306	0.	0.
	N2	0.	0.	0.	0.	0.	0.	0.46094	0.29020	0.	0.
3B2	H1	0.	0.								
0.1497	H2	0.	0.								
	C1	0.	0.	0.	0.	0.	0.	0.30248	0.68060	0.	0.
	N1	0.	0.	0.	0.	0.	0.	-0.55429	-0.76842	0.	0.
	N2	0.	0.	0.	0.	0.	0.	0.38907	0.58969	0.	0.

Center		S1	S2	S3	S4	X1	X2	Basis Function Type Y1	Y2	Z1	Z2
4B2	H1	0.	0.								
0.4887	H2	0.	0.								
	C1	0.	0.	0.	0.	0.	0.	-0.96543	1.08130	0.	0.
	N1	0.	0.	0.	0.	0.	0.	0.02491	-0.09570	0.	0.
	N2	0.	0.	0.	0.	0.	0.	-0.25404	0.17701	0.	0.
5B2	H1	0.	0.								
0.6564	H2	0.	0.								
	C1	0.	0.	0.	0.	0.	0.	0.23727	-0.60018	0.	0.
	N1	0.	0.	0.	0.	0.	0.	-0.44575	0.62560	0.	0.
	N2	0.	0.	0.	0.	0.	0.	-0.72310	0.80023	0.	0.

Overlap Matrix

First Function	Second Function and Overlap

H1S1 with

H1S2	.68301	H2S1	.00875	H2S2	.10229	C1S1	.00234	C1S2	.03581	C1S3	.29023	C1S4	.37554
C1X1	.39026	C1X2	.38721	C1Z1	-.19458	C1Z2	-.19306	N1S2	.00005	N1S3	.00601	N1S4	.06606
N1X1	.01011	N1X2	.10704	N1Z1	-.01885	N1Z2	-.19956	N2S3	.00001	N2S4	.00324	N2X1	.00004
N2X2	.00854	N2Z1	-.00011	N2Z2	-.02582						

H1S2 with

H2S1	.10229	H2S2	.30580	C1S1	.00777	C1S2	.09909	C1S3	.50124	C1S4	.71003	C1X1	.32279
C1X2	.54261	C1Z1	-.16094	C1Z2	-.27054	N1S1	.00091	N1S2	.01230	N1S3	.09214	N1S4	.23820
N1X1	.05455	N1X2	.21398	N1Z1	-.10169	N1Z2	-.39892	N2S1	.00003	N2S2	.00048	N2S3	.00651
N2S4	.03900	N2X1	.00416	N2X2	.04244	N2Z1	-.01257	N2Z2	-.12829		

C1S1 with

C1S2	.35436	C1S3	.00805	C1S4	.01419	N1S3	.00070	N1S4	.00503	N1Z1	-.00316	N1Z2	-.01118
N2S4	.00022	N2Z2	-.00172								

C1S2 with

C1S3	.26226	C1S4	.17745	N1S2	.00001	N1S3	.01320	N1S4	.06547	N1Z1	-.04532	N1Z2	-.13898
N2S4	.00336	N2Z1	-.00008	N2Z2	-.02356						

C1S3 with

C1S4	.79970	N1S1	.00149	N1S2	.02164	N1S3	.18566	N1S4	.36843	N1Z1	-.28060	N1Z2	-.58335
N2S2	.00003	N2S3	.00288	N2S4	.04263	N2Z1	-.01064	N2Z2	-.17352		

C1S4 with

N1S1	.00447	N1S2	.05753	N1S3	.32308	N1S4	.57149	N1Z1	-.22140	N1Z2	-.58800	N2S1	.00048
N2S2	.00659	N2S3	.05239	N2S4	.15572	N2Z1	-.06852	N2Z2	-.33581		

C1X1 with

C1X2	.53875	N1X1	.14849	N1X2	.32404	N2X1	.00440	N2X2	.05476

C1X2 with

N1X1	.21600	N1X2	.62940	N2X1	.05077	N2X2	.22741

C1Y1 with

C1Y2	.53875	N1Y1	.14849	N1Y2	.32404	N2Y1	.00440	N2Y2	.05476

C1Y2 with

N1Y1	.21600	N1Y2	.62940	N2Y1	.05077	N2Y2	.22741

C1Z1 with

C1Z2	.53875	N1S1	.00455	N1S2	.05958	N1S3	.33150	N1S4	.36236	N1Z1	-.32671	N1Z2	-.15708
N2S1	.00004	N2S2	.00067	N2S3	.01537	N2S4	.08356	N2Z1	-.03717	N2Z2	-.21827		

C1Z2 with

N1S1	.00766	N1S2	.09666	N1S3	.47241	N1S4	.64389	N1Z1	-.04329	N1Z2	.09919	N2S1	.00253
N2S2	.03302	N2S3	.20114	N2S4	.39185	N2Z1	-.15672	N2Z2	-.42717		

N1S1 with

N1S2	.35584	N1S3	.00894	N1S4	.01446	N2S3	.00178	N2S4	.00570	N2Z1	-.00540	N2Z2	-.01000

N1S2 with

N1S3	.27246	N1S4	.17925	N2S2	.00003	N2S3	.02706	N2S4	.07302	N2Z1	-.07107	N2Z2	-.12427

N1S3 with

N1S4	.79352	N2S1	.00178	N2S2	.02706	N2S3	.23437	N2S4	.39137	N2Z1	-.35821	N2Z2	-.54889

N1S4 with

N2S1	.00570	N2S2	.07302	N2S3	.39137	N2S4	.62772	N2Z1	-.29519	N2Z2	-.62844

N1X1 with

N1X2	.52739	N2X1	.18485	N2X2	.29355

N1X2 with

N2X1	.29355	N2X2	.69046

N1Y1 with

N1Y2	.52739	N2Y1	.18485	N2Y2	.29355

N1Y2 with

N2Y1	.29355	N2Y2	.69046

Overlap Matrix (continued)

First Function   Second Function and Overlap

N1Z1 with
     N1Z2 .52739   N2S1 .00540   N2S2 .07107   N2S3 .35821   N2S4 .29519   N2Z1 -.34929   N2Z2 -.04987

N1Z2 with
     N2S1 .01000   N2S2 .12427   N2S3 .54889   N2S4 .62844   N2Z1 -.04987   N2Z2 .17896

N2S1 with
     N2S2 .35584   N2S3 .00894   N2S4 .01446

N2S2 with
     N2S3 .27246   N2S4 .17925

N2S3 with
     N2S4 .79352

N2X1 with
     N2X2 .52739

N2Y1 with
     N2Y2 .52739

N2Z1 with
     N2Z2 .52739

NNO   Nitrous Oxide

Molecular Geometry   Symmetry C-ᵥ

Center		Coordinates	
	X	Y	Z
N1	0.	0.	-2.13162401
N2	0.	0.	0.
O1	0.	0.	2.23744929
Mass	0.	0.	0.13496402

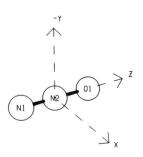

MO Symmetry Types and Occupancy

Representation	SIGMA	PI-X	PI-Y
Number of MOs	18	6	6
Number Occupied	7	2	2

Energy Expectation Values

Total	-183.5761	Kinetic	183.6301
Electronic	-244.4091	Potential	-367.2062
Nuclear Repulsion	60.8330	Virial	1.9997

Electronic Potential	-428.0392
One-Electron Potential	-553.5657
Two-Electron Potential	125.5265

Moments of the Charge Distribution

First

x	y	z	Dipole Moment
0.	0.	-0.3984	0.3894
0.	0.	0.0090	

Second

x^2	y^2	z^2	xy	xz	yz	r^2
-11.4665	-11.4665	-87.2748	0.	0.	0.	-110.2078
0.	0.	71.4530	0.	0.	0.	71.4530

Third

x^3	xy^2	xz^2	x^2y	y^3	yz^2	x^2z	y^2z	z^3	xyz
0.	0.	0.	0.	0.	0.	0.8952	0.8952	12.1906	0.
0.	0.	0.	0.	0.	0.	0.	0.	-7.1768	0.

Fourth (even)

x^4	y^4	z^4	x^2y^2	x^2z^2	y^2z^2	r^2x^2	r^2y^2	r^2z^2	r^4
-32.2449	-32.2449	-635.0220	-10.7483	-51.9370	-51.9370	-94.9301	-94.9301	-738.8960	-928.7562
0.	0.	341.0766	0.	0.	0.	0.	0.	341.0766	341.0766

Expectation Values at Atomic Centers

Center			Expectation Value		
	r^{-1}	δ	xr^{-3}	yr^{-3}	zr^{-3}
N1	-18.2000	189.7978	0.	0.	-0.2553
N2	-18.0722	189.2004	0.	0.	0.0134
O1	-22.2494	287.5964	0.	0.	0.3092

Center	$\dfrac{3x^2-r^2}{r^5}$	$\dfrac{3y^2-r^2}{r^5}$	$\dfrac{3z^2-r^2}{r^5}$	$\dfrac{3xy}{r^5}$	$\dfrac{3xz}{r^5}$	$\dfrac{3yz}{r^5}$
N1	-0.0200	-0.0200	0.0400	0.	0.	0.
N2	0.0117	0.0117	-0.0235	0.	0.	0.
O1	-1.5337	-1.5337	3.0673	0.	0.	0.

Mulliken Charge

Center				Basis Function Type							Total
	S1	S2	S3	S4	X1	X2	Y1	Y2	Z1	Z2	
N1	0.0435	1.9559	0.9732	0.8312	0.8410	0.2348	0.8410	0.2348	0.9462	0.0672	6.9687
N2	0.0433	1.9561	0.9755	0.3014	0.9329	0.2666	0.9329	0.2666	1.0645	-0.0024	6.7373
O1	0.0427	1.9566	0.9868	0.8875	1.2612	0.4635	1.2612	0.4635	0.8026	0.1684	8.2940

Overlap Populations

Pair of Atomic Centers

	N2 N1	O1 N1	O1 N2
1SIGMA	0.	-0.0001	-0.0018
2SIGMA	-0.0010	0.	0.0001
3SIGMA	-0.0044	0.	0.
4SIGMA	0.2418	0.0184	0.4409
5SIGMA	0.4901	-0.0308	0.1754
6SIGMA	-0.5920	0.0128	-0.1592
7SIGMA	-0.0167	0.0089	0.0347
Total SIGMA	0.1179	0.0092	0.4901
1PI-X	0.2433	0.0104	0.2625
2PI-X	0.1729	-0.0782	-0.2111
Total PI-X	0.4162	-0.0678	0.0514
1PI-Y	0.2433	0.0104	0.2625
2PI-Y	0.1729	-0.0782	-0.2111
Total PI-Y	0.4162	-0.0678	0.0514
Total	0.9503	-0.1263	0.5930
Distance	2.1316	4.3691	2.2374

Molecular Orbitals and Eigenvalues

	Center	S1	S2	S3	S4	X1	X2	Y1	Y2	Z1	Z2
* 1SIGMA	N1	-0.00000	0.00001	0.00013	-0.00300	0.	0.	0.	0.	-0.00024	-0.00051
-20.6577	N2	0.00000	0.00004	-0.00013	-0.00042	0.	0.	0.	0.	-0.00040	-0.00386
	O1	0.05113	0.97939	0.00270	0.00561	0.	0.	0.	0.	-0.00237	-0.00046
* 2SIGMA	N1	-0.00056	-0.01052	-0.00099	0.00251	0.	0.	0.	0.	-0.00078	0.00113
-15.8632	N2	-0.05193	-0.97894	-0.00379	-0.00295	0.	0.	0.	0.	0.00011	0.00133
	O1	0.00001	0.00030	-0.00020	-0.00038	0.	0.	0.	0.	0.00062	-0.00012
* 3SIGMA	N1	0.05196	0.97912	0.00215	0.00898	0.	0.	0.	0.	0.00339	0.00154
-15.7428	N2	-0.00059	-0.01110	0.00104	-0.00577	0.	0.	0.	0.	-0.00006	0.00527
	O1	-0.00000	0.00003	0.00046	-0.00237	0.	0.	0.	0.	0.00010	-0.00016
* 4SIGMA	N1	0.00408	0.08561	-0.20366	-0.05174	0.	0.	0.	0.	-0.14197	0.01124
-1.6483	N2	0.00876	0.18443	-0.43888	-0.13758	0.	0.	0.	0.	-0.05063	0.02395
	O1	0.00658	0.14394	-0.31401	-0.23347	0.	0.	0.	0.	0.16125	0.02066
* 5SIGMA	N1	-0.00686	-0.14807	0.31703	0.16944	0.	0.	0.	0.	0.19389	-0.01069
-1.4674	N2	-0.00273	-0.05787	0.13746	0.03726	0.	0.	0.	0.	-0.43576	0.02890
	O1	0.00622	0.13658	-0.29693	-0.22710	0.	0.	0.	0.	0.09945	-0.00820
* 6SIGMA	N1	-0.00430	-0.09400	0.20425	0.38228	0.	0.	0.	0.	-0.08978	0.03096
-0.8221	N2	0.00544	0.11189	-0.31277	-0.32497	0.	0.	0.	0.	-0.24759	0.10245
	O1	-0.00651	-0.14237	0.32899	0.40833	0.	0.	0.	0.	0.40174	0.10402
* 7SIGMA	N1	-0.00634	-0.13524	0.32891	0.40744	0.	0.	0.	0.	-0.47984	-0.10629
-0.6960	N2	0.00181	0.03734	-0.10759	-0.06568	0.	0.	0.	0.	0.35328	-0.06333
	O1	0.00172	0.03694	-0.09226	-0.08796	0.	0.	0.	0.	-0.22313	-0.07266
8SIGMA	N1	-0.00396	-0.06098	0.38694	-0.52779	0.	0.	0.	0.	0.41069	0.09704
0.3269	N2	0.00354	0.10090	-0.02970	-1.96445	0.	0.	0.	0.	-0.15434	-1.34988
	O1	-0.00474	-0.11375	0.18615	1.69311	0.	0.	0.	0.	-0.43131	-1.00285
9SIGMA	N1	0.00338	0.09707	-0.02232	-3.69730	0.	0.	0.	0.	0.05728	-2.29957
0.4161	N2	-0.00352	-0.08247	0.15234	2.84680	0.	0.	0.	0.	-0.09076	-2.83203
	O1	-0.00080	-0.03055	-0.04817	1.27087	0.	0.	0.	0.	0.16182	-0.10597
10SIGMA	N1	-0.00270	-0.02736	0.36240	1.22666	0.	0.	0.	0.	0.33702	-0.57591
0.6138	N2	-0.00045	0.01643	0.20558	0.26912	0.	0.	0.	0.	-0.38346	2.66450
	O1	0.00095	0.03952	0.07649	-1.74430	0.	0.	0.	0.	-0.40538	0.65402
* 1PI-X	N1	0.	0.	0.	0.	0.30995	0.05105	0.	0.	0.	0.
-0.7849	N2	0.	0.	0.	0.	0.55288	0.17224	0.	0.	0.	0.
	O1	0.	0.	0.	0.	0.37310	0.10645	0.	0.	0.	0.
* 2PI-X	N1	0.	0.	0.	0.	0.45060	0.19692	0.	0.	0.	0.
-0.4870	N2	0.	0.	0.	0.	0.16812	0.07460	0.	0.	0.	0.
	O1	0.	0.	0.	0.	-0.59603	-0.33980	0.	0.	0.	0.
3PI-X	N1	0.	0.	0.	0.	0.49474	0.62546	0.	0.	0.	0.
0.1213	N2	0.	0.	0.	0.	-0.61332	-0.62808	0.	0.	0.	0.
	O1	0.	0.	0.	0.	0.32202	0.35338	0.	0.	0.	0.
4PI-X	N1	0.	0.	0.	0.	-0.57493	0.42386	0.	0.	0.	0.
0.5703	N2	0.	0.	0.	0.	-0.58779	0.89707	0.	0.	0.	0.
	O1	0.	0.	0.	0.	-0.19009	-0.05420	0.	0.	0.	0.
* 1PI-Y	N1	0.	0.	0.	0.	0.	0.	0.30995	0.05105	0.	0.
-0.7849	N2	0.	0.	0.	0.	0.	0.	0.55288	0.17224	0.	0.
	O1	0.	0.	0.	0.	0.	0.	0.37310	0.10645	0.	0.

		S1	S2	S3	S4	X1	X2	Y1	Y2	Z1	Z2
• 2PI-Y	N1	0.	0.	0.	0.	0.	0.	0.45060	0.19692	0.	0.
-0.4870	N2	0.	0.	0.	0.	0.	0.	0.16812	0.07460	0.	0.
	O1	0.	0.	0.	0.	0.	0.	-0.59603	-0.33980	0.	0.
3PI-Y	N1	0.	0.	0.	0.	0.	0.	0.49474	0.62546	0.	0.
0.1213	N2	0.	0.	0.	0.	0.	0.	-0.61332	-0.62808	0.	0.
	O1	0.	0.	0.	0.	0.	0.	0.32202	0.35338	0.	0.
4PI-Y	N1	0.	0.	0.	0.	0.	0.	-0.57493	0.42386	0.	0.
0.5703	N2	0.	0.	0.	0.	0.	0.	-0.58779	0.89707	0.	0.
	O1	0.	0.	0.	0.	0.	0.	-0.19009	-0.05420	0.	0.

Overlap Matrix

First Function   Second Function and Overlap

N1S1 with
   N1S2 .35584   N1S3 .00894   N1S4 .01446   N2S3 .00170   N2S4 .00563   N2Z1 -.00525   N2Z2 -.00996
   O1S4 .00008   O1Z2 -.00101

N1S2 with
   N1S3 .27246   N1S4 .17925   N2S2 .00003   N2S3 .02603   N2S4 .07208   N2Z1 -.06923   N2Z2 -.12387
   O1S4 .00131   O1Z1 -.00001   O1Z2 -.01398

N1S3 with
   N1S4 .79352   N2S1 .00170   N2S2 .02603   N2S3 .22946   N2S4 .38743   N2Z1 -.35327   N2Z2 -.54815
   O1S3 .00061   O1S4 .02014   O1Z1 -.00310   O1Z2 -.11249

N1S4 with
   N2S1 .00563   N2S2 .07208   N2S3 .38743   N2S4 .62355   N2Z1 -.29433   N2Z2 -.62920   O1S1 .00022
   O1S2 .00310   O1S3 .02741   O1S4 .09989   O1Z1 -.04165   O1Z2 -.26640

N1X1 with
   N1X2 .52739   N2X1 .18107   N2X2 .29110   O1X1 .00120   O1X2 .03047

N1X2 with
   N2X1 .29110   N2X2 .68680   O1X1 .02988   O1X2 .16526

N1Y1 with
   N1Y2 .52739   N2Y1 .18107   N2Y2 .29110   O1Y1 .00120   O1Y2 .03047

N1Y2 with
   N2Y1 .29110   N2Y2 .68680   O1Y1 .02988   O1Y2 .16526

N1Z1 with
   N1Z2 .52739   N2S1 .00525   N2S2 .06923   N2S3 .35327   N2S4 .29433   N2Z1 -.34853   N2Z2 -.05431
   O1S2 .00008   O1S3 .00403   O1S4 .04250   O1Z1 -.01302   O1Z2 -.15191

N1Z2 with
   N2S1 .00996   N2S2 .12387   N2S3 .54815   N2S4 .62920   N2Z1 -.05431   N2Z2 .17073   O1S1 .00149
   O1S2 .01979   O1S3 .12879   O1S4 .29020   O1Z1 -.12645   O1Z2 -.42298

N2S1 with
   N2S2 .35584   N2S3 .00894   N2S4 .01446   O1S3 .00045   O1S4 .00443   O1Z1 -.00242   O1Z2 -.01048

N2S2 with
   N2S3 .27246   N2S4 .17925   O1S3 .00901   O1S4 .05757   O1Z1 -.03526   O1Z2 -.12988

N2S3 with
   N2S4 .79352   O1S1 .00100   O1S2 .01529   O1S3 .14887   O1S4 .32850   O1Z1 -.24504   O1Z2 -.55884

N2S4 with
   O1S1 .00413   O1S2 .05340   O1S3 .29717   O1S4 .53950   O1Z1 -.21947   O1Z2 -.60179

N2X1 with
   N2X2 .52739   O1X1 .11864   O1X2 .28193

N2X2 with
   O1X1 .20691   O1X2 .61431

N2Y1 with
   N2Y2 .52739   O1Y1 .11864   O1Y2 .28193

N2Y2 with
   O1Y1 .20691   O1Y2 .61431

N2Z1 with
   N2Z2 .52739   O1S1 .00346   O1S2 .04620   O1S3 .27529   O1S4 .32646   O1Z1 -.30778   O1Z2 -.16795

N2Z2 with
   O1S1 .00782   O1S2 .09886   O1S3 .47268   O1S4 .64568   O1Z1 -.07857   O1Z2 .04086

O1S1 with
   O1S2 .35461   O1S3 .00974   O1S4 .01469

O1S2 with
   O1S3 .28286   O1S4 .18230

O1S3 with
   O1S4 .79029

Overlap Matrix (continued)

01X1 with
    01X2 .50607

01Y1 with
    01Y2 .50607

01Z1 with
    01Z2 .50607

NH2CN   Cyanamide

Molecular Geometry   Symmetry C$_{2v}$

Center	Coordinates		
	X	Y	Z
H1	1.53674001	0.	-3.39682999
H2	-1.53674001	0.	-3.39682999
C1	0.	0.	0.
N1	0.	0.	-2.50959000
N2	0.	0.	2.22613001
Mass	0.	0.	-0.25739006

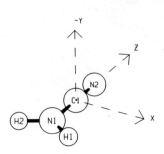

MO Symmetry Types and Occupancy

Representation	A1	A2	B1	B2
Number of MOs	20	0	8	6
Number Occupied	7	0	2	2

Energy Expectation Values

Total	-147.8437	Kinetic	147.9541
Electronic	-207.6285	Potential	-295.7977
Nuclear Repulsion	59.7849	Virial	1.9993

Electronic Potential	-355.5826
One-Electron Potential	-465.8260
Two-Electron Potential	110.2434

Moments of the Charge Distribution

First

x	y	z	Dipole Moment
0.	0.	1.1845	1.9308
0.	0.	-3.1153	

Second

x^2	y^2	z^2	xy	xz	yz	r^2
-16.0161	-14.1416	-112.4733	0.	0.	0.	-142.6310
4.7231	0.	98.7916	0.	0.	0.	103.5147

Third

x^3	xy^2	xz^2	x^2y	y^3	yz^2	x^2z	y^2z	z^3	xyz
0.	0.	0.	0.	0.	0.	8.2206	0.8308	7.9854	0.
0.	0.	0.	0.	0.	0.	-14.8280	0.	-34.5251	0.

Fourth (even)

x^4	y^4	z^4	x^2y^2	x^2z^2	y^2z^2	r^2x^2	r^2y^2	r^2z^2	r^4
-57.5047	-49.8165	-1005.6471	-16.9051	-100.1960	-79.1596	-174.6058	-145.8811	-1185.0028	-1505.4896
11.1540	0.	640.7146	0.	46.5517	0.	57.7057	0.	687.2663	744.9719

Expectation Values at Atomic Centers

Center			Expectation Value		
	r^{-1}	δ	xr^{-3}	yr^{-3}	zr^{-3}
H1	-0.9641	0.4414	0.1297	0.	-0.0777
C1	-14.6380	116.6077	0.	0.	-0.0032
N1	-18.3013	188.3462	0.	0.	-0.0295
N2	-18.3416	189.5962	0.	0.	0.2053

	$\frac{3x^2-r^2}{r^5}$	$\frac{3y^2-r^2}{r^5}$	$\frac{3z^2-r^2}{r^5}$	$\frac{3xy}{r^5}$	$\frac{3xz}{r^5}$	$\frac{3yz}{r^5}$
H1	0.4268	-0.3883	-0.0385	0.	-0.4150	0.
C1	-0.0906	0.1926	-0.1020	0.	0.	0.
N1	0.4370	-1.4151	0.9781	0.	0.	0.
N2	0.6838	-0.2287	-0.4551	0.	0.	0.

Mulliken Charge

Center					Basis Function Type						Total
	S1	S2	S3	S4	X1	X2	Y1	Y2	Z1	Z2	
H1	0.5166	0.0968									0.6133
C1	0.0429	1.9565	0.8200	0.1468	0.8178	0.2379	0.7111	0.1612	0.9483	0.0764	5.9189
N1	0.0431	1.9560	0.8345	0.6395	1.0329	0.2445	1.2803	0.5996	0.9235	0.1689	7.7227
N2	0.0434	1.9560	0.9554	0.7233	0.7206	0.2518	0.9109	0.3369	1.0525	0.1809	7.1318

Overlap Populations

	Pair of Atomic Centers						
	H2	C1	N1	N1	N2	N2	N2
	H1	H1	H1	C1	H1	C1	N1
1A1	0.	0.	-0.0001	-0.0007	0.	0.	0.
2A1	0.	0.	0.	0.	0.	-0.0020	0.
3A1	0.	0.	0.	-0.0004	0.	0.0006	0.
4A1	0.0013	0.0056	0.1059	0.3173	0.0002	0.1330	0.0353
5A1	0.0011	0.0033	0.0479	0.0615	0.	0.6192	0.0027
6A1	0.0212	-0.0552	0.1739	0.3293	-0.0011	0.0475	0.0003
7A1	-0.0001	-0.0002	0.0011	-0.0344	0.	-0.0220	-0.0036
Total A1	0.0236	-0.0465	0.3286	0.6726	-0.0009	0.7763	0.0347
1B1	-0.0443	0.0082	0.3315	0.0804	0.0003	0.0088	0.0029
2B1	-0.0066	-0.0240	0.0429	-0.1943	-0.0014	0.5258	-0.0219
Total B1	-0.0509	-0.0157	0.3743	-0.1139	-0.0012	0.5347	-0.0190
1B2	0.	0.	0.	0.3137	0.	0.1813	0.0255
2B2	0.	0.	0.	-0.3642	0.	0.3336	-0.0794
Total B2	0.	0.	0.	-0.0504	0.	0.5149	-0.0539
Total	-0.0273	-0.0623	0.7029	0.5083	-0.0020	1.8258	-0.0382
Distance	3.0735	3.7283	1.7745	2.5096	5.8292	2.2261	4.7357

Molecular Orbitals and Eigenvalues

Center		S1	S2	S3	S4	X1	X2	Y1	Y2	Z1	Z2
* 1A1	H1	-0.00039	0.00054								
-15.6140	H2	-0.00039	0.00054								
	C1	-0.00000	0.00002	-0.00009	0.00129	0.	0.	0.	0.	0.00007	-0.00098
	N1	-0.05193	-0.97885	-0.00405	-0.00345	0.	0.	0.	0.	-0.00051	-0.00020
	N2	0.00005	0.00092	-0.00001	0.00012	0.	0.	0.	0.	-0.00002	0.00028
* 2A1	H1	0.00009	-0.00051								
-15.5940	H2	0.00009	-0.00051								
	C1	0.00001	0.00002	0.00002	-0.00194	0.	0.	0.	0.	-0.00019	-0.00332
	N1	0.00005	0.00093	0.00017	-0.00115	0.	0.	0.	0.	-0.00003	-0.00026
	N2	0.05196	0.97917	0.00209	0.00604	0.	0.	0.	0.	-0.00256	-0.00011
* 3A1	H1	0.00007	-0.00035								
-11.3393	H2	0.00007	-0.00035								
	C1	0.05190	0.97897	0.00471	0.00104	0.	0.	0.	0.	-0.00057	-0.00055
	N1	-0.00002	-0.00034	0.00052	-0.00056	0.	0.	0.	0.	0.00070	-0.00065
	N2	-0.00004	-0.00085	0.00100	0.00039	0.	0.	0.	0.	-0.00154	0.00016
* 4A1	H1	-0.11022	0.00190								
-1.3028	H2	-0.11022	0.00190								
	C1	0.00615	0.13221	-0.28435	-0.03853	0.	0.	0.	0.	0.05185	0.00346
	N1	0.00868	0.18755	-0.41009	-0.30807	0.	0.	0.	0.	-0.05335	-0.00319
	N2	0.00356	0.07612	-0.17308	-0.11436	0.	0.	0.	0.	0.10276	0.00859
* 5A1	H1	0.07900	0.00290								
-1.2151	H2	0.07900	0.00290								
	C1	0.00457	0.09787	-0.21711	-0.07318	0.	0.	0.	0.	-0.32956	-0.01315
	N1	-0.00492	-0.10605	0.23349	0.12961	0.	0.	0.	0.	-0.03087	-0.03718
	N2	0.00814	0.17496	-0.38706	-0.15947	0.	0.	0.	0.	0.19441	-0.03494
* 6A1	H1	0.15140	0.05786								
-0.8284	H2	0.15140	0.05786								
	C1	0.00498	0.10589	-0.25517	-0.08057	0.	0.	0.	0.	0.33120	0.02664
	N1	-0.00147	-0.03159	0.07323	0.10528	0.	0.	0.	0.	-0.51770	-0.10285
	N2	-0.00253	-0.05473	0.12110	0.04656	0.	0.	0.	0.	0.00348	0.03838
* 7A1	H1	0.01977	-0.00883								
-0.5507	H2	0.01977	-0.00883								
	C1	-0.00369	-0.07603	0.21733	0.09920	0.	0.	0.	0.	0.28897	-0.08184
	N1	0.00058	0.01323	-0.02407	-0.05436	0.	0.	0.	0.	-0.10418	-0.04314
	N2	0.00582	0.12476	-0.29838	-0.43500	0.	0.	0.	0.	-0.54155	-0.14195
8A1	H1	-0.03049	-1.12332								
0.2112	H2	-0.03049	-1.12332								
	C1	0.00125	0.04311	0.03709	-1.31871	0.	0.	0.	0.	0.11966	0.07602
	N1	-0.00541	-0.13032	0.19680	2.09261	0.	0.	0.	0.	-0.01984	-0.09455
	N2	-0.00181	-0.03987	0.08977	0.48225	0.	0.	0.	0.	-0.14372	-0.30090
9A1	H1	0.02280	0.66894								
0.3397	H2	0.02280	0.66894								
	C1	0.00188	0.06297	0.04139	-2.88520	0.	0.	0.	0.	0.07402	-1.52676
	N1	-0.00045	0.00126	0.09932	-0.84946	0.	0.	0.	0.	0.36160	0.88554
	N2	-0.00413	-0.10471	0.11392	2.26755	0.	0.	0.	0.	-0.12043	-0.81823
10A1	H1	0.10468	1.67012								
0.4283	H2	0.10468	1.67012								
	C1	0.00209	0.05414	-0.06315	-2.22063	0.	0.	0.	0.	0.18896	3.39352
	N1	-0.00164	-0.04219	0.04142	0.99832	0.	0.	0.	0.	0.13677	2.48583
	N2	0.00083	0.03709	0.08750	-1.68478	0.	0.	0.	0.	-0.23138	0.00281

NH2CN

Center		S1	S2	S3	S4	X1	X2	Y1	Y2	Z1	Z2
11A1	H1	-0.11563	-0.07066								
0.5616	H2	-0.11563	-0.07066								
	C1	-0.00453	-0.02018	0.79016	0.10214	0.	0.	0.	0.	-0.41351	0.85479
	N1	0.00034	0.01060	0.00566	0.28417	0.	0.	0.	0.	0.48330	-1.39865
	N2	-0.00071	-0.00060	0.14074	-0.57351	0.	0.	0.	0.	-0.18244	0.10577
12A1	H1	0.06830	0.34280								
0.6475	H2	0.06830	0.34280								
	C1	0.00072	-0.01558	-0.24944	-1.92886	0.	0.	0.	0.	-0.58268	0.88869
	N1	0.00068	0.01158	-0.05779	0.38147	0.	0.	0.	0.	-0.16987	0.47492
	N2	-0.00143	-0.04815	-0.03556	0.86398	0.	0.	0.	0.	0.43900	-1.90634
* 1B1	H1	0.23279	0.07952								
-0.7923	H2	-0.23279	-0.07952								
	C1	0.	0.	0.	0.	0.13777	0.00745	0.	0.	0.	0.
	N1	0.	0.	0.	0.	0.56516	0.16060	0.	0.	0.	0.
	N2	0.	0.	0.	0.	0.05228	0.01081	0.	0.	0.	0.
* 2B1	H1	-0.07961	-0.03375								
-0.4695	H2	0.07961	0.03375								
	C1	0.	0.	0.	0.	0.50935	0.20128	0.	0.	0.	0.
	N1	0.	0.	0.	0.	-0.12074	-0.11878	0.	0.	0.	0.
	N2	0.	0.	0.	0.	0.48797	0.16993	0.	0.	0.	0.
3B1	H1	-0.11304	-0.83777								
0.1903	H2	0.11304	0.83777								
	C1	0.	0.	0.	0.	0.48035	0.49475	0.	0.	0.	0.
	N1	0.	0.	0.	0.	0.03112	0.65161	0.	0.	0.	0.
	N2	0.	0.	0.	0.	-0.46732	-0.59233	0.	0.	0.	0.
4B1	H1	0.04189	-1.92261								
0.2954	H2	-0.04189	1.92261								
	C1	0.	0.	0.	0.	0.07225	-1.29499	0.	0.	0.	0.
	N1	0.	0.	0.	0.	0.38355	1.92972	0.	0.	0.	0.
	N2	0.	0.	0.	0.	0.13649	0.51670	0.	0.	0.	0.
5B1	H1	-0.04389	0.91680								
0.5073	H2	0.04389	-0.91680								
	C1	0.	0.	0.	0.	0.90398	-1.07633	0.	0.	0.	0.
	N1	0.	0.	0.	0.	-0.26263	-0.33274	0.	0.	0.	0.
	N2	0.	0.	0.	0.	0.08827	-0.04730	0.	0.	0.	0.
* 1B2	H1	0.	0.								
-0.5809	H2	0.	0.								
	C1	0.	0.	0.	0.	0.	0.	0.39631	0.07655	0.	0.
	N1	0.	0.	0.	0.	0.	0.	0.50418	0.23433	0.	0.
	N2	0.	0.	0.	0.	0.	0.	0.27925	0.08531	0.	0.
* 2B2	H1	0.	0.								
-0.4059	H2	0.	0.								
	C1	0.	0.	0.	0.	0.	0.	0.26239	0.13351	0.	0.
	N1	0.	0.	0.	0.	0.	0.	-0.48087	-0.35086	0.	0.
	N2	0.	0.	0.	0.	0.	0.	0.49179	0.21358	0.	0.
3B2	H1	0.	0.								
0.2415	H2	0.	0.								
	C1	0.	0.	0.	0.	0.	0.	-0.53202	-1.04040	0.	0.
	N1	0.	0.	0.	0.	0.	0.	0.26428	0.46188	0.	0.
	N2	0.	0.	0.	0.	0.	0.	0.44347	0.73880	0.	0.

NH2CN

Center		S1	S2	S3	S4	X1	X2	Y1	Y2	Z1	Z2
4B2	H1	0.	0.								
0.4695	H2	0.	0.								
	C1	0.	0.	0.	0.	0.	0.	0.92452	-1.20057	0.	0.
	N1	0.	0.	0.	0.	0.	0.	0.13470	0.06436	0.	0.
	N2	0.	0.	0.	0.	0.	0.	0.05200	0.03898	0.	0.

Overlap Matrix

First Function   Second Function and Overlap

H1S1 with
    H1S2 .68301   H2S1 .03221   H2S2 .17790   C1S2 -.00016   C1S3 .01907   C1S4 .11393   C1X1 .02576
    C1X2 .12447   C1Z1 -.05693   C1Z2 -.27513   N1S1 .00358   N1S2 .04941   N1S3 .32493   N1S4 .43973
    N1X1 .37766   N1X2 .43972   N1Z1 -.21804   N1Z2 -.25387   N2S3 .00001   N2S4 .00316   N2X1 .00003
    N2X2 .00704   N2Z1 -.00011   N2Z2 -.02576

H1S2 with
    H2S1 .17790   H2S2 .43227   C1S1 .00138   C1S2 .01893   C1S3 .14411   C1S4 .32360   C1X1 .08032
    C1X2 .23174   C1Z1 -.17754   C1Z2 -.51225   N1S1 .00735   N1S2 .09335   N1S3 .47918   N1S4 .73622
    N1X1 .23236   N1X2 .49375   N1Z1 -.13415   N1Z2 -.28507   N2S1 .00003   N2S2 .00047   N2S3 .00637
    N2S4 .03843   N2X1 .00343   N2X2 .03524   N2Z1 -.01254   N2Z2 -.12893

C1S1 with
    C1S2 .35436   C1S3 .00805   C1S4 .01419   N1S3 .00066   N1S4 .00495   N1Z1 .00306   N1Z2 .01111
    N2S3 .00164   N2S4 .00654   N2Z1 -.00555   N2Z2 -.01230

C1S2 with
    C1S3 .26226   C1S4 .17745   N1S2 .00001   N1S3 .01264   N1S4 .06450   N1Z1 .04397   N1Z2 .13815
    N2S2 .00006   N2S3 .02739   N2S4 .08411   N2Z1 -.07508   N2Z2 -.15169

C1S3 with
    C1S4 .79970   N1S1 .00144   N1S2 .02094   N1S3 .18187   N1S4 .36446   N1Z1 .27671   N1Z2 .58153
    N2S1 .00271   N2S2 .03758   N2S3 .26186   N2S4 .44151   N2Z1 -.35036   N2Z2 -.60690

C1S4 with
    N1S1 .00442   N1S2 .05690   N1S3 .32014   N1S4 .56777   N1Z1 .22074   N1Z2 .58808   N2S1 .00539
    N2S2 .06905   N2S3 .37653   N2S4 .63763   N2Z1 -.22982   N2Z2 -.57932

C1X1 with
    C1X2 .53875   N1X1 .14563   N1X2 .32113   N2X1 .20684   N2X2 .37696

C1X2 with
    N1X1 .21443   N1X2 .62618   N2X1 .24407   N2X2 .68575

C1Y1 with
    C1Y2 .53875   N1Y1 .14563   N1Y2 .32113   N2Y1 .20684   N2Y2 .37696

C1Y2 with
    N1Y1 .21443   N1Y2 .62618   N2Y1 .24407   N2Y2 .68575

C1Z1 with
    C1Z2 .53875   N1S1 -.00445   N1S2 -.05832   N1S3 -.32688   N1S4 -.36070   N1Z1 -.32521   N1Z2 -.16141
    N2S1 .00661   N2S2 .08534   N2S3 .41601   N2S4 .38650   N2Z1 -.33985   N2Z2 -.06956

C1Z2 with
    N1S1 -.00764   N1S2 -.09643   N1S3 -.47177   N1S4 -.64419   N1Z1 -.04609   N1Z2 .09227   N2S1 .00791
    N2S2 .09945   N2S3 .47705   N2S4 .63100   N2Z1 .01066   N2Z2 .22568

N1S1 with
    N1S2 .35584   N1S3 .00894   N1S4 .01446   N2S4 .00014   N2Z2 -.00115

N1S2 with
    N1S3 .27246   N1S4 .17925   N2S4 .00201   N2Z1 -.00003   N2Z2 -.01558

N1S3 with
    N1S4 .79352   N2S3 .00055   N2S4 .02285   N2Z1 -.00324   N2Z2 -.11453

N1S4 with
    N2S1 .00014   N2S2 .00201   N2S3 .02285   N2S4 .09717   N2Z1 -.04121   N2Z2 -.26922

N1X1 with
    N1X2 .52739   N2X1 .00109   N2X2 .02857

N1X2 with
    N2X1 .02857   N2X2 .15655

N1Y1 with
    N1Y2 .52739   N2Y1 .00109   N2Y2 .02857

N1Y2 with
    N2Y1 .02857   N2Y2 .15655

NH2CN   T-151

Overlap Matrix (continued)

First Function   Second Function and Overlap

N1Z1 with
       N1Z2 .52739   N2S2 .00003   N2S3 .00324   N2S4 .04121   N2Z1 -.01211   N2Z2 -.13689

N1Z2 with
       N2S1 .00115   N2S2 .01558   N2S3 .11453   N2S4 .26922   N2Z1 -.13689   N2Z2 -.42406

N2S1 with
       N2S2 .35584   N2S3 .00894   N2S4 .01446

N2S2 with
       N2S3 .27246   N2S4 .17925

N2S3 with
       N2S4 .79352

N2X1 with
       N2X2 .52739

N2Y1 with
       N2Y2 .52739

N2Z1 with
       N2Z2 .52739

BH3CO   Borine Carbonyl

Molecular Geometry   Symmetry C₃ᵥ

Center	Coordinates		
	X	Y	Z
H1	2.18385017	0.	-3.47763470
H2	-1.09192508	1.89126951	-3.47763470
H3	-1.09192508	-1.89126951	-3.47763470
B1	0.	0.	-2.91021499
C1	0.	0.	0.
O1	0.	0.	2.13730726
Mass	0.	0.	-0.19910233

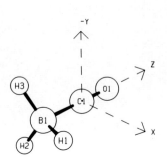

MO Symmetry Types and Occupancy

Representation	A1	A2	E
Number of MOs	20	0	16
Number Occupied	7	0	4

Energy Expectation Values

Total	-139.0670	Kinetic	139.1403
Electronic	-195.5662	Potential	-278.2072
Nuclear Repulsion	56.4993	Virial	1.9995

Electronic Potential	-334.7065
One-Electron Potential	-438.3049
Two-Electron Potential	103.5984

Moments of the Charge Distribution

First

x	y	z	Dipole Moment
0.	0.	4.1436	0.6383
0.	0.	-3.5053	

Second

x^2	y^2	z^2	xy	xz	yz	r^2
-21.2823	-21.2823	-131.9406	0.	0.	0.	-174.5052
7.1538	7.1538	112.9053	0.	0.	0.	127.2129

Third

x^3	xy^2	xz^2	x^2y	y^3	yz^2	x^2z	y^2z	z^3	xyz
-7.4633	7.4633	-0.0001	0.	0.	0.	34.8189	34.8189	136.5696	0.
7.8114	-7.8114	0.	0.	0.	0.	-23.4540	-23.4540	-103.2763	0.

Fourth (even)

x^4	y^4	z^4	x^2y^2	x^2z^2	y^2z^2	r^2x^2	r^2y^2	r^2z^2	r^4
-107.2967	-107.2966	-1410.4585	-35.7655	-181.1531	-181.1530	-324.2153	-324.2151	-1772.7645	-2421.1949
25.5884	25.5884	855.1289	8.5295	76.8946	76.8946	111.0125	111.0125	1008.9181	1230.9431

Fourth (odd)

x^3y	xy^3	xyz^2	x^3z	xy^2z	xz^3	x^2yz	y^3z	yz^3
0.	0.	0.0001	24.0396	-24.0394	0.0002	0.	0.	-0.0001
0.	0.	0.	-25.6100	25.6100	0.	0.	0.	0.

Expectation Values at Atomic Centers

Center	r^{-1}	δ	xr^{-3}	yr^{-3}	zr^{-3}
			Expectation Value		
H1	-1.1384	0.3946	0.0525	0.	-0.0331
B1	-11.4585	65.4570	0.	0.	-0.0462
C1	-14.5252	117.2932	0.	0.	-0.1193
O1	-22.1809	287.5467	0.	0.	0.2951

	$\dfrac{3x^2-r^2}{r^5}$	$\dfrac{3y^2-r^2}{r^5}$	$\dfrac{3z^2-r^2}{r^5}$	$\dfrac{3xy}{r^5}$	$\dfrac{3xz}{r^5}$	$\dfrac{3yz}{r^5}$
H1	0.2157	-0.1351	-0.0805	0.	-0.0968	0.
B1	-0.1241	-0.1241	0.2483	0.	0.	0.
C1	0.4134	0.4134	-0.8268	0.	0.	0.
O1	-0.0413	-0.0413	0.0825	0.	0.	0.

Mulliken Charge

Center	S1	S2	S3	S4	X1	X2	Y1	Y2	Z1	Z2	Total
					Basis Function Type						Total
H1	0.4937	0.4485									0.9422
B1	0.0435	1.9556	0.6676	0.3541	0.8770	0.0963	0.8770	0.0963	0.5224	-0.0083	5.4815
C1	0.0431	1.9565	0.9750	0.3261	0.5042	0.0685	0.5042	0.0685	1.0523	0.0208	5.5192
O1	0.0427	1.9566	0.9954	0.7970	1.1041	0.4208	1.1041	0.4208	1.1759	0.1551	8.1726

Overlap Populations

	H2 / H1	B1 / H1	C1 / H1	C1 / B1	O1 / H1	O1 / B1	O1 / C1
			Pair of Atomic Centers				
1A1	0.	0.	0.	0.	0.	0.	-0.0009
2A1	0.	0.	0.	-0.0007	0.	0.	-0.0012
3A1	0.	-0.0003	0.	-0.0017	0.	0.	0.
4A1	0.0002	-0.0014	-0.0008	0.0088	-0.0003	0.0076	0.6412
5A1	0.0020	0.0355	0.0204	0.4393	0.0003	0.0067	-0.0252
6A1	0.0025	0.0420	-0.0003	0.0528	-0.0005	0.0071	0.0219
7A1	0.0339	0.2179	-0.0670	0.0086	0.0009	-0.0017	-0.0082
Total A1	0.0386	0.2937	-0.0477	0.5072	0.0004	0.0198	0.6276
1E	-0.0003	0.0065	0.0040	0.0280	0.0006	0.0064	0.3773
2E	0.	0.0001	0.	0.0279	0.	0.0064	0.3773
3E	-0.0664	0.4842	0.0158	0.0473	-0.0036	-0.0198	-0.0280
4E	0.0123	0.0105	0.0003	0.0473	-0.0001	-0.0198	-0.0280
Total E	-0.0543	0.5012	0.0202	0.1505	-0.0031	-0.0268	0.6985
Total	-0.0157	0.7949	-0.0275	0.6576	-0.0027	-0.0070	1.3261
Distance	3.7825	2.2564	4.1065	2.9102	6.0247	5.0475	2.1373

Molecular Orbitals and Eigenvalues

Center		S1	S2	S3	S4	X1	X2	Y1	Y2	Z1	Z2
* 1A1	H1	0.00002	-0.00032								
-20.7271	H2	0.00002	-0.00032								
	H3	0.00002	-0.00032								
	B1	0.00000	-0.00002	-0.00045	0.00017	-0.00000	-0.00000	-0.00000	-0.00000	-0.00024	0.00001
	C1	0.00000	-0.00005	-0.00031	-0.00079	-0.00000	0.00000	0.00000	0.00000	-0.00035	-0.00169
	O1	0.05113	0.97935	0.00325	0.00375	0.00000	-0.00000	-0.00000	-0.00000	-0.00219	0.00025
* 2A1	H1	-0.00008	0.00021								
-11.4729	H2	-0.00008	0.00021								
	H3	-0.00008	0.00021								
	B1	-0.00000	-0.00008	0.00090	-0.00249	0.00000	0.00000	-0.00000	-0.00000	0.00085	-0.00119
	C1	0.05191	0.97919	0.00373	0.00402	0.00000	-0.00000	-0.00000	0.00000	0.00304	-0.00115
	O1	-0.00003	-0.00070	0.00137	-0.00185	-0.00000	0.00000	0.00000	-0.00000	-0.00142	0.00151
* 3A1	H1	0.00105	-0.00097								
-7.5703	H2	0.00105	-0.00097								
	H3	0.00105	-0.00097								
	B1	0.05292	0.97849	0.00575	0.00418	0.00000	0.00000	-0.00000	-0.00000	0.00203	0.00026
	C1	-0.00002	-0.00040	0.00085	-0.00365	0.00000	-0.00000	-0.00000	0.00000	-0.00078	0.00141
	O1	0.00000	0.00001	-0.00013	0.00066	-0.00000	0.00000	0.00000	-0.00000	0.00002	-0.00071
* 4A1	H1	0.00058	0.01275								
-1.6076	H2	0.00058	0.01274								
	H3	0.00058	0.01274								
	B1	0.00024	0.00356	-0.02693	-0.02281	-0.00000	-0.00000	0.00000	0.00000	-0.03103	0.00433
	C1	0.00574	0.12122	-0.27024	-0.03255	-0.00000	0.00000	0.00000	-0.00000	-0.25032	0.03089
	O1	0.00976	0.21283	-0.47497	-0.28668	0.00000	-0.00000	-0.00000	0.00000	0.21292	-0.02356
* 5A1	H1	-0.05837	-0.02690								
-0.9053	H2	-0.05837	-0.02691								
	H3	-0.05837	-0.02691								
	B1	0.00571	0.12402	-0.24242	-0.07669	-0.00000	-0.00000	0.00000	0.00000	-0.12443	0.01600
	C1	0.00777	0.16316	-0.41674	-0.18302	-0.00000	0.00000	0.00000	-0.00000	0.20412	0.02989
	O1	-0.00397	-0.08649	0.20189	0.21533	-0.00000	-0.00000	-0.00000	0.00000	0.22006	0.07605
* 6A1	H1	0.05603	0.03171								
-0.7962	H2	0.05603	0.03171								
	H3	0.05603	0.03171								
	B1	-0.00427	-0.09353	0.17891	0.14163	0.00000	0.00000	-0.00000	-0.00000	0.04562	0.02300
	C1	0.00144	0.02615	-0.10107	-0.01624	-0.00000	-0.00000	-0.00000	0.00000	-0.41251	0.07187
	O1	-0.00401	-0.08859	0.19761	0.35158	0.00000	0.00000	-0.00000	-0.00000	0.54933	0.09821
* 7A1	H1	0.15256	0.13048								
-0.5615	H2	0.15255	0.13049								
	H3	0.15255	0.13049								
	B1	-0.00492	-0.10641	0.23530	0.13843	0.00000	0.00001	-0.00000	-0.00000	-0.30218	0.00061
	C1	0.00429	0.09103	-0.23703	-0.22388	0.00000	-0.00001	0.00000	0.00000	0.33928	-0.04163
	O1	-0.00039	-0.00811	0.02221	0.00046	-0.00000	0.00000	0.00000	-0.00000	-0.14864	-0.01113
8A1	H1	-0.01937	0.09863								
0.1742	H2	-0.01937	0.09866								
	H3	-0.01937	0.09866								
	B1	-0.00128	-0.00597	0.21917	-1.94795	0.00000	0.00004	-0.00000	0.00000	0.02270	-1.58038
	C1	-0.00182	-0.04782	0.04052	1.43586	-0.00000	-0.00001	-0.00000	-0.00000	-0.11895	-1.01847
	O1	-0.00166	-0.04015	0.06015	0.50350	0.00000	0.00000	0.00000	0.00000	-0.00412	0.09537

Center		Basis Function Type									
		S1	S2	S3	S4	X1	X2	Y1	Y2	Z1	Z2
9A1	H1	-0.03389	-0.45986								
0.2483	H2	-0.03389	-0.45980								
	H3	-0.03389	-0.45980								
	B1	0.00387	0.05996	-0.38214	1.95301	-0.00003	0.00011	-0.00000	0.00000	-0.20110	-0.57396
	C1	-0.00184	-0.04672	0.06191	0.79106	-0.00002	-0.00002	0.00000	-0.00000	-0.01841	1.38468
	O1	0.00294	0.07308	-0.09570	-1.12311	0.00000	0.00001	-0.00000	0.00000	0.10582	0.18799
10A1	H1	0.03075	-0.21357								
0.4546	H2	0.03075	-0.21364								
	H3	0.03074	-0.21367								
	B1	-0.00936	-0.08191	1.38692	-1.11629	-0.00005	-0.00004	-0.00001	-0.00001	-0.67940	0.03789
	C1	-0.00202	-0.03164	0.19981	0.72639	-0.00003	0.00006	-0.00001	0.00002	-0.08068	0.35495
	O1	0.00129	0.03065	-0.05429	-0.43983	-0.00001	0.00000	-0.00000	-0.00000	-0.00623	0.18430
11A1	H1	0.06299	2.03059								
0.5126	H2	0.06297	2.02934								
	H3	0.06297	2.02927								
	B1	0.00363	0.11579	0.02078	-6.03542	-0.00074	-0.00039	-0.00004	-0.00001	-0.08350	-0.54028
	C1	-0.00306	-0.07264	0.13922	3.15435	0.00005	0.00028	0.00001	-0.00000	0.06344	-0.07340
	O1	0.00239	0.05674	-0.10118	-0.87781	-0.00007	-0.00004	-0.00000	-0.00000	0.07166	0.61023
12A1	H1	0.01201	-0.00173								
0.5808	H2	0.01201	-0.00176								
	H3	0.01201	-0.00176								
	B1	-0.00544	-0.04638	0.81843	1.05810	-0.00002	-0.00002	0.00000	0.00000	1.12240	-0.09963
	C1	0.00328	0.09684	-0.02300	-0.96954	-0.00000	0.00002	-0.00000	-0.00000	0.00230	1.82173
	O1	0.00003	0.01546	0.10377	-0.78595	-0.00000	-0.00001	0.00000	0.00000	-0.49093	0.27086
* 1E	H1	-0.03389	-0.01379								
-0.7017	H2	0.02003	0.00815								
	H3	0.01386	0.00564								
	B1	-0.00000	-0.00000	0.00000	-0.00001	-0.06909	-0.00830	0.00726	0.00087	-0.00000	-0.00000
	C1	-0.00000	-0.00000	-0.00000	0.00001	-0.38325	-0.04247	0.04026	0.00446	0.00000	-0.00000
	O1	0.00000	0.00000	0.00000	-0.00000	-0.61411	-0.25687	0.06452	0.02698	0.00000	0.00000
* 2E	H1	0.00356	0.00145								
-0.7017	H2	0.02757	0.01122								
	H3	-0.03114	-0.01267								
	B1	0.00000	0.00000	-0.00000	0.00000	0.00726	0.00087	0.06909	0.00830	0.00000	0.00000
	C1	-0.00000	-0.00000	0.00000	-0.00000	0.04026	0.00446	0.38324	0.04248	-0.00000	0.00000
	O1	-0.00000	-0.00000	-0.00000	0.00000	0.06451	0.02698	0.61412	0.25686	0.00000	-0.00000
* 3E	H1	0.25431	0.23428								
-0.4700	H2	-0.15951	-0.14694								
	H3	-0.09480	-0.08732								
	B1	0.00000	0.00000	0.00000	-0.00002	0.47973	0.06934	-0.07048	-0.01018	0.00000	-0.00000
	C1	-0.00000	-0.00000	0.00000	0.00001	0.04620	0.02997	-0.00679	-0.00440	-0.00000	-0.00000
	O1	0.00000	0.00000	0.00000	-0.00000	-0.18513	-0.11427	0.02720	0.01679	0.00000	0.00000
* 4E	H1	-0.03736	-0.03442								
-0.4700	H2	-0.20156	-0.18568								
	H3	0.23892	0.22010								
	B1	-0.00000	-0.00000	0.00000	0.00000	-0.07048	-0.01019	-0.47972	-0.06934	-0.00000	-0.00000
	C1	0.00000	0.00000	-0.00000	0.00000	-0.00679	-0.00440	-0.04620	-0.02997	0.00000	-0.00000
	O1	0.00000	0.00000	-0.00000	0.00000	0.02720	0.01679	0.18513	0.11427	-0.00000	0.00000

Center		S1	S2	S3	S4	X1	X2	Y1	Y2	Z1	Z2
5E 0.0885	H1	0.00297	0.00789								
	H2	0.09929	0.26348								
	H3	-0.10226	-0.27138								
	B1	-0.00000	-0.00000	-0.00000	0.00001	0.00056	0.00181	0.02188	0.07107	-0.00000	0.00000
	C1	0.00000	0.00000	0.00000	-0.00000	-0.01583	-0.01775	-0.61951	-0.69468	0.00000	0.00000
	O1	0.00000	0.00000	-0.00000	-0.00000	0.01055	0.01217	0.41275	0.47638	0.00000	-0.00000
6E 0.0885	H1	-0.11636	-0.30881								
	H2	0.06076	0.16125								
	H3	0.05561	0.14758								
	B1	-0.00000	0.00000	0.00000	-0.00002	-0.02187	-0.07107	0.00056	0.00182	-0.00000	-0.00001
	C1	-0.00000	-0.00000	-0.00000	0.00003	0.61951	0.69468	-0.01583	-0.01775	0.00000	0.00000
	O1	0.00000	0.00000	0.00000	-0.00001	-0.41275	-0.47638	0.01055	0.01217	0.00000	0.00001
7E 0.2137	H1	0.00051	0.01134								
	H2	0.03424	0.76398								
	H3	-0.03475	-0.77533								
	B1	0.00000	0.00000	-0.00000	0.00001	0.00274	-0.02380	0.21505	-1.86512	-0.00000	-0.00000
	C1	-0.00000	-0.00000	0.00000	0.00000	0.00198	0.00641	0.15535	0.50235	-0.00000	0.00000
	O1	0.00000	0.00000	-0.00000	-0.00000	-0.00016	-0.00186	-0.01217	-0.14585	0.00000	0.00000
8E 0.2137	H1	0.03984	0.88864								
	H2	-0.02036	-0.45425								
	H3	-0.01948	-0.43460								
	B1	0.00000	0.00000	-0.00002	0.00020	0.21505	-1.86511	-0.00274	0.02380	-0.00002	-0.00005
	C1	-0.00000	-0.00000	0.00001	0.00001	0.15536	0.50235	-0.00198	-0.00641	-0.00000	0.00007
	O1	0.00000	0.00000	-0.00001	-0.00004	-0.01217	-0.14585	0.00016	0.00186	0.00001	-0.00000
9E 0.4658	H1	-0.04002	-0.15363								
	H2	0.01974	0.07594								
	H3	0.02029	0.07807								
	B1	-0.00000	0.00000	0.00005	-0.00043	0.09305	0.77675	0.00075	0.00623	-0.00003	-0.00004
	C1	-0.00000	-0.00001	0.00001	0.00025	0.92684	-1.56196	0.00744	-0.01253	-0.00000	0.00003
	O1	0.00000	0.00001	-0.00000	-0.00009	0.06413	0.08696	0.00051	0.00070	0.00000	0.00005
10E 0.4658	H1	0.00032	0.00122								
	H2	-0.03482	-0.13379								
	H3	0.03450	0.13253								
	B1	-0.00000	-0.00000	0.00002	0.00002	-0.00075	-0.00623	0.09306	0.77675	-0.00001	0.00000
	C1	-0.00000	-0.00000	0.00000	-0.00001	-0.00744	0.01253	0.92684	-1.56196	-0.00000	0.00000
	O1	0.00000	0.00000	-0.00000	0.00000	-0.00051	-0.00070	0.06414	0.08695	-0.00000	-0.00000
11E 0.4954	H1	0.01593	1.43150								
	H2	-0.00685	-0.61359								
	H3	-0.00917	-0.82151								
	B1	-0.00000	-0.00006	-0.00008	0.00359	-1.25703	-0.53359	-0.10532	-0.04470	0.00006	0.00032
	C1	0.00000	0.00004	-0.00007	-0.00190	0.18965	0.28837	0.01589	0.02416	-0.00003	-0.00002
	O1	-0.00000	-0.00003	0.00005	0.00057	-0.11536	-0.06406	-0.00967	-0.00537	-0.00004	-0.00036
12E 0.4954	H1	-0.00134	-0.12000								
	H2	0.01449	1.30082								
	H3	-0.01315	-1.18069								
	B1	0.00000	0.00000	-0.00001	-0.00012	0.10532	0.04471	-1.25703	-0.53357	0.00001	-0.00001
	C1	-0.00000	-0.00000	-0.00000	0.00006	-0.01589	-0.02416	0.18965	0.28835	0.00000	0.00000
	O1	0.00000	0.00000	-0.00000	-0.00002	0.00967	0.00537	-0.11536	-0.06405	0.00000	0.00001

Overlap Matrix

First Function   Second Function and Overlap

H1S1 with
    H1S2 .68301   H2S1 .00636   H2S2 .08922   B1S1 .00172   B1S2 .03092   B1S3 .29370   B1S4 .32262
    B1X1 .44126   B1X2 .31335   B1Z1 -.11465   B1Z2 -.08142   C1S2 .00003   C1S3 .00838   C1S4 .07929
    C1X1 .01783   C1X2 .13201   C1Z1 -.02840   C1Z2 -.21022   O1S4 .00057   O1X2 .00299   O1Z1 -.00001
    O1Z2 -.00770

H1S2 with
    H2S1 .08922   H2S2 .28074   B1S1 .00883   B1S2 .11218   B1S3 .54857   B1S4 .67900   B1X1 .46958
    B1X2 .54899   B1Z1 -.12201   B1Z2 -.14264   C1S1 .00082   C1S2 .01144   C1S3 .09853   C1S4 .25476
    C1X1 .07904   C1X2 .26792   C1Z1 -.12587   C1Z2 -.42664   O1S1 .00002   O1S2 .00024   O1S3 .00291
    O1S4 .01836   O1X1 .00206   O1X2 .02694   O1Z1 -.00529   O1Z2 -.06926

B1S1 with
    B1S2 .35489   B1S3 .00732   B1S4 .01371   C1S3 .00088   C1S4 .00543   C1Z1 -.00396   C1Z2 -.01171
    O1S4 .00002   O1Z2 -.00051

B1S2 with
    B1S3 .24693   B1S4 .17217   C1S2 .00001   C1S3 .01635   C1S4 .07071   C1Z1 -.05557   C1Z2 -.14535
    O1S4 .00053   O1Z2 -.00823

B1S3 with
    B1S4 .80606   C1S1 .00201   C1S2 .02828   C1S3 .21840   C1S4 .40345   C1Z1 -.31205   C1Z2 -.60502
    O1S1 .00001   O1S2 .00013   O1S3 .00271   O1S4 .02579   O1Z1 -.00709   O1Z2 -.11561

B1S4 with
    C1S1 .00455   C1S2 .05907   C1S3 .33777   C1S4 .59002   C1Z1 -.21891   C1Z2 -.57301   O1S1 .00057
    O1S2 .00757   O1S3 .05000   O1S4 .13121   O1Z1 -.04410   O1Z2 -.23811

B1X1 with
    B1X2 .55377   C1X1 .18420   C1X2 .37963   O1X1 .00446   O1X2 .04688

B1X2 with
    C1X1 .21805   C1X2 .64204   O1X1 .03517   O1X2 .18001

B1Y1 with
    B1Y2 .55377   C1Y1 .18420   C1Y2 .37963   O1Y1 .00446   O1Y2 .04688

B1Y2 with
    C1Y1 .21805   C1Y2 .64204   O1Y1 .03517   O1Y2 .18001

B1Z1 with
    B1Z2 .55377   C1S1 .00560   C1S2 .07300   C1S3 .38877   C1S4 .41602   C1Z1 -.33079   C1Z2 -.13019
    O1S1 .00016   O1S2 .00223   O1S3 .02109   O1S4 .08176   O1Z1 -.03390   O1Z2 -.22053

B1Z2 with
    C1S1 .00690   C1S2 .08806   C1S3 .44858   C1S4 .63098   C1Z1 -.00316   C1Z2 .16896   O1S1 .00232
    O1S2 .03017   O1S3 .17662   O1S4 .36510   O1Z1 -.08037   O1Z2 -.30425

C1S1 with
    C1S2 .35436   C1S3 .00805   C1S4 .01419   O1S3 .00086   O1S4 .00635   O1Z1 -.00400   O1Z2 -.01390

C1S2 with
    C1S3 .26226   C1S4 .17745   O1S2 .00004   O1S3 .01804   O1S4 .08221   O1Z1 -.05834   O1Z2 -.16958

C1S3 with
    C1S4 .79970   O1S1 .00263   O1S2 .03568   O1S3 .23493   O1S4 .43362   O1Z1 -.29478   O1Z2 -.62463

C1S4 with
    O1S1 .00461   O1S2 .05940   O1S3 .32622   O1S4 .59411   O1Z1 -.17393   O1Z2 -.52879

C1X1 with
    C1X2 .53875   O1X1 .17810   O1X2 .40208

C1X2 with
    O1X1 .18791   O1X2 .63112

C1Y1 with
    C1Y2 .53875   O1Y1 .17810   O1Y2 .40208

C1Y2 with
    O1Y1 .18791   O1Y2 .63112

Overlap Matrix (continued)

First Function   Second Function and Overlap

C1Z1 with
    C1Z2 .53875   01S1 .00600   01S2 .07763   01S3 .39555   01S4 .44219   01Z1 -.31210   01Z2 -.12640

C1Z2 with
    01S1 .00641   01S2 .08150   01S3 .40788   01S4 .61195   01Z1 .01534   01Z2 .20097

01S1 with
    01S2 .35461   01S3 .00974   01S4 .01469

01S2 with
    01S3 .28286   01S4 .18230

01S3 with
    01S4 .79029

01X1 with
    01X2 .50607

01Y1 with
    01Y2 .50607

01Z1 with
    01Z2 .50607

CH3CN   Methyl Cyanide

Molecular Geometry   Symmetry C₃ᵥ

Center	X	Coordinates Y	Z
H1	1.96393999	0.	-3.45140100
H2	-0.98197080	1.70082200	-3.45140100
H3	-0.98197080	-1.70082200	-3.45140100
C1	0.	0.	-2.75593501
C2	0.	0.	0.
N1	0.	0.	2.18663001
Mass	-0.00000004	0.	-0.31410739

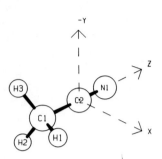

MO Symmetry Types and Occupancy

Representation	A1	A2	E
Number of MOs	20	0	16
Number Occupied	7	0	4

Energy Expectation Values

Total	-131.8674	Kinetic	131.7387
Electronic	-190.2071	Potential	-263.6061
Nuclear Repulsion	58.3397	Virial	2.0010

Electronic Potential	-321.9457
One-Electron Potential	-424.2900
Two-Electron Potential	102.3442

Moments of the Charge Distribution

First

x	y	z	Dipole Moment
0.	0.	3.0273	1.6457
0.	0.	-4.6730	

Second

x^2	y^2	z^2	xy	xz	yz	r^2
-19.0665	-19.0665	-125.3210	0.	0.	0.	-163.4541
5.7856	5.7856	109.6708	0.	0.	0.	121.2419

Third

x^3	xy^2	xz^2	x^2y	y^3	yz^2	x^2z	y^2z	z^3	xyz
-4.5222	4.5221	0.	0.	0.	0.	18.1478	18.1478	61.5332	0.
5.6813	-5.6813	0.	0.	0.	0.	-18.1511	-18.1511	-70.3364	0.

Fourth (even)

x^4	y^4	z^4	x^2y^2	x^2z^2	y^2z^2	r^2x^2	r^2y^2	r^2z^2	r^4
-80.9497	-80.9500	-1255.2073	-26.9833	-138.1588	-138.1589	-246.0918	-246.0922	-1531.5250	-2023.7089
16.7365	16.7365	777.7597	5.5789	56.9453	56.9453	79.2607	79.2607	891.6503	1050.1718

Fourth (odd)

x^3y	xy^3	xyz^2	x^3z	xy^2z	xz^3	x^2yz	y^3z	yz^3
0.	0.	0.	14.0087	-14.0086	-0.0002	0.	0.	-0.0001
0.	0.	0.	-17.8238	17.8238	0.	0.	0.	0.

Expectation Values at Atomic Centers

Center		Expectation Value			
	r^{-1}	δ	xr^{-3}	yr^{-3}	zr^{-3}
H1	-1.0474	0.4059	0.0527	0.	-0.0254
C1	-14.6699	116.2560	0.	0.	-0.0289
C2	-14.6804	116.7677	0.	0.	-0.0393
N1	-18.3381	189.6295	0.	0.	0.2058

	$\dfrac{3x^2-r^2}{r^5}$	$\dfrac{3y^2-r^2}{r^5}$	$\dfrac{3z^2-r^2}{r^5}$	$\dfrac{3xy}{r^5}$	$\dfrac{3xz}{r^5}$	$\dfrac{3yz}{r^5}$
H1	0.2532	-0.1622	-0.0911	0.	-0.1497	0.
C1	-0.1321	-0.1321	0.2643	0.	0.	0.
C2	0.1141	0.1141	-0.2282	0.	0.	0.
N1	0.3521	0.3521	-0.7042	0.	0.	0.

Mulliken Charge

Center	Basis Function Type										Total
	S1	S2	S3	S4	X1	X2	Y1	Y2	Z1	Z2	
H1	0.5019	0.2756									0.7775
C1	0.0427	1.9563	0.7493	0.6829	0.9470	0.1762	0.9470	0.1761	0.7985	0.0779	6.5540
C2	0.0429	1.9564	0.8313	0.2556	0.7395	0.2153	0.7395	0.2153	0.9893	0.0498	6.0348
N1	0.0434	1.9560	0.9588	0.7051	0.7966	0.2767	0.7966	0.2767	1.1055	0.1634	7.0787

Overlap Populations

	Pair of Atomic Centers						
	H2	C1	C2	C2	N1	N1	N1
	H1	H1	H1	C1	H1	C1	C2
1A1	0.	0.	0.	0.	0.	-0.0001	-0.0020
2A1	0.	0.	0.	-0.0004	0.	0.	0.0002
3A1	0.	-0.0003	0.	-0.0015	0.	0.	0.
4A1	0.0003	-0.0045	-0.0030	0.0561	-0.0005	0.0144	0.7214
5A1	0.0036	0.1316	0.0051	0.2697	0.0004	0.0051	0.0590
6A1	0.0164	0.1358	-0.0695	0.1718	-0.0003	-0.0038	-0.0011
7A1	0.0002	0.0004	0.0032	-0.0190	0.0003	-0.0004	0.0839
Total A1	0.0204	0.2630	-0.0642	0.4767	-0.0001	0.0153	0.8613
1E	0.0043	0.2807	0.0211	0.1221	0.0011	0.0079	0.0456
2E	-0.0520	0.1191	0.0090	0.1221	0.0005	0.0079	0.0456
3E	0.0014	0.0004	-0.0002	-0.1899	0.	-0.0287	0.4948
4E	-0.0158	0.1105	-0.0488	-0.1899	-0.0042	-0.0287	0.4948
Total E	-0.0620	0.5107	-0.0189	-0.1357	-0.0027	-0.0415	1.0806
Total	-0.0416	0.7737	-0.0831	0.3409	-0.0028	-0.0262	1.9419
Distance	3.4016	2.0834	3.9710	2.7559	5.9703	4.9426	2.1866

Molecular Orbitals and Eigenvalues

Center					Basis Function Type						
		S1	S2	S3	S4	X1	X2	Y1	Y2	Z1	Z2
* 1A1	H1	-0.00003	0.00040								
-15.6001	H2	-0.00003	0.00040								
	H3	-0.00003	0.00040								
	C1	0.00000	-0.00000	0.00002	0.00138	-0.00000	0.00000	-0.00000	0.00000	0.00011	0.00056
	C2	-0.00001	-0.00003	0.00017	0.00142	-0.00000	-0.00000	0.00000	-0.00000	0.00003	0.00376
	N1	-0.05196	-0.97915	-0.00213	-0.00620	-0.00000	0.00000	-0.00000	0.00000	0.00252	0.00001
* 2A1	H1	-0.00008	0.00035								
-11.3038	H2	-0.00008	0.00035								
	H3	-0.00008	0.00035								
	C1	-0.00166	-0.03130	-0.00062	0.00083	-0.00000	-0.00000	-0.00000	0.00000	-0.00057	0.00093
	C2	-0.05187	-0.97843	-0.00460	-0.00220	-0.00000	0.00000	0.00000	-0.00000	-0.00043	0.00097
	N1	0.00004	0.00095	-0.00110	0.00021	0.00000	-0.00000	-0.00000	0.00000	0.00151	-0.00064
* 3A1	H1	0.00052	-0.00092								
-11.2914	H2	0.00052	-0.00092								
	H3	0.00052	-0.00092								
	C1	0.05188	0.97843	0.00405	0.00555	-0.00000	0.00000	-0.00000	-0.00000	0.00073	0.00053
	C2	-0.00168	-0.03161	0.00031	-0.00292	0.00000	-0.00000	0.00000	0.00000	-0.00029	0.00159
	N1	0.00000	0.00004	-0.00003	0.00007	-0.00000	0.00000	0.00000	-0.00000	0.00005	-0.00054
* 4A1	H1	-0.01064	0.01886								
-1.2532	H2	-0.01064	0.01885								
	H3	-0.01064	0.01885								
	C1	0.00140	0.03043	-0.06656	-0.08254	0.00000	-0.00001	0.00000	0.00000	-0.04788	0.00386
	C2	0.00737	0.15906	-0.33800	-0.11545	-0.00000	0.00001	0.00000	-0.00000	-0.22913	0.02182
	N1	0.00869	0.18667	-0.41441	-0.19707	0.00000	-0.00000	-0.00000	0.00000	0.22482	-0.03160
* 5A1	H1	-0.11350	-0.01877								
-1.0431	H2	-0.11350	-0.01879								
	H3	-0.11350	-0.01879								
	C1	0.00897	0.19543	-0.41510	-0.33062	0.00000	-0.00002	0.00000	0.00000	-0.04512	-0.00874
	C2	0.00243	0.05182	-0.11999	0.03035	0.00000	0.00001	0.00000	-0.00000	0.26060	-0.01259
	N1	-0.00257	-0.05553	0.12101	0.03116	0.00000	-0.00000	0.00000	0.00000	-0.04197	0.03570
* 6A1	H1	0.11283	0.07071								
-0.6942	H2	0.11283	0.07073								
	H3	0.11283	0.07073								
	C1	-0.00207	-0.04569	0.09744	0.17464	-0.00000	0.00002	-0.00000	0.00000	-0.43912	-0.04545
	C2	0.00580	0.12483	-0.29594	-0.22402	-0.00000	-0.00001	-0.00000	0.00000	0.29476	0.06582
	N1	-0.00327	-0.07062	0.16024	0.13283	-0.00001	0.00001	-0.00000	-0.00000	0.07771	0.04162
* 7A1	H1	-0.02869	-0.00305								
-0.5513	H2	-0.02868	-0.00305								
	H3	-0.02868	-0.00305								
	C1	-0.00029	-0.00719	0.00773	0.07401	-0.00000	-0.00000	0.00000	0.00000	0.16258	0.08016
	C2	0.00255	0.05304	-0.14507	-0.07862	-0.00000	-0.00000	-0.00000	-0.00000	-0.34204	0.11820
	N1	-0.00520	-0.11103	0.26946	0.38553	0.00000	0.00000	-0.00000	0.00000	0.55683	0.13566
8A1	H1	-0.01666	0.90699								
0.2437	H2	-0.01667	0.90693								
	H3	-0.01667	0.90689								
	C1	0.00627	0.15564	-0.21826	-3.50759	0.00001	-0.00009	-0.00000	-0.00003	0.05877	-1.01471
	C2	-0.00150	-0.05984	-0.09808	2.78364	-0.00001	0.00003	0.00001	0.00002	-0.12987	-0.83431
	N1	0.00145	0.03136	-0.07590	-0.39531	0.00002	-0.00001	-0.00001	-0.00001	0.12824	0.45687

CH3CN

Center		S1	S2	S3	S4	X1	X2	Y1	Y2	Z1	Z2
9A1	H1	0.03349	1.05402								
0.2709	H2	0.03349	1.05409								
	H3	0.03349	1.05408								
	C1	0.00168	0.02947	-0.14247	-0.35321	0.00002	0.00004	-0.00000	-0.00001	0.00397	2.21526
	C2	0.00286	0.06124	-0.16435	-1.42993	-0.00001	-0.00004	0.00000	0.00001	0.19469	1.20352
	N1	0.00078	0.02177	-0.00722	-0.58202	0.00002	0.00002	-0.00000	-0.00000	-0.02762	-0.04507
10A1	H1	0.02996	0.26717								
0.3242	H2	0.02997	0.26780								
	H3	0.02997	0.26757								
	C1	-0.00050	0.01043	0.17187	-2.10769	0.00004	0.00044	-0.00001	-0.00017	0.40210	-0.20021
	C2	0.00142	0.04101	-0.01256	-1.38646	-0.00003	-0.00024	0.00001	0.00009	0.05029	-2.69486
	N1	-0.00406	-0.10760	0.07894	2.59105	0.00003	0.00010	-0.00001	-0.00004	-0.06652	-0.58080
11A1	H1	0.00344	0.33674								
0.5647	H2	0.00344	0.33672								
	H3	0.00344	0.33672								
	C1	-0.00222	-0.03288	0.23135	0.55872	-0.00001	0.00001	-0.00000	-0.00000	0.83906	0.01093
	C2	0.00133	0.06901	0.19240	0.18018	0.00002	-0.00003	0.00000	0.00000	0.20486	1.81158
	N1	0.00031	0.03369	0.16790	-1.69034	0.00000	0.00000	-0.00000	-0.00000	-0.51731	0.96664
12A1	H1	-0.01102	0.26282								
0.6733	H2	-0.01101	0.26281								
	H3	-0.01102	0.26281								
	C1	-0.00032	-0.01342	-0.02157	1.44396	-0.00000	-0.00001	-0.00000	-0.00000	0.49643	0.37663
	C2	0.00454	0.06380	-0.49652	-2.08807	0.00000	0.00001	0.00000	0.00000	-0.48534	1.99805
	N1	-0.00174	-0.04423	0.05710	0.14939	0.00000	-0.00001	-0.00000	-0.00000	0.13513	-1.72485
* 1E	H1	0.20277	0.15025								
-0.6294	H2	0.01300	0.00965								
	H3	-0.21577	-0.15987								
	C1	-0.00000	-0.00000	0.00000	-0.00001	0.42248	0.08527	0.27520	0.05553	-0.00000	0.00001
	C2	0.00000	0.00000	-0.00000	-0.00000	0.17531	0.03634	0.11419	0.02368	-0.00000	0.00001
	N1	-0.00000	-0.00000	0.00000	-0.00000	0.11254	0.02919	0.07331	0.01901	0.00000	0.00000
* 2E	H1	0.13208	0.09787								
-0.6294	H2	-0.24165	-0.17905								
	H3	0.10956	0.08120								
	C1	-0.00000	-0.00000	0.00000	-0.00001	0.27520	0.05554	-0.42248	-0.08525	-0.00000	0.00000
	C2	0.00000	0.00000	-0.00000	-0.00000	0.11419	0.02368	-0.17531	-0.03635	-0.00000	0.00000
	N1	-0.00000	-0.00000	0.00000	-0.00000	0.07331	0.01901	-0.11254	-0.02918	0.00000	0.00000
* 3E	H1	0.00731	0.00615								
-0.4678	H2	0.10073	0.08479								
	H3	-0.10804	-0.09094								
	C1	0.00000	0.00000	-0.00000	-0.00000	0.01186	0.00647	0.19557	0.10666	0.00000	0.00000
	C2	-0.00000	-0.00000	0.00000	-0.00000	-0.02680	-0.01065	-0.44196	-0.17559	-0.00000	0.00000
	N1	0.00000	0.00000	-0.00000	0.00000	-0.03045	-0.01106	-0.50229	-0.18235	0.00000	-0.00000
* 4E	H1	0.12053	0.10145								
-0.4678	H2	-0.06659	-0.05605								
	H3	-0.05393	-0.04540								
	C1	0.00000	0.00000	-0.00000	-0.00001	0.19557	0.10666	-0.01186	-0.00647	0.00001	0.00000
	C2	-0.00000	-0.00000	0.00000	-0.00000	-0.44196	-0.17558	0.02680	0.01065	-0.00000	-0.00000
	N1	0.00000	0.00000	-0.00000	0.00000	-0.50229	-0.18236	0.03045	0.01106	0.00000	-0.00000

		S1	S2	S3	S4	X1	X2	Y1	Y2	Z1	Z2
5E	H1	0.12064	0.79445								
0.1922	H2	-0.05802	-0.38209								
	H3	-0.06264	-0.41250								
	C1	-0.00000	-0.00001	-0.00000	0.00014	-0.06507	-0.53888	-0.00144	-0.01191	-0.00001	-0.00000
	C2	-0.00000	0.00000	0.00001	-0.00009	-0.43524	-0.60033	-0.00962	-0.01326	-0.00000	0.00001
	N1	-0.00000	-0.00000	-0.00000	0.00004	0.44063	0.60820	0.00973	0.01344	-0.00000	-0.00002
6E	H1	-0.00267	-0.01754								
0.1922	H2	0.10582	0.69681								
	H3	-0.10315	-0.67923								
	C1	0.00000	0.00000	0.00000	-0.00005	0.00144	0.01190	-0.06507	-0.53883	0.00000	-0.00000
	C2	-0.00000	-0.00000	-0.00000	0.00002	0.00962	0.01326	-0.43524	-0.60036	-0.00000	-0.00001
	N1	0.00000	0.00000	-0.00000	0.00000	-0.00974	-0.01344	0.44063	0.60821	0.00000	0.00000
7E	H1	0.05131	2.13765								
0.3413	H2	-0.01665	-0.69400								
	H3	-0.03465	-1.44354								
	C1	-0.00000	-0.00000	0.00002	-0.00024	-0.21017	-2.79822	-0.04254	-0.56649	0.00005	0.00004
	C2	0.00000	0.00001	-0.00001	-0.00028	0.22590	1.46318	0.04573	0.29621	0.00002	-0.00036
	N1	-0.00000	-0.00002	0.00001	0.00038	-0.17679	-0.64324	-0.03579	-0.13022	-0.00001	-0.00009
8E	H1	-0.01039	-0.43279								
0.3413	H2	0.04963	2.06756								
	H3	-0.03925	-1.63488								
	C1	-0.00000	-0.00000	-0.00001	0.00019	0.04255	0.56648	-0.21013	-2.79827	-0.00004	0.00001
	C2	-0.00000	-0.00000	0.00000	0.00012	-0.04573	-0.29621	0.22590	1.46317	-0.00001	0.00023
	N1	0.00000	0.00001	-0.00001	-0.00022	0.03579	0.13022	-0.17678	-0.64323	0.00001	0.00005
9E	H1	0.08097	-0.22625								
0.4229	H2	-0.03351	0.09357								
	H3	-0.04747	0.13263								
	C1	-0.00000	-0.00000	0.00002	0.00004	0.84902	-0.82518	0.08456	-0.08216	-0.00000	-0.00001
	C2	-0.00000	-0.00000	0.00001	-0.00002	0.35722	-0.21131	0.03557	-0.02106	-0.00001	0.00002
	N1	-0.00000	-0.00000	-0.00000	-0.00000	0.05682	-0.07272	0.00566	-0.00724	-0.00000	-0.00001
10E	H1	0.00806	-0.02254								
0.4229	H2	-0.07414	0.20729								
	H3	0.06608	-0.18476								
	C1	-0.00000	-0.00000	0.00000	0.00000	0.08456	-0.08217	-0.84903	0.82502	-0.00000	-0.00000
	C2	-0.00000	-0.00000	0.00000	0.00000	0.03558	-0.02105	-0.35720	0.21139	-0.00000	-0.00000
	N1	0.00000	0.00000	-0.00000	-0.00000	0.00566	-0.00724	-0.05683	0.07268	0.00000	0.00000
11E	H1	-0.05461	0.14195								
0.5186	H2	0.02543	-0.06611								
	H3	0.02918	-0.07584								
	C1	0.00000	0.00000	-0.00002	-0.00000	-0.43264	0.88214	-0.01712	0.03490	-0.00002	0.00001
	C2	-0.00000	-0.00000	-0.00001	-0.00001	0.85968	-1.43607	0.03402	-0.05682	-0.00000	-0.00004
	N1	-0.00000	-0.00000	-0.00000	0.00003	0.11876	0.04555	0.00470	0.00180	0.00001	-0.00002
12E	H1	0.00216	-0.00562								
0.5186	H2	-0.04837	0.12575								
	H3	0.04621	-0.12014								
	C1	-0.00000	-0.00000	0.00000	0.00000	0.01712	-0.03491	-0.43264	0.88212	-0.00000	-0.00000
	C2	-0.00000	-0.00000	0.00000	0.00000	-0.03402	0.05682	0.85968	-1.43606	0.00000	0.00000
	N1	0.00000	0.00000	-0.00000	-0.00000	-0.00470	-0.00180	0.11876	0.04555	-0.00000	0.00000

Overlap Matrix

First Function   Second Function and Overlap

H1S1 with
    H1S2 .68301   H2S1 .01578   H2S2 .13155   C1S1 .00208   C1S2 .03242   C1S3 .27627   C1S4 .36755
    C1X1 .39875   C1X2 .40919   C1Z1 -.14120   C1Z2 -.14490   C2S2 .00006   C2S3 .01135   C2S4 .09064
    C2X1 .02088   C2X2 .13226   C2Z1 -.03670   C2Z2 -.23243   N1S3 .00001   N1S4 .00242   N1X1 .00002
    N1X2 .00720   N1Z1 -.00007   N1Z2 -.02068

H1S2 with
    H2S1 .13155   H2S2 .35794   C1S1 .00753   C1S2 .09618   C1S3 .49013   C1S4 .70005   C1X1 .33949
    C1X2 .57636   C1Z1 -.12022   C1Z2 -.20410   C2S1 .00099   C2S2 .01378   C2S3 .11338   C2S4 .27828
    C2X1 .08140   C2X2 .26001   C2Z1 -.14305   C2Z2 -.45694   N1S1 .00002   N1S2 .00035   N1S3 .00504
    N1S4 .03276   N1X1 .00349   N1X2 .03905   N1Z1 -.01003   N1Z2 -.11210

C1S1 with
    C1S2 .35436   C1S3 .00805   C1S4 .01419   C2S3 .00099   C2S4 .00462   C2Z1 -.00384   C2Z2 -.00915
    N1S4 .00011   N1Z2 -.00109

C1S2 with
    C1S3 .26226   C1S4 .17745   C2S2 .00001   C2S3 .01599   C2S4 .05999   C2Z1 -.05195   C2Z2 -.11536
    N1S4 .00180   N1Z1 -.00002   N1Z2 -.01528

C1S3 with
    C1S4 .79970   C2S1 .00099   C2S2 .01599   C2S3 .17883   C2S4 .34216   C2Z1 -.30706   C2Z2 -.53513
    N1S2 .00001   N1S3 .00119   N1S4 .02703   N1Z1 -.00512   N1Z2 -.12646

C1S4 with
    C2S1 .00462   C2S2 .05999   C2S3 .34216   C2S4 .57004   C2Z1 -.29378   C2Z2 -.63606   N1S1 .00030
    N1S2 .00417   N1S3 .03571   N1S4 .11842   N1Z1 -.05048   N1Z2 -.28106

C1X1 with
    C1X2 .53875   C2X1 .15109   C2X2 .27164   N1X1 .00219   N1X2 .03779

C1X2 with
    C2X1 .27164   C2X2 .64713   N1X1 .03745   N1X2 .18352

C1Y1 with
    C1Y2 .53875   C2Y1 .15109   C2Y2 .27164   N1Y1 .00219   N1Y2 .03779

C1Y2 with
    C2Y1 .27164   C2Y2 .64713   N1Y1 .03745   N1Y2 .18352

C1Z1 with
    C1Z2 .53875   C2S1 .00384   C2S2 .05195   C2S3 .30706   C2S4 .29378   C2Z1 -.33938   C2Z2 -.09964
    N1S1 .00001   N1S2 .00025   N1S3 .00780   N1S4 .05817   N1Z1 -.02131   N1Z2 -.17807

C1Z2 with
    C2S1 .00915   C2S2 .11536   C2S3 .53513   C2S4 .63606   C2Z1 -.09964   C2Z2 .08386   N1S1 .00189
    N1S2 .02480   N1S3 .15823   N1S4 .33239   N1Z1 -.13828   N1Z2 -.42343

C2S1 with
    C2S2 .35436   C2S3 .00805   C2S4 .01419   N1S3 .00184   N1S4 .00678   N1Z1 -.00599   N1Z2 -.01243

C2S2 with
    C2S3 .26226   C2S4 .17745   N1S2 .00007   N1S3 .03029   N1S4 .08706   N1Z1 -.08045   N1Z2 -.15319

C2S3 with
    C2S4 .79970   N1S1 .00294   N1S2 .04055   N1S3 .27447   N1S4 .45263   N1Z1 -.36043   N1Z2 -.60884

C2S4 with
    N1S1 .00553   N1S2 .07081   N1S3 .38454   N1S4 .64732   N1Z1 -.23048   N1Z2 -.57684

C2X1 with
    C2X2 .53875   N1X1 .21671   N1X2 .38489

C2X2 with
    N1X1 .24821   N1X2 .69388

C2Y1 with
    C2Y2 .53875   N1Y1 .21671   N1Y2 .38489

C2Y2 with
    N1Y1 .24821   N1Y2 .69388

CH3CN   T-167

Overlap Matrix (continued)

First Function   Second Function and Overlap

C2Z1 with
    C2Z2 .53875   N1S1 .00697   N1S2 .08973   N1S3 .42855   N1S4 .38909   N1Z1 -.33921   N1Z2 -.05509

C2Z2 with
    N1S1 .00792   N1S2 .09962   N1S3 .47662   N1S4 .62783   N1Z1 .01918   N1Z2 .24473

N1S1 with
    N1S2 .35584   N1S3 .00894   N1S4 .01446

N1S2 with
    N1S3 .27246   N1S4 .17925

N1S3 with
    N1S4 .79352

N1X1 with
    N1X2 .52739

N1Y1 with
    N1Y2 .52739

N1Z1 with
    N1Z2 .52739

CH3CN   T-169

CH3NC   Methyl Isocyanide

Molecular Geometry   Symmetry C_{3v}

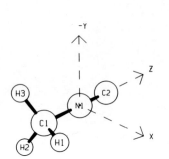

Center	Coordinates		
	X	Y	Z
H1	1.96689001	0.	-3.37197000
H2	-0.98344500	1.70337500	-3.37197000
H3	-0.98344500	-1.70337500	-3.37197000
C1	0.	0.	-2.69086999
C2	0.	0.	2.20375001
N1	0.	0.	0.
Mass	0.	0.	-0.39097393

MO Symmetry Types and Occupancy

Representation	A1	A2	E
Number of MOs	20	0	16
Number Occupied	7	0	4

Energy Expectation Values

Total	-131.8400	Kinetic	131.8186
Electronic	-191.8140	Potential	-263.6586
Nuclear Repulsion	59.9740	Virial	2.0002

Electronic Potential	-323.6326
One-Electron Potential	-427.5105
Two-Electron Potential	103.8779

Moments of the Charge Distribution

First

x	y	z	Dipole Moment
0.	0.	3.0013	1.4359
0.	0.	-4.4372	

Second

x^2	y^2	z^2	xy	xz	yz	r^2
-19.1155	-19.1154	-115.6920	0.	0.	0.	-153.9229
5.8030	5.8030	99.8617	0.	0.	0.	111.4677

Third

x^3	xy^2	xz^2	x^2y	y^3	yz^2	x^2z	y^2z	z^3	xyz
-4.6401	4.6401	0.	0.	0.	0.	16.3780	16.3779	28.2642	0.
5.7069	-5.7069	0.	0.	0.	0.	-17.2987	-17.2986	-47.2289	0.

Fourth (even)

x^4	y^4	z^4	x^2y^2	x^2z^2	y^2z^2	r^2x^2	r^2y^2	r^2z^2	r^4
-80.1222	-80.1215	-1188.4379	-26.7073	-132.5978	-132.5973	-239.4273	-239.4261	-1453.6332	-1932.4865
16.8373	16.8372	676.9056	5.6124	51.5673	51.5672	74.0170	74.0168	780.0401	928.0739

Fourth (odd)

x^3y	xy^3	xyz^2	x^3z	xy^2z	xz^3	x^2yz	y^3z	yz^3
0.	0.	0.	13.7279	-13.7281	-0.0006	0.	-0.0001	-0.0001
0.	0.	0.	-17.0123	17.0123	0.	0.	0.	0.

Expectation Values at Atomic Centers

Center	r^{-1}	δ	xr^{-3}	yr^{-3}	zr^{-3}
			Expectation Value		
H1	-1.0625	0.4146	0.0567	0.	-0.0236
C1	-14.6489	116.2480	0.	0.	-0.0373
C2	-14.6909	117.2703	0.	0.	0.1528
N1	-18.3337	188.9265	0.	0.	-0.0466

	$\dfrac{3x^2-r^2}{r^5}$	$\dfrac{3y^2-r^2}{r^5}$	$\dfrac{3z^2-r^2}{r^5}$	$\dfrac{3xy}{r^5}$	$\dfrac{3xz}{r^5}$	$\dfrac{3yz}{r^5}$
H1	0.2567	-0.1656	-0.0911	0.	-0.1465	0.
C1	-0.2249	-0.2249	0.4498	0.	0.	0.
C2	0.3789	0.3789	-0.7578	0.	0.	0.
N1	-0.1320	-0.1320	0.2640	0.	0.	0.

Mulliken Charge

Center	S1	S2	S3	S4	X1	X2	Y1	Y2	Z1	Z2	Total
					Basis Function Type						Total
H1	0.5111	0.2762									0.7873
C1	0.0427	1.9564	0.7624	0.6543	0.9602	0.1590	0.9602	0.1590	0.6952	0.0177	6.3669
C2	0.0431	1.9564	0.9059	0.9492	0.4797	0.0597	0.4797	0.0597	1.0085	0.1454	6.0873
N1	0.0433	1.9560	0.8889	0.4666	1.0123	0.4540	1.0123	0.4540	1.0623	-0.1658	7.1838

Overlap Populations

	H2 / H1	C1 / H1	C2 / H1	C2 / C1	N1 / H1	N1 / C1	N1 / C2
			Pair of Atomic Centers				
1A1	0.	0.	0.	0.	0.	-0.0006	-0.0001
2A1	0.	-0.0005	0.	-0.0002	0.	-0.0020	0.
3A1	0.	0.	0.	-0.0005	0.	0.	-0.0059
4A1	0.	0.0015	0.	-0.0335	0.0028	0.0349	0.6367
5A1	0.0016	0.1124	-0.0025	-0.1901	-0.0070	0.0829	0.0822
6A1	0.0147	0.1336	0.0057	-0.0344	-0.0425	0.2203	-0.1644
7A1	0.0008	-0.0081	0.0013	0.0007	0.0055	-0.1082	0.1459
Total A1	0.0170	0.2389	0.0045	-0.2579	-0.0412	0.2273	0.6946
1E	-0.0116	0.0086	0.	0.0061	0.0011	0.1711	0.0621
2E	-0.0271	0.3286	0.0012	0.0061	0.0402	0.1711	0.0621
3E	0.0008	0.	0.	-0.0203	0.	-0.2680	0.3834
4E	-0.0303	0.1861	-0.0045	-0.0203	-0.0823	-0.2680	0.3834
Total E	-0.0682	0.5233	-0.0033	-0.0285	-0.0410	-0.1939	0.8910
Total	-0.0511	0.7622	0.0012	-0.2864	-0.0822	0.0334	1.5856
Distance	3.4068	2.0815	5.9125	4.8946	3.9037	2.6909	2.2038

CH3NC

Molecular Orbitals and Eigenvalues

Center		S1	S2	S3	S4	X1	X2	Y1	Y2	Z1	Z2
· 1A1	H1	-0.00005	-0.00015								
-15.5963	H2	-0.00005	-0.00015								
	H3	-0.00005	-0.00015								
	C1	-0.00000	-0.00000	-0.00030	0.00176	-0.00000	-0.00000	-0.00000	0.00000	-0.00009	0.00064
	C2	-0.00001	-0.00004	-0.00055	0.00048	-0.00000	-0.00000	0.00000	0.00000	0.00061	-0.00057
	N1	-0.05193	-0.97891	-0.00383	-0.00301	-0.00000	0.00000	0.00000	-0.00000	-0.00025	0.00077
· 2A1	H1	-0.00057	0.00135								
-11.3102	H2	-0.00057	0.00135								
	H3	-0.00057	0.00135								
	C1	-0.05191	-0.97897	-0.00394	-0.00712	0.00000	-0.00000	-0.00000	0.00000	-0.00125	-0.00058
	C2	0.00016	0.00303	0.00011	0.00232	-0.00000	-0.00000	-0.00000	0.00000	-0.00011	-0.00070
	N1	0.00001	0.00021	-0.00048	0.00121	-0.00000	0.00000	0.00000	-0.00000	0.00046	-0.00332
· 3A1	H1	-0.00010	0.00005								
-11.3019	H2	-0.00011	0.00005								
	H3	-0.00011	0.00005								
	C1	-0.00016	-0.00303	-0.00011	0.00473	-0.00000	0.00000	-0.00000	0.00000	0.00033	0.00193
	C2	-0.05192	-0.97931	-0.00379	-0.00910	0.00000	0.00000	-0.00000	0.00000	0.00376	0.00178
	N1	0.00004	0.00088	-0.00158	0.00451	0.00000	-0.00000	0.00000	-0.00000	-0.00131	0.00762
· 4A1	H1	0.02560	-0.00337								
-1.2869	H2	0.02560	-0.00337								
	H3	0.02560	-0.00337								
	C1	-0.00304	-0.06574	0.14299	-0.04196	-0.00000	0.00000	0.00000	-0.00000	0.07595	-0.05717
	C2	-0.00561	-0.11992	0.26144	0.21180	0.00000	0.00000	0.00000	-0.00000	-0.22247	-0.02022
	N1	-0.00957	-0.20565	0.45838	0.21429	0.00000	-0.00000	-0.00000	0.00000	0.12256	-0.16598
· 5A1	H1	-0.10244	-0.00808								
-1.0356	H2	-0.10244	-0.00808								
	H3	-0.10244	-0.00807								
	C1	0.00825	0.17862	-0.38972	-0.40520	-0.00000	0.00000	-0.00000	0.00000	-0.04868	-0.03710
	C2	-0.00425	-0.09231	0.19591	0.22172	-0.00000	-0.00000	0.00000	0.00000	-0.14557	-0.03820
	N1	-0.00028	-0.00602	0.01449	-0.03503	-0.00000	-0.00000	0.00000	-0.00000	0.36033	-0.12970
· 6A1	H1	0.12298	0.06287								
-0.7352	H2	0.12298	0.06287								
	H3	0.12298	0.06287								
	C1	-0.00304	-0.06635	0.14883	0.14132	0.00000	0.00000	0.00000	-0.00000	-0.41146	-0.05728
	C2	-0.00384	-0.08262	0.19074	0.18658	0.00000	0.00000	0.00000	-0.00000	-0.07367	-0.01584
	N1	0.00369	0.08086	-0.17607	-0.28159	0.00000	-0.00000	0.00000	0.00000	0.39280	0.00232
· 7A1	H1	0.01681	0.01748								
-0.4654	H2	0.01682	0.01746								
	H3	0.01682	0.01746								
	C1	0.00038	0.00824	-0.01766	-0.15302	-0.00001	-0.00002	0.00000	0.00000	-0.11408	-0.08668
	C2	0.00674	0.14775	-0.33262	-0.41352	0.00001	-0.00000	-0.00000	-0.00000	-0.50112	-0.15650
	N1	-0.00233	-0.04865	0.13114	0.09336	0.00001	0.00001	-0.00000	-0.00000	0.23638	-0.11742
8A1	H1	0.03056	0.12871								
0.2473	H2	0.03056	0.12874								
	H3	0.03056	0.12872								
	C1	-0.00212	-0.05551	0.05727	1.60779	0.00001	-0.00000	-0.00000	-0.00002	0.10907	2.09640
	C2	-0.00018	-0.00542	0.00197	-1.12791	-0.00001	-0.00000	-0.00000	-0.00000	-0.09792	0.39717
	N1	0.00309	0.07840	-0.08354	-1.29677	0.00000	0.00001	0.00000	0.00001	0.17171	1.48309

		S1	S2	S3	S4	X1	X2	Y1	Y2	Z1	Z2
9A1	H1	0.00220	1.30206								
0.2585	H2	0.00222	1.30243								
	H3	0.00222	1.30244								
	C1	0.00568	0.13503	-0.24093	-3.68113	0.00004	0.00028	-0.00000	0.00001	0.13284	-0.06023
	C2	0.00056	0.00430	-0.08647	1.66048	0.00001	0.00005	-0.00000	0.00000	0.03947	-1.00614
	N1	0.00074	0.01607	-0.03954	-0.41841	-0.00002	-0.00009	-0.00000	-0.00000	-0.02280	-1.41819
10A1	H1	-0.01970	-0.77539								
0.3773	H2	-0.01970	-0.77550								
	H3	-0.01970	-0.77551								
	C1	-0.00213	-0.05082	0.09038	0.75325	-0.00005	-0.00003	0.00000	-0.00000	-0.15870	-0.81486
	C2	-0.00136	-0.04113	-0.00082	2.27587	0.00001	-0.00003	0.00000	-0.00000	0.09034	-2.06541
	N1	0.00206	0.05903	-0.01388	-1.90876	0.00001	0.00001	0.00000	0.00000	-0.09591	-1.08693
11A1	H1	-0.01711	-0.05991								
0.5098	H2	-0.01710	-0.05987								
	H3	-0.01710	-0.05987								
	C1	0.00459	0.06696	-0.47895	0.96838	0.00000	0.00003	-0.00000	0.00000	-0.89804	0.74983
	C2	0.00649	0.08219	-0.76694	-0.67492	-0.00001	0.00001	-0.00000	0.00000	0.35942	0.46521
	N1	-0.00384	-0.09325	0.14121	0.79946	-0.00001	0.00000	-0.00000	-0.00000	-0.26985	1.00897
* 1E	H1	-0.03540	-0.02569								
-0.6433	H2	0.20725	0.15042								
	H3	-0.17185	-0.12473								
	C1	0.00000	-0.00000	-0.00000	0.00001	-0.07683	-0.01324	0.47502	0.08182	0.00000	0.00000
	C2	0.00000	0.00000	-0.00000	-0.00000	-0.02108	-0.00143	0.13032	0.00886	0.00000	0.00000
	N1	-0.00000	-0.00000	0.00000	0.00000	-0.04773	-0.01831	0.29513	0.11323	-0.00000	0.00000
* 2E	H1	-0.21887	-0.15885								
-0.6433	H2	0.07878	0.05717								
	H3	0.14010	0.10167								
	C1	-0.00000	-0.00000	-0.00000	0.00002	-0.47502	-0.08184	-0.07683	-0.01323	-0.00000	-0.00000
	C2	0.00000	-0.00000	-0.00000	-0.00001	-0.13032	-0.00886	-0.02108	-0.00143	-0.00000	0.00000
	N1	0.00000	0.00000	-0.00000	-0.00000	-0.29513	-0.11323	-0.04773	-0.01831	0.00000	0.00001
* 3E	H1	0.00256	0.00234								
-0.4749	H2	0.14028	0.12797								
	H3	-0.14284	-0.13031								
	C1	-0.00000	-0.00000	-0.00000	0.00000	0.00413	0.00163	0.26359	0.10374	0.00000	0.00000
	C2	-0.00000	-0.00000	0.00000	-0.00000	-0.00547	-0.00083	-0.34873	-0.05289	0.00000	0.00000
	N1	0.00000	0.00000	-0.00000	-0.00000	-0.00815	-0.00468	-0.51983	-0.29811	-0.00000	0.00000
* 4E	H1	0.16346	0.14911								
-0.4749	H2	-0.08395	-0.07657								
	H3	-0.07951	-0.07252								
	C1	0.00000	0.00000	-0.00000	0.00001	0.26359	0.10376	-0.00413	-0.00163	-0.00000	0.00001
	C2	0.00000	0.00000	-0.00001	-0.00003	-0.34873	-0.05290	0.00547	0.00083	-0.00001	0.00001
	N1	-0.00000	-0.00000	0.00000	0.00001	-0.51983	-0.29811	0.00815	0.00468	0.00001	0.00002
5E	H1	0.00113	0.00599								
0.1832	H2	0.06478	0.34471								
	H3	-0.06590	-0.35070								
	C1	-0.00000	-0.00000	-0.00000	0.00001	-0.00136	-0.00224	-0.09130	-0.15038	0.00000	0.00000
	C2	-0.00000	-0.00000	-0.00000	-0.00000	0.00670	0.01277	0.44879	0.85614	-0.00000	0.00000
	N1	0.00000	0.00000	0.00000	-0.00000	-0.00567	-0.00887	-0.37979	-0.59433	0.00000	0.00000

		S1	S2	S3	S4	X1	X2	Y1	Y2	Z1	Z2
6E	H1	0.07545	0.40147								
0.1832	H2	-0.03870	-0.20595								
	H3	-0.03675	-0.19557								
	C1	-0.00000	-0.00000	-0.00000	0.00009	-0.09130	-0.15036	0.00136	0.00224	0.00000	0.00003
	C2	0.00000	0.00000	-0.00000	-0.00004	0.44879	0.85614	-0.00670	-0.01277	-0.00000	0.00001
	N1	0.00000	0.00000	0.00000	-0.00002	-0.37979	-0.59433	0.00567	0.00887	0.00000	0.00005
7E	H1	0.00210	0.04132								
0.3057	H2	0.08505	1.67220								
	H3	-0.08715	-1.71350								
	C1	0.00000	0.00000	-0.00000	-0.00003	-0.00091	-0.05212	-0.04287	-2.46578	0.00000	-0.00001
	C2	0.00000	0.00000	-0.00000	0.00001	0.00291	-0.00999	0.13760	-0.47263	0.00000	-0.00000
	N1	-0.00000	-0.00000	0.00000	0.00001	0.00264	0.01571	0.12473	0.74333	-0.00000	-0.00001
8E	H1	0.09944	1.95487								
0.3057	H2	-0.05154	-1.01299								
	H3	-0.04790	-0.94143								
	C1	0.00000	0.00002	-0.00005	-0.00033	-0.04289	-2.46574	0.00091	0.05212	0.00001	0.00004
	C2	0.00000	0.00000	-0.00003	0.00006	0.13762	-0.47265	-0.00291	0.00999	0.00002	-0.00003
	N1	0.00000	-0.00000	-0.00001	0.00004	0.12473	0.74332	-0.00264	-0.01571	-0.00000	-0.00005
9E	H1	-0.07188	0.82995								
0.4333	H2	0.03681	-0.42496								
	H3	0.03507	-0.40480								
	C1	0.00000	0.00001	-0.00004	-0.00005	-0.96750	0.31081	0.01357	-0.00436	0.00002	0.00005
	C2	0.00000	0.00000	-0.00001	-0.00008	-0.17854	-0.03160	0.00250	0.00044	-0.00000	0.00008
	N1	0.00000	-0.00000	-0.00001	0.00007	0.13332	-0.00039	-0.00187	0.00000	0.00001	0.00004
10E	H1	0.00101	-0.01164								
0.4333	H2	0.06174	-0.71292								
	H3	-0.06275	0.72456								
	C1	0.00000	-0.00000	-0.00000	-0.00000	0.01357	-0.00436	0.96750	-0.31076	-0.00000	0.00000
	C2	0.00000	0.00000	-0.00000	-0.00000	0.00250	0.00044	0.17854	0.03160	-0.00000	0.00000
	N1	0.00000	0.00000	-0.00000	0.00000	-0.00187	0.00001	-0.13332	0.00038	0.00000	-0.00000
11E	H1	0.00003	0.00133								
0.5602	H2	-0.01352	-0.51933								
	H3	0.01348	0.51800								
	C1	-0.00000	-0.00000	0.00000	-0.00000	0.00022	-0.00226	-0.09926	1.01806	-0.00000	0.00000
	C2	0.00000	0.00000	-0.00000	0.00000	-0.00228	0.00216	1.02734	-0.97451	-0.00000	-0.00000
	N1	0.00000	0.00000	-0.00000	-0.00000	0.00014	0.00064	-0.06167	-0.28707	0.00000	-0.00000
12E	H1	0.01560	0.59892								
0.5602	H2	-0.00778	-0.29832								
	H3	-0.00784	-0.30062								
	C1	0.00000	0.00000	-0.00002	0.00010	0.09925	-1.01810	0.00022	-0.00226	0.00001	0.00001
	C2	-0.00000	-0.00000	0.00002	-0.00009	-1.02733	0.97450	-0.00228	0.00216	0.00002	-0.00000
	N1	0.00000	0.00000	-0.00001	0.00002	0.06167	0.28708	0.00014	0.00064	-0.00000	0.00008

Overlap Matrix

First Function   Second Function and Overlap

H1S1 with
    H1S2 .68301   H2S1 .01560   H2S2 .13091   C1S1 .00210   C1S2 .03258   C1S3 .27691   C1S4 .36792
    C1X1 .40030   C1X2 .41014   C1Z1 -.13862 C1Z2 -.14202 C2S3 .00006   C2S4 .00872   C2X1 .00023
    C2X2 .01997   C2Z1 -.00064 C2Z2 -.05661 N1S2 .00005   N1S3 .00542   N1S4 .06288   N1X1 .00992
    N1X2 .11060   N1Z1 -.01701 N1Z2 -.18962

H1S2 with
    H2S1 .13091   H2S2 .35684   C1S1 .00754   C1S2 .09631   C1S3 .49064   C1S4 .70051   C1X1 .34035
    C1X2 .57755   C1Z1 -.11786 C1Z2 -.20000 C2S1 .00003   C2S2 .00053   C2S3 .00961   C2S4 .05913
    C2X1 .00774   C2X2 .06843   C2Z1 -.02195 C2Z2 -.19398 N1S1 .00086   N1S2 .01166   N1S3 .08825
    N1S4 .23128   N1X1 .05631   N1X2 .22443   N1Z1 -.09654 N1Z2 -.38476

C1S1 with
    C1S2 .35436   C1S3 .00805   C1S4 .01419   C2S4 .00041   C2Z1 -.00002 C2Z2 -.00250 N1S3 .00035
    N1S4 .00407   N1Z1 -.00199 N1Z2 -.01019

C1S2 with
    C1S3 .26226   C1S4 .17745   C2S3 .00002   C2S4 .00581   C2Z1 -.00042 C2Z2 -.03309 N1S3 .00736
    N1S4 .05353   N1Z1 -.03006 N1Z2 -.12751

C1S3 with
    C1S4 .79970   C2S2 .00002   C2S3 .00381   C2S4 .05467   C2Z1 -.01689 C2Z2 -.21202 N1S1 .00092
    N1S2 .01389   N1S3 .14055   N1S4 .31851   N1Z1 -.23125 N1Z2 -.55623

C1S4 with
    C2S1 .00041   C2S2 .00581   C2S3 .05467   C2S4 .16985   C2Z1 -.08788 C2Z2 -.39263 N1S1 .00384
    N1S2 .04967   N1S3 .28567   N1S4 .52333   N1Z1 -.21152 N1Z2 -.58580

C1X1 with
    C1X2 .53875   C2X1 .00602   C2X2 .06268   N1X1 .11462   N1X2 .28699

C1X2 with
    C2X1 .06268   C2X2 .25341   N1X1 .19579   N1X2 .58747

C1Y1 with
    C1Y2 .53875   C2Y1 .00602   C2Y2 .06268   N1Y1 .11462   N1Y2 .28699

C1Y2 with
    C2Y1 .06268   C2Y2 .25341   N1Y1 .19579   N1Y2 .58747

C1Z1 with
    C1Z2 .53875   C2S1 .00002   C2S2 .00042   C2S3 .01689   C2S4 .08788   C2Z1 -.04801 C2Z2 -.20550
    N1S1 .00337   N1S2 .04472   N1S3 .27355   N1S4 .33882   N1Z1 -.30265 N1Z2 -.20818

C1Z2 with
    C2S1 .00250   C2S2 .03309   C2S3 .21202   C2S4 .39263   C2Z1 -.20550 C2Z2 -.44233 N1S1 .00736
    N1S2 .09301   N1S3 .46139   N1S4 .64425   N1Z1 -.07763 N1Z2 .01159

C2S1 with
    C2S2 .35436   C2S3 .00805   C2S4 .01419   N1S3 .00175   N1S4 .00667   N1Z1 .00580   N1Z2 .01238

C2S2 with
    C2S3 .26226   C2S4 .17745   N1S2 .00007   N1S3 .02900   N1S4 .08577   N1Z1 .07808   N1Z2 .15255

C2S3 with
    C2S4 .79970   N1S1 .00284   N1S2 .03924   N1S3 .26896   N1S4 .44780   N1Z1 .35609   N1Z2 .60805

C2S4 with
    N1S1 .00547   N1S2 .07005   N1S3 .38107   N1S4 .64312   N1Z1 .23021   N1Z2 .57795

C2X1 with
    C2X2 .53875   N1X1 .21239   N1X2 .38145

C2X2 with
    N1X1 .24642   N1X2 .69036

C2Y1 with
    C2Y2 .53875   N1Y1 .21239   N1Y2 .38145

C2Y2 with
    N1Y1 .24642   N1Y2 .69036

CH3NC   T-175

Overlap Matrix (continued)

First Function   Second Function and Overlap

C2Z1 with
    C2Z2 .53875   N1S1 -.00681   N1S2 -.08781   N1S3 -.42312   N1S4 -.38800   N1Z1 -.33958   N1Z2 -.06141

C2Z2 with
    N1S1 -.00792   N1S2 -.09956   N1S3 -.47685   N1S4 -.62925   N1Z1 .01547   N1Z2 .23646

N1S1 with
    N1S2 .35584   N1S3 .00894   N1S4 .01446

N1S2 with
    N1S3 .27246   N1S4 .17925

N1S3 with
    N1S4 .79352

N1X1 with
    N1X2 .52739

N1Y1 with
    N1Y2 .52739

N1Z1 with
    N1Z2 .52739

C3H4*   Cyclopropene

Molecular Geometry   Symmetry C$_{2v}$

Center		Coordinates	
	X	Y	Z
H1	2.97770551	0.	-1.01407170
H2	-2.97770551	0.	-1.01407170
H3	0.	1.72956550	3.69430920
H4	0.	-1.72956550	3.69430920
C1	1.22803010	0.	0.
C2	-1.22803010	0.	0.
C3	0.	0.	2.58607799
Mass	0.	0.	0.91017072

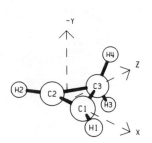

MO Symmetry Types and Occupancy

Representation	A1	A2	B1	B2
Number of MOs	18	2	12	6
Number Occupied	6	0	3	2

Energy Expectation Values

Total	-115.7655	Kinetic	115.8291
Electronic	-179.4674	Potential	-231.5947
Nuclear Repulsion	63.7019	Virial	1.9995

Electronic Potential	-295.2966
One-Electron Potential	-396.5699
Two-Electron Potential	101.2734

Moments of the Charge Distribution

First

x	y	z	Dipole Moment
0.	0.	-0.9962	0.1430
0.	0.	0.8532	

Second

x^2	y^2	z^2	xy	xz	yz	r^2
-48.6021	-20.7779	-63.6170	0.	0.	0.	-132.9970
35.8302	5.9828	49.7012	0.	0.	0.	91.5141

Third

x^3	xy^2	xz^2	x^2y	y^3	yz^2	x^2z	y^2z	z^3	xyz
0.	0.	0.	0.	0.	0.	43.1725	-14.2178	-49.2573	0.
0.	0.	0.	0.	0.	0.	-50.5946	16.6569	48.1067	0.

Fourth (even)

x^4	y^4	z^4	x^2y^2	x^2z^2	y^2z^2	r^2x^2	r^2y^2	r^2z^2	r^4
-336.8825	-95.5354	-435.7243	-45.0937	-140.7453	-97.1383	-522.7215	-237.7674	-673.6079	-1434.0967
184.5287	17.8969	203.1561	0.	80.6533	46.3752	265.1820	64.2721	330.1847	659.6388

Fourth (odd)

x^3y	xy^3	xyz^2	x^3z	xy^2z	xz^3	x^2yz	y^3z	yz^3
0.	0.	0.	0.0001	0.	0.	0.	0.	0.
0.	0.	0.	0.	0.	0.	0.	0.	0.

Expectation Values at Atomic Centers

Center	r^{-1}	δ	xr^{-3}	yr^{-3}	zr^{-3}
			Expectation Value		
H1	-1.0572	0.4064	0.0568	0.	-0.0333
H3	-1.1244	0.4268	0.	0.0553	0.0386
C1	-14.7098	116.4395	0.0484	0.	-0.0311
C3	-14.7141	116.2568	0.	0.	0.0543

Center	$\dfrac{3x^2-r^2}{r^5}$	$\dfrac{3y^2-r^2}{r^5}$	$\dfrac{3z^2-r^2}{r^5}$	$\dfrac{3xy}{r^5}$	$\dfrac{3xz}{r^5}$	$\dfrac{3yz}{r^5}$
H1	0.2316	-0.1827	-0.0489	0.	-0.2376	0.
H3	-0.1825	0.1928	-0.0103	0.	0.	0.2272
C1	-0.0926	0.1682	-0.0756	0.	-0.1466	0.
C3	-0.1489	-0.1300	0.2789	0.	0.	0.

Mulliken Charge

Center	S1	S2	S3	S4	X1	X2	Y1	Y2	Z1	Z2	Total
					Basis Function Type						
H1	0.5049	0.2635									0.7683
H3	0.5243	0.3055									0.8298
C1	0.0428	1.9564	0.7819	0.4595	0.9551	0.0528	0.7223	0.2539	0.8751	0.1225	6.2221
C3	0.0427	1.9563	0.7565	0.5429	0.8763	0.1983	0.9390	0.1939	0.7956	0.0580	6.3596

Overlap Populations

Pair of Atomic Centers

	H2 H1	H4 H3	C1 H1	C1 H2	C1 H3	C2 C1	C3 H1	C3 H3	C3 C1
1A1	0.	0.	-0.0001	0.	0.	0.0016	0.	0.	-0.0004
2A1	0.	0.	0.	0.	0.	0.	0.	-0.0003	-0.0005
3A1	0.	0.0003	0.0114	0.0015	0.0025	0.3973	0.0011	0.0183	0.2591
4A1	0.0001	0.0146	0.0611	0.0063	-0.0074	0.2108	-0.0081	0.2348	-0.1075
5A1	0.0030	0.0088	0.2682	-0.0165	-0.0073	0.0921	0.0004	0.0703	0.0344
6A1	0.0009	0.0299	0.0174	-0.0312	-0.0176	0.3476	0.0248	0.0404	-0.0134
Total A1	0.0040	0.0536	0.3580	-0.0399	-0.0298	1.0495	0.0181	0.3635	0.1718
1B1	0.	0.	0.0001	0.	0.	-0.0032	0.	0.	0.
2B1	-0.0018	0.	0.3265	-0.0130	0.	-0.0812	0.0060	0.	0.0471
3B1	-0.0028	0.	0.0227	0.0399	0.	-0.5591	-0.0443	0.	0.2638
Total B1	-0.0046	0.	0.3493	0.0269	0.	-0.6436	-0.0382	0.	0.3109
1B2	0.	-0.0484	0.	0.	0.0118	0.0457	0.	0.2885	0.0978
2B2	0.	-0.0687	0.	0.	-0.0586	0.5259	0.	0.1314	-0.1721
Total B2	0.	-0.1171	0.	0.	-0.0468	0.5717	0.	0.4199	-0.0743
Total	-0.0006	-0.0635	0.7072	-0.0130	-0.0765	0.9775	-0.0201	0.7834	0.4084
Distance	5.9554	3.4591	2.0223	4.3263	4.2600	2.4561	4.6720	2.0542	2.8628

C3H4*

Molecular Orbitals and Eigenvalues

Center		Basis Function Type									
		S1	S2	S3	S4	X1	X2	Y1	Y2	Z1	Z2
* 1A1	H1	0.00074	-0.00059								
-11.2597	H2	0.00074	-0.00059								
	H3	0.00012	-0.00015								
	H4	0.00012	-0.00015								
	C1	0.03667	0.69163	0.00350	0.00158	-0.00112	0.00041	0.	0.	0.00043	0.00015
	C2	0.03667	0.69163	0.00350	0.00158	0.00112	-0.00041	0.	0.	0.00043	0.00015
	C3	0.00163	0.03069	0.00047	-0.00133	0.	0.	0.	0.	-0.00009	0.00076
* 2A1	H1	0.00001	0.00004								
-11.2460	H2	0.00001	0.00004								
	H3	0.00071	-0.00096								
	H4	0.00071	-0.00096								
	C1	-0.00116	-0.02191	0.00013	-0.00106	0.00007	0.00006	0.	0.	0.00009	-0.00051
	C2	-0.00116	-0.02191	0.00013	-0.00106	-0.00007	-0.00006	0.	0.	0.00009	-0.00051
	C3	0.05188	0.97849	0.00414	0.00413	0.	0.	0.	0.	-0.00116	-0.00004
* 3A1	H1	-0.04755	-0.00020								
-1.1743	H2	-0.04755	-0.00020								
	H3	-0.04547	-0.00368								
	H4	-0.04547	-0.00368								
	C1	0.00655	0.14220	-0.29884	-0.11512	0.15960	-0.02149	0.	0.	-0.07595	0.00873
	C2	0.00655	0.14220	-0.29884	-0.11512	-0.15960	0.02149	0.	0.	-0.07595	0.00873
	C3	0.00559	0.12078	-0.25758	-0.12039	0.	0.	0.	0.	0.13303	-0.01182
* 5A1	H1	-0.18146	-0.11404								
-0.6843	H2	-0.18146	-0.11404								
	H3	-0.08473	-0.05323								
	H4	-0.08473	-0.05323								
	C1	0.00097	0.01972	-0.05636	-0.07078	-0.33201	-0.02777	0.	0.	0.17611	0.01889
	C2	0.00097	0.01972	-0.05636	-0.07078	0.33201	0.02777	0.	0.	0.17611	0.01889
	C3	0.00131	0.02823	-0.06646	-0.09965	0.	0.	0.	0.	-0.23491	-0.01269
* 6A1	H1	0.06645	0.06584								
-0.4951	H2	0.06645	0.06584								
	H3	-0.10739	-0.11199								
	H4	-0.10739	-0.11199								
	C1	0.00039	0.00927	-0.01552	-0.03788	0.31762	0.06087	0.	0.	0.31320	0.05963
	C2	0.00039	0.00927	-0.01552	-0.03788	-0.31762	-0.06087	0.	0.	0.31320	0.05963
	C3	-0.00040	-0.00900	0.02249	0.09679	0.	0.	0.	0.	-0.37634	-0.09429
7A1	H1	0.01500	-0.05080								
0.2352	H2	0.01500	-0.05080								
	H3	-0.00566	-0.12183								
	H4	-0.00566	-0.12183								
	C1	0.00224	0.06229	-0.03056	-1.30769	-0.02910	0.30572	0.	0.	-0.02312	-1.01942
	C2	0.00224	0.06229	-0.03056	-1.30769	0.02910	-0.30572	0.	0.	-0.02312	-1.01942
	C3	-0.00436	-0.11726	0.08700	2.83303	0.	0.	0.	0.	-0.07745	-1.55859
8A1	H1	0.04755	0.51731								
0.3333	H2	0.04755	0.51731								
	H3	-0.03904	-1.51390								
	H4	-0.03904	-1.51390								
	C1	-0.00034	0.00838	0.12535	0.03814	-0.20830	-0.08368	0.	0.	0.08576	0.09604
	C2	-0.00034	0.00838	0.12535	0.03814	0.20830	0.08368	0.	0.	0.08576	0.09604
	C3	-0.00262	-0.05389	0.17041	1.44299	0.	0.	0.	0.	0.02630	1.90020

Center		S1	S2	S3	S4	X1	X2	Y1	Y2	Z1	Z2
9A1	H1	0.06025	1.22670								
0.3610	H2	0.06025	1.22670								
	H3	0.03768	0.98250								
	H4	0.03768	0.98250								
	C1	0.00148	0.04214	-0.01429	-0.94656	-0.37272	-0.30955	0.	0.	0.12695	0.27231
	C2	0.00148	0.04214	-0.01429	-0.94656	0.37272	0.30955	0.	0.	0.12695	0.27231
	C3	0.00194	0.05432	-0.02629	-1.46916	0.	0.	0.	0.	-0.28702	-0.47795
10A1	H1	0.12905	2.02375								
0.4024	H2	0.12905	2.02375								
	H3	0.06982	-0.40532								
	H4	0.06982	-0.40532								
	C1	-0.00122	-0.01387	0.15277	-1.52157	0.22308	-2.45120	0.	0.	-0.07347	1.04270
	C2	-0.00122	-0.01387	0.15277	-1.52157	-0.22308	2.45120	0.	0.	-0.07347	1.04270
	C3	-0.00291	-0.04393	0.29366	-1.38413	0.	0.	0.	0.	-0.06483	0.74688
11A1	H1	-0.03479	-0.38154								
0.4462	H2	-0.03479	-0.38154								
	H3	-0.06916	0.24680								
	H4	-0.06916	0.24680								
	C1	0.00088	0.00786	-0.12391	0.62282	0.21745	-0.11546	0.	0.	0.29476	-0.54465
	C2	0.00088	0.00786	-0.12391	0.62282	-0.21745	0.11546	0.	0.	0.29476	-0.54465
	C3	0.00220	0.04253	-0.15756	-0.13590	0.	0.	0.	0.	-0.54230	0.98496
1A2	H1	0.	0.								
0.1450	H2	0.	0.								
	H3	0.	0.								
	H4	0.	0.								
	C1	0.	0.	0.	0.	0.	0.	0.40593	0.85444	0.	0.
	C2	0.	0.	0.	0.	0.	0.	-0.40593	-0.85444	0.	0.
	C3	0.	0.	0.	0.	0.	0.	0.	0.	0.	0.
2A2	H1	0.	0.								
0.6052	H2	0.	0.								
	H3	0.	0.								
	H4	0.	0.								
	C1	0.	0.	0.	0.	0.	0.	-0.80678	1.21202	0.	0.
	C2	0.	0.	0.	0.	0.	0.	0.80678	-1.21202	0.	0.
	C3	0.	0.	0.	0.	0.	0.	0.	0.	0.	0.
* 1B1	H1	-0.00050	-0.00014								
-11.2577	H2	0.00050	0.00014								
	H3	0.	0.								
	H4	0.	0.								
	C1	-0.03671	-0.69247	-0.00261	-0.00370	0.00025	0.00138	0.	0.	-0.00053	-0.00016
	C2	0.03671	0.69247	0.00261	0.00370	0.00025	0.00138	0.	0.	0.00053	0.00016
	C3	0.	0.	0.	0.	-0.00016	0.00023	0.	0.	0.	0.
* 2B1	H1	-0.19290	-0.08457								
-0.7667	H2	0.19290	0.08457								
	H3	0.	0.								
	H4	0.	0.								
	C1	0.00567	0.12334	-0.27403	-0.21559	-0.21333	-0.03245	0.	0.	0.02553	0.01138
	C2	-0.00567	-0.12334	0.27403	0.21559	-0.21333	-0.03245	0.	0.	-0.02553	-0.01138
	C3	0.	0.	0.	0.	-0.14093	-0.01410	0.	0.	0.	0.

Center		S1	S2	S3	S4	X1	X2	Y1	Y2	Z1	Z2
• 3B1	H1	0.12580	0.11431								
-0.4227	H2	-0.12580	-0.11431								
	H3	0.	0.								
	H4	0.	0.								
	C1	0.00110	0.02370	-0.06228	-0.20784	0.07469	0.06663	0.	0.	-0.33518	-0.13651
	C2	-0.00110	-0.02370	0.06228	0.20784	0.07469	0.06663	0.	0.	0.33518	0.13651
	C3	0.	0.	0.	0.	-0.51039	-0.16539	0.	0.	0.	0.
4B1	H1	0.04323	-0.27564								
0.2309	H2	-0.04323	0.27564								
	H3	0.	0.								
	H4	0.	0.								
	C1	0.00361	0.09993	-0.05259	-2.94608	-0.08401	1.60549	0.	0.	-0.06618	-0.89498
	C2	-0.00361	-0.09993	0.05259	2.94608	-0.08401	1.60549	0.	0.	0.06618	0.89498
	C3	0.	0.	0.	0.	0.12822	0.91184	0.	0.	0.	0.
5B1	H1	-0.02247	-0.35838								
0.2970	H2	0.02247	0.35838								
	H3	0.	0.								
	H4	0.	0.								
	C1	0.00020	0.01441	0.05587	-1.34565	0.07637	1.69400	0.	0.	0.28771	1.04445
	C2	-0.00020	-0.01441	-0.05587	1.34565	0.07637	1.69400	0.	0.	-0.28771	-1.04445
	C3	0.	0.	0.	0.	-0.31426	-1.31712	0.	0.	0.	0.
6B1	H1	-0.11415	-2.69458								
0.4494	H2	0.11415	2.69458								
	H3	0.	0.								
	H4	0.	0.								
	C1	-0.00299	-0.06006	0.20179	-0.08830	-0.05571	2.20921	0.	0.	-0.09575	-1.97154
	C2	0.00299	0.06006	-0.20179	0.08830	-0.05571	2.20921	0.	0.	0.09575	1.97154
	C3	0.	0.	0.	0.	-0.23623	0.31045	0.	0.	0.	0.
7B1	H1	0.05712	0.30718								
0.4829	H2	-0.05712	-0.30718								
	H3	0.	0.								
	H4	0.	0.								
	C1	-0.00022	-0.00412	0.01320	0.09217	0.20469	-0.63060	0.	0.	-0.41796	1.13737
	C2	0.00022	0.00412	-0.01320	-0.09217	0.20469	-0.63060	0.	0.	0.41796	-1.13737
	C3	0.	0.	0.	0.	-0.54746	0.67881	0.	0.	0.	0.
• 1B2	H1	0.	0.								
-0.6037	H2	0.	0.								
	H3	-0.20087	-0.12418								
	H4	0.20087	0.12418								
	C1	0.	0.	0.	0.	0.	0.	-0.18209	-0.03217	0.	0.
	C2	0.	0.	0.	0.	0.	0.	-0.18209	-0.03217	0.	0.
	C3	0.	0.	0.	0.	0.	0.	-0.49946	-0.12282	0.	0.
• 2B2	H1	0.	0.								
-0.3574	H2	0.	0.								
	H3	-0.14562	-0.17480								
	H4	0.14562	0.17480								
	C1	0.	0.	0.	0.	0.	0.	0.45159	0.20603	0.	0.
	C2	0.	0.	0.	0.	0.	0.	0.45159	0.20603	0.	0.
	C3	0.	0.	0.	0.	0.	0.	-0.19394	-0.08614	0.	0.

Center		S1	S2	S3	S4	X1	X2	Y1	Y2	Z1	Z2
3B2	H1	0.	0.								
0.4167	H2	0.	0.								
	H3	0.03560	2.09163								
	H4	-0.03560	-2.09163								
	C1	0.	0.	0.	0.	0.	0.	-0.27068	0.95232	0.	0.
	C2	0.	0.	0.	0.	0.	0.	-0.27068	0.95232	0.	0.
	C3	0.	0.	0.	0.	0.	0.	-0.45731	-2.80529	0.	0.
4B2	H1	0.	0.								
0.4747	H2	0.	0.								
	H3	0.11582	1.21019								
	H4	-0.11582	-1.21019								
	C1	0.	0.	0.	0.	0.	0.	0.21902	0.34884	0.	0.
	C2	0.	0.	0.	0.	0.	0.	0.21902	0.34884	0.	0.
	C3	0.	0.	0.	0.	0.	0.	0.62105	-2.77555	0.	0.
5B2	H1	0.	0.								
0.5236	H2	0.	0.								
	H3	0.01832	-1.01542								
	H4	-0.01832	1.01542								
	C1	0.	0.	0.	0.	0.	0.	-0.53927	0.46507	0.	0.
	C2	0.	0.	0.	0.	0.	0.	-0.53927	0.46507	0.	0.
	C3	0.	0.	0.	0.	0.	0.	0.60312	0.35967	0.	0.

Basis Function Type

Overlap Matrix

First Function   Second Function and Overlap

H1S1 with
    H1S2 .68301   H2S1 .00001   H2S2 .00449   C1S1 .00245   C1S2 .03738   C1S3 .29646   C1S4 .37905
    C1X1 .38221   C1X2 .37371   C1Z1 -.22152 C1Z2 -.21659 C3S3 .00213   C3S4 .04323   C3X1 .00745
    C3X2 .11028   C3Z1 -.00900 C3Z2 -.13333

H1S2 with
    H2S1 .00449   H2S2 .04289   C1S1 .00787   C1S2 .10038   C1S3 .50611   C1S4 .71438   C1X1 .31218
    C1X2 .52254   C1Z1 -.18093 C1Z2 -.30285 C3S1 .00034   C3S2 .00492   C3S3 .05207   C3S4 .17064
    C3X1 .05840   C3X2 .25851   C3Z1 -.07061 C3Z2 -.31255

H2S1 with
    C1S2 .00001   C1S3 .00502   C1S4 .06322   C1X1 -.02202 C1X2 -.21173 C1Z1 -.00531 C1Z2 -.05105

H2S2 with
    C1S1 .00059   C1S2 .00835   C1S3 .07766   C1S4 .21936   C1X1 -.12104 C1X2 -.45348 C1Z1 -.02919
    C1Z2 -.10934

H3S1 with
    H3S2 .68301   H4S1 .01384   H4S2 .12439   C1S2 .00002   C1S3 .00588   C1S4 .06777   C1X1 -.00737
    C1X2 -.06540 C1Y1 .01038   C1Y2 .09210   C1Z1 .02217   C1Z2 .19673   C3S1 .00226   C3S2 .03473
    C3S3 .28584   C3S4 .37305   C3Y1 .36375   C3Y2 .36470   C3Z1 .23308   C3Z2 .23369

H3S2 with
    H4S1 .12439   H4S2 .34563   C1S1 .00065   C1S2 .00920   C1S3 .08355   C1S4 .22967   C1X1 -.03792
    C1X2 -.13777 C1Y1 .05340   C1Y2 .19403   C1Z1 .11407   C1Z2 .41444   C3S1 .00769   C3S2 .09818
    C3S3 .49778   C3S4 .70693   C3Y1 .30358   C3Y2 .51188   C3Z1 .19452   C3Z2 .32799

C1S1 with
    C1S2 .35436   C1S3 .00805   C1S4 .01419   C2S3 .00207   C2S4 .00581   C2X1 .00609   C2X2 .00976
    C3S3 .00074   C3S4 .00422   C3X1 .00138   C3X2 .00381   C3Z1 -.00290 C3Z2 -.00802

C1S2 with
    C1S3 .26226   C1S4 .17745   C2S2 .00004   C2S3 .03093   C2S4 .07499   C2X1 .08008   C2X2 .12237
    C3S3 .01241   C3S4 .05506   C3X1 .01886   C3X2 .04807   C3Z1 -.03971 C3Z2 -.10124

C1S3 with
    C1S4 .79970   C2S1 .00207   C2S2 .03093   C2S3 .25636   C2S4 .40753   C2X1 .38867   C2X2 .55028
    C3S1 .00074   C3S2 .01241   C3S3 .15560   C3S4 .31992   C3X1 .11972   C3X2 .22569   C3Z1 -.25212
    C3Z2 -.47527

C1S4 with
    C2S1 .00581   C2S2 .07499   C2S3 .40753   C2S4 .63994   C2X1 .31081   C2X2 .62706   C3S1 .00422
    C3S2 .05506   C3S3 .31992   C3S4 .54526   C3X1 .12256   C3X2 .27264   C3Z1 -.25810 C3Z2 -.57415

C1X1 with
    C1X2 .53875   C2S1 -.00609 C2S2 -.08008 C2S3 -.38867 C2S4 -.31081 C2X1 -.35329 C2X2 -.02702
    C3S1 -.00138 C3S2 -.01886 C3S3 -.11972 C3S4 -.12256 C3X1 .04815   C3X2 .18753   C3Z1 .17864
    C3Z2 .14703

C1X2 with
    C2S1 -.00976 C2S2 -.12237 C2S3 -.55028 C2S4 -.62706 C2X1 -.02702 C2X2 .21849   C3S1 -.00381
    C3S2 -.04807 C3S3 -.22569 C3S4 -.27264 C3X1 .18753   C3X2 .51718   C3Z1 .14703   C3Z2 .22755

C1Y1 with
    C1Y2 .53875   C2Y1 .21213   C2Y2 .31268   C3Y1 .13298   C3Y2 .25735

C1Y2 with
    C2Y1 .31268   C2Y2 .70776   C3Y1 .25735   C3Y2 .62524

C1Z1 with
    C1Z2 .53875   C2Z1 .21213   C2Z2 .31268   C3S1 .00290   C3S2 .03971   C3S3 .25212   C3S4 .25810
    C3X1 .17864   C3X2 .14703   C3Z1 -.24321 C3Z2 -.05228

C1Z2 with
    C2Z1 .31268   C2Z2 .70776   C3S1 .00802   C3S2 .10124   C3S3 .47527   C3S4 .57415   C3X1 .14703
    C3X2 .22755   C3Z1 -.05228 C3Z2 .14604

C3S1 with
    C3S2 .35436   C3S3 .00805   C3S4 .01419

Overlap Matrix (continued)

First Function   Second Function and Overlap

C3S2 with
     C3S3 .26226   C3S4 .17745

C3S3 with
     C3S4 .79970

C3X1 with
     C3X2 .53875

C3Y1 with
     C3Y2 .53875

C3Z1 with
     C3Z2 .53875

CH2N2*   Diazirine

Molecular Geometry   Symmetry C$_{2v}$

Center	Coordinates		
	X	Y	Z
H1	0.	1.75629330	-1.07625790
H2	0.	-1.75629330	-1.07625790
C1	0.	0.	0.
N1	1.16028270	0.	2.54895490
N2	-1.16028270	0.	2.54895490
Mass	0.	0.	1.64717212

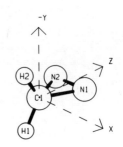

MO Symmetry Types and Occupancy

Representation	A1	A2	B1	B2
Number of MOs	16	2	10	6
Number Occupied	6	0	3	2

Energy Expectation Values

Total	-147.7287	Kinetic	147.8475
Electronic	-211.6275	Potential	-295.5762
Nuclear Repulsion	63.8988	Virial	1.9992
Electronic Potential	-359.4750		
One-Electron Potential	-473.3660		
Two-Electron Potential	113.8910		

Moments of the Charge Distribution

First

x	y	z	Dipole Moment
0.	0.	1.7840	0.9210
0.	0.	-2.7049	

Second

x^2	y^2	z^2	xy	xz	yz	r^2
-34.5087	-18.1320	-55.4103	0.	0.	0.	-108.0509
18.8476	6.1691	42.4982	0.	0.	0.	67.5149

Third

x^3	xy^2	xz^2	x^2y	y^3	yz^2	x^2z	y^2z	z^3	xyz
0.	0.	0.	0.	0.	0.	-18.0182	16.3838	57.9545	0.
0.	0.	0.	0.	0.	0.	16.9964	-16.8012	-56.9474	0.

Fourth (even)

x^4	y^4	z^4	x^2y^2	x^2z^2	y^2z^2	r^2x^2	r^2y^2	r^2z^2	r^4
-147.1758	-74.9914	-346.8623	-29.0461	-74.0473	-81.4698	-250.2692	-185.5073	-502.3795	-938.1559
25.3737	19.0291	163.4522	0.	15.3271	45.7569	40.7008	64.7860	224.5362	330.0229

Expectation Values at Atomic Centers

Center			Expectation Value		
	r^{-1}	δ	xr^{-3}	yr^{-3}	zr^{-3}
H1	-1.0578	0.4177	0.	0.0516	-0.0343
C1	-14.6439	116.3028	0.	0.	-0.0905
N1	-18.2564	189.7805	0.2391	0.	0.1259

	$\dfrac{3x^2-r^2}{r^5}$	$\dfrac{3y^2-r^2}{r^5}$	$\dfrac{3z^2-r^2}{r^5}$	$\dfrac{3xy}{r^5}$	$\dfrac{3xz}{r^5}$	$\dfrac{3yz}{r^5}$
H1	-0.1771	0.1906	-0.0135	0.	0.	-0.2099
C1	-0.1860	-0.2549	0.4409	0.	0.	0.
N1	-0.4865	0.7161	-0.2297	0.	-0.2166	0.

Mulliken Charge

Center				Basis Function Type							Total
	S1	S2	S3	S4	X1	X2	Y1	Y2	Z1	Z2	
H1	0.5156	0.2795									0.7951
C1	0.0427	1.9563	0.7952	0.5422	0.8652	0.1399	0.9924	0.1946	0.7383	0.0367	6.3034
N1	0.0434	1.9559	0.9804	0.8488	1.0333	0.0746	0.7152	0.2519	0.9492	0.2005	7.0532

Overlap Populations

	Pair of Atomic Centers				
	H2	C1	N1	N1	N2
	H1	H1	H1	C1	N1
1A1	0.	0.	0.	-0.0006	0.0011
2A1	0.	-0.0003	0.	-0.0005	0.
3A1	0.	0.0019	0.0006	0.1320	0.5614
4A1	0.0073	0.2049	-0.0025	-0.0954	0.2048
5A1	0.0315	0.1444	-0.0209	0.0634	0.1857
6A1	0.0067	0.0039	-0.0031	0.0569	-0.0013
Total A1	0.0455	0.3548	-0.0259	0.1558	0.9517
1B1	0.	0.	0.	0.	-0.0034
2B1	0.	0.	0.	0.1174	-0.8095
3B1	0.	0.	0.	0.0568	-0.2281
Total B1	0.	0.	0.	0.1742	-1.0411
1B2	-0.0220	0.2101	0.0114	0.1304	0.0966
2B2	-0.0781	0.1825	-0.0450	-0.1501	0.3520
Total B2	-0.1001	0.3926	-0.0337	-0.0197	0.4486
Total	-0.0546	0.7474	-0.0596	0.3104	0.3592
Distance	3.5126	2.0598	4.1920	2.8006	2.3206

Molecular Orbitals and Eigenvalues

Center		S1	S2	S3	S4	X1	X2	Y1	Y2	Z1	Z2
* 1A1	H1	0.00011	0.00025								
-15.6895	H2	0.00011	0.00025								
	C1	-0.00000	0.00006	0.00047	-0.00268	0.	0.	0.	0.	-0.00036	-0.00093
	N1	0.03673	0.69224	0.00202	0.00239	-0.00189	0.00070	0.	0.	-0.00090	-0.00028
	N2	0.03673	0.69224	0.00202	0.00239	0.00189	-0.00070	0.	0.	-0.00090	-0.00028
* 2A1	H1	-0.00079	0.00115								
-11.3193	H2	-0.00079	0.00115								
	C1	-0.05190	-0.97895	-0.00451	-0.00391	0.	0.	0.	0.	-0.00188	0.00032
	N1	0.00001	0.00015	-0.00034	0.00104	-0.00001	-0.00008	0.	0.	0.00020	-0.00057
	N2	0.00001	0.00015	-0.00034	0.00104	0.00001	0.00008	0.	0.	0.00020	-0.00057
* 3A1	H1	0.01949	-0.00094								
-1.5253	H2	0.01949	-0.00094								
	C1	-0.00357	-0.07535	0.17135	0.00625	0.	0.	0.	0.	0.12315	-0.03365
	N1	-0.00706	-0.15072	0.33939	0.17238	-0.20861	0.00918	0.	0.	-0.06721	-0.00870
	N2	-0.00706	-0.15072	0.33939	0.17238	0.20861	-0.00918	0.	0.	-0.06721	-0.00870
* 4A1	H1	0.14367	0.03480								
-0.9385	H2	0.14367	0.03480								
	C1	-0.00861	-0.18499	0.42622	0.34449	0.	0.	0.	0.	-0.03757	0.00700
	N1	0.00232	0.04946	-0.11843	-0.12409	0.12648	0.00222	0.	0.	-0.20389	-0.01892
	N2	0.00232	0.04946	-0.11843	-0.12409	-0.12648	-0.00222	0.	0.	-0.20389	-0.01892
* 5A1	H1	-0.13654	-0.11049								
-0.6687	H2	-0.13654	-0.11049								
	C1	0.00116	0.02780	-0.03816	-0.06933	0.	0.	0.	0.	0.44013	0.02055
	N1	0.00148	0.03220	-0.07229	-0.11464	-0.04843	-0.00960	0.	0.	-0.36223	-0.08583
	N2	0.00148	0.03220	-0.07229	-0.11464	0.04843	0.00960	0.	0.	-0.36223	-0.08583
* 6A1	H1	-0.05840	-0.05203								
-0.5777	H2	-0.05840	-0.05203								
	C1	0.00104	0.01724	-0.08436	0.12918	0.	0.	0.	0.	0.10690	0.08830
	N1	-0.00258	-0.05742	0.11600	0.21446	0.51048	0.13048	0.	0.	-0.03527	0.02026
	N2	-0.00258	-0.05742	0.11600	0.21446	-0.51048	-0.13048	0.	0.	-0.03527	0.02026
7A1	H1	-0.01114	-0.09832								
0.2411	H2	-0.01114	-0.09832								
	C1	0.00424	0.10081	-0.17876	-1.46709	0.	0.	0.	0.	-0.13280	-1.84056
	N1	-0.00229	-0.05861	0.05818	1.13460	0.00833	0.02231	0.	0.	-0.14653	-0.51047
	N2	-0.00229	-0.05861	0.05818	1.13460	-0.00833	-0.02231	0.	0.	-0.14653	-0.51047
8A1	H1	-0.04752	-1.77268								
0.2932	H2	-0.04752	-1.77268								
	C1	-0.00545	-0.12007	0.29462	2.49455	0.	0.	0.	0.	-0.19329	-1.09418
	N1	-0.00003	0.00301	0.02653	0.00123	-0.00474	0.02970	0.	0.	0.04447	0.19327
	N2	-0.00003	0.00301	0.02653	0.00123	0.00474	-0.02970	0.	0.	0.04447	0.19327
9A1	H1	0.09037	0.04603								
0.5059	H2	0.09037	0.04603								
	C1	0.00230	0.04216	-0.18455	-0.58300	0.	0.	0.	0.	-0.98330	0.85629
	N1	-0.00250	-0.04179	0.21878	-0.09181	-0.12309	0.10992	0.	0.	-0.03355	-0.49043
	N2	-0.00250	-0.04179	0.21878	-0.09181	0.12309	-0.10992	0.	0.	-0.03355	-0.49043
10A1	H1	0.08943	0.16760								
0.6019	H2	0.08943	0.16760								
	C1	-0.00373	-0.04092	0.48056	-1.88055	0.	0.	0.	0.	-0.09011	-0.34199
	N1	-0.00169	-0.04547	0.03174	0.25726	0.45998	-0.91822	0.	0.	0.18494	-0.51033
	N2	-0.00169	-0.04547	0.03174	0.25726	-0.45998	0.91822	0.	0.	0.18494	-0.51033

Center		Basis Function Type									
		S1	S2	S3	S4	X1	X2	Y1	Y2	Z1	Z2
1A2	H1	0.	0.								
0.0937	H2	0.	0.								
	C1	0.	0.	0.	0.	0.	0.	0.	0.	0.	0.
	N1	0.	0.	0.	0.	0.	0.	-0.51065	-0.57504	0.	0.
	N2	0.	0.	0.	0.	0.	0.	0.51065	0.57504	0.	0.
* 1B1	H1	0.	0.								
-15.6879	H2	0.	0.								
	C1	0.	0.	0.	0.	0.00019	0.00020	0.	0.	0.	0.
	N1	-0.03674	-0.69247	-0.00153	-0.00509	0.00185	0.00131	0.	0.	0.00101	-0.00025
	N2	0.03674	0.69247	0.00153	0.00509	0.00185	0.00131	0.	0.	-0.00101	0.00025
* 2B1	H1	0.	0.								
-0.8540	H2	0.	0.								
	C1	0.	0.	0.	0.	-0.21656	-0.01899	0.	0.	0.	0.
	N1	0.00728	0.15616	-0.36828	-0.46263	-0.14565	0.01003	0.	0.	0.08640	0.02324
	N2	-0.00728	-0.15616	0.36828	0.46263	-0.14565	0.01003	0.	0.	-0.08640	-0.02324
* 3B1	H1	0.	0.								
-0.4180	H2	0.	0.								
	C1	0.	0.	0.	0.	0.49777	0.16523	0.	0.	0.	0.
	N1	0.00234	0.04821	-0.13273	-0.06061	-0.13368	-0.13745	0.	0.	-0.38199	-0.19017
	N2	-0.00234	-0.04821	0.13273	0.06061	-0.13368	-0.13745	0.	0.	0.38199	0.19017
4B1	H1	0.	0.								
0.2042	H2	0.	0.								
	C1	0.	0.	0.	0.	-0.26177	-1.17256	0.	0.	0.	0.
	N1	-0.00186	-0.04584	0.06044	0.73004	0.00391	-0.06323	0.	0.	-0.35848	-0.66222
	N2	0.00186	0.04584	-0.06044	-0.73004	0.00391	-0.06323	0.	0.	0.35848	0.66222
5B1	H1	0.	0.								
0.3886	H2	0.	0.								
	C1	0.	0.	0.	0.	-0.49240	0.11721	0.	0.	0.	0.
	N1	0.00256	0.06894	-0.04767	-1.95919	0.19588	1.42041	0.	0.	-0.23329	-0.24238
	N2	-0.00256	-0.06894	0.04767	1.95919	0.19588	1.42041	0.	0.	0.23329	0.24238
6B1	H1	0.	0.								
0.5668	H2	0.	0.								
	C1	0.	0.	0.	0.	-0.92541	1.22217	0.	0.	0.	0.
	N1	-0.00281	-0.06208	0.14565	1.29100	0.02326	-1.34827	0.	0.	-0.07297	-0.05150
	N2	0.00281	0.06208	-0.14565	-1.29100	0.02326	-1.34827	0.	0.	0.07297	0.05150
* 1B2	H1	0.16900	0.07817								
-0.6970	H2	-0.16900	-0.07817								
	C1	0.	0.	0.	0.	0.	0.	0.47170	0.11588	0.	0.
	N1	0.	0.	0.	0.	0.	0.	0.26984	0.07461	0.	0.
	N2	0.	0.	0.	0.	0.	0.	0.26984	0.07461	0.	0.
* 2B2	H1	0.17822	0.18443								
-0.4920	H2	-0.17822	-0.18443								
	C1	0.	0.	0.	0.	0.	0.	0.30323	0.05331	0.	0.
	N1	0.	0.	0.	0.	0.	0.	-0.41856	-0.18805	0.	0.
	N2	0.	0.	0.	0.	0.	0.	-0.41856	-0.18805	0.	0.
3B2	H1	-0.09424	-1.51867								
0.3973	H2	0.09424	1.51867								
	C1	0.	0.	0.	0.	0.	0.	-0.48929	2.98099	0.	0.
	N1	0.	0.	0.	0.	0.	0.	-0.14089	-0.40026	0.	0.
	N2	0.	0.	0.	0.	0.	0.	-0.14089	-0.40026	0.	0.

CH2N2*

Center		S1	S2	S3	S4	X1	X2	Y1	Y2	Z1	Z2
4B2	H1	0.01484	1.94979								
0.4294	H2	-0.01484	-1.94979								
	C1	0.	0.	0.	0.	0.	0.	-0.84887	-1.68623	0.	0.
	N1	0.	0.	0.	0.	0.	0.	0.13596	0.24218	0.	0.
	N2	0.	0.	0.	0.	0.	0.	0.13596	0.24218	0.	0.

Basis Function Type

Overlap Matrix

First Function   Second Function and Overlap

H1S1 with
        H1S2 .68301   H2S1 .01222   H2S2 .11798   C1S1 .00222   C1S2 .03427   C1S3 .28397   C1S4 .37199
        C1Y1 .36687   C1Y2 .36948   C1Z1 -.22482   C1Z2 -.22642   N1S2 .00001   N1S3 .00249   N1S4 .04327
        N1X1 -.00290   N1X2 -.04770   N1Y1 .00439   N1Y2 .07221   N1Z1 -.00907   N1Z2 -.14904

H1S2 with
        H2S1 .11798   H2S2 .33438   C1S1 .00766   C1S2 .09779   C1S3 .49630   C1S4 .70560   C1Y1 .30737
        C1Y2 .51894   C1Z1 -.18835   C1Z2 -.31801   N1S1 .00057   N1S2 .00781   N1S3 .06363   N1S4 .18495
        N1X1 -.02413   N1X2 -.10841   N1Y1 .03653   N1Y2 .16410   N1Z1 -.07540   N1Z2 -.33872

C1S1 with
        C1S2 .35436   C1S3 .00805   C1S4 .01419   N1S3 .00023   N1S4 .00359   N1X1 -.00062   N1X2 -.00398
        N1Z1 -.00137   N1Z2 -.00874

C1S2 with
        C1S3 .26226   C1S4 .17745   N1S3 .00521   N1S4 .04752   N1X1 -.00974   N1X2 -.04996   N1Z1 -.02140
        N1Z2 -.10976

C1S3 with
        C1S4 .79970   N1S1 .00069   N1S2 .01068   N1S3 .11913   N1S4 .29221   N1X1 -.08507   N1X2 -.22295
        N1Z1 -.18688   N1Z2 -.48978

C1S4 with
        N1S1 .00352   N1S2 .04554   N1S3 .26560   N1S4 .49676   N1X1 -.08488   N1X2 -.24098   N1Z1 -.18647
        N1Z2 -.52940

C1X1 with
        C1X2 .53875   N1S1 .00117   N1S2 .01562   N1S3 .10089   N1S4 .13427   N1X1 .03262   N1X2 .18149
        N1Z1 -.14493   N1Z2 -.18810

C1X2 with
        N1S1 .00296   N1S2 .03748   N1S3 .18759   N1S4 .26570   N1X1 .13678   N1X2 .46120   N1Z1 -.10536
        N1Z2 -.22582

C1Y1 with
        C1Y2 .53875   N1Y1 .09860   N1Y2 .26711

C1Y2 with
        N1Y1 .18474   N1Y2 .56399

C1Z1 with
        C1Z2 .53875   N1S1 .00256   N1S2 .03433   N1S3 .22163   N1S4 .29497   N1X1 -.14493   N1X2 -.18810
        N1Z1 -.21980   N1Z2 -.14611

C1Z2 with
        N1S1 .00650   N1S2 .08234   N1S3 .41210   N1S4 .58370   N1X1 -.10536   N1X2 -.22582   N1Z1 -.04672
        N1Z2 .06791

N1S1 with
        N1S2 .35584   N1S3 .00894   N1S4 .01446   N2S3 .00096   N2S4 .00473   N2X1 .00363   N2X2 .00944

N1S2 with
        N1S3 .27246   N1S4 .17925   N2S2 .00001   N2S3 .01568   N2S4 .06090   N2X1 .04914   N2X2 .11781

N1S3 with
        N1S4 .79352   N2S1 .00096   N2S2 .01568   N2S3 .17371   N2S4 .33923   N2X1 .29243   N2X2 .53403

N1S4 with
        N2S1 .00473   N2S2 .06090   N2S3 .33923   N2S4 .57134   N2X1 .28100   N2X2 .63390

N1X1 with
        N1X2 .52739   N2S1 -.00363   N2S2 -.04914   N2S3 -.29243   N2S4 -.28100   N2X1 -.33036   N2X2 -.10574

N1X2 with
        N2S1 -.00944   N2S2 -.11781   N2S3 -.53403   N2S4 -.63390   N2X1 -.10574   N2X2 .07014

N1Y1 with
        N1Y2 .52739   N2Y1 .13858   N2Y2 .26083

N1Y2 with
        N2Y1 .26083   N2Y2 .64066

Overlap Matrix (continued)

First Function   Second Function and Overlap

N1Z1 with
    N1Z2 .52739   N2Z1 .13858   N2Z2 .26083

N1Z2 with
    N2Z1 .26083   N2Z2 .64066

03   Ozone

Molecular Geometry   Symmetry C_{2v}

Center	Coordinates		
	X	Y	Z
01	-2.05700001	0.	-1.26549999
02	0.	0.	0.
03	2.05700001	0.	-1.26549999
Mass	0.	0.	-0.84366667

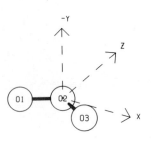

MO Symmetry Types and Occupancy

Representation	A1	A2	B1	B2
Number of MOs	14	2	10	4
Number Occupied	6	1	4	1

Energy Expectation Values

Total	-224.1905	Kinetic	224.3859
Electronic	-292.7468	Potential	-448.5763
Nuclear Repulsion	68.5564	Virial	1.9991

Electronic Potential	-517.1327
One-Electron Potential	-668.4351
Two-Electron Potential	151.3023

Moments of the Charge Distribution

First

x	y	z	Dipole Moment
0.	0.	0.2775	0.2775
0.	0.	0.	

Second

x^2	y^2	z^2	xy	xz	yz	r^2
-81.4093	-11.0321	-20.5956	0.	0.	0.	-113.0370
67.7000	0.	8.5413	0.	0.	0.	76.2413

Third

x^3	xy^2	xz^2	x^2y	y^3	yz^2	x^2z	y^2z	z^3	xyz
0.	0.	0.	0.	0.	0.	28.4516	0.3976	-2.7014	0.
0.	0.	0.	0.	0.	0.	-28.5581	0.	3.6030	0.

Fourth (even)

x^4	y^4	z^4	x^2y^2	x^2z^2	y^2z^2	r^2x^2	r^2y^2	r^2z^2	r^4
-523.5114	-27.9845	-59.6389	-42.9503	-58.2278	-13.3581	-624.6895	-84.2928	-131.2247	-840.2070
286.4555	0.	4.5596	0.	12.0468	0.	298.5023	0.	16.6064	315.1086

Expectation Values at Atomic Centers

Center			Expectation Value		
	r^{-1}	δ	xr^{-3}	yr^{-3}	zr^{-3}
01	-22.1985	288.1050	-0.2151	0.	-0.1214
02	-22.0193	287.9004	0.	0.	0.2493

Center	$\dfrac{3x^2-r^2}{r^5}$	$\dfrac{3y^2-r^2}{r^5}$	$\dfrac{3z^2-r^2}{r^5}$	$\dfrac{3xy}{r^5}$	$\dfrac{3xz}{r^5}$	$\dfrac{3yz}{r^5}$
01	1.5126	0.5442	-2.0568	0.	3.1267	0.
02	0.8982	1.4201	-2.3183	0.	0.	0.

Mulliken Charge

Center				Basis Function Type							Total
	S1	S2	S3	S4	X1	X2	Y1	Y2	Z1	Z2	
01	0.0428	1.9566	1.0209	0.9485	0.9026	0.1988	0.9734	0.4009	1.2944	0.3704	8.1092
02	0.0427	1.9566	1.0328	0.8295	1.0106	0.0411	0.8891	0.3624	1.3370	0.2798	7.7815

Overlap Populations

	Pair of Atomic Centers	
	02	03
	01	01
1A1	-0.0004	0.
2A1	-0.0004	0.
3A1	0.2768	0.0078
4A1	-0.3627	0.0595
5A1	0.0291	-0.0018
6A1	-0.2370	0.0941
Total A1	-0.2946	0.1597
1A2	0.	-0.0928
Total A2	0.	-0.0928
1B1	-0.0004	0.
2B1	0.1679	-0.0922
3B1	-0.0256	-0.0032
4B1	0.0128	-0.1979
Total B1	0.1548	-0.2933
1B2	0.1974	0.0149
Total B2	0.1974	0.0149
Total	0.0575	-0.2115
Distance	2.4151	4.1140

Molecular Orbitals and Eigenvalues

Center		S1	S2	S3	S4	X1	X2	Y1	Y2	Z1	Z2
* 1A1	01	0.00010	0.00189	0.00032	-0.00108	0.00004	-0.00059	0.	0.	0.00007	-0.00039
-20.8980	02	0.05113	0.97944	0.00309	0.00346	0.	0.	0.	0.	-0.00177	0.00003
	03	0.00010	0.00189	0.00032	-0.00108	-0.00004	0.00059	0.	0.	0.00007	-0.00039
* 2A1	01	0.03615	0.69260	0.00224	0.00203	0.00116	0.00018	0.	0.	0.00065	-0.00001
-20.7155	02	-0.00015	-0.00276	0.00070	-0.00222	0.	0.	0.	0.	-0.00003	0.00089
	03	0.03615	0.69260	0.00224	0.00203	-0.00116	-0.00018	0.	0.	0.00065	-0.00001
* 3A1	01	-0.00451	-0.09776	0.22256	0.10502	0.10661	-0.00843	0.	0.	0.05426	-0.00136
-1.7847	02	-0.00880	-0.19127	0.43325	0.32446	0.	0.	0.	0.	-0.14215	-0.03496
	03	-0.00451	-0.09776	0.22256	0.10502	-0.10661	0.00843	0.	0.	0.05426	-0.00136
* 4A1	01	-0.00628	-0.13701	0.31347	0.34688	-0.05936	0.00431	0.	0.	-0.09063	-0.01362
-1.0977	02	0.00714	0.15390	-0.37484	-0.41151	0.	0.	0.	0.	-0.19981	-0.03598
	03	-0.00628	-0.13701	0.31347	0.34688	0.05936	-0.00431	0.	0.	-0.09063	-0.01362
* 5A1	01	0.00251	0.05629	-0.11682	-0.23019	0.28973	0.06339	0.	0.	-0.12701	-0.05306
-0.8369	02	0.00060	0.01254	-0.03478	0.01543	0.	0.	0.	0.	-0.57096	-0.21720
	03	0.00251	0.05629	-0.11682	-0.23019	-0.28973	-0.06339	0.	0.	-0.12701	-0.05306
* 6A1	01	-0.00014	-0.00467	-0.00441	0.06357	0.09322	0.07776	0.	0.	-0.49742	-0.23016
-0.5615	02	-0.00155	-0.03482	0.07293	0.15172	0.	0.	0.	0.	0.39914	0.17744
	03	-0.00014	-0.00467	-0.00441	0.06357	-0.09322	-0.07776	0.	0.	-0.49742	-0.23016
7A1	01	0.00243	0.05627	-0.10842	-0.54590	-0.40784	-0.61709	0.	0.	-0.16402	-0.24842
0.2533	02	-0.00563	-0.12723	0.27627	1.29141	0.	0.	0.	0.	-0.39599	-0.62306
	03	0.00243	0.05627	-0.10842	-0.54590	0.40784	0.61709	0.	0.	-0.16402	-0.24842
* 1A2	01	0.	0.	0.	0.	0.	0.	-0.52968	-0.29273	0.	0.
-0.4936	02	0.	0.	0.	0.	0.	0.	0.	0.	0.	0.
	03	0.	0.	0.	0.	0.	0.	0.52968	0.29273	0.	0.
* 1B1	01	-0.03615	-0.69260	-0.00207	-0.00239	-0.00109	-0.00014	0.	0.	-0.00062	-0.00033
-20.7156	02	0.	0.	0.	0.	0.00001	-0.00157	0.	0.	0.	0.
	03	0.03615	0.69260	0.00207	0.00239	-0.00109	-0.00014	0.	0.	0.00062	0.00033
* 3B1	01	0.00370	0.08136	-0.18492	-0.26366	0.20266	0.06355	0.	0.	0.29224	0.08763
-0.7940	02	0.	0.	0.	0.	-0.50430	-0.12054	0.	0.	0.	0.
	03	-0.00370	-0.08136	0.18492	0.26366	0.20266	0.06355	0.	0.	-0.29224	-0.08763
* 4B1	01	-0.00036	-0.00724	0.02153	-0.01445	-0.41234	-0.19665	0.	0.	0.39197	0.16306
-0.5790	02	0.	0.	0.	0.	0.13372	-0.03526	0.	0.	0.	0.
	03	0.00036	0.00724	-0.02153	0.01445	-0.41234	-0.19665	0.	0.	-0.39197	-0.16306
5B1	01	0.00317	0.07427	-0.13749	-0.91031	-0.23771	-0.62257	0.	0.	-0.26272	-0.54746
0.3425	02	0.	0.	0.	0.	-0.59869	-1.31425	0.	0.	0.	0.
	03	-0.00317	-0.07427	0.13749	0.91031	-0.23771	-0.62257	0.	0.	0.26272	0.54746
* 1B2	01	0.	0.	0.	0.	0.	0.	-0.30141	-0.10534	0.	0.
-0.7909	02	0.	0.	0.	0.	0.	0.	-0.57471	-0.23957	0.	0.
	03	0.	0.	0.	0.	0.	0.	-0.30141	-0.10534	0.	0.
2B2	01	0.	0.	0.	0.	0.	0.	0.43130	0.32735	0.	0.
-0.0692	02	0.	0.	0.	0.	0.	0.	-0.54955	-0.38892	0.	0.
	03	0.	0.	0.	0.	0.	0.	0.43130	0.32735	0.	0.

Overlap Matrix

First Function			Second Function and Overlap				

O1S1 with
| O1S2 .35461 | O1S3 .00974 | O1S4 .01469 | O2S3 .00017 | O2S4 .00283 | O2X1 -.00099 | O2X2 -.00652 |
| O2Z1 -.00061 | O2Z2 -.00401 | O3S4 .00012 | O3X2 -.00122 | | | |

O1S2 with
| O1S3 .28286 | O1S4 .18230 | O2S3 .00338 | O2S4 .03727 | O2X1 -.01488 | O2X2 -.08247 | O2Z1 -.00915 |
| O2Z2 -.05073 | O3S4 .00183 | O3X1 -.00002 | O3X2 -.01647 | | | |

O1S3 with
| O1S4 .79029 | O2S1 .00017 | O2S2 .00338 | O2S3 .07229 | O2S4 .22540 | O2X1 -.13263 | O2X2 -.39393 |
| O2Z1 -.08160 | O2Z2 -.24235 | O3S3 .00042 | O3S4 .02059 | O3X1 -.00271 | O3X2 -.11571 | |

O1S4 with
| O2S1 .00283 | O2S2 .03727 | O2S3 .22540 | O2S4 .43873 | O2X1 -.19471 | O2X2 -.52481 | O2Z1 -.11979 |
| O2Z2 -.32287 | O3S1 .00012 | O3S2 .00183 | O3S3 .02059 | O3S4 .09157 | O3X1 -.03816 | O3X2 -.27220 |

O1X1 with
| O1X2 .50607 | O2S1 .00099 | O2S2 .01488 | O2S3 .13263 | O2S4 .19471 | O2X1 -.15354 | O2X2 -.08214 |
| O2Z1 -.13335 | O2Z2 -.16575 | O3S2 .00002 | O3S3 .00271 | O3S4 .03816 | O3X1 -.01071 | O3X2 -.13484 |

O1X2 with
| O2S1 .00652 | O2S2 .08247 | O2S3 .39393 | O2S4 .52481 | O2X1 -.08214 | O2X2 .05129 | O2Z1 -.16575 |
| O2Z2 -.29830 | O3S1 .00122 | O3S2 .01647 | O3S3 .11571 | O3S4 .27220 | O3X1 -.13484 | O3X2 -.42891 |

O1Y1 with
| O1Y2 .50607 | O2Y1 .06321 | O2Y2 .18728 | O3Y1 .00094 | O3Y2 .02865 | | |

O1Y2 with
| O2Y1 .18728 | O2Y2 .53616 | O3Y1 .02865 | O3Y2 .16387 | | | |

O1Z1 with
| O1Z2 .50607 | O2S1 .00061 | O2S2 .00915 | O2S3 .08160 | O2S4 .11979 | O2X1 -.13335 | O2X2 -.16575 |
| O2Z1 -.01883 | O2Z2 .08531 | O3Z1 .00094 | O3Z2 .02865 | | | |

O1Z2 with
| O2S1 .00401 | O2S2 .05073 | O2S3 .24235 | O2S4 .32287 | O2X1 -.16575 | O2X2 -.29830 | O2Z1 .08531 |
| O2Z2 .35264 | O3Z1 .02865 | O3Z2 .16387 | | | | |

O2S1 with
| O2S2 .35461 | O2S3 .00974 | O2S4 .01469 | | | | |

O2S2 with
| O2S3 .28286 | O2S4 .18230 | | | | | |

O2S3 with
| O2S4 .79029 | | | | | | |

O2X1 with
| O2X2 .50607 | | | | | | |

O2Y1 with
| O2Y2 .50607 | | | | | | |

O2Z1 with
| O2Z2 .50607 | | | | | | |

CF2   Difluorocarbene

Molecular Geometry   Symmetry C_{2v}

Center	Coordinates		
	X	Y	Z
C1	0.	0.	0.
F1	-1.94770300	0.	-1.49722201
F2	1.94770300	0.	-1.49722201
Mass	0.	0.	-1.13786647

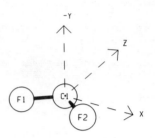

MO Symmetry Types and Occupancy

Representation	A1	A2	B1	B2
Number of MOs	14	2	10	4
Number Occupied	6	1	4	1

Energy Expectation Values

Total	-236.6114	Kinetic	236.7642
Electronic	-301.3671	Potential	-473.3756
Nuclear Repulsion	64.7557	Virial	1.9994
Electronic Potential	-538.1313		
One-Electron Potential	-693.3869		
Two-Electron Potential	155.2555		

Moments of the Charge Distribution

First

x	y	z	Dipole Moment
0.	0.	-0.3834	0.0246
0.	0.	0.3588	

Second

x^2	y^2	z^2	xy	xz	yz	r^2
-81.5421	-10.3390	-23.1087	0.	0.	0.	-114.9898
68.2838	0.	10.0929	0.	0.	0.	78.3767

Third

x^3	xy^2	xz^2	x^2y	y^3	yz^2	x^2z	y^2z	z^3	xyz
0.	0.	0.	0.	0.	0.	23.3874	-1.8108	-20.2433	0.
0.	0.	0.	0.	0.	0.	-24.5382	0.	8.0041	0.

Fourth (even)

x^4	y^4	z^4	x^2y^2	x^2z^2	y^2z^2	r^2x^2	r^2y^2	r^2z^2	r^4
-460.5444	-25.2022	-97.8493	-36.5843	-52.2861	-16.6174	-549.4149	-78.4039	-166.7528	-794.5717
259.0380	0.	10.3583	0.	8.8179	0.	267.8559	0.	19.1762	287.0321

Expectation Values at Atomic Centers

Center	r^{-1}	δ	xr^{-3}	yr^{-3}	zr^{-3}
C1	-14.4794	117.3462	0.	0.	0.2150
F1	-26.4999	413.7151	-0.2265	0.	-0.1488

Center	$\dfrac{3x^2-r^2}{r^5}$	$\dfrac{3y^2-r^2}{r^5}$	$\dfrac{3z^2-r^2}{r^5}$	$\dfrac{3xy}{r^5}$	$\dfrac{3xz}{r^5}$	$\dfrac{3yz}{r^5}$
C1	0.1813	0.8336	-1.0149	0.	0.	0.
F1	0.8419	-0.3652	-0.4767	0.	2.1331	0.

Mulliken Charge

Center	S1	S2	S3	S4	X1	X2	Y1	Y2	Z1	Z2	Total
C1	0.0431	1.9562	1.0058	0.7021	0.5412	-0.0222	0.2206	0.0255	0.9539	0.0680	5.4942
F1	0.0436	1.9557	1.0336	0.8967	1.2732	0.3903	1.3400	0.5369	1.3557	0.4271	9.2529

Overlap Populations

Pair of Atomic Centers

	F1 C1	F2 F1
1A1	-0.0001	0.
2A1	-0.0007	0.
3A1	0.1780	0.0514
4A1	0.0845	0.0031
5A1	0.0840	0.0789
6A1	-0.3864	0.0300
Total A1	-0.0407	0.1633
1A2	0.	-0.0590
Total A2	0.	-0.0590
1B1	-0.0001	0.
2B1	0.0857	-0.0609
3B1	0.1381	-0.0254
4B1	0.0182	-0.1947
Total B1	0.2420	-0.2810
1B2	0.1219	0.0421
Total B2	0.1219	0.0421
Total	0.3231	-0.1346
Distance	2.4567	3.8954

CF2

Molecular Orbitals and Eigenvalues

Center		S1	S2	S3	S4	X1	X2	Y1	Y2	Z1	Z2
* 1A1	C1	0.00000	-0.00003	-0.00005	0.00057	0.	0.	0.	0.	-0.00014	-0.00032
-26.3790	F1	-0.03658	-0.69219	-0.00311	-0.00111	-0.00071	0.00009	0.	0.	-0.00047	0.00012
	F2	-0.03658	-0.69219	-0.00311	-0.00111	0.00071	-0.00009	0.	0.	-0.00047	0.00012
* 2A1	C1	0.05193	0.97944	0.00318	0.00268	0.	0.	0.	0.	-0.00429	0.00103
-11.5111	F1	-0.00001	-0.00021	0.00074	-0.00137	0.00051	-0.00075	0.	0.	0.00046	-0.00042
	F2	-0.00001	-0.00021	0.00074	-0.00137	-0.00051	0.00075	0.	0.	0.00046	-0.00042
* 3A1	C1	-0.00416	-0.08839	0.19400	0.01373	0.	0.	0.	0.	-0.12406	0.02421
-1.7865	F1	-0.00748	-0.16053	0.35982	0.29509	0.08659	0.01553	0.	0.	0.04871	0.00602
	F2	-0.00748	-0.16053	0.35982	0.29509	-0.08659	-0.01553	0.	0.	0.04871	0.00602
* 4A1	C1	-0.00715	-0.14592	0.41043	0.08743	0.	0.	0.	0.	-0.15788	0.05058
-0.9751	F1	0.00355	0.07576	-0.18001	-0.21413	0.32701	0.11327	0.	0.	0.18917	0.05921
	F2	0.00355	0.07576	-0.18001	-0.21413	-0.32701	-0.11327	0.	0.	0.18917	0.05921
* 5A1	C1	0.00259	0.05396	-0.14194	-0.07581	0.	0.	0.	0.	-0.19117	-0.02339
-0.8124	F1	0.00011	0.00243	-0.00568	-0.01585	0.28667	0.12648	0.	0.	-0.39640	-0.18327
	F2	0.00011	0.00243	-0.00568	-0.01585	-0.28667	-0.12648	0.	0.	-0.39640	-0.18327
* 6A1	C1	-0.00622	-0.13404	0.32875	0.55941	0.	0.	0.	0.	0.52064	0.08490
-0.4901	F1	0.00076	0.01662	-0.03660	-0.09139	-0.05543	-0.02032	0.	0.	-0.29307	-0.16656
	F2	0.00076	0.01662	-0.03660	-0.09139	0.05543	0.02032	0.	0.	-0.29307	-0.16656
7A1	C1	-0.00304	-0.07924	0.08875	1.23822	0.	0.	0.	0.	-0.23715	-1.30779
0.2984	F1	0.00274	0.06587	-0.09402	-0.80486	-0.17633	-0.36141	0.	0.	-0.00269	-0.04781
	F2	0.00274	0.06587	-0.09402	-0.80486	0.17633	0.36141	0.	0.	-0.00269	-0.04781
8A1	C1	0.00069	-0.01125	-0.22382	1.03629	0.	0.	0.	0.	-1.11227	0.75163
0.5238	F1	0.00197	0.03903	-0.12688	-0.20587	-0.18234	-0.23391	0.	0.	-0.08933	-0.24622
	F2	0.00197	0.03903	-0.12688	-0.20587	0.18234	0.23391	0.	0.	-0.08933	-0.24622
9A1	C1	-0.01246	-0.14126	1.58647	-1.64321	0.	0.	0.	0.	-0.30956	0.33900
0.6172	F1	0.00054	0.00626	-0.06551	0.32847	-0.26249	0.01347	0.	0.	-0.20690	0.00786
	F2	0.00054	0.00626	-0.06551	0.32847	0.26249	-0.01347	0.	0.	-0.20690	0.00786
* 1A2	C1	0.	0.	0.	0.	0.	0.	0.	0.	0.	0.
-0.7114	F1	0.	0.	0.	0.	0.	0.	-0.54280	-0.27213	0.	0.
	F2	0.	0.	0.	0.	0.	0.	0.54280	0.27213	0.	0.
* 1B1	C1	0.	0.	0.	0.	0.00011	0.00028	0.	0.	0.	0.
-26.3790	F1	0.03658	0.69219	0.00319	0.00089	0.00072	-0.00022	0.	0.	0.00045	-0.00005
	F2	-0.03658	-0.69219	-0.00319	-0.00089	0.00072	-0.00022	0.	0.	-0.00045	0.00005
* 2B1	C1	0.	0.	0.	0.	0.14647	-0.02870	0.	0.	0.	0.
-1.6902	F1	0.00816	0.17534	-0.39195	-0.34587	-0.05607	-0.01343	0.	0.	-0.05030	-0.00648
	F2	-0.00816	-0.17534	0.39195	0.34587	-0.05607	-0.01343	0.	0.	0.05030	0.00648
* 3B1	C1	0.	0.	0.	0.	0.36501	-0.05207	0.	0.	0.	0.
-0.8277	F1	-0.00185	-0.03908	0.09730	0.11739	-0.21473	-0.07115	0.	0.	-0.42507	-0.16111
	F2	0.00185	0.03908	-0.09730	-0.11739	-0.21473	-0.07115	0.	0.	0.42507	0.16111
* 4B1	C1	0.	0.	0.	0.	0.08651	-0.04693	0.	0.	0.	0.
-0.6991	F1	-0.00022	-0.00396	0.01638	-0.01314	-0.49592	-0.22726	0.	0.	0.27527	0.12950
	F2	0.00022	0.00396	-0.01638	0.01314	-0.49592	-0.22726	0.	0.	-0.27527	-0.12950
5B1	C1	0.	0.	0.	0.	0.35893	1.27850	0.	0.	0.	0.
0.2679	F1	-0.00226	-0.05336	0.08311	0.61362	0.02099	0.04147	0.	0.	0.18031	0.24446
	F2	0.00226	0.05336	-0.08311	-0.61362	0.02099	0.04147	0.	0.	-0.18031	-0.24446

Center		Basis Function Type									
		S1	S2	S3	S4	X1	X2	Y1	Y2	Z1	Z2
6B1	C1	0.	0.	0.	0.	-1.25321	0.88751	0.	0.	0.	0.
0.5981	F1	0.00152	0.02719	-0.11917	0.04789	-0.24480	-0.23530	0.	0.	-0.21492	-0.13634
	F2	-0.00152	-0.02719	0.11917	-0.04789	-0.24480	-0.23530	0.	0.	0.21492	0.13634
* 1B2	C1	0.	0.	0.	0.	0.	0.	0.23426	0.02589	0.	0.
-0.7941	F1	0.	0.	0.	0.	0.	0.	0.48696	0.22650	0.	0.
	F2	0.	0.	0.	0.	0.	0.	0.48696	0.22650	0.	0.
2B2	C1	0.	0.	0.	0.	0.	0.	-0.66839	-0.58650	0.	0.
0.0595	F1	0.	0.	0.	0.	0.	0.	0.25847	0.23712	0.	0.
	F2	0.	0.	0.	0.	0.	0.	0.25847	0.23712	0.	0.
3B2	C1	0.	0.	0.	0.	0.	0.	-0.95103	1.19845	0.	0.
0.4267	F1	0.	0.	0.	0.	0.	0.	0.00503	-0.12578	0.	0.
	F2	0.	0.	0.	0.	0.	0.	0.00503	-0.12578	0.	0.

Overlap Matrix

First Function		Second Function and Overlap				

C1S1 with

C1S2	.35436	C1S3	.00805	C1S4	.01419	F1S3	.00005	F1S4	.00319	F1X1	.00047	F1X2	.00879
F1Z1	.00036	F1Z2	.00676										

C1S2 with

C1S3	.26226	C1S4	.17745	F1S3	.00263	F1S4	.04397	F1X1	.01053	F1X2	.10936	F1Z1	.00809
F1Z2	.08406												

C1S3 with

C1S4	.79970	F1S1	.00110	F1S2	.01522	F1S3	.11984	F1S4	.29666	F1X1	.13191	F1X2	.43434
F1Z1	.10140	F1Z2	.33388										

C1S4 with

F1S1	.00312	F1S2	.04004	F1S3	.22936	F1S4	.46167	F1X1	.09975	F1X2	.37389	F1Z1	.07668
F1Z2	.28741												

C1X1 with

C1X2	.53875	F1S1	-.00259	F1S2	-.03354	F1S3	-.19758	F1S4	-.31522	F1X1	-.11513	F1X2	-.08562
F1Z1	-.15486	F1Z2	-.30081										

C1X2 with

F1S1	-.00415	F1S2	-.05266	F1S3	-.27808	F1S4	-.46968	F1X1	.02617	F1X2	.18928	F1Z1	-.07342
F1Z2	-.23008												

C1Y1 with

C1Y2	.53875	F1Y1	.08633	F1Y2	.30571

C1Y2 with

F1Y1	.12167	F1Y2	.48858

C1Z1 with

C1Z2	.53875	F1S1	-.00199	F1S2	-.02579	F1S3	-.15188	F1S4	-.24231	F1X1	-.15486	F1X2	-.30081
F1Z1	-.03271	F1Z2	.07447										

C1Z2 with

F1S1	-.00319	F1S2	-.04048	F1S3	-.21376	F1S4	-.36105	F1X1	-.07342	F1X2	-.23008	F1Z1	.06524
F1Z2	.31172												

F1S1 with

F1S2	.35647	F1S3	.01003	F1S4	.01474	F2S4	.00007	F2X2	-.00078

F1S2 with

F1S3	.28653	F1S4	.18096	F2S4	.00098	F2X2	-.01055

F1S3 with

F1S4	.78531	F2S3	.00013	F2S4	.01234	F2X1	-.00094	F2X2	-.08029

F1S4 with

F2S1	.00007	F2S2	.00098	F2S3	.01234	F2S4	.06643	F2X1	-.02380	F2X2	-.21717

F1X1 with

F1X2	.49444	F2S3	.00094	F2S4	.02380	F2X1	-.00426	F2X2	-.10091

F1X2 with

F2S1	.00078	F2S2	.01055	F2S3	.08029	F2S4	.21717	F2X1	-.10091	F2X2	-.39572

F1Y1 with

F1Y2	.49444	F2Y1	.00032	F2Y2	.01813

F1Y2 with

F2Y1	.01813	F2Y2	.12575

F1Z1 with

F1Z2	.49444	F2Z1	.00032	F2Z2	.01813

F1Z2 with

F2Z1	.01813	F2Z2	.12575

FNO   Nitrozylfluoride

Molecular Geometry   Symmetry C₁ₕ

Center	Coordinates		
	X	Y	Z
N1	0.	0.	0.
O1	0.	0.	2.13541999
F1	-2.69918701	0.	-0.98242610
Mass	-1.04661489	0.	0.31617155

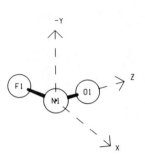

MO Symmetry Types and Occupancy

Representation	A*	A**
Number of MOs	24	6
Number Occupied	10	2

Energy Expectation Values

Total	-228.5521	Kinetic	228.5536
Electronic	-294.1684	Potential	-457.1057
Nuclear Repulsion	65.6163	Virial	2.0000
Electronic Potential	-522.7220		
One-Electron Potential	-674.0634		
Two-Electron Potential	151.3414		

Moments of the Charge Distribution

First

x	y	z	Dipole Moment
-0.1232	0.	-0.3636	0.7602
0.8261	0.	0.6534	

Second

x^2	y^2	z^2	xy	xz	yz	r^2
-53.5630	-10.5672	-55.3923	0.	-33.0900	0.	-119.5225
41.0100	0.	42.3543	0.	32.2303	0.	83.3643

Third

x^3	xy^2	xz^2	x^2y	y^3	yz^2	x^2z	y^2z	z^3	xyz
20.0586	-0.9195	-5.1478	0.	0.	0.	19.2514	-0.3464	-31.3089	0.
-23.4215	0.	3.3625	0.	0.	0.	-18.4000	0.	28.2385	0.

Fourth (even)

x^4	y^4	z^4	x^2y^2	x^2z^2	y^2z^2	r^2x^2	r^2y^2	r^2z^2	r^4
-227.0211	-25.3089	-261.8871	-27.2275	-122.9424	-27.6277	-377.1910	-80.1641	-412.4571	-869.8122
85.1236	0.	113.2952	0.	71.2187	0.	156.3423	0.	184.5139	340.8562

Fourth (odd)

x^3y	xy^3	xyz^2	x^3z	xy^2z	xz^3	x^2yz	y^3z	yz^3
0.	0.	0.	-110.3290	-13.7849	-128.0452	0.	0.	0.
0.	0.	0.	66.8954	0.	82.7533	0.	0.	0.

Expectation Values at Atomic Centers

Center	r^{-1}	δ	Expectation Value xr^{-3}	yr^{-3}	zr^{-3}
N1	-18.0597	190.2342	0.1209	0.	-0.2367
O1	-22.1537	287.7438	0.0093	0.	0.3249
F1	-26.5753	413.8890	-0.1545	0.	-0.0709

	$\dfrac{3x^2-r^2}{r^5}$	$\dfrac{3y^2-r^2}{r^5}$	$\dfrac{3z^2-r^2}{r^5}$	$\dfrac{3xy}{r^5}$	$\dfrac{3xz}{r^5}$	$\dfrac{3yz}{r^5}$
N1	0.2690	0.6359	-0.9049	0.	0.9512	0.
O1	-2.6273	1.7198	0.9075	0.	0.3301	0.
F1	2.7936	-1.4288	-1.3647	0.	1.7261	0.

Mulliken Charge

Center	S1	S2	S3	S4	X1	X2	Y1	Y2	Z1	Z2	Total
N1	0.0436	1.9559	1.0771	0.7345	0.7530	0.1111	0.6566	0.1640	1.0331	0.0119	6.5409
O1	0.0427	1.9566	1.0048	0.8593	1.3866	0.4115	0.8806	0.3438	1.0698	0.1150	8.0708
F1	0.0436	1.9558	1.0314	0.9516	1.0807	0.4449	1.3984	0.5565	1.3939	0.5316	9.3883

Overlap Populations

Pair of Atomic Centers

	O1 N1	F1 N1	F1 O1
1A*	0.	-0.0001	0.
2A*	-0.0011	0.	0.
3A*	-0.0014	-0.0002	0.
4A*	0.5880	0.0617	0.0205
5A*	0.0464	0.0710	-0.0330
6A*	-0.4374	0.0138	-0.0065
7A*	0.2778	0.0571	0.0258
8A*	0.0643	0.1062	-0.0028
9A*	0.0010	-0.0164	-0.0728
10A*	-0.3018	-0.2526	-0.0786
Total A*	0.2358	0.0404	-0.1474
1A**	0.3364	0.0877	0.0219
2A**	0.0641	-0.0579	-0.0420
Total A**	0.4004	0.0299	-0.0201
Total	0.6363	0.0702	-0.1675
Distance	2.1354	2.8724	4.1239

Molecular Orbitals and Eigenvalues

Center		S1	S2	S3	S4	X1	X2	Y1	Y2	Z1	Z2
* 1A*	N1	-0.00000	0.00002	0.00019	-0.00058	-0.00003	0.00040	0.	0.	-0.00002	0.00001
-26.3001	O1	0.00000	-0.00000	-0.00004	0.00022	0.00002	-0.00013	0.	0.	-0.00002	-0.00008
	F1	0.05174	0.97897	0.00422	0.00141	0.00068	-0.00003	0.	0.	0.00031	-0.00004
* 2A*	N1	-0.00000	-0.00008	-0.00033	0.00226	0.00001	-0.00023	0.	0.	0.00020	0.00129
-20.7617	O1	-0.05112	-0.97936	-0.00333	-0.00383	0.00009	-0.00001	0.	0.	0.00232	0.00020
	F1	-0.00000	-0.00000	0.00009	-0.00030	-0.00002	-0.00009	0.	0.	0.00002	-0.00024
* 3A*	N1	0.05196	0.97930	0.00243	0.00423	-0.00143	0.00006	0.	0.	0.00280	-0.00006
-15.8916	O1	-0.00002	-0.00043	0.00096	-0.00266	0.00010	0.00007	0.	0.	-0.00043	0.00144
	F1	-0.00000	-0.00003	0.00032	-0.00088	0.00014	-0.00044	0.	0.	0.00005	-0.00016
* 4A*	N1	0.00683	0.14326	-0.34272	-0.09756	0.03679	0.00277	0.	0.	-0.22034	0.02969
-1.7576	O1	0.00813	0.17701	-0.39538	-0.24601	0.00709	-0.00048	0.	0.	0.21225	0.00141
	F1	0.00301	0.06457	-0.14419	-0.11362	-0.03857	-0.01012	0.	0.	-0.02300	-0.01043
* 5A*	N1	-0.00044	-0.00843	0.02652	-0.01853	0.08156	-0.00117	0.	0.	0.12526	-0.01935
-1.5677	O1	-0.00342	-0.07466	0.16687	0.13399	0.01662	0.00409	0.	0.	-0.07164	-0.00419
	F1	0.01089	0.23443	-0.52077	-0.48591	-0.04970	-0.02679	0.	0.	-0.02067	-0.01065
* 6A*	N1	-0.00807	-0.16705	0.44964	0.32468	-0.02074	0.01201	0.	0.	-0.08500	-0.00915
-0.9953	O1	0.00664	0.14588	-0.33060	-0.42950	-0.00813	-0.00410	0.	0.	-0.27377	-0.05216
	F1	0.00259	0.05527	-0.13065	-0.14970	0.17693	0.06871	0.	0.	0.04026	0.01485
* 7A*	N1	-0.00061	-0.01127	0.04335	-0.01372	0.39826	0.06635	0.	0.	0.09584	-0.02204
-0.8179	O1	0.00112	0.02443	-0.05762	-0.06069	0.50463	0.17723	0.	0.	-0.16905	-0.04860
	F1	-0.00208	-0.04460	0.10451	0.13499	-0.23455	-0.09854	0.	0.	-0.18576	-0.08288
* 8A*	N1	0.00260	0.05849	-0.11523	-0.26378	-0.12024	-0.05169	0.	0.	0.48450	0.02879
-0.7832	O1	0.00241	0.05340	-0.11636	-0.18799	-0.18228	-0.06533	0.	0.	-0.44196	-0.12284
	F1	-0.00100	-0.02069	0.05613	0.04371	-0.21973	-0.10463	0.	0.	0.13969	0.06368
* 9A*	N1	0.00094	0.02180	-0.03789	-0.13880	-0.09274	-0.03793	0.	0.	0.13149	-0.03672
-0.6355	O1	-0.00023	-0.00459	0.01564	0.03860	-0.19374	-0.06241	0.	0.	-0.13293	-0.04941
	F1	-0.00020	-0.00348	0.01597	-0.01229	0.19736	0.08676	0.	0.	-0.69904	-0.34218
* 10A*	N1	-0.00330	-0.07084	0.17007	0.33477	0.26419	0.08211	0.	0.	-0.20768	-0.07568
-0.5188	O1	0.00037	0.00706	-0.02631	-0.00750	-0.48757	-0.25244	0.	0.	0.08642	0.03437
	F1	-0.00002	0.00120	0.01250	-0.05196	-0.48387	-0.29650	0.	0.	-0.11483	-0.06891
11A*	N1	-0.00320	-0.06681	0.18464	0.44965	-0.67622	-0.68948	0.	0.	-0.14045	-0.06669
0.1500	O1	0.00028	0.00722	-0.00767	-0.09263	0.28917	0.36150	0.	0.	0.05817	0.09200
	F1	0.00255	0.05823	-0.10602	-0.46365	-0.41864	-0.42889	0.	0.	-0.13415	-0.15764
12A*	N1	-0.00228	-0.06254	0.03857	1.94330	0.00397	0.00749	0.	0.	0.16441	1.87616
0.3979	O1	0.00437	0.10905	-0.14561	-2.10038	-0.04197	0.01311	0.	0.	0.14807	1.09410
	F1	-0.00017	-0.00397	0.00698	0.07099	0.02060	0.03403	0.	0.	-0.13912	-0.19393
13A*	N1	0.00089	0.02215	-0.03453	-0.48009	-0.78277	1.60625	0.	0.	0.06295	-0.02026
0.5833	O1	0.00038	0.00268	-0.06067	0.20761	-0.15496	-0.19409	0.	0.	0.04997	-0.14948
	F1	-0.00086	-0.02846	-0.02447	0.72524	-0.17634	0.36343	0.	0.	-0.05853	0.20688
* 1A**	N1	0.	0.	0.	0.	0.	0.	0.45157	0.12133	0.	0.
-0.7571	O1	0.	0.	0.	0.	0.	0.	0.47982	0.19159	0.	0.
	F1	0.	0.	0.	0.	0.	0.	0.29364	0.12955	0.	0.
* 2A**	N1	0.	0.	0.	0.	0.	0.	-0.15180	-0.01838	0.	0.
-0.6269	O1	0.	0.	0.	0.	0.	0.	-0.30248	-0.14545	0.	0.
	F1	0.	0.	0.	0.	0.	0.	0.68760	0.35081	0.	0.

FNO

| Center | | S1 | S2 | S3 | S4 | Basis Function Type | | | | | |
						X1	X2	Y1	Y2	Z1	Z2
3A**	N1	0.	0.	0.	0.	0.	0.	0.61824	0.52564	0.	0.
0.0276	O1	0.	0.	0.	0.	0.	0.	-0.49940	-0.47896	0.	0.
	F1	0.	0.	0.	0.	0.	0.	-0.14757	-0.12151	0.	0.
4A**	N1	0.	0.	0.	0.	0.	0.	0.86004	-1.20393	0.	0.
0.5938	O1	0.	0.	0.	0.	0.	0.	0.18590	0.02009	0.	0.
	F1	0.	0.	0.	0.	0.	0.	0.11416	0.11459	0.	0.

Overlap Matrix

First Function				Second Function and Overlap			

N1S1 with

N1S2	.35584	N1S3	.00894	N1S4	.01446	O1S3	.00068	O1S4	.00503	O1Z1	-.00318	O1Z2	-.01100
F1S4	.00114	F1X1	.00006	F1X2	.00526	F1Z1	.00002	F1Z2	.00192				

N1S2 with

N1S3	.27246	N1S4	.17925	O1S2	.00001	O1S3	.01281	O1S4	.06491	O1Z1	-.04495	O1Z2	-.13589
F1S3	.00019	F1S4	.01621	F1X1	.00177	F1X2	.06779	F1Z1	.00064	F1Z2	.02467		

N1S3 with

N1S4	.79352	O1S1	.00136	O1S2	.02016	O1S3	.17675	O1S4	.35876	O1Z1	-.27593	O1Z2	-.57305
F1S1	.00009	F1S2	.00160	F1S3	.02907	F1S4	.13901	F1X1	.05972	F1X2	.35490	F1Z1	.02174
F1Z2	.12918												

N1S4 with

O1S1	.00453	O1S2	.05844	O1S3	.32045	O1S4	.56911	O1Z1	-.22560	O1Z2	-.60200	F1S1	.00177
F1S2	.02303	F1S3	.14263	F1S4	.32018	F1X1	.11575	F1X2	.45388	F1Z1	.04213	F1Z2	.16520

N1X1 with

N1X2	.52739	O1X1	.13939	O1X2	.30272	F1S1	-.00062	F1S2	-.00889	F1S3	-.08005	F1S4	-.19707
F1X1	-.11377	F1X2	-.26240	F1Z1	-.05091	F1Z2	-.14976						

N1X2 with

O1X1	.22003	O1X2	.64041	F1S1	-.00464	F1S2	-.05901	F1S3	-.31429	F1S4	-.52967	F1X1	-.10779
F1X2	-.19814	F1Z1	-.07535	F1Z2	-.21600								

N1Y1 with

N1Y2	.52739	O1Y1	.13939	O1Y2	.30272	F1Y1	.02610	F1Y2	.14907

N1Y2 with

O1Y1	.22003	O1Y2	.64041	F1Y1	.09923	F1Y2	.39530

N1Z1 with

N1Z2	.52739	O1S1	.00421	O1S2	.05558	O1S3	.31076	O1S4	.33998	O1Z1	-.32324	O1Z2	-.13756
F1S1	-.00023	F1S2	-.00323	F1S3	-.02914	F1S4	-.07173	F1X1	-.05091	F1X2	-.14976	F1Z1	.00757
F1Z2	.09456												

N1Z2 with

O1S1	.00804	O1S2	.10141	O1S3	.48015	O1S4	.64559	O1Z1	-.05655	O1Z2	.09588	F1S1	-.00169
F1S2	-.02148	F1S3	-.11439	F1S4	-.19279	F1X1	-.07535	F1X2	-.21600	F1Z1	.07180	F1Z2	.31669

O1S1 with

O1S2	.35461	O1S3	.00974	O1S4	.01469	F1S4	.00004	F1X2	.00039	F1Z2	.00045

O1S2 with

O1S3	.28286	O1S4	.18230	F1S4	.00065	F1X2	.00549	F1Z2	.00634

O1S3 with

O1S4	.79029	F1S3	.00015	F1S4	.01093	F1X1	.00060	F1X2	.04761	F1Z1	.00070	F1Z2	.05499

O1S4 with

F1S1	.00010	F1S2	.00145	F1S3	.01483	F1S4	.06763	F1X1	.01626	F1X2	.13509	F1Z1	.01878
F1Z2	.15604												

O1X1 with

O1X2	.50607	F1S2	-.00001	F1S3	-.00088	F1S4	-.01644	F1X1	-.00194	F1X2	-.03588	F1Z1	-.00269
F1Z2	-.06262												

O1X2 with

F1S1	-.00066	F1S2	-.00878	F1S3	-.06064	F1S4	-.15293	F1X1	-.03111	F1X2	-.09543	F1Z1	-.05861
F1Z2	-.25767												

O1Y1 with

O1Y2	.50607	F1Y1	.00039	F1Y2	.01833

O1Y2 with

F1Y1	.01963	F1Y2	.12764

O1Z1 with

O1Z2	.50607	F1S2	-.00001	F1S3	-.00102	F1S4	-.01899	F1X1	-.00269	F1X2	-.06262	F1Z1	-.00272
F1Z2	-.05400												

Overlap Matrix (continued)

First Function   Second Function and Overlap

01Z2 with
     F1S1 -.00076   F1S2 -.01015   F1S3 -.07004   F1S4 -.17665   F1X1 -.05861   F1X2 -.25767   F1Z1 -.04807
     F1Z2 -.16999

F1S1 with
     F1S2 .35647   F1S3 .01003   F1S4 .01474

F1S2 with
     F1S3 .28653   F1S4 .18096

F1S3 with
     F1S4 .78531

F1X1 with
     F1X2 .49444

F1Y1 with
     F1Y2 .49444

F1Z1 with
     F1Z2 .49444

FNO   T-209

CHONH2   Formamide

Molecular Geometry   Symmetry C₁

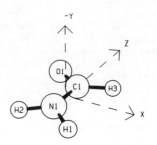

Center	Coordinates		
	X	Y	Z
H1	1.50399999	0.39400000	-2.38100001
H2	-1.66599999	0.23400000	-2.21700001
H3	1.91400000	0.	2.12200001
C1	0.	0.	1.30000000
N1	0.	0.	-1.30000000
O1	-1.87300000	0.	2.55500001
Mass	-0.62620918	0.01405776	0.79446065

MO Symmetry Types and Occupancy

Representation	A
Number of MOs	36
Number Occupied	12

Energy Expectation Values

Total	-168.8685	Kinetic	169.0049
Electronic	-240.2535	Potential	-337.8734
Nuclear Repulsion	71.3851	Virial	1.9992
Electronic Potential	-409.2584		
One-Electron Potential	-540.5026		
Two-Electron Potential	131.2441		

Moments of the Charge Distribution

First

x	y	z	Dipole Moment
-0.6449	0.0001	1.1479	1.7284
1.7970	0.2906	-2.4031	

Second

x^2	y^2	z^2	xy	xz	yz	r^2
-42.0277	-14.3064	-90.5476	-0.2892	26.8722	0.9677	-146.8817
29.6053	0.1971	77.9516	0.5707	-25.1026	-1.9220	107.7540

Third

x^3	xy^2	xz^2	x^2y	y^3	yz^2	x^2z	y^2z	z^3	xyz
-10.4078	-2.0899	-2.0784	-0.7241	0.1553	-1.9115	-13.1712	4.0379	64.6441	0.6040
12.6207	0.2573	5.8011	1.6247	0.0654	4.9992	8.2348	-0.6033	-76.8768	-1.5796

Fourth (even)

x^4	y^4	z^4	x^2y^2	x^2z^2	y^2z^2	r^2x^2	r^2y^2	r^2z^2	r^4
-206.9718	-46.5726	-654.8088	-30.9110	-171.9700	-62.7820	-409.8528	-140.2655	-889.5608	-1439.6791
84.7278	0.0232	398.9818	0.7121	118.1218	1.9059	203.5617	2.6412	519.0096	725.2125

Fourth (odd)

x^3y	xy^3	xyz^2	x^3z	xy^2z	xz^3	x^2yz	y^3z	yz^3
-2.2516	-0.2677	-3.3937	72.9785	15.7046	167.2867	4.2159	0.9643	9.9266
3.3681	0.1058	6.1750	-35.7026	-0.8292	-128.0859	-6.5550	-0.2062	-17.9258

Expectation Values at Atomic Centers

Center			Expectation Value		
	r^{-1}	δ	xr^{-3}	yr^{-3}	zr^{-3}
H1	-0.9829	0.4066	0.0660	0.0069	-0.0375
H2	-0.9929	0.3986	-0.0615	-0.0013	-0.0266
H3	-1.0534	0.4227	0.0618	-0.0019	0.0368
C1	-14.5838	116.5186	0.0486	-0.0007	-0.0043
N1	-18.3291	188.5130	-0.0009	-0.0492	-0.0198
O1	-22.3623	287.2086	-0.2244	-0.0006	0.1599

	$\dfrac{3x^2-r^2}{r^5}$	$\dfrac{3y^2-r^2}{r^5}$	$\dfrac{3z^2-r^2}{r^5}$	$\dfrac{3xy}{r^5}$	$\dfrac{3xz}{r^5}$	$\dfrac{3yz}{r^5}$
H1	0.2304	-0.2452	0.0148	0.1139	-0.2984	-0.0835
H2	0.2840	-0.2453	-0.0387	-0.0616	0.2569	-0.0429
H3	0.2336	-0.1578	-0.0759	0.0003	0.1669	-0.0008
C1	-0.3478	0.4161	-0.0683	-0.0038	-0.0388	0.0208
N1	0.4723	-1.0871	0.6148	0.0886	0.0114	-0.1982
O1	0.0720	1.0100	-1.0821	0.0034	-1.4656	0.0633

Mulliken Charge

Center			Basis Function Type								Total
	S1	S2	S3	S4	X1	X2	Y1	Y2	Z1	Z2	
H1	0.4937	0.1486									0.6423
H2	0.4861	0.1448									0.6310
H3	0.5211	0.3271									0.8482
C1	0.0428	1.9565	0.8475	0.2694	0.9524	0.0833	0.5925	0.1180	0.8436	0.0363	5.7424
N1	0.0432	1.9560	0.8405	0.7033	0.9784	0.2900	1.2037	0.6181	0.9550	0.1703	7.7585
O1	0.0426	1.9566	0.9599	0.8402	1.1599	0.2901	0.9935	0.4572	1.2655	0.4119	8.3775

Overlap Populations

	Pair of Atomic Centers									
	H2 H1	H3 H1	H3 H2	C1 H1	C1 H2	C1 H3	N1 H1	N1 H2	N1 H3	N1 C1
1A	0.	0.	0.	0.	0.	0.	0.	0.	0.	0.
2A	0.	0.	0.	0.	0.	0.	-0.0001	-0.0001	0.	-0.0006
3A	0.	0.	0.	0.	0.	-0.0002	0.	0.	0.	-0.0003
4A	-0.0006	-0.0001	0.	-0.0019	0.0038	0.0089	-0.0046	0.0088	0.0021	0.0576
5A	0.0016	0.0003	0.	0.0057	0.0035	0.0058	0.1401	0.1118	0.0078	0.2932
6A	0.0109	-0.0030	-0.0015	-0.0266	-0.0259	0.2760	0.0837	0.1624	0.0142	0.1094
7A	-0.0130	0.0085	-0.0005	-0.0243	0.0144	0.1078	0.3101	0.0313	0.0024	0.1858
8A	-0.0167	-0.0017	0.0011	0.0012	-0.0335	0.0989	0.0448	0.2776	-0.0278	-0.0168
9A	-0.0235	-0.0094	0.0030	-0.0014	-0.0146	0.0966	0.0896	0.0821	-0.0093	-0.0717
10A	-0.0003	-0.0002	0.	-0.0012	0.0005	0.0016	0.0015	0.0009	-0.0003	0.2145
11A	0.0068	0.0120	0.0028	0.0183	0.0119	0.1728	-0.0390	-0.0259	-0.1524	-0.1382
12A	0.0001	0.0001	0.0002	0.0001	0.0005	0.0053	0.0003	-0.0014	-0.0046	-0.1372
Total A	-0.0349	0.0065	0.0052	-0.0301	-0.0394	0.7734	0.6264	0.6476	-0.1680	0.4957
Total	-0.0349	0.0065	0.0052	-0.0301	-0.0394	0.7734	0.6264	0.6476	-0.1680	0.4957
Distance	3.1783	4.5388	5.6301	3.9959	3.8987	2.0830	1.8936	1.9160	3.9209	2.6000

Overlap Populations (continued)

| | | | Pair of Atomic Centers | | |
| | 01 | 01 | 01 | 01 | 01 |
	H1	H2	H3	C1	N1
1A	0.	0.	0.	-0.0008	0.0001
2A	0.	0.	0.	0.	0.
3A	0.	0.	0.	0.0001	0.
4A	-0.0006	0.0026	0.0038	0.5661	0.0230
5A	0.	-0.0001	-0.0019	0.0414	-0.0197
6A	0.	0.0028	-0.0046	-0.0185	-0.0073
7A	-0.0011	0.0018	0.0107	0.0849	0.0165
8A	0.	-0.0091	0.0048	0.0609	-0.0220
9A	0.0033	-0.0072	-0.0114	0.1299	-0.0143
10A	0.	0.0001	-0.0005	0.2856	0.0395
11A	0.0024	0.0146	-0.1175	-0.0362	-0.1015
12A	0.0001	0.0013	-0.0034	0.0991	-0.1120
Total A	0.0040	0.0068	-0.1201	1.2125	-0.1977
Total	0.0040	0.0068	-0.1201	1.2125	-0.1977
Distance	5.9936	4.7822	3.8117	2.2546	4.2859

Molecular Orbitals and Eigenvalues

Center		Basis Function Type									
		S1	S2	S3	S4	X1	X2	Y1	Y2	Z1	Z2
* 1A	H1	0.00009	0.00027								
-20.5383	H2	-0.00011	-0.00027								
	H3	0.00008	-0.00083								
	C1	0.00000	-0.00002	-0.00016	-0.00080	0.00035	0.00169	0.00000	0.00000	-0.00016	-0.00021
	N1	0.00000	-0.00000	-0.00005	0.00021	0.00006	-0.00099	0.00000	-0.00002	0.00002	0.00014
	O1	0.05112	0.97934	0.00337	0.00348	0.00163	-0.00022	-0.00000	0.00000	-0.00115	0.00002
* 2A	H1	0.00023	-0.00042								
-15.5878	H2	0.00022	-0.00041								
	H3	0.00006	-0.00009								
	C1	0.00000	-0.00000	0.00021	-0.00143	0.00004	0.00014	-0.00001	0.00007	-0.00009	0.00087
	N1	0.05194	0.97892	0.00373	0.00340	0.00000	-0.00003	0.00050	-0.00017	0.00015	0.00038
	O1	0.00000	0.00001	-0.00000	0.00016	0.00003	-0.00001	0.00000	-0.00002	0.00003	-0.00018
* 3A	H1	0.00006	-0.00041								
-11.3852	H2	0.00004	-0.00007								
	H3	0.00086	-0.00090								
	C1	0.05191	0.97906	0.00442	0.00183	-0.00071	0.00061	-0.00001	-0.00000	-0.00024	-0.00012
	N1	-0.00001	-0.00029	0.00046	-0.00044	0.00002	0.00008	-0.00001	0.00005	0.00064	-0.00053
	O1	-0.00002	-0.00049	0.00073	-0.00012	0.00103	-0.00020	0.00000	-0.00000	-0.00066	0.00013
* 4A	H1	0.01488	-0.02015								
-1.4288	H2	0.02039	0.01694								
	H3	0.03344	0.00455								
	C1	-0.00616	-0.13039	0.29444	0.05163	-0.17307	0.01552	0.00085	-0.00020	0.08728	-0.00380
	N1	-0.00216	-0.04651	0.10043	0.08032	-0.01045	0.02132	0.00493	0.00305	0.04367	0.01061
	O1	-0.00907	-0.19888	0.43424	0.30182	0.14887	-0.00761	0.00068	0.00018	-0.10603	0.00198
* 5A	H1	-0.12284	-0.00561								
-1.2200	H2	-0.11427	0.00319								
	H3	-0.02167	-0.01318								
	C1	0.00219	0.04692	-0.10537	-0.01727	-0.08496	0.00443	-0.00374	0.00122	0.20595	-0.00747
	N1	0.00948	0.20496	-0.44744	-0.35442	-0.00575	0.00201	-0.03252	-0.01040	-0.03421	0.01976
	O1	-0.00333	-0.07294	0.16196	0.11718	0.04219	-0.01533	-0.00031	-0.00003	-0.00342	0.00814
* 6A	H1	-0.09462	-0.03064								
-0.8585	H2	-0.13975	-0.06056								
	H3	0.17720	0.06460								
	C1	-0.00677	-0.14240	0.35854	0.16874	0.22029	0.05536	-0.00317	0.00427	-0.06180	-0.00540
	N1	0.00229	0.04987	-0.10983	-0.13400	0.08435	0.00190	-0.03946	-0.01230	0.33126	0.07067
	O1	0.00329	0.07215	-0.16332	-0.16333	0.09952	0.03738	-0.00222	-0.00246	-0.02443	-0.01808
* 7A	H1	-0.21628	-0.10575								
-0.7561	H2	0.07737	0.01919								
	H3	-0.11890	-0.06590								
	C1	-0.00044	-0.00806	0.03011	-0.01257	-0.24497	-0.02387	-0.00906	0.00012	-0.26921	-0.03153
	N1	0.00016	0.00389	-0.00547	-0.03316	-0.35161	-0.10953	-0.04547	-0.01300	0.27564	0.05130
	O1	0.00105	0.02306	-0.05172	-0.07774	-0.07281	-0.02104	-0.00466	-0.00128	-0.20940	-0.05924
* 8A	H1	0.08429	0.03746								
-0.6753	H2	-0.20514	-0.08595								
	H3	-0.11218	-0.02856								
	C1	0.00205	0.04105	-0.12508	-0.08035	-0.01277	-0.08900	-0.00458	0.00189	-0.34342	-0.02036
	N1	-0.00033	-0.00693	0.01710	-0.01083	0.34817	0.15375	-0.01536	-0.00897	0.23681	0.07071
	O1	-0.00294	-0.06498	0.14474	0.22619	-0.32996	-0.08686	-0.00341	-0.00253	-0.00194	-0.00787

Center		S1	S2	S3	S4	X1	X2	Y1	Y2	Z1	Z2
* 9A	H1	-0.13416	-0.07303								
-0.6119	H2	0.11542	0.07987								
	H3	0.11994	0.12109								
	C1	0.00149	0.02672	-0.11130	0.03427	0.39555	-0.02908	-0.04040	-0.00643	-0.09053	0.04112
	N1	-0.00028	-0.00614	0.01316	0.01686	-0.25085	-0.07007	-0.05646	-0.02601	0.04767	0.01351
	O1	-0.00310	-0.06865	0.15031	0.26584	-0.33904	-0.09590	-0.03815	-0.01568	0.37291	0.11834
* 10A	H1	-0.01084	-0.01214								
-0.5829	H2	0.01555	0.00428								
	H3	0.01481	0.01588								
	C1	0.00000	-0.00073	-0.00593	0.01281	0.04210	-0.00502	0.41332	0.08901	-0.02662	-0.00091
	N1	0.00059	0.01226	-0.03267	-0.02924	-0.04107	-0.01595	0.34808	0.16271	0.04188	0.01577
	O1	-0.00027	-0.00602	0.01306	0.02455	-0.05370	-0.01717	0.41809	0.18196	0.02111	0.00649
* 11A	H1	0.04467	0.05767								
-0.4401	H2	0.04301	0.04021								
	H3	0.17477	0.23983								
	C1	-0.00055	-0.01328	0.02052	0.10892	0.11949	-0.00624	-0.03229	-0.01437	0.07776	-0.07229
	N1	0.00184	0.04072	-0.08911	-0.24120	-0.00568	-0.02472	0.09646	0.05992	-0.22163	-0.10282
	O1	0.00012	0.00242	-0.00773	-0.00144	-0.38664	-0.20852	-0.08550	-0.04687	-0.52435	-0.27871
* 12A	H1	0.00765	-0.00563								
-0.4198	H2	0.00589	-0.01888								
	H3	-0.02946	-0.04206								
	C1	-0.00009	-0.00173	0.00568	-0.02325	-0.01647	0.00282	-0.08786	-0.05436	-0.03407	0.01023
	N1	0.00048	0.00941	-0.03082	0.00874	-0.02535	-0.02143	0.55360	0.37460	0.09677	0.04809
	O1	-0.00001	-0.00034	0.00048	0.00402	0.05907	0.03611	-0.42399	-0.24244	0.10350	0.05732
13A	H1	-0.02866	-0.07417								
0.1710	H2	-0.01747	-0.11798								
	H3	-0.00020	-0.02240								
	C1	0.00014	0.00186	-0.01450	0.08087	-0.00063	0.03051	0.54314	0.87195	0.01396	-0.02689
	N1	-0.00057	-0.01266	0.02685	0.10216	0.00695	-0.02732	-0.23635	-0.41999	-0.04046	-0.08690
	O1	-0.00005	-0.00104	0.00294	0.00292	-0.00419	-0.00983	-0.39409	-0.53774	-0.00993	-0.00245
14A	H1	-0.02571	-1.24805								
0.2362	H2	-0.05425	-1.12815								
	H3	0.05334	0.10637								
	C1	0.00089	0.02481	-0.01450	-0.48851	0.01185	0.01794	-0.00440	-0.15655	0.13608	0.49698
	N1	-0.00492	-0.12222	0.15188	2.04915	0.01778	-0.02399	0.08345	0.23437	-0.18087	-0.32608
	O1	-0.00033	-0.00658	0.02105	0.01832	-0.05382	-0.03152	0.01785	0.05096	-0.10161	-0.18895
15A	H1	-0.03308	-1.92133								
0.2679	H2	0.03840	1.08567								
	H3	-0.04472	-0.29471								
	C1	-0.00248	-0.05397	0.14084	1.20264	0.29052	-0.88191	0.02269	-0.08762	0.11079	-0.62407
	N1	0.00039	0.01294	0.00992	-0.31663	0.29786	1.52316	0.02341	0.16613	-0.08367	-0.73855
	O1	0.00020	0.01110	0.03655	-0.43902	-0.00298	0.03293	-0.01319	0.01658	-0.03665	0.11460
16A	H1	0.05206	0.71904								
0.3257	H2	-0.06634	-1.15920								
	H3	-0.01796	-0.95474								
	C1	-0.00357	-0.08623	0.14863	2.28559	0.16841	-0.43288	-0.00947	-0.00180	0.04988	0.38685
	N1	0.00099	0.02286	-0.04447	-0.22900	-0.18370	-0.70244	-0.01033	0.00472	-0.08804	-0.63026
	O1	0.00272	0.07123	-0.06385	-1.12170	-0.15710	-0.35734	0.00898	0.00477	0.04580	0.20832

Center		S1	S2	S3	S4	X1	X2	Y1	Y2	Z1	Z2
17A	H1	0.08081	0.16878								
0.3449	H2	-0.03374	1.53285								
	H3	-0.02008	0.89365								
	C1	0.00177	0.04237	-0.07491	-1.65744	-0.17451	-1.24873	-0.01425	0.11068	0.00141	1.82065
	N1	-0.00038	-0.01695	-0.03941	0.73539	0.23091	0.85205	-0.02435	-0.15929	0.12287	1.23476
	O1	0.00197	0.05228	-0.03949	-0.86046	0.01781	-0.06897	-0.01376	-0.02727	-0.04585	-0.19537
18A	H1	-0.05324	0.36833								
0.4534	H2	0.08716	0.59043								
	H3	-0.08134	-2.72995								
	C1	-0.00166	-0.03590	0.09623	0.94802	0.11891	2.53208	-0.42844	0.57284	-0.08107	1.74579
	N1	0.00024	0.00147	-0.03889	0.33929	0.02202	-0.44579	-0.04879	-0.13228	0.01851	0.35443
	O1	-0.00077	-0.02185	0.00695	0.44796	-0.14223	-0.34804	0.01985	-0.05678	-0.14521	-0.48491
19A	H1	-0.03528	0.14986								
0.4617	H2	0.01654	0.32700								
	H3	-0.05876	-1.24897								
	C1	-0.00075	-0.01793	0.03290	0.74254	-0.04540	1.22454	0.85018	-1.07831	-0.16600	0.88560
	N1	0.00081	0.01489	-0.06125	0.05018	0.01897	-0.19153	0.05485	0.09228	0.04415	-0.08045
	O1	-0.00015	-0.00497	-0.00456	0.12406	-0.06689	-0.20285	-0.04011	0.09445	-0.05620	-0.21402
20A	H1	-0.05523	-1.29956								
0.5366	H2	-0.16950	-0.56231								
	H3	-0.10946	0.37593								
	C1	0.00052	-0.01774	-0.22066	2.75152	-0.61668	0.00728	-0.16756	0.13145	-0.66090	-0.77497
	N1	0.00237	0.04870	-0.14663	-0.49117	0.15959	0.08461	0.05108	0.16012	-0.09159	-2.11073
	O1	0.00114	0.02908	-0.03397	-0.53691	0.02969	-0.24748	0.00347	0.00846	0.11290	0.35306
21A	H1	0.04844	-0.01918								
0.6152	H2	-0.07898	-0.23216								
	H3	0.23480	0.95344								
	C1	-0.00577	-0.02684	0.99931	-2.26244	0.36528	-0.69147	-0.02419	0.02722	-0.49568	0.06830
	N1	0.00028	0.01005	0.01563	0.40532	0.06910	-0.13785	-0.02340	0.03789	0.29769	-0.77536
	O1	-0.00202	-0.04505	0.10413	0.57604	0.00749	0.54263	-0.00145	-0.00178	0.00923	-0.29715

Overlap Matrix

First Function   Second Function and Overlap

H1S1 with
     H1S2 .68301   H2S1 .02583   H2S2 .16209   H3S1 .00080   H3S2 .03659   C1S2 .00005   C1S3 .01075
     C1S4 .08847   C1X1 .01524   C1X2 .09933   C1Y1 .00399   C1Y2 .02602   C1Z1 -.03731   C1Z2 -.24310
     N1S1 .00269   N1S2 .03812   N1S3 .27922   N1S4 .40976   N1X1 .31579   N1X2 .40571   N1Y1 .08273
     N1Y2 .10628   N1Z1 -.22698   N1Z2 -.29160   O1S4 .00061   O1X1 .00001   O1X2 .00492   O1Y2 .00057
     O1Z1 -.00001   O1Z2 -.00719

H1S2 with
     H2S1 .16209   H2S2 .40784   H3S1 .03659   H3S2 .16056   C1S1 .00096   C1S2 .01332   C1S3 .11054
     C1S4 .27387   C1X1 .06082   C1X2 .19640   C1Y1 .01593   C1Y2 .05145   C1Z1 -.14887   C1Z2 -.48067
     N1S1 .00680   N1S2 .08659   N1S3 .45078   N1S4 .70604   N1X1 .21407   N1X2 .46548   N1Y1 .05608
     N1Y2 .12194   N1Z1 -.15386   N1Z2 -.33457   O1S1 .00002   O1S2 .00026   O1S3 .00308   O1S4 .01912
     O1X1 .00336   O1X2 .04319   O1Y1 .00039   O1Y2 .00504   O1Z1 -.00491   O1Z2 -.06313

H2S1 with
     H2S2 .68301   H3S1 .00002   H3S2 .00763   C1S2 .00008   C1S3 .01329   C1S4 .09718   C1X1 -.02034
     C1X2 -.11869   C1Y1 .00286   C1Y2 .01667   C1Z1 -.04293   C1Z2 -.25055   N1S1 .00254   N1S2 .03624
     N1S3 .27108   N1S4 .40415   N1X1 -.33934   N1X2 -.44426   N1Y1 .04766   N1Y2 .06240   N1Z1 -.18678
     N1Z2 -.24453   O1S3 .00013   O1S4 .00818   O1X1 .00003   O1X2 .00253   O1Y1 .00004   O1Y2 .00286
     O1Z1 -.00080   O1Z2 -.05838

H2S2 with
     H3S1 .00763   H3S2 .05994   C1S1 .00110   C1S2 .01518   C1S3 .12198   C1S4 .29137   C1X1 -.07410
     C1X2 -.22949   C1Y1 .01041   C1Y2 .03223   C1Z1 -.15643   C1Z2 -.48447   N1S1 .00670   N1S2 .08532
     N1S3 .44543   N1S4 .70029   N1X1 -.23434   N1X2 -.51186   N1Y1 .03291   N1Y2 .07189   N1Z1 -.12899
     N1Z2 -.28174   O1S1 .00018   O1S2 .00249   O1S3 .02146   O1S4 .07937   O1X1 .00136   O1X2 .00939
     O1Y1 .00154   O1Y2 .01062   O1Z1 -.03138   O1Z2 -.21647

H3S1 with
     H3S2 .68301   C1S1 .00209   C1S2 .03246   C1S3 .27640   C1S4 .36762   C1X1 .38880   C1X2 .39885
     C1Z1 .16698   C1Z2 .17129   N1S2 .00004   N1S3 .00519   N1S4 .06154   N1X1 .00927   N1X2 .10570
     N1Z1 .01658   N1Z2 .18898   O1S2 .00004   O1S3 .00317   O1S4 .04337   O1X1 .01142   O1X2 .18177
     O1Z1 -.00131   O1Z2 -.02078

H3S2 with
     C1S1 .00753   C1S2 .09621   C1S3 .49023   C1S4 .70014   C1X1 .33093   C1X2 .56177   C1Z1 .14212
     C1Z2 .24126   N1S1 .00084   N1S2 .01139   N1S3 .08660   N1S4 .22832   N1X1 .05380   N1X2 .21589
     N1Z1 .09618   N1Z2 .38599   O1S1 .00079   O1S2 .01058   O1S3 .07415   O1S4 .19706   O1X1 .08366
     O1X2 .38578   O1Z1 -.00957   O1Z2 -.04411

C1S1 with
     C1S2 .35436   C1S3 .00805   C1S4 .01419   N1S3 .00048   N1S4 .00450   N1Z1 .00248   N1Z2 .01066
     O1S3 .00053   O1S4 .00549   O1X1 .00243   O1X2 .01092   O1Z1 -.00163   O1Z2 -.00731

C1S2 with
     C1S3 .26226   C1S4 .17745   N1S3 .00969   N1S4 .05887   N1Z1 .03651   N1Z2 .13299   O1S2 .00002
     O1S3 .01230   O1S4 .07175   O1X1 .03702   O1X2 .13384   O1Z1 -.02480   O1Z2 -.08968

C1S3 with
     C1S4 .79970   N1S1 .00116   N1S2 .01712   N1S3 .16031   N1S4 .34118   N1Z1 .25370   N1Z2 .56969
     O1S1 .00206   O1S2 .02837   O1S3 .20065   O1S4 .39639   O1X1 .22194   O1X2 .50805   O1Z1 -.14871
     O1Z2 -.34042

C1S4 with
     N1S1 .00413   N1S2 .05324   N1S3 .30276   N1S4 .54554   N1Z1 .21643   N1Z2 .58768   O1S1 .00427
     O1S2 .05513   O1S3 .30553   O1S4 .56513   O1X1 .14288   O1X2 .44299   O1Z1 -.09574   O1Z2 -.29683

C1X1 with
     C1X2 .53875   N1X1 .12943   N1X2 .30392   O1S1 -.00427   O1S2 -.05550   O1S3 -.29567   O1S4 -.35441
     O1X1 -.16364   O1X2 -.00300   O1Z1 .21199   O1Z2 .25220

C1X2 with
     N1X1 .20508   N1X2 .60690   O1S1 -.00529   O1S2 -.06739   O1S3 -.33920   O1S4 -.51422   O1X1 .05260
     O1X2 .28945   O1Z1 .08432   O1Z2 .21300

C1Y1 with
     C1Y2 .53875   N1Y1 .12943   N1Y2 .30392   O1Y1 .15274   O1Y2 .37338

Overlap Matrix   (continued)

First Function   Second Function and Overlap

C1Y2 with
    N1Y1 .20508   N1Y2 .60690   O1Y1 .17843   O1Y2 .60733

C1Z1 with
    C1Z2 .53875   N1S1 -.00389   N1S2 -.05121   N1S3 -.29976   N1S4 -.35022   N1Z1 -.31494   N1Z2 -.18596
    O1S1 .00286   O1S2 .03719   O1S3 .19811   O1S4 .23747   O1X1 .21199   O1X2 .25220   O1Z1 .01070
    O1Z2 .20440

C1Z2 with
    N1S1 -.00751   N1S2 -.09485   N1S3 -.46723   N1S4 -.64500   N1Z1 -.06233   N1Z2 .05148   O1S1 .00355
    O1S2 .04515   O1S3 .22728   O1S4 .34455   O1X1 .08432   O1X2 .21300   O1Z1 .12194   O1Z2 .46461

N1S1 with
    N1S2 .35584   N1S3 .00894   N1S4 .01446   O1S4 .00010   O1X2 .00051   O1Z2 -.00104

N1S2 with
    N1S3 .27246   N1S4 .17925   O1S4 .00159   O1X1 .00001   O1X2 .00695   O1Z1 -.00001   O1Z2 -.01430

N1S3 with
    N1S4 .79352   O1S3 .00080   O1S4 .02324   O1X1 .00171   O1X2 .05418   O1Z1 -.00352   O1Z2 -.11150

N1S4 with
    O1S1 .00026   O1S2 .00358   O1S3 .03096   O1S4 .10888   O1X1 .02011   O1X2 .12320   O1Z1 -.04138
    O1Z2 -.25358

N1X1 with
    N1X2 .52739   O1S2 -.00005   O1S3 -.00220   O1S4 -.02089   O1X1 -.00178   O1X2 -.00343   O1Z1 .00674
    O1Z2 .07728

N1X2 with
    O1S1 -.00072   O1S2 -.00953   O1S3 -.06106   O1S4 -.13411   O1X1 .00126   O1X2 .06112   O1Z1 .06528
    O1Z2 .23795

N1Y1 with
    N1Y2 .52739   O1Y1 .00150   O1Y2 .03411

N1Y2 with
    O1Y1 .03298   O1Y2 .17673

N1Z1 with
    N1Z2 .52739   O1S2 .00011   O1S3 .00453   O1S4 .04299   O1X1 .00674   O1X2 .07728   O1Z1 -.01238
    O1Z2 -.12495

N1Z2 with
    O1S1 .00148   O1S2 .01962   O1S3 .12568   O1S4 .27601   O1X1 .06528   O1X2 .23795   O1Z1 -.10138
    O1Z2 -.31301

O1S1 with
    O1S2 .35461   O1S3 .00974   O1S4 .01469

O1S2 with
    O1S3 .28286   O1S4 .18230

O1S3 with
    O1S4 .79029

O1X1 with
    O1X2 .50607

O1Y1 with
    O1Y2 .50607

O1Z1 with
    O1Z2 .50607

CHONH2

T-217

CHOOH   Formic Acid

Molecular Geometry   Symmetry C_{1h}

Center	Coordinates		
	X	Y	Z
H1	0.	0.	-2.07300001
H2	-2.21869999	0.	2.74010000
C1	0.	0.	0.
O1	1.88079999	0.	1.27339999
O2	-2.36939999	0.	0.90960000
Mass	-0.21847692	0.	0.77358686

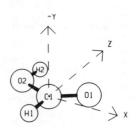

MO Symmetry Types and Occupancy

Representation	A*	A**
Number of MOs	28	6
Number Occupied	10	2

Energy Expectation Values

Total	-188.6888	Kinetic	188.8443
Electronic	-258.8997	Potential	-377.5331
Nuclear Repulsion	70.2108	Virial	1.9992
Electronic Potential	-447.7439		
One-Electron Potential	-586.3418		
Two-Electron Potential	138.5979		

Moments of the Charge Distribution

First

x	y	z	Dipole Moment
0.3187	0.	0.2762	0.5872
-0.8841	0.	-0.4350	

Second

x^2	y^2	z^2	xy	xz	yz	r^2
-94.1724	-12.6809	-27.6404	0.	-3.1150	0.	-134.4936
76.6025	0.	17.7074	0.	0.4841	0.	94.3098

Third

x^3	xy^2	xz^2	x^2y	y^3	yz^2	x^2z	y^2z	z^3	xyz
11.9140	0.5220	-2.9267	0.	0.	0.	-26.6542	0.7447	19.0498	0.
-13.5276	0.	-1.3033	0.	0.	0.	30.1657	0.	-17.2199	0.

Fourth (even)

x^4	y^4	z^4	x^2y^2	x^2z^2	y^2z^2	r^2x^2	r^2y^2	r^2z^2	r^4
-651.2430	-36.5705	-150.8263	-54.3101	-79.0390	-20.6072	-784.5921	-111.4877	-250.4725	-1146.5523
342.6275	0.	83.2653	0.	25.5224	0.	368.1499	0.	108.7877	476.9375

Fourth (odd)

x^3y	xy^3	xyz^2	x^3z	xy^2z	xz^3	x^2yz	y^3z	yz^3
0.	0.	0.	-25.3143	-1.3852	5.7359	0.	0.	0.
0.	0.	0.	10.3485	0.	-18.8040	0.	0.	0.

Expectation Values at Atomic Centers

Center	r^{-1}	δ	xr^{-3}	yr^{-3}	zr^{-3}
			Expectation Value		
H1	-1.0235	0.4166	0.0035	0.	-0.0701
H2	-0.9067	0.3935	-0.0248	0.	0.0951
C1	-14.5401	116.5526	-0.0028	0.	-0.0615
O1	-22.3201	287.2801	0.2353	0.	0.1407
O2	-22.2735	286.9959	-0.2211	0.	-0.1101

Center	$\dfrac{3x^2-r^2}{r^5}$	$\dfrac{3y^2-r^2}{r^5}$	$\dfrac{3z^2-r^2}{r^5}$	$\dfrac{3xy}{r^5}$	$\dfrac{3xz}{r^5}$	$\dfrac{3yz}{r^5}$
H1	-0.1543	-0.1560	0.3103	0.	-0.0048	0.
H2	-0.2184	-0.2731	0.4915	0.	0.0123	0.
C1	-0.0208	0.4224	-0.4016	0.	-0.1157	0.
O1	0.0683	1.0929	-1.1612	0.	1.4128	0.
O2	0.4072	-1.4832	1.0760	0.	-0.9620	0.

Mulliken Charge

Center	S1	S2	S3	S4	X1	X2	Y1	Y2	Z1	Z2	Total
					Basis Function Type						
H1	0.5144	0.2927									0.8071
H2	0.4716	0.1014									0.5730
C1	0.0429	1.9565	0.8636	0.2413	0.8011	0.0257	0.5820	0.1130	0.9677	0.1033	5.6970
O1	0.0426	1.9567	0.9641	0.8494	1.1554	0.3104	0.9777	0.4564	1.2673	0.3955	8.3755
O2	0.0426	1.9567	0.9406	0.8790	1.1255	0.3539	1.3034	0.5677	1.0585	0.3196	8.5473

Overlap Populations

	H2 H1	C1 H1	C1 H2	O1 H1	O1 H2	O1 C1	O2 H1	O2 H2	O2 C1	O2 O1
				Pair of Atomic Centers						
1A*	0.	0.	0.	0.	0.	0.	0.	0.	-0.0006	0.
2A*	0.	0.	0.	0.	0.	-0.0007	0.	0.	0.	0.
3A*	0.	-0.0003	0.	0.	0.	0.0001	0.	0.	0.	0.
4A*	0.0001	0.0092	0.0062	0.0039	0.0022	0.2523	0.0058	0.0806	0.2279	0.0467
5A*	0.	-0.0016	0.0013	-0.0032	0.0015	0.3396	0.0029	0.0564	0.0805	-0.0382
6A*	-0.0024	0.2561	-0.0369	-0.0021	0.0052	-0.0120	0.0200	0.2129	0.1477	-0.0004
7A*	0.0018	0.2213	-0.0440	0.0112	-0.0110	0.1867	-0.0206	0.0756	0.0893	0.0134
8A*	0.0002	0.0073	0.0068	0.0007	-0.0014	-0.0604	-0.0044	0.0726	0.0571	-0.0175
9A*	0.0051	0.1137	-0.0034	-0.0320	-0.0053	0.1096	-0.0456	0.0612	-0.0911	-0.0118
10A*	0.0032	0.1532	0.0116	-0.0908	0.0202	-0.0418	-0.0850	-0.0212	-0.0719	-0.1069
Total A*	0.0080	0.7590	-0.0584	-0.1123	0.0115	0.7735	-0.1269	0.5382	0.4389	-0.1148
1A**	0.	0.	0.	0.	0.	0.1710	0.	0.	0.2498	0.0310
2A**	0.	0.	0.	0.	0.	0.2220	0.	0.	-0.1851	-0.0737
Total A**	0.	0.	0.	0.	0.	0.3930	0.	0.	0.0646	-0.0427
Total	0.0080	0.7590	-0.0584	-0.1123	0.0115	1.1664	-0.1269	0.5382	0.5036	-0.1576
Distance	5.2999	2.0730	3.5257	3.8387	4.3540	2.2713	3.8092	1.8367	2.5380	4.2657

Molecular Orbitals and Eigenvalues

Center		S1	S2	S3	S4	X1	X2	Y1	Y2	Z1	Z2
* 1A*	H1	0.00003	0.00024								
-20.6248	H2	-0.00009	0.00003								
	C1	-0.00000	0.00001	0.00001	-0.00133	0.00019	0.00086	0.	0.	-0.00015	-0.00001
	O1	-0.00002	-0.00045	-0.00001	0.00004	0.00003	-0.00030	0.	0.	-0.00002	0.00009
	O2	0.05112	0.97932	0.00359	0.00292	0.00147	-0.00005	0.	0.	0.00085	-0.00061
* 2A*	H1	-0.00007	0.00058								
-20.5820	H2	0.00011	-0.00009								
	C1	-0.00000	0.00002	0.00015	0.00091	0.00025	0.00069	0.	0.	0.00025	0.00110
	O1	-0.05112	-0.97936	-0.00335	-0.00333	0.00168	-0.00007	0.	0.	0.00102	-0.00013
	O2	-0.00002	-0.00045	0.00008	-0.00022	0.00003	-0.00030	0.	0.	0.00004	-0.00036
* 3A*	H1	-0.00097	0.00100								
-11.4302	H2	0.00001	-0.00006								
	C1	-0.05191	-0.97908	-0.00448	-0.00126	0.00021	-0.00003	0.	0.	-0.00102	0.00080
	O1	0.00002	0.00047	-0.00070	0.00004	0.00095	-0.00012	0.	0.	0.00071	-0.00013
	O2	0.00001	0.00022	-0.00033	0.00020	-0.00057	0.00010	0.	0.	0.00023	-0.00006
* 4A*	H1	-0.03086	-0.00982								
-1.5114	H2	-0.08478	-0.00492								
	C1	0.00587	0.12484	-0.27788	-0.02698	-0.00916	-0.00554	0.	0.	-0.11534	0.01215
	O1	0.00580	0.12723	-0.27675	-0.19643	0.11059	0.00485	0.	0.	0.05930	0.00098
	O2	0.00730	0.16073	-0.34447	-0.30225	-0.11217	-0.02605	0.	0.	-0.03110	-0.00105
* 5A*	H1	-0.01036	0.01124								
-1.4041	H2	0.08603	-0.01608								
	C1	0.00198	0.04112	-0.09967	-0.02615	-0.25266	0.01618	0.	0.	-0.06427	0.01666
	O1	0.00772	0.16913	-0.37180	-0.26356	0.10945	-0.00896	0.	0.	0.08399	-0.01357
	O2	-0.00700	-0.15420	0.33329	0.29595	0.06312	-0.00303	0.	0.	0.03559	0.02826
* 6A*	H1	0.16992	0.05768								
-0.9083	H2	-0.18142	-0.06678								
	C1	-0.00666	-0.13859	0.36364	0.15252	-0.15740	-0.00631	0.	0.	-0.19314	-0.03875
	O1	0.00281	0.06179	-0.13812	-0.13752	-0.04013	-0.02651	0.	0.	-0.08054	-0.02872
	O2	0.00173	0.03815	-0.08393	-0.11068	0.18864	0.04634	0.	0.	-0.33854	-0.07875
* 7A*	H1	0.16783	0.05726								
-0.7413	H2	0.08767	0.05557								
	C1	-0.00023	-0.00648	0.00234	0.07881	0.25704	-0.00165	0.	0.	-0.41145	-0.08579
	O1	0.00002	0.00025	-0.00137	0.00809	0.17045	0.06151	0.	0.	-0.25439	-0.07336
	O2	-0.00163	-0.03579	0.08136	0.12635	-0.37031	-0.12043	0.	0.	0.12867	0.04010
* 8A*	H1	-0.03724	-0.00554								
-0.7137	H2	-0.13974	-0.04806								
	C1	0.00366	0.07336	-0.22152	-0.03764	-0.23814	0.01401	0.	0.	-0.12438	0.03069
	O1	-0.00456	-0.10055	0.22440	0.35269	0.35708	0.10604	0.	0.	0.22399	0.05524
	O2	-0.00237	-0.05132	0.12394	0.13306	-0.12978	-0.08943	0.	0.	-0.34708	-0.14375
* 9A*	H1	-0.13856	-0.15454								
-0.6019	H2	-0.15301	-0.06732								
	C1	0.00026	0.00884	0.00716	-0.03831	0.23081	-0.06251	0.	0.	0.25027	-0.03500
	O1	0.00152	0.03368	-0.07404	-0.13149	-0.32311	-0.08699	0.	0.	-0.21167	-0.06873
	O2	-0.00280	-0.06081	0.14604	0.19663	-0.34076	-0.16362	0.	0.	-0.34565	-0.15573
* 10A*	H1	0.15980	0.19462								
-0.4810	H2	0.02506	0.04125								
	C1	-0.00062	-0.01488	0.02380	0.11331	0.02506	-0.06038	0.	0.	-0.13619	0.03033
	O1	0.00021	0.00423	-0.01331	-0.00598	-0.36123	-0.18746	0.	0.	0.55352	0.29277
	O2	0.00049	0.01264	-0.01370	-0.13297	-0.29407	-0.15794	0.	0.	0.02391	0.01249

CHOOH

Center		S1	S2	S3	S4	X1	X2	Y1	Y2	Z1	Z2
11A*	H1	0.07221	0.13997								
0.2162	H2	-0.10043	-1.44321								
	C1	0.00124	0.03003	-0.05143	-0.41166	0.09523	0.06192	0.	0.	0.04870	-0.41544
	O1	-0.00084	-0.02091	0.02537	0.25794	-0.06447	-0.14174	0.	0.	0.10169	0.09236
	O2	-0.00343	-0.08568	0.10761	1.10346	0.07041	0.17570	0.	0.	0.27569	0.63915
12A*	H1	-0.04798	-1.16574								
0.3045	H2	-0.04291	-0.75493								
	C1	-0.00425	-0.10012	0.19244	2.22646	0.02727	0.43888	0.	0.	-0.32739	0.42704
	O1	0.00243	0.06482	-0.04831	-1.08292	0.07268	0.30917	0.	0.	0.14251	0.23024
	O2	-0.00002	0.00060	0.00684	-0.02664	-0.14484	-0.35020	0.	0.	0.12910	0.31038
13A*	H1	-0.00425	0.67597								
0.3284	H2	-0.03249	0.90987								
	C1	0.00218	0.04903	-0.11381	-0.98168	0.12849	1.93358	0.	0.	0.11676	0.19464
	O1	0.00175	0.04685	-0.03217	-0.81075	-0.07862	-0.14576	0.	0.	0.04043	0.20969
	O2	-0.00198	-0.05223	0.04330	0.84219	0.04575	0.32337	0.	0.	-0.21088	-0.70563
14A*	H1	0.03683	-2.13181								
0.4491	H2	0.12786	1.09116								
	C1	-0.00206	-0.01153	0.33991	-0.74088	0.06933	0.44721	0.	0.	-0.65265	-2.17939
	O1	-0.00240	-0.05920	0.08369	0.91234	-0.16235	-0.58112	0.	0.	0.16457	0.20610
	O2	-0.00125	-0.02967	0.05235	0.44409	0.15675	0.52452	0.	0.	-0.09321	-0.30101
15A*	H1	0.18622	2.27241								
0.4631	H2	0.03606	-0.55950								
	C1	0.00021	0.03891	0.21780	-2.63379	0.26804	-0.29349	0.	0.	-0.56859	2.58793
	O1	-0.00099	-0.01961	0.06830	0.11993	-0.04273	-0.09007	0.	0.	-0.13219	-0.54183
	O2	-0.00227	-0.04843	0.13023	0.37685	0.08328	0.33895	0.	0.	-0.06194	-0.38359
16A*	H1	-0.16889	-0.76425								
0.6570	H2	0.10569	-0.11690								
	C1	0.00839	0.08404	-1.14769	2.00014	0.90174	-0.44069	0.	0.	-0.21489	-0.35556
	O1	0.00090	0.02185	-0.03317	-0.44508	-0.11055	0.33916	0.	0.	0.04058	0.35230
	O2	-0.00209	-0.03762	0.16937	-0.24344	0.34542	0.31888	0.	0.	-0.04249	-0.16916
* 1A**	H1	0.	0.								
-0.6521	H2	0.	0.								
	C1	0.	0.	0.	0.	0.	0.	-0.36455	-0.07111	0.	0.
	O1	0.	0.	0.	0.	0.	0.	-0.30525	-0.12548	0.	0.
	O2	0.	0.	0.	0.	0.	0.	-0.52044	-0.24559	0.	0.
* 2A**	H1	0.	0.								
-0.5000	H2	0.	0.								
	C1	0.	0.	0.	0.	0.	0.	0.21884	0.06432	0.	0.
	O1	0.	0.	0.	0.	0.	0.	0.51348	0.27458	0.	0.
	O2	0.	0.	0.	0.	0.	0.	-0.48506	-0.29718	0.	0.
3A**	H1	0.	0.								
0.1392	H2	0.	0.								
	C1	0.	0.	0.	0.	0.	0.	-0.57251	-0.77273	0.	0.
	O1	0.	0.	0.	0.	0.	0.	0.40952	0.51638	0.	0.
	O2	0.	0.	0.	0.	0.	0.	0.25011	0.31427	0.	0.
4A**	H1	0.	0.								
0.4298	H2	0.	0.								
	C1	0.	0.	0.	0.	0.	0.	-0.95368	1.23710	0.	0.
	O1	0.	0.	0.	0.	0.	0.	0.03816	-0.11622	0.	0.
	O2	0.	0.	0.	0.	0.	0.	-0.01650	-0.16079	0.	0.

Basis Function Type

CHOOH

Overlap Matrix

First Function   Second Function and Overlap

H1S1 with

H1S2 .68301	H2S1 .00007	H2S2 .01270	C1S1 .00214	C1S2 .03323	C1S3 .27966	C1S4 .36951
C1Z1 -.42623	C1Z2 -.43377	O1S2 .00004	O1S3 .00293	O1S4 .04161	O1X1 -.00527	O1X2 -.08726
O1Z1 -.00938	O1Z2 -.15525	O2S2 .00005	O2S3 .00320	O2S4 .04354	O2X1 .00719	O2X2 .11408
O2Z1 -.00905	O2Z2 -.14361					

H1S2 with

H2S1 .01270	H2S2 .08258	C1S1 .00759	C1S2 .09689	C1S3 .49286	C1S4 .70250	C1Z1 -.36031
C1Z2 -.61021	O1S1 .00076	O1S2 .01021	O1S3 .07190	O1S4 .19266	O1X1 -.04032	O1X2 -.18779
O1Z1 -.07174	O1Z2 -.33412	O2S1 .00079	O2S2 .01061	O2S3 .07436	O2S4 .19747	O2X1 .05249
O2X2 .24181	O2Z1 -.06607	O2Z2 -.30439				

H2S1 with

H2S2 .68301	C1S1 .00001	C1S2 .00037	C1S3 .02869	C1S4 .13637	C1X1 -.05342	C1X2 -.20777
C1Z1 .06597	C1Z2 .25660	O1S3 .00058	O1S4 .01787	O1X1 -.00265	O1X2 -.09515	O1Z1 .00095
O1Z2 .03404	O2S1 .00249	O2S2 .03447	O2S3 .24332	O2S4 .41274	O2X1 .02788	O2X2 .04561
O2Z1 .33867	O2Z2 .55397					

H2S2 with

C1S1 .00179	C1S2 .02430	C1S3 .17398	C1S4 .36434	C1X1 -.13922	C1X2 -.37064	C1Z1 .17193
C1Z2 .45774	O1S1 .00036	O1S2 .00491	O1S3 .03839	O1S4 .12159	O1X1 -.04757	O1X2 -.27180
O1Z1 .01702	O1Z2 .09724	O2S1 .00570	O2S2 .07318	O2S3 .38877	O2S4 .66505	O2X1 .01696
O2X2 .04530	O2Z1 .20602	O2Z2 .55019				

C1S1 with

C1S2 .35436	C1S3 .00805	C1S4 .01419	O1S3 .00050	O1S4 .00537	O1X1 -.00231	O1X2 -.01079
O1Z1 -.00156	O1Z2 -.00730	O2S3 .00015	O2S4 .00374	O2X1 .00116	O2X2 .01033	O2Z1 -.00045
O2Z2 -.00397						

C1S2 with

C1S3 .26226	C1S4 .17745	O1S2 .00002	O1S3 .01163	O1S4 .07033	O1X1 -.03546	O1X2 -.13234
O1Z1 -.02401	O1Z2 -.08960	O2S3 .00451	O2S4 .05012	O2X1 .02027	O2X2 .12848	O2Z1 -.00778
O2Z2 -.04932						

C1S3 with

C1S4 .79970	O1S1 .00199	O1S2 .02743	O1S3 .19603	O1S4 .39119	O1X1 -.21793	O1X2 -.50460
O1Z1 -.14755	O1Z2 -.34164	O2S1 .00108	O2S2 .01551	O2S3 .13201	O2S4 .31275	O2X1 .18786
O2X2 .52788	O2Z1 -.07212	O2Z2 -.20265				

C1S4 with

O1S1 .00422	O1S2 .05452	O1S3 .30259	O1S4 .56098	O1X1 -.14212	O1X2 -.44190	O1Z1 -.09622
O1Z2 -.29919	O2S1 .00349	O2S2 .04527	O2S3 .25704	O2S4 .49529	O2X1 .15243	O2X2 .49764
O2Z1 -.05852	O2Z2 -.19104					

C1X1 with

C1X2 .53875	O1S1 .00416	O1S2 .05412	O1S3 .29010	O1S4 .35125	O1X1 -.16174	O1X2 -.00600
O1Z1 -.21064	O1Z2 -.25413	O2S1 -.00322	O2S2 -.04229	O2S3 -.24905	O2S4 -.35468	O2X1 -.22447
O2X2 -.18670	O2Z1 .12579	O2Z2 .18964				

C1X2 with

O1S1 .00527	O1S2 .06709	O1S3 .33800	O1S4 .51318	O1X1 .05116	O1X2 .28518	O1Z1 -.08526
O1Z2 -.21580	O2S1 -.00573	O2S2 -.07313	O2S3 -.37373	O2S4 -.58231	O2X1 -.01998	O2X2 .08912
O2Z1 .06743	O2Z2 .17647					

C1Y1 with

C1Y2 .53875	O1Y1 .14937	O1Y2 .36934	O2Y1 .10320	O2Y2 .30729

C1Y2 with

O1Y1 .17708	O1Y2 .60391	O2Y1 .15566	O2Y2 .54881

C1Z1 with

C1Z2 .53875	O1S1 .00282	O1S2 .03664	O1S3 .19642	O1S4 .23781	O1X1 -.21064	O1X2 -.25413
O1Z1 .00676	O1Z2 .19728	O2S1 .00124	O2S2 .01623	O2S3 .09561	O2S4 .13616	O2X1 .12579
O2X2 .18964	O2Z1 .05491	O2Z2 .23449				

C1Z2 with

O1S1 .00357	O1S2 .04542	O1S3 .22884	O1S4 .34745	O1X1 -.08526	O1X2 -.21580	O1Z1 .11936
O1Z2 .45780	O2S1 .00220	O2S2 .02807	O2S3 .14347	O2S4 .22355	O2X1 .06743	O2X2 .17647
O2Z1 .12978	O2Z2 .48106					

Overlap Matrix (continued)

First Function   Second Function and Overlap

O1S1 with
     O1S2 .35461   O1S3 .00974   O1S4 .01469   O2S4 .00009   O2X2 .00096   O2Z2 .00008

O1S2 with
     O1S3 .28286   O1S4 .18230   O2S4 .00129   O2X1 .00001   O2X2 .01307   O2Z2 .00112

O1S3 with
     O1S4 .79029   O2S3 .00023   O2S4 .01565   O2X1 .00167   O2X2 .09592   O2Z1 .00014   O2Z2 .00821

O1S4 with
     O2S1 .00009   O2S2 .00129   O2S3 .01565   O2S4 .07652   O2X1 .03027   O2X2 .24090   O2Z1 .00259
     O2Z2 .02062

O1X1 with
     O1X2 .50607   O2S2 -.00001 O2S3 -.00167 O2S4 -.03027 O2X1 -.00721 O2X2 -.11760 O2Z1 -.00067
     O2Z2 -.01205

O1X2 with
     O2S1 -.00096 O2S2 -.01307 O2S3 -.09592 O2S4 -.24090 O2X1 -.11760 O2X2 -.40924 O2Z1 -.01205
     O2Z2 -.04727

O1Y1 with
     O1Y2 .50607   O2Y1 .00059   O2Y2 .02312

O1Y2 with
     O2Y1 .02312   O2Y2 .14305

O1Z1 with
     O1Z2 .50607   O2S3 -.00014 O2S4 -.00259 O2X1 -.00067 O2X2 -.01205 O2Z1 .00054   O2Z2 .02209

O1Z2 with
     O2S1 -.00008 O2S2 -.00112 O2S3 -.00821 O2S4 -.02062 O2X1 -.01205 O2X2 -.04727 O2Z1 .02209
     O2Z2 .13900

O2S1 with
     O2S2 .35461   O2S3 .00974   O2S4 .01469

O2S2 with
     O2S3 .28286   O2S4 .18230

O2S3 with
     O2S4 .79029

O2X1 with
     O2X2 .50607

O2Y1 with
     O2Y2 .50607

O2Z1 with
     O2Z2 .50607

CHOOH   T-223

CHOF   Formyl Fluoride

Molecular Geometry   Symmetry C_{1h}

Center	X	Coordinates Y	Z
H1	0.	0.	-2.06930000
C1	0.	0.	0.
O1	1.77540000	0.	1.35250001
F1	-2.37750000	0.	0.86150000
Mass	-0.34939694	0.	0.74820761

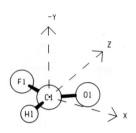

MO Symmetry Types and Occupancy

Representation	A*	A**
Number of MOs	26	6
Number Occupied	10	2

Energy Expectation Values

Total	-212.6841	Kinetic	212.7743
Electronic	-280.1218	Potential	-425.4584
Nuclear Repulsion	67.4377	Virial	1.9996

Electronic Potential	-492.8961
One-Electron Potential	-639.9450
Two-Electron Potential	147.0489

Moments of the Charge Distribution

First

x	y	z	Dipole Moment
-1.3509	0.	0.3608	1.1036
1.1912	0.	-1.4528	

Second

x^2	y^2	z^2	xy	xz	yz	r^2
-89.7471	-11.3454	-25.0142	0.	-6.4765	0.	-126.1067
73.9915	0.	14.3341	0.	5.6511	0.	88.3256

Third

x^3	xy^2	xz^2	x^2y	y^3	yz^2	x^2z	y^2z	z^3	xyz
-12.7107	-2.4995	-12.6430	0.	0.	0.	-25.7104	0.9569	21.4444	0.
1.9642	0.	9.9202	0.	0.	0.	25.1278	0.	-23.1011	0.

Fourth (even)

x^4	y^4	z^4	x^2y^2	x^2z^2	y^2z^2	r^2x^2	r^2y^2	r^2z^2	r^4
-573.8458	-30.0245	-129.0387	-44.7226	-67.7322	-17.7432	-686.3005	-92.4902	-214.5140	-993.3047
315.4348	0.	65.9660	0.	15.0435	0.	330.4783	0.	81.0095	411.4878

Fourth (odd)

x^3y	xy^3	xyz^2	x^3z	xy^2z	xz^3	x^2yz	y^3z	yz^3
0.	0.	0.	-49.4110	-2.8999	-6.6075	0.	0.	0.
0.	0.	0.	37.5582	0.	-4.9683	0.	0.	0.

Expectation Values at Atomic Centers

Center	r^{-1}	δ	xr^{-3}	yr^{-3}	zr^{-3}
H1	-0.9857	0.4132	0.0032	0.	-0.0716
C1	-14.4969	116.5554	0.0052	0.	-0.0619
O1	-22.2829	287.3592	0.2270	0.	0.1672
F1	-26.5379	413.6237	-0.2325	0.	0.0882

Center	$\dfrac{3x^2-r^2}{r^5}$	$\dfrac{3y^2-r^2}{r^5}$	$\dfrac{3z^2-r^2}{r^5}$	$\dfrac{3xy}{r^5}$	$\dfrac{3xz}{r^5}$	$\dfrac{3yz}{r^5}$
H1	-0.1515	-0.1594	0.3110	0.	-0.0040	0.
C1	0.1223	0.2880	-0.4103	0.	-0.2047	0.
O1	-0.2224	1.4250	-1.2027	0.	1.4180	0.
F1	2.3633	-1.0327	-1.3306	0.	-1.4750	0.

Mulliken Charge

Center	S1	S2	S3	S4	X1	X2	Y1	Y2	Z1	Z2	Total
H1	0.5107	0.2852									0.7959
C1	0.0429	1.9565	0.8645	0.2506	0.7341	0.0080	0.6191	0.1072	0.9593	0.0792	5.6213
O1	0.0427	1.9566	0.9756	0.8291	1.1843	0.3062	0.9329	0.4187	1.2732	0.3587	8.2780
F1	0.0436	1.9557	1.0256	0.9134	1.1616	0.3752	1.3867	0.5354	1.4103	0.4971	9.3047

Basis Function Type

Overlap Populations

Pair of Atomic Centers

	C1 H1	O1 H1	O1 C1	F1 H1	F1 C1	F1 O1
1A*	0.	0.	0.	0.	-0.0002	0.
2A*	0.	0.	-0.0007	0.	0.	0.
3A*	-0.0002	0.	0.0002	0.	-0.0001	0.
4A*	0.0043	0.0014	0.0286	0.0068	0.2416	0.0164
5A*	0.0031	0.0010	0.5899	-0.0002	-0.0130	-0.0093
6A*	0.2944	-0.0029	-0.0096	0.0159	0.1188	-0.0055
7A*	0.1507	0.0115	0.1128	-0.0017	0.1752	0.0208
8A*	0.0116	0.0015	-0.0573	-0.0046	0.0469	-0.0127
9A*	0.1297	-0.0132	0.1206	-0.0448	-0.1309	-0.0349
10A*	0.1750	-0.0870	-0.0212	-0.0803	-0.1232	-0.0656
Total A*	0.7685	-0.0877	0.7634	-0.1088	0.3152	-0.0907
1A**	0.	0.	0.0697	=0.	0.1958	0.0154
2A**	0.	0.	0.3474	0.	-0.1653	-0.0421
Total A**	0.	0.	0.4171	0.	0.0305	-0.0268
Total	0.7685	-0.0877	1.1805	-0.1088	0.3456	-0.1175
Distance	2.0693	3.8550	2.2319	3.7739	2.5288	4.1818

Molecular Orbitals and Eigenvalues

Center		S1	S2	S3	S4	X1	X2	Y1	Y2	Z1	Z2
* 1A*	H1	0.00003	-0.00012								
-26.3384	C1	-0.00000	0.00002	0.00002	-0.00031	0.00010	0.00031	0.	0.	-0.00006	-0.00026
	O1	0.00000	-0.00000	-0.00004	0.00011	0.00002	-0.00016	0.	0.	-0.00002	0.00012
	F1	0.05174	0.97891	0.00439	0.00146	0.00106	-0.00021	0.	0.	-0.00039	0.00009
* 2A*	H1	0.00006	-0.00052								
-20.6209	C1	0.00000	-0.00003	-0.00016	-0.00085	-0.00025	-0.00063	0.	0.	-0.00027	-0.00100
	O1	0.05112	0.97936	0.00337	0.00326	-0.00164	0.00006	0.	0.	-0.00121	0.00017
	F1	0.00000	0.00001	-0.00007	0.00017	-0.00003	0.00022	0.	0.	-0.00006	0.00029
* 3A*	H1	0.00098	-0.00092								
-11.4729	C1	0.05191	0.97910	0.00436	0.00120	-0.00042	0.00011	0.	0.	0.00107	-0.00074
	O1	-0.00002	-0.00053	0.00080	-0.00003	-0.00099	0.00011	0.	0.	-0.00085	0.00010
	F1	-0.00001	-0.00016	0.00024	-0.00015	0.00052	-0.00010	0.	0.	-0.00020	0.00003
* 4A*	H1	0.01669	0.01173								
-1.6883	C1	-0.00368	-0.07847	0.16989	0.00224	-0.08492	0.01040	0.	0.	0.05796	-0.00324
	O1	-0.00169	-0.03704	0.08046	0.05831	-0.04060	-0.00310	0.	0.	-0.01910	-0.00787
	F1	-0.01074	-0.23095	0.51467	0.44812	0.10286	0.02410	0.	0.	-0.03708	-0.01183
* 5A*	H1	0.02500	-0.00053								
-1.4850	C1	-0.00487	-0.10237	0.23777	0.04889	0.21714	-0.02012	0.	0.	0.12911	-0.01555
	O1	-0.00947	-0.20704	0.45793	0.30101	-0.14848	0.01310	0.	0.	-0.11690	0.01116
	F1	0.00298	0.06395	-0.14489	-0.12914	0.01037	0.01155	0.	0.	0.01482	-0.00569
* 6A*	H1	-0.18807	-0.05807								
-0.9218	C1	0.00711	0.14779	-0.39106	-0.13636	0.15324	-0.01494	0.	0.	0.21079	0.06063
	O1	-0.00333	-0.07282	0.16634	0.15627	0.05629	0.04073	0.	0.	0.08795	0.02458
	F1	-0.00319	-0.06789	0.16263	0.19083	-0.29024	-0.10980	0.	0.	0.23174	0.07694
* 7A*	H1	-0.13312	-0.06612								
-0.8073	C1	-0.00028	-0.00323	0.03387	-0.06477	-0.19088	0.02099	0.	0.	0.38322	0.02929
	O1	0.00103	0.02280	-0.04967	-0.08585	-0.20804	-0.06625	0.	0.	0.16363	0.05217
	F1	0.00177	0.03783	-0.09010	-0.12312	0.46372	0.17750	0.	0.	0.18236	0.08258
* 8A*	H1	-0.04578	-0.00111								
-0.7322	C1	0.00277	0.05481	-0.17235	-0.06777	-0.33665	0.02155	0.	0.	-0.05804	0.03113
	O1	-0.00424	-0.09375	0.20800	0.35213	0.28861	0.06814	0.	0.	0.34151	0.08462
	F1	0.00076	0.01608	-0.03947	-0.05906	0.11256	0.02830	0.	0.	-0.41997	-0.19106
* 9A*	H1	0.13639	0.12409								
-0.6454	C1	0.00039	0.00490	-0.04181	0.06646	-0.12552	0.05361	0.	0.	-0.29644	0.02369
	O1	-0.00161	-0.03543	0.07875	0.13337	0.38180	0.11030	0.	0.	0.10327	0.01884
	F1	0.00002	0.00033	-0.00224	0.00279	0.21065	0.09837	0.	0.	0.53913	0.27134
* 10A*	H1	-0.17031	-0.18558								
-0.5197	C1	0.00106	0.02351	-0.05236	-0.12326	0.02675	0.05132	0.	0.	0.12206	-0.02085
	O1	-0.00008	-0.00166	0.00421	0.00424	0.36496	0.18456	0.	0.	-0.55659	-0.27562
	F1	-0.00041	-0.01058	0.00759	0.11204	0.26661	0.15531	0.	0.	-0.10669	-0.07004
11A*	H1	0.09411	1.11393								
0.2557	C1	0.00498	0.11504	-0.23894	1.96735	0.16353	0.21370	0.	0.	0.23052	-0.42478
	O1	-0.00190	-0.04824	0.05514	0.68577	-0.13527	-0.33644	0.	0.	-0.02629	-0.05344
	F1	-0.00254	-0.06103	0.08607	0.69412	0.20405	0.36532	0.	0.	-0.06274	-0.05410
12A*	H1	0.01005	0.27145								
0.2957	C1	0.00024	0.00470	-0.01962	-0.49652	-0.19283	-1.60964	0.	0.	0.16593	-0.18270
	O1	-0.00212	-0.05811	0.03078	1.06233	0.08070	0.04018	0.	0.	-0.16890	-0.32093
	F1	0.00221	0.05406	-0.06786	-0.68936	-0.03914	-0.08501	0.	0.	0.12059	0.16358

Center		S1	S2	S3	S4	X1	X2	Y1	Y2	Z1	Z2
13A*	H1	-0.04120	2.39642								
0.3999	C1	0.00293	0.04058	-0.32327	-0.35074	-0.08277	0.19407	0.	0.	0.54305	2.12807
	O1	0.00225	0.05674	-0.06817	-0.91515	0.09343	0.40610	0.	0.	-0.18984	-0.22977
	F1	0.00150	0.03325	-0.07011	-0.32231	-0.07761	-0.22244	0.	0.	-0.14922	-0.21460
14A*	H1	-0.19543	-1.61261								
0.4404	C1	0.00034	-0.03540	-0.30843	2.43516	-0.32275	0.32721	0.	0.	0.68106	-1.90399
	O1	0.00138	0.02949	-0.08041	-0.29965	0.06629	0.16929	0.	0.	0.09001	0.45699
	F1	0.00152	0.03129	-0.08964	-0.19057	-0.18002	-0.27714	0.	0.	0.11834	0.32318
15A*	H1	0.16411	0.90881								
0.6107	C1	-0.00774	-0.08017	1.03161	-2.18811	-0.90867	0.52134	0.	0.	0.18627	0.48769
	O1	-0.00095	-0.02282	0.03685	0.44222	0.10164	-0.32893	0.	0.	-0.07597	-0.38994
	F1	0.00136	0.02039	-0.13206	0.32932	-0.42597	-0.10934	0.	0.	0.09053	-0.01771
* 1A**	H1	0.	0.					0.27306	0.05073	0.	0.
-0.7405	C1	0.	0.	0.	0.	0.	0.	0.27306	0.05073	0.	0.
	O1	0.	0.	0.	0.	0.	0.	0.18027	0.05968	0.	0.
	F1	0.	0.	0.	0.	0.	0.	0.65666	0.30365	0.	0.
* 2A**	H1	0.	0.								
-0.5641	C1	0.	0.	0.	0.	0.	0.	0.35422	0.07393	0.	0.
	O1	0.	0.	0.	0.	0.	0.	0.55324	0.27101	0.	0.
	F1	0.	0.	0.	0.	0.	0.	-0.35097	-0.20925	0.	0.
3A**	H1	0.	0.								
0.1114	C1	0.	0.	0.	0.	0.	0.	0.57666	0.71389	0.	0.
	O1	0.	0.	0.	0.	0.	0.	-0.43461	-0.52632	0.	0.
	F1	0.	0.	0.	0.	0.	0.	-0.21023	-0.21990	0.	0.
4A**	H1	0.	0.								
0.4076	C1	0.	0.	0.	0.	0.	0.	0.94009	-1.21860	0.	0.
	O1	0.	0.	0.	0.	0.	0.	-0.04862	0.10889	0.	0.
	F1	0.	0.	0.	0.	0.	0.	0.00214	0.14747	0.	0.

Overlap Matrix

First Function   Second Function and Overlap

H1S1 with
      H1S2 .68301   C1S1 .00217   C1S2 .03352   C1S3 .28087   C1S4 .37021   C1Z1 -.42737   C1Z2 -.43365
      O1S2 .00004   O1S3 .00279   O1S4 .04058   O1X1 -.00476 O1X2 -.08069 O1Z1 -.00918   O1Z2 -.15552
      F1S2 .00004   F1S3 .00195   F1S4 .02972   F1X1 .00405   F1X2 .08670   F1Z1 -.00500   F1Z2 -.10687

H1S2 with
      C1S1 .00761   C1S2 .09714   C1S3 .49382   C1S4 .70337   C1Z1 -.36036   C1Z2 -.60977   O1S1 .00074
      O1S2 .00999   O1S3 .07058   O1S4 .19006   O1X1 -.03737 O1X2 -.17513   O1Z1 -.07203   O1Z2 -.33753
      F1S1 .00070   F1S2 .00920   F1S3 .06256   F1S4 .16872   F1X1 .03954   F1X2 .20418   F1Z1 -.04875
      F1Z2 -.25170

C1S1 with
      C1S2 .35436   C1S3 .00805   C1S4 .01419   O1S3 .00058   O1S4 .00565   O1X1 -.00247 O1X2 -.01058
      O1Z1 -.00188 O1Z2 -.00806 F1S3 .00004   F1S4 .00280   F1X1 .00041   F1X2 .00973   F1Z1 -.00015
      F1Z2 -.00353

C1S2 with
      C1S3 .26226   C1S4 .17745   O1S2 .00002   O1S3 .01327   O1S4 .07371   O1X1 -.03739 O1X2 -.12952
      O1Z1 -.02848 O1Z2 -.09867 F1S3 .00192   F1S4 .03908   F1X1 .00980   F1X2 .12174   F1Z1 -.00355
      F1Z2 -.04411

C1S3 with
      C1S4 .79970   O1S1 .00216   O1S2 .02969   O1S3 .20701   O1S4 .40349   O1X1 -.21679 O1X2 -.48874
      O1Z1 -.16515 O1Z2 -.37232 F1S1 .00093   F1S2 .01294   F1S3 .10622   F1S4 .27601   F1X1 .14332
      F1X2 .49878   F1Z1 -.05193 F1Z2 -.18074

C1S4 with
      O1S1 .00433   O1S2 .05595   O1S3 .30951   O1S4 .57074   O1X1 -.13718 O1X2 -.42366   O1Z1 -.10450
      O1Z2 -.32275 F1S1 .00296   F1S2 .03800   F1S3 .21878   F1S4 .44463   F1X1 .11620   F1X2 .44091
      F1Z1 -.04211 F1Z2 -.15977

C1X1 with
      C1X2 .53875   O1S1 .00421   O1S2 .05473   O1S3 .28915   O1S4 .34192   O1X1 -.13667 O1X2 .03563
      O1Z1 -.22403 O1Z2 -.26149 F1S1 -.00276 F1S2 -.03588 F1S3 -.21546 F1S4 -.35833 F1X1 -.18841
      F1X2 -.26020 F1Z1 .09630   F1Z2 .19855

C1X2 with
      O1S1 .00508   O1S2 .06461   O1S3 .32487   O1S4 .49149   O1X1 .06604   O1X2 .32417   O1Z1 -.08702
      O1Z2 -.21924 F1S1 -.00486 F1S2 -.06172 F1S3 -.32694 F1S4 -.55571 F1X1 -.01987 F1X2 .04139
      F1Z1 .04968   F1Z2 .15698

C1Y1 with
      C1Y2 .53875   O1Y1 .15741   O1Y2 .37889   F1Y1 .07734   F1Y2 .28773

C1Y2 with
      O1Y1 .18027   O1Y2 .61196   F1Y1 .11723   F1Y2 .47460

C1Z1 with
      C1Z2 .53875   O1S1 .00321   O1S2 .04170   O1S3 .22028   O1S4 .26048   O1X1 -.22403 O1X2 -.26149
      O1Z1 -.01325 O1Z2 .17968   F1S1 .00100   F1S2 .01300   F1S3 .07807   F1S4 .12984   F1X1 .09630
      F1X2 .19855   F1Z1 .04245   F1Z2 .21579

C1Z2 with
      O1S1 .00387   O1S2 .04922   O1S3 .24749   O1S4 .37442   O1X1 -.08702 O1X2 -.21924   O1Z1 .11398
      O1Z2 .44494   F1S1 .00176   F1S2 .02237   F1S3 .11847   F1S4 .20137   F1X1 .04968   F1X2 .15698
      F1Z1 .09923   F1Z2 .41772

O1S1 with
      O1S2 .35461   O1S3 .00974   O1S4 .01469   F1S4 .00003   F1X2 .00052   F1Z2 .00006

O1S2 with
      O1S3 .28286   O1S4 .18230   F1S4 .00056   F1X2 .00744   F1Z2 .00088

O1S3 with
      O1S4 .79029   F1S3 .00012   F1S4 .00966   F1X1 .00074   F1X2 .06620   F1Z1 .00009   F1Z2 .00783

O1S4 with
      F1S1 .00009   F1S2 .00127   F1S3 .01329   F1S4 .06269   F1X1 .02248   F1X2 .19441   F1Z1 .00266
      F1Z2 .02299

Overlap Matrix (continued)

First Function   Second Function and Overlap

O1X1 with
    O1X2 .50607   F1S2 -.00001   F1S3 -.00109   F1S4 -.02249   F1X1 -.00420   F1X2 -.09971   F1Z1 -.00053
    F1Z2 -.01376

O1X2 with
    F1S1 -.00092   F1S2 -.01222   F1S3 -.08553   F1S4 -.22063   F1X1 -.09224   F1X2 -.37796   F1Z1 -.01304
    F1Z2 -.05893

O1Y1 with
    O1Y2 .50607   F1Y1 .00032   F1Y2 .01663

O1Y2 with
    F1Y1 .01802   F1Y2 .12049

O1Z1 with
    O1Z2 .50607   F1S3 -.00013   F1S4 -.00266   F1X1 -.00053   F1X2 -.01376   F1Z1 .00026   F1Z2 .01500

O1Z2 with
    F1S1 -.00011   F1S2 -.00145   F1S3 -.01011   F1S4 -.02608   F1X1 -.01304   F1X2 -.05893   F1Z1 .01648
    F1Z2 .11352

F1S1 with
    F1S2 .35647   F1S3 .01003   F1S4 .01474

F1S2 with
    F1S3 .28653   F1S4 .18096

F1S3 with
    F1S4 .78531

F1X1 with
    F1X2 .49444

F1Y1 with
    F1Y2 .49444

F1Z1 with
    F1Z2 .49444

CHOO-   Formate Anion

Molecular Geometry   Symmetry C$_{2v}$

Center		Coordinates	
	X	Y	Z
H1	0.	0.	-2.05980000
C1	0.	0.	0.
O1	2.10470000	0.	1.07240000
O2	-2.10470000	0.	1.07240000
Mass	0.	0.	0.71626031

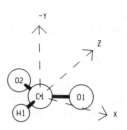

MO Symmetry Types and Occupancy

Representation	A1	A2	B1	B2
Number of MOs	16	2	10	4
Number Occupied	6	1	4	1

Energy Expectation Values

Total	-188.1261	Kinetic	188.3570
Electronic	-251.1237	Potential	-376.4830
Nuclear Repulsion	62.9977	Virial	1.9988

Electronic Potential	-439.4807
One-Electron Potential	-578.3219
Two-Electron Potential	138.8412

Moments of the Charge Distribution

First

x	y	z	Dipole Moment
-0.0003	0.	0.4018	0.9736
0.	0.	-1.3754	

Second

x^2	y^2	z^2	xy	xz	yz	r^2
-93.0908	-13.5010	-26.8722	0.	-0.0003	0.	-133.4640
70.8762	0.	12.8141	0.	0.	0.	83.6902

Third

x^3	xy^2	xz^2	x^2y	y^3	yz^2	x^2z	y^2z	z^3	xyz
-0.0036	-0.0002	-0.0003	0.	0.	0.	-26.5895	0.8167	24.9077	0.
0.	0.	0.	0.	0.	0.	25.2418	0.	-22.8758	0.

Fourth (even)

x^4	y^4	z^4	x^2y^2	x^2z^2	y^2z^2	r^2x^2	r^2y^2	r^2z^2	r^4
-652.1620	-42.1549	-152.7226	-57.8466	-75.2702	-22.3952	-785.2789	-122.3968	-250.3881	-1158.0637
313.9647	0.	61.2269	0.	8.9896	0.	322.9543	0.	70.2165	393.1708

Fourth (odd)

x^3y	xy^3	xyz^2	x^3z	xy^2z	xz^3	x^2yz	y^3z	yz^3
0.	0.	0.	-0.0026	-0.0002	-0.0007	0.	0.	0.
0.	0.	0.	0.	0.	0.	0.	0.	0.

Expectation Values at Atomic Centers

Center			Expectation Value		
	r^{-1}	δ	xr^{-3}	yr^{-3}	zr^{-3}
H1	-1.3076	0.4464	0.	0.	-0.0911
C1	-14.8325	116.5526	0.	0.	-0.0217
O1	-22.6266	287.0556	0.2099	0.	0.1155

	$\dfrac{3x^2-r^2}{r^5}$	$\dfrac{3y^2-r^2}{r^5}$	$\dfrac{3z^2-r^2}{r^5}$	$\dfrac{3xy}{r^5}$	$\dfrac{3xz}{r^5}$	$\dfrac{3yz}{r^5}$
H1	-0.1798	-0.1735	0.3533	0.	0.	0.
C1	-0.2304	0.4489	-0.2186	0.	0.0001	0.
O1	0.6695	0.4198	-1.0893	0.	1.2548	0.

Mulliken Charge

Center				Basis Function Type							Total
	S1	S2	S3	S4	X1	X2	Y1	Y2	Z1	Z2	
H1	0.5435	0.4249									0.9684
C1	0.0428	1.9566	0.8546	0.2312	0.8672	0.0659	0.5842	0.1726	0.8997	0.1197	5.7946
O1	0.0426	1.9567	0.9383	0.8639	1.0920	0.3461	1.0691	0.5525	1.2582	0.4992	8.6185

Overlap Populations

	Pair of Atomic Centers			
	C1	O1	O1	O2
	H1	H1	C1	O1
1A1	0.	0.	0.	0.
2A1	-0.0002	0.	0.0002	0.
3A1	0.0188	0.0071	0.2971	0.0394
4A1	0.5226	-0.0011	0.0192	-0.0016
5A1	0.0404	0.0072	0.0997	0.0101
6A1	0.1279	-0.2165	-0.0180	0.0695
Total A1	0.7095	-0.2034	0.3982	0.1174
1A2	0.	0.	-0.0001	-0.0999
Total A2	0.	0.	-0.0001	-0.0999
1B1	0.	0.	-0.0006	0.
2B1	0.	0.	0.2530	-0.0310
3B1	0.	0.	0.0741	-0.0089
4B1	0.	-0.0001	0.0444	-0.2159
Total B1	0.	-0.0002	0.3709	-0.2557
1B2	0.	0.	0.2640	0.0303
Total B2	0.	0.	0.2640	0.0303
Total	0.7095	-0.2035	1.0330	-0.2079
Distance	2.0598	3.7737	2.3622	4.2094

Molecular Orbitals and Eigenvalues

Center		S1	S2	S3	S4	X1	X2	Y1	Y2	Z1	Z2
• 1A1	H1	0.00008	-0.00040								
-20.2646	C1-	-0.00000	-0.00001	-0.00011	-0.00109	0.00023	0.00064	0.	0.	-0.00021	-0.00087
	O1	0.00550	0.10538	0.00030	0.00053	-0.00013	-0.00028	0.	0.	-0.00013	0.00028
	O2	0.05083	0.97367	0.00334	0.00318	0.00143	0.00006	0.	0.	-0.00080	0.00011
• 2A1	H1	-0.00081	0.00091								
-11.1372	C1	-0.05191	-0.97909	-0.00436	-0.00087	-0.00000	0.00000	0.	0.	-0.00030	0.00069
	O1	0.00002	0.00036	-0.00046	-0.00021	0.00087	0.00006	0.	0.	0.00047	-0.00008
	O2	0.00002	0.00036	-0.00046	-0.00021	-0.00087	-0.00007	0.	0.	0.00047	-0.00008
• 3A1	H1	0.03841	0.01458								
-1.1710	C1	-0.00668	-0.14190	0.31760	0.05929	0.00010	0.00003	0.	0.	0.10601	-0.00754
	O1	-0.00640	-0.14079	0.30294	0.23756	-0.11504	-0.00341	0.	0.	-0.05133	-0.00756
	O2	-0.00640	-0.14071	0.30273	0.23750	0.11499	0.00345	0.	0.	-0.05129	-0.00759
• 4A1	H1	-0.24752	-0.11840								
-0.5686	C1	0.00626	0.13148	-0.33434	-0.14083	-0.00001	0.00005	0.	0.	0.32767	0.09222
	O1	-0.00320	-0.07030	0.15718	0.17420	0.05687	0.04268	0.	0.	0.15363	0.03868
	O2	-0.00320	-0.07030	0.15717	0.17426	-0.05684	-0.04267	0.	0.	0.15358	0.03864
• 5A1	H1	0.08219	0.09744								
-0.4203	C1	0.00356	0.06934	-0.22925	0.01829	-0.00010	0.00008	0.	0.	-0.36017	0.01606
	O1	-0.00301	-0.06696	0.14355	0.22948	0.36762	0.14928	0.	0.	-0.05042	-0.03097
	O2	-0.00301	-0.06694	0.14348	0.22948	-0.36741	-0.14921	0.	0.	-0.05051	-0.03103
• 6A1	H1	0.20312	0.35280								
-0.1811	C1	0.00015	0.00069	-0.02301	0.08427	-0.00016	0.00039	0.	0.	-0.17608	0.09194
	O1	0.00003	0.00140	0.00382	-0.04903	-0.02628	-0.01900	0.	0.	0.46651	0.26917
	O2	0.00002	0.00137	0.00384	-0.04859	0.02835	0.02012	0.	0.	0.46868	0.27042
7A1	H1	-0.02165	-1.08040								
0.5326	C1	-0.00405	-0.09562	0.18242	2.22623	-0.00011	-0.00300	0.	0.	-0.39061	0.69386
	O1	0.00198	0.05400	-0.03005	-0.87917	0.09196	0.41922	0.	0.	0.04612	0.04257
	O2	0.00199	0.05418	-0.03006	-0.88262	-0.09184	-0.41929	0.	0.	0.04649	0.04353
8A1	H1	-0.02987	3.00195								
0.6680	C1	0.00256	0.05017	-0.18176	-1.18128	0.00000	0.00063	0.	0.	0.37089	2.83169
	O1	0.00141	0.03650	-0.03468	-0.57039	0.02016	0.31319	0.	0.	-0.21166	-0.47010
	O2	0.00141	0.03646	-0.03462	-0.56972	-0.02029	-0.31312	0.	0.	-0.21175	-0.47027
• 1A2	H1	0.	0.								
-0.1594	C1	0.	0.	0.	0.	0.	0.	-0.00006	-0.00002	0.	0.
	O1	0.	0.	0.	0.	0.	0.	0.49804	0.33138	0.	0.
	O2	0.	0.	0.	0.	0.	0.	-0.49812	-0.33143	0.	0.
• 1B1	H1	-0.00006	0.00032								
-20.2648	C1	0.00000	0.00001	0.00009	0.00088	0.00028	0.00080	0.	0.	0.00017	0.00070
	O1	-0.05083	-0.97367	-0.00335	-0.00314	0.00144	0.00000	0.	0.	0.00079	-0.00005
	O2	0.00550	0.10538	0.00042	0.00015	0.00018	-0.00027	0.	0.	-0.00004	-0.00027
• 2B1	H1	-0.00001	-0.00001								
-1.0560	C1	0.00000	0.00004	-0.00010	-0.00005	0.30298	-0.00935	0.	0.	-0.00004	-0.00002
	O1	-0.00734	-0.16143	0.34957	0.27305	-0.08806	0.01512	0.	0.	-0.06311	0.00837
	O2	0.00734	0.16150	-0.34973	-0.27311	-0.08813	0.01514	0.	0.	0.06314	-0.00836
• 3B1	H1	0.00001	0.00001								
-0.3639	C1	-0.00000	-0.00003	0.00008	0.00013	-0.44973	0.05165	0.	0.	0.00008	0.00006
	O1	-0.00276	-0.06140	0.13272	0.24139	0.29310	0.09512	0.	0.	0.28930	0.11691
	O2	0.00277	0.06145	-0.13278	-0.24169	0.29326	0.09512	0.	0.	-0.28922	-0.11687

Center		Basis Function Type									
		S1	S2	S3	S4	X1	X2	Y1	Y2	Z1	Z2
* 4B1	H1	-0.00059	-0.00102								
-0.1880	C1	-0.00000	-0.00001	0.00010	-0.00011	-0.07876	0.13009	0.	0.	0.00056	-0.00023
	O1	0.00012	0.00339	-0.00146	-0.06010	0.37192	0.19375	0.	0.	-0.37837	-0.21920
	O2	-0.00012	-0.00338	0.00144	0.06019	0.37180	0.19358	0.	0.	0.37571	0.21770
5B1	H1	-0.00001	0.00251								
0.5584	C1	0.00001	0.00016	-0.00043	-0.00319	-0.09107	-1.92295	0.	0.	0.00087	-0.00023
	O1	-0.00215	-0.06080	0.01658	1.10845	0.06376	-0.05007	0.	0.	-0.12923	-0.30150
	O2	0.00214	0.06067	-0.01640	-1.10660	0.06399	-0.04924	0.	0.	0.12899	0.30116
* 1B2	H1	0.	0.								
-0.3397	C1	0.	0.	0.	0.	0.	0.	0.41606	0.12736	0.	0.
	O1	0.	0.	0.	0.	0.	0.	0.37969	0.16849	0.	0.
	O2	0.	0.	0.	0.	0.	0.	0.37956	0.16842	0.	0.
2B2	H1	0.	0.								
0.4173	C1	0.	0.	0.	0.	0.	0.	-0.46863	-0.93904	0.	0.
	O1	0.	0.	0.	0.	0.	0.	0.32915	0.49287	0.	0.
	O2	0.	0.	0.	0.	0.	0.	0.32915	0.49290	0.	0.
3B2	H1	0.	0.								
0.6744	C1	0.	0.	0.	0.	0.	0.	-1.01310	1.16506	0.	0.
	O1	0.	0.	0.	0.	0.	0.	0.05470	-0.11212	0.	0.
	O2	0.	0.	0.	0.	0.	0.	0.05470	-0.11213	0.	0.

Overlap Matrix

First Function   Second Function and Overlap

H1S1 with
      H1S2 .68301   C1S1 .00222   C1S2 .03427   C1S3 .28398   C1S4 .37199   C1Z1 -.43029   C1Z2 -.43334
      O1S2 .00005   O1S3 .00355   O1S4 .04595   O1X1 -.00702   O1X2 -.10593   O1Z1 -.01045   O1Z2 -.15764
H1S2 with
      C1S1 .00766   C1S2 .09779   C1S3 .49630   C1S4 .70560   C1Z1 -.36049   C1Z2 -.60863   O1S1 .00083
      O1S2 .01112   O1S3 .07740   O1S4 .20336   O1X1 -.04849   O1X2 -.22049   O1Z1 -.07217   O1Z2 -.32813
C1S1 with
      C1S2 .35436   C1S3 .00805   C1S4 .01419   O1S3 .00034   O1S4 .00477   O1X1 -.00191   O1X2 -.01103
      O1Z1 -.00097   O1Z2 -.00562
C1S2 with
      C1S3 .26226   C1S4 .17745   O1S2 .00001   O1S3 .00851   O1S4 .06293   O1X1 -.03057   O1X2 -.13596
      O1Z1 -.01557   O1Z2 -.06927
C1S3 with
      C1S4 .79970   O1S1 .00163   O1S2 .02275   O1S3 .17223   O1S4 .36351   O1X1 -.21535   O1X2 -.53128
      O1Z1 -.10973   O1Z2 -.27070
C1S4 with
      O1S1 .00397   O1S2 .05129   O1S3 .28682   O1S4 .53851   O1X1 -.15086   O1X2 -.47664   O1Z1 -.07687
      O1Z2 -.24286
C1X1 with
      C1X2 .53875   O1S1 .00396   O1S2 .05160   O1S3 .28576   O1S4 .36549   O1X1 -.20745   O1X2 -.09430
      O1Z1 -.17300   O1Z2 -.22520
C1X2 with
      O1S1 .00562   O1S2 .07159   O1S3 .36238   O1S4 .55490   O1X1 .01860   O1X2 .19844   O1Z1 -.07701
      O1Z2 -.19708
C1Y1 with
      C1Y2 .53875   O1Y1 .13208   O1Y2 .34769
C1Y2 with
      O1Y1 .16974   O1Y2 .58524
C1Z1 with
      C1Z2 .53875   O1S1 .00202   O1S2 .02629   O1S3 .14560   O1S4 .18623   O1X1 -.17300   O1X2 -.22520
      O1Z1 .04393   O1Z2 .23294
C1Z2 with
      O1S1 .00286   O1S2 .03648   O1S3 .18464   O1S4 .28274   O1X1 -.07701   O1X2 -.19708   O1Z1 .13050
      O1Z2 .48482
O1S1 with
      O1S2 .35461   O1S3 .00974   O1S4 .01469   O2S4 .00010   O2X2 .00105
O1S2 with
      O1S3 .28286   O1S4 .18230   O2S4 .00147   O2X1 .00001   O2X2 .01429
O1S3 with
      O1S4 .79029   O2S3 .00029   O2S4 .01735   O2X1 .00201   O2X2 .10318
O1S4 with
      O2S1 .00010   O2S2 .00147   O2S3 .01735   O2S4 .08185   O2X1 .03311   O2X2 .25286
O1X1 with
      O1X2 .50607   O2S2 -.00001   O2S3 -.00201   O2S4 -.03311   O2X1 -.00841   O2X2 -.12458
O1X2 with
      O2S1 -.00105   O2S2 -.01429   O2S3 -.10318   O2S4 -.25286   O2X1 -.12458   O2X2 -.41956
O1Y1 with
      O1Y2 .50607   O2Y1 .00071   O2Y2 .02506
O1Y2 with
      O2Y1 .02506   O2Y2 .15054
O1Z1 with
      O1Z2 .50607   O2Z1 .00071   O2Z2 .02506

Overlap Matrix (continued)

First Function   Second Function and Overlap

0122 with
    0221 .02506   0222 .15054

C3H6   Cyclopropane

Molecular Geometry   Symmetry D_{3h}

Center	Coordinates		
	X	Y	Z
H1	2.75169519	0.	1.73661330
H2	2.75169519	0.	-1.73661330
H3	-1.37584750	-2.38303789	1.73661330
H4	-1.37584750	-2.38303789	-1.73661330
H5	-1.37584750	2.38303789	1.73661330
H6	-1.37584750	2.38303789	-1.73661330
C1	1.64748000	0.	0.
C2	-0.82374000	-1.42675950	0.
C3	-0.82374000	1.42675950	0.
Mass	0.00000001	0.	0.

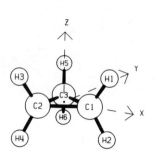

MO Symmetry Types and Occupancy

Representation	A1*	A1**	A2*	A2**	E*	E**
Number of MOs	8	0	2	4	20	8
Number Occupied	3	0	0	1	6	2

Energy Expectation Values

Total	-117.0099	Kinetic	117.0667
Electronic	-192.5399	Potential	-234.0765
Nuclear Repulsion	75.5301	Virial	1.9995
Electronic Potential	-309.6066		
One-Electron Potential	-422.9223		
Two-Electron Potential	113.3158		

Moments of the Charge Distribution

First

x	y	z	Dipole Moment
-0.0003	0.	0.	0.0003
0.	0.	0.	

Second

x^2	y^2	z^2	xy	xz	yz	r^2
-62.8578	-62.8563	-31.8027	0.	0.	0.	-157.5168
47.1432	47.1432	18.0950	0.	0.	0.	112.3813

Third

x^3	xy^2	xz^2	x^2y	y^3	yz^2	x^2z	y^2z	z^3	xyz
-48.0967	48.0883	-0.0009	0.	0.	0.	0.	0.	0.	0.
51.3751	-51.3751	0.	0.	0.	0.	0.	0.	0.	0.

Fourth (even)

x^4	y^4	z^4	x^2y^2	x^2z^2	y^2z^2	r^2x^2	r^2y^2	r^2z^2	r^4
-413.2779	-413.2531	-158.6553	-137.7532	-118.8323	-118.8276	-669.8635	-669.8339	-396.3153	-1736.0128
178.7243	178.7243	54.5712	59.5748	68.5059	68.5059	306.8050	306.8050	191.5831	805.1932

Expectation Values at Atomic Centers

Center			Expectation Value		
	r^{-1}	δ	xr^{-3}	yr^{-3}	zr^{-3}
H1	-1.1118	0.4195	0.0358	0.	0.0487
C1	-14.7328	116.2735	0.0456	0.	0.

	$\dfrac{3x^2-r^2}{r^5}$	$\dfrac{3y^2-r^2}{r^5}$	$\dfrac{3z^2-r^2}{r^5}$	$\dfrac{3xy}{r^5}$	$\dfrac{3xz}{r^5}$	$\dfrac{3yz}{r^5}$
H1	-0.0115	-0.1802	0.1917	0.	0.2240	0.
C1	0.2118	-0.1285	-0.0833	0.	0.	0.

Mulliken Charge

Center			Basis Function Type							Total
	S1	S2	S3	S4	X1	X2	Y1	Y2	Z1	Z2	
H1	0.5174	0.2973									0.8148
C1	0.0427	1.9563	0.7565	0.5475	0.8303	0.0510	0.8772	0.1610	0.9329	0.2154	6.3707

Overlap Populations

	Pair of Atomic Centers								
	H2	H3	H3	C1	C1	C2	C2	C2	C3
	H1	H1	H2	H1	H3	H3	H5	C1	C2
1A1*	0.	0.	0.	-0.0001	0.	-0.0001	0.	0.0004	0.0004
2A1*	0.0005	0.0001	0.	0.0254	0.0041	0.0254	0.0041	0.2665	0.2665
3A1*	0.0187	0.0056	0.0016	0.0865	-0.0152	0.0865	-0.0152	0.1134	0.1134
Total A1*	0.0192	0.0057	0.0016	0.1118	-0.0111	0.1118	-0.0111	0.3803	0.3803
1A2**	-0.0137	0.0038	-0.0011	0.0933	0.0116	0.0933	0.0116	0.0968	0.0968
Total A2**	-0.0137	0.0038	-0.0011	0.0933	0.0116	0.0933	0.0116	0.0968	0.0968
1E*	0.	0.	0.	0.	0.	-0.0002	0.	-0.0001	-0.0020
2E*	0.	0.	0.	-0.0002	0.	-0.0001	0.	-0.0013	0.0005
3E*	0.	0.	0.	0.	0.0044	0.1908	-0.0269	0.0504	-0.2731
4E*	0.0174	-0.0023	-0.0006	0.2544	-0.0194	0.0636	0.0119	-0.1653	0.1582
5E*	0.0236	-0.0038	-0.0011	-0.0063	0.0325	-0.0016	-0.0352	-0.1571	0.4867
6E*	0.	0.	0.	0.	-0.0379	-0.0046	0.0296	0.2720	-0.3700
Total E*	0.0410	-0.0061	-0.0017	0.2479	-0.0205	0.2481	-0.0206	-0.0014	0.0004
1E**	0.	0.	0.	0.	0.	0.2392	-0.0394	0.	-0.2442
2E**	-0.0647	-0.0095	0.0027	0.3189	-0.0263	0.0797	0.0131	-0.1629	0.0815
Total E**	-0.0647	-0.0095	0.0027	0.3189	-0.0263	0.3189	-0.0263	-0.1629	-0.1627
Total	-0.0181	-0.0060	0.0015	0.7720	-0.0463	0.7721	-0.0464	0.3127	0.3147
Distance	3.4732	4.7661	5.8974	2.0579	4.2232	2.0579	4.2232	2.8535	2.8535

Molecular Orbitals and Eigenvalues

Center		S1	S2	S3	S4	X1	X2	Y1	Y2	Z1	Z2
* 1A1*	H1	0.00044	-0.00051								
-11.2296	H2	0.00044	-0.00051								
	H3	0.00044	-0.00051								
	H4	0.00044	-0.00051								
	H5	0.00044	-0.00051								
	H6	0.00044	-0.00051								
	C1	0.02996	0.56506	0.00279	0.00108	-0.00066	0.00040	0.	0.	0.	0.
	C2	0.02996	0.56506	0.00279	0.00108	0.00033	-0.00020	0.00057	-0.00035	0.	0.
	C3	0.02996	0.56506	0.00279	0.00108	0.00033	-0.00020	-0.00057	0.00035	0.	0.
* 2A1*	H1	-0.05367	-0.00619								
-1.1403	H2	-0.05367	-0.00619								
	H3	-0.05367	-0.00619								
	H4	-0.05367	-0.00619								
	H5	-0.05367	-0.00619								
	H6	-0.05367	-0.00619								
	C1	0.00601	0.13002	-0.27783	-0.12985	0.12878	-0.01047	0.	0.	0.	0.
	C2	0.00601	0.13002	-0.27783	-0.12985	-0.06439	0.00523	-0.11153	0.00907	0.	0.
	C3	0.00601	0.13002	-0.27783	-0.12985	-0.06439	0.00523	0.11153	-0.00907	0.	0.
* 3A1*	H1	0.10470	0.08346								
-0.6228	H2	0.10470	0.08346								
	H3	0.10470	0.08346								
	H4	0.10470	0.08346								
	H5	0.10470	0.08346								
	H6	0.10470	0.08346								
	C1	-0.00042	-0.01010	0.01386	0.04759	0.35577	0.03397	0.	0.	0.	0.
	C2	-0.00042	-0.01010	0.01386	0.04759	-0.17789	-0.01698	-0.30811	-0.02942	0.	0.
	C3	-0.00042	-0.01010	0.01386	0.04759	-0.17789	-0.01698	0.30811	0.02942	0.	0.
4A1*	H1	-0.02439	-0.83698								
0.2668	H2	-0.02439	-0.83698								
	H3	-0.02439	-0.83698								
	H4	-0.02439	-0.83698								
	H5	-0.02439	-0.83698								
	H6	-0.02439	-0.83698								
	C1	-0.00216	-0.04732	0.11594	1.31020	0.08461	0.81274	0.	0.	0.	0.
	C2	-0.00216	-0.04732	0.11594	1.31020	-0.04230	-0.40637	-0.07327	-0.70386	0.	0.
	C3	-0.00216	-0.04732	0.11594	1.31020	-0.04230	-0.40637	0.07327	0.70386	0.	0.
5A1*	H1	0.06299	-0.24566								
0.4090	H2	0.06299	-0.24566								
	H3	0.06299	-0.24566								
	H4	0.06299	-0.24566								
	H5	0.06299	-0.24566								
	H6	0.06299	-0.24566								
	C1	-0.00211	-0.04623	0.11258	-0.09473	0.42076	-0.65886	0.	0.	0.	0.
	C2	-0.00211	-0.04623	0.11258	-0.09473	-0.21038	0.32943	-0.36439	0.57059	0.	0.
	C3	-0.00211	-0.04623	0.11258	-0.09473	-0.21038	0.32943	0.36439	-0.57059	0.	0.
1A2*	H1	0.	0.								
0.2826	H2	0.	0.								
	H3	0.	0.								
	H4	0.	0.								
	H5	0.	0.								
	H6	0.	0.								
	C1	0.	0.	0.	0.	0.	0.	-0.33376	-1.47099	0.	0.
	C2	0.	0.	0.	0.	-0.28904	-1.27392	0.16688	0.73550	0.	0.
	C3	0.	0.	0.	0.	0.28904	1.27392	0.16688	0.73550	0.	0.

Center		S1	S2	S3	S4	X1	X2	Y1	Y2	Z1	Z2
						Basis Function Type					
* 1A2**	H1	0.11542	0.06446								
-0.6720	H2	-0.11542	-0.06446								
	H3	0.11542	0.06446								
	H4	-0.11542	-0.06446								
	H5	0.11542	0.06446								
	H6	-0.11542	-0.06446								
	C1	0.	0.	0.	0.	0.	0.	0.	0.	0.30701	0.05884
	C2	0.	0.	0.	0.	0.	0.	0.	0.	0.30701	0.05884
	C3	0.	0.	0.	0.	0.	0.	0.	0.	0.30701	0.05884
2A2**	H1	-0.06362	-0.99578								
0.3367	H2	0.06362	0.99578								
	H3	-0.06362	-0.99578								
	H4	0.06362	0.99578								
	H5	-0.06362	-0.99578								
	H6	0.06362	0.99578								
	C1	0.	0.	0.	0.	0.	0.	0.	0.	0.34167	0.62957
	C2	0.	0.	0.	0.	0.	0.	0.	0.	0.34167	0.62957
	C3	0.	0.	0.	0.	0.	0.	0.	0.	0.34167	0.62957
3A2**	H1	-0.09899	-0.47414								
0.4778	H2	0.09899	0.47414								
	H3	-0.09899	-0.47414								
	H4	0.09899	0.47414								
	H5	-0.09899	-0.47414								
	H6	0.09899	0.47414								
	C1	0.	0.	0.	0.	0.	0.	0.	0.	-0.40757	0.84158
	C2	0.	0.	0.	0.	0.	0.	0.	0.	-0.40757	0.84158
	C3	0.	0.	0.	0.	0.	0.	0.	0.	-0.40757	0.84158
* 1E*	H1	0.	0.								
-11.2287	H2	0.	0.								
	H3	-0.00042	0.00069								
	H4	-0.00042	0.00069								
	H5	0.00042	-0.00069								
	H6	0.00042	-0.00069								
	C1	0.	0.	0.	0.	0.	0.	0.00014	-0.00056	0.	0.
	C2	-0.03671	-0.69229	-0.00281	-0.00397	-0.00037	0.00004	-0.00049	-0.00050	0.	0.
	C3	0.03671	0.69229	0.00281	0.00397	0.00037	-0.00004	-0.00049	-0.00050	0.	0.
* 2E*	H1	0.00048	-0.00079								
-11.2286	H2	0.00048	-0.00079								
	H3	-0.00024	0.00040								
	H4	-0.00024	0.00040								
	H5	-0.00024	0.00040								
	H6	-0.00024	0.00040								
	C1	0.04239	0.79938	0.00325	0.00458	-0.00071	-0.00048	0.	0.	0.	0.
	C2	-0.02119	-0.39969	-0.00162	-0.00229	-0.00007	-0.00054	-0.00037	0.00003	0.	0.
	C3	-0.02119	-0.39969	-0.00162	-0.00229	-0.00007	-0.00054	0.00037	-0.00003	0.	0.
* 3E*	H1	0.	0.								
-0.8156	H2	0.	0.								
	H3	0.13101	0.05809								
	H4	0.13101	0.05809								
	H5	-0.13101	-0.05809								
	H6	-0.13101	-0.05809								
	C1	0.	0.	0.	0.	0.	0.	-0.14398	-0.00483	0.	0.
	C2	-0.00618	-0.13453	0.29447	0.27734	0.00221	0.00514	-0.14015	0.00408	0.	0.
	C3	0.00618	0.13453	-0.29447	-0.27734	0.00221	-0.00514	-0.14015	0.00408	0.	0.

Center		S1	S2	S3	S4	X1	X2	Y1	Y2	Z1	Z2
* 4E *	H1	0.15127	0.06713								
-0.8156	H2	0.15127	0.06713								
	H3	-0.07563	-0.03356								
	H4	-0.07563	-0.03356								
	H5	-0.07563	-0.03356								
	H6	-0.07563	-0.03356								
	C1	-0.00713	-0.15534	0.34003	0.32019	0.13887	-0.00712	0.	0.	0.	0.
	C2	0.00357	0.07767	-0.17002	-0.16009	0.14272	0.00187	-0.00222	-0.00519	0.	0.
	C3	0.00357	0.07767	-0.17002	-0.16009	0.14272	0.00187	0.00222	0.00519	0.	0.
* 5E *	H1	-0.08986	-0.10189								
-0.4188	H2	-0.08986	-0.10189								
	H3	0.04493	0.05095								
	H4	0.04493	0.05095								
	H5	0.04493	0.05095								
	H6	0.04493	0.05095								
	C1	-0.00054	-0.01347	0.01879	0.18654	-0.29045	-0.13995	0.	0.	0.	0.
	C2	0.00027	0.00674	-0.00940	-0.09327	0.30840	0.08369	-0.34574	-0.12912	0.	0.
	C3	0.00027	0.00674	-0.00940	-0.09327	0.30840	0.08369	0.34574	0.12912	0.	0.
* 6E *	H1	0.	0.								
-0.4188	H2	0.	0.								
	H3	0.07779	0.08826								
	H4	0.07779	0.08826								
	H5	-0.07779	-0.08826								
	H6	-0.07779	-0.08826								
	C1	0.	0.	0.	0.	0.	0.	0.50808	0.15839	0.	0.
	C2	0.00046	0.01165	-0.01641	-0.16068	-0.34577	-0.12894	-0.09081	-0.06494	0.	0.
	C3	-0.00046	-0.01165	0.01641	0.16068	0.34577	0.12894	-0.09081	-0.06494	0.	0.
7E *	H1	0.	0.								
0.2405	H2	0.	0.								
	H3	-0.00108	-0.13676								
	H4	-0.00108	-0.13676								
	H5	0.00108	0.13676								
	H6	0.00108	0.13676								
	C1	0.	0.	0.	0.	0.	0.	0.00737	0.78314	0.	0.
	C2	-0.00363	-0.10154	0.04567	2.58678	0.01633	0.33704	0.03566	1.36690	0.	0.
	C3	0.00363	0.10154	-0.04567	-2.58678	-0.01633	-0.33704	0.03566	1.36690	0.	0.
8E *	H1	-0.00118	-0.15633								
0.2406	H2	-0.00118	-0.15633								
	H3	0.00059	0.07816								
	H4	0.00059	0.07816								
	H5	0.00059	0.07816								
	H6	0.00059	0.07816								
	C1	-0.00419	-0.11721	0.05304	2.98469	-0.04485	-1.56325	0.	0.	0.	0.
	C2	0.00209	0.05860	-0.02652	-1.49234	-0.01693	-0.97779	-0.01612	-0.33802	0.	0.
	C3	0.00209	0.05860	-0.02652	-1.49234	-0.01693	-0.97779	0.01612	0.33802	0.	0.
9E *	H1	0.06515	1.73431								
0.4050	H2	0.06515	1.73431								
	H3	-0.03257	-0.86716								
	H4	-0.03257	-0.86716								
	H5	-0.03257	-0.86716								
	H6	-0.03257	-0.86716								
	C1	0.00216	0.05386	-0.07752	-1.28420	-0.15474	-2.04229	0.	0.	0.	0.
	C2	-0.00108	-0.02693	0.03876	0.64210	-0.11446	-0.36435	-0.02326	-0.96876	0.	0.
	C3	-0.00108	-0.02693	0.03876	0.64210	-0.11446	-0.36435	0.02326	0.96876	0.	0.

Center — Basis Function Type

Center		S1	S2	S3	S4	X1	X2	Y1	Y2	Z1	Z2
10E* 0.4051	H1	0.	0.								
	H2	0.	0.								
	H3	0.05647	1.50228								
	H4	0.05647	1.50228								
	H5	-0.05647	-1.50228								
	H6	-0.05647	-1.50228								
	C1	0.	0.	0.	0.	0.	0.	0.10032	-0.19351	0.	0.
	C2	0.00187	0.04658	-0.06679	-1.11167	0.02389	0.96819	0.14170	1.48344	0.	0.
	C3	-0.00187	-0.04658	0.06679	1.11167	-0.02389	-0.96819	0.14170	1.48344	0.	0.
11E* 0.4808	H1	-0.03189	0.36706								
	H2	-0.03189	0.36706								
	H3	0.01594	-0.18353								
	H4	0.01594	-0.18353								
	H5	0.01594	-0.18353								
	H6	0.01594	-0.18353								
	C1	0.00100	0.02244	-0.04974	-0.49153	-0.50796	0.59790	0.	0.	0.	0.
	C2	-0.00050	-0.01122	0.02487	0.24576	0.23913	-0.39646	-0.43133	0.57410	0.	0.
	C3	-0.00050	-0.01122	0.02487	0.24576	0.23913	-0.39646	0.43133	-0.57410	0.	0.
12E* 0.4808	H1	0.	0.								
	H2	0.	0.								
	H3	-0.02774	0.31590								
	H4	-0.02774	0.31590								
	H5	0.02774	-0.31590								
	H6	0.02774	-0.31590								
	C1	0.	0.	0.	0.	0.	0.	-0.48821	0.72798	0.	0.
	C2	0.00086	0.01941	-0.04318	-0.42448	0.43127	-0.57547	0.25877	-0.26876	0.	0.
	C3	-0.00086	-0.01941	0.04318	0.42448	-0.43127	0.57547	0.25877	-0.26876	0.	0.
* 1E** -0.5106	H1	0.	0.								
	H2	0.	0.								
	H3	0.18376	0.13023								
	H4	-0.18376	-0.13023								
	H5	-0.18376	-0.13023								
	H6	0.18376	0.13023								
	C1	0.	0.	0.	0.	0.	0.	0.	0.	0.	0.
	C2	0.	0.	0.	0.	0.	0.	0.	0.	0.37853	0.14554
	C3	0.	0.	0.	0.	0.	0.	0.	0.	-0.37853	-0.14554
* 2E** -0.5105	H1	-0.21219	-0.15029								
	H2	0.21219	0.15029								
	H3	0.10610	0.07514								
	H4	-0.10610	-0.07514								
	H5	0.10610	0.07514								
	H6	-0.10610	-0.07514								
	C1	0.	0.	0.	0.	0.	0.	0.	0.	-0.43708	-0.16821
	C2	0.	0.	0.	0.	0.	0.	0.	0.	0.21854	0.08411
	C3	0.	0.	0.	0.	0.	0.	0.	0.	0.21854	0.08411
3E** 0.4366	H1	-0.06153	-2.23510								
	H2	0.06153	2.23510								
	H3	0.03077	1.11755								
	H4	-0.03077	-1.11755								
	H5	0.03077	1.11755								
	H6	-0.03077	-1.11755								
	C1	0.	0.	0.	0.	0.	0.	0.	0.	-0.24840	4.65794
	C2	0.	0.	0.	0.	0.	0.	0.	0.	0.12420	-2.32897
	C3	0.	0.	0.	0.	0.	0.	0.	0.	0.12420	-2.32897

Center		S1	S2	S3	S4	X1	X2	Y1	Y2	Z1	Z2
4E**	H1	0.	0.								
0.4371	H2	0.	0.								
	H3	0.05353	1.93238								
	H4	-0.05353	-1.93238								
	H5	-0.05353	-1.93238								
	H6	0.05353	1.93238								
	C1	0.	0.	0.	0.	0.	0.	0.	0.	0.	0.
	C2	0.	0.	0.	0.	0.	0.	0.	0.	0.21637	-4.03061
	C3	0.	0.	0.	0.	0.	0.	0.	0.	-0.21637	4.03061
5E**	H1	0.04837	-1.68523								
0.5313	H2	-0.04837	1.68523								
	H3	-0.02418	0.84262								
	H4	0.02418	-0.84262								
	H5	-0.02418	0.84262								
	H6	0.02418	-0.84262								
	C1	0.	0.	0.	0.	0.	0.	0.	0.	0.79032	1.66614
	C2	0.	0.	0.	0.	0.	0.	0.	0.	-0.39516	-0.83307
	C3	0.	0.	0.	0.	0.	0.	0.	0.	-0.39516	-0.83307
6E**	H1	0.	0.								
0.5313	H2	0.	0.								
	H3	-0.04171	1.46315								
	H4	0.04171	-1.46315								
	H5	0.04171	-1.46315								
	H6	-0.04171	1.46315								
	C1	0.	0.	0.	0.	0.	0.	0.	0.	0.	0.
	C2	0.	0.	0.	0.	0.	0.	0.	0.	-0.68405	-1.45077
	C3	0.	0.	0.	0.	0.	0.	0.	0.	0.68405	1.45077

Center   Basis Function Type

Overlap Matrix

First Function   Second Function and Overlap

H1S1 with
        H1S2 .68301   H2S1 .01339   H2S2 .12268   H3S1 .00040   H3S2 .02713   C1S1 .00223   C1S2 .03442
        C1S3 .28459   C1S4 .37234   C1X1 .23118   C1X2 .23248   C1Z1 .36358   C1Z2 .36563

H1S2 with
        H2S1 .12268   H2S2 .34264   H3S1 .02713   H3S2 .13307   C1S1 .00767   C1S2 .09792   C1S3 .49679
        C1S4 .70604   C1X1 .19344   C1X2 .32645   C1Z1 .30422   C1Z2 .51341

H2S1 with
        H3S1 .00001   H3S2 .00494

H2S2 with
        H3S1 .00494   H3S2 .04560

H3S1 with
        C1S2 .00002   C1S3 .00641   C1S4 .07041   C1X1 -.01955   C1X2 -.16604   C1Y1 -.01541   C1Y2 -.13087
        C1Z1 .01123   C1Z2 .09537   C2S1 .00223   C2S2 .03442   C2S3 .28459   C2S4 .37234   C2X1 -.11559
        C2X2 -.11624   C2Y1 -.20021   C2Y2 -.20133   C2Z1 .36358   C2Z2 .36563

H3S2 with
        C1S1 .00069   C1S2 .00970   C1S3 .08697   C1S4 .23553   C1X1 -.09703   C1X2 -.34662   C1Y1 -.07648
        C1Y2 -.27321   C1Z1 .05573   C1Z2 .19910   C2S1 .00767   C2S2 .09792   C2S3 .49679   C2S4 .70604
        C2X1 -.09672   C2X2 -.16322   C2Y1 -.16752   C2Y2 -.28271   C2Z1 .30422   C2Z2 .51341

H5S1 with
        C2S2 .00002   C2S3 .00641   C2S4 .07041   C2X1 -.00357   C2X2 -.03032   C2Y1 .02463   C2Y2 .20923
        C2Z1 .01123   C2Z2 .09537

H5S2 with
        C2S1 .00069   C2S2 .00970   C2S3 .08697   C2S4 .23553   C2X1 -.01772   C2X2 -.06330   C2Y1 .12227
        C2Y2 .43679   C2Z1 .05573   C2Z2 .19910

C1S1 with
        C1S2 .35436   C1S3 .00805   C1S4 .01419   C2S3 .00076   C2S4 .00426   C2X1 .00282   C2X2 .00771
        C2Y1 .00163   C2Y2 .00445

C1S2 with
        C1S3 .26226   C1S4 .17745   C2S3 .01269   C2S4 .05548   C2X1 .03864   C2X2 .09732   C2Y1 .02231
        C2Y2 .05619

C1S3 with
        C1S4 .79970   C2S1 .00076   C2S2 .01269   C2S3 .15753   C2S4 .32183   C2X1 .24378   C2X2 .45638
        C2Y1 .14075   C2Y2 .26349

C1S4 with
        C2S1 .00426   C2S2 .05548   C2S3 .32183   C2S4 .54741   C2X1 .24809   C2X2 .55054   C2Y1 .14323
        C2Y2 .31785

C1X1 with
        C1X2 .53875   C2S1 -.00282   C2S2 -.03864   C2S3 -.24378   C2S4 -.24809   C2X1 -.21323   C2X2 -.02552
        C2Y1 -.20075   C2Y2 -.16403

C1X2 with
        C2S1 -.00771   C2S2 -.09732   C2S3 -.45638   C2S4 -.55054   C2X1 -.02552   C2X2 .18824   C2Y1 -.16403
        C2Y2 -.25341

C1Y1 with
        C1Y2 .53875   C2S1 -.00163   C2S2 -.02231   C2S3 -.14075   C2S4 -.14323   C2X1 -.20075   C2X2 -.16403
        C2Y1 .01858   C2Y2 .16389

C1Y2 with
        C2S1 -.00445   C2S2 -.05619   C2S3 -.26349   C2S4 -.31785   C2X1 -.16403   C2X2 -.25341   C2Y1 .16389
        C2Y2 .48085

C1Z1 with
        C1Z2 .53875   C2Z1 .13448   C2Z2 .25859

C1Z2 with
        C2Z1 .25859   C2Z2 .62715

C2S1 with
        C3S3 .00076   C3S4 .00426   C3Y1 -.00326   C3Y2 -.00890

                                    C3H6   T-243

Overlap Matrix (continued)

First Function   Second Function and Overlap

C2S2 with
     C3S3 .01269   C3S4 .05548   C3Y1 -.04462   C3Y2 -.11238

C2S3 with
     C3S1 .00076   C3S2 .01269   C3S3 .15753   C3S4 .32183   C3Y1 -.28150   C3Y2 -.52698

C2S4 with
     C3S1 .00426   C3S2 .05548   C3S3 .32183   C3S4 .54741   C3Y1 -.28646   C3Y2 -.63571

C2X1 with
     C3X1 .13448   C3X2 .25859

C2X2 with
     C3X1 .25859   C3X2 .62715

C2Y1 with
     C3S1 .00326   C3S2 .04462   C3S3 .28150   C3S4 .28646   C3Y1 -.32913   C3Y2 -.12022

C2Y2 with
     C3S1 .00890   C3S2 .11238   C3S3 .52698   C3S4 .63571   C3Y1 -.12022   C3Y2 .04193

C2Z1 with
     C3Z1 .13448   C3Z2 .25859

C2Z2 with
     C3Z1 .25859   C3Z2 .62715

C2H5N   Ethylenimine

Molecular Geometry   Symmetry C₁ₕ

Center	Coordinates		
	X	Y	Z
H1	-2.40366319	1.74220780	0.37784637
H2	-2.40366319	-1.74220780	0.37784637
H3	2.40366319	1.74220780	0.37784637
H4	2.40366319	-1.74220780	0.37784637
H5	0.	1.75214569	-3.14748049
C1	-1.39841710	0.	0.
C2	1.39841710	0.	0.
N1	0.	0.	-2.43956769
Mass	0.	0.04102536	-0.83198275

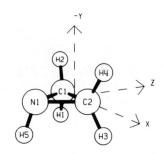

MO Symmetry Types and Occupancy

Representation	A*	A**
Number of MOs	24	16
Number Occupied	8	4

Energy Expectation Values

Total	-132.9726	Kinetic	133.1411
Electronic	-208.9681	Potential	-266.1138
Nuclear Repulsion	75.9954	Virial	1.9987

Electronic Potential	-342.1092
One-Electron Potential	-462.4017
Two-Electron Potential	120.2924

Moments of the Charge Distribution

First

x	y	z	Dipole Moment
0.	-0.0843	-0.6052	0.9426
0.	0.7675	1.2545	

Second

x^2	y^2	z^2	xy	xz	yz	r^2
-60.2727	-28.1210	-54.3764	0.	0.	2.0778	-142.7701
46.5772	15.1078	37.6129	0.	0.	-4.1086	99.2980

Third

x^3	xy^2	xz^2	x^2y	y^3	yz^2	x^2z	y^2z	z^3	xyz
0.	0.	0.	1.6298	-2.1343	-3.1075	-46.4515	-8.4591	25.5493	0.
0.	0.	0.	-1.9108	3.5142	7.8511	47.4836	7.9151	-27.5023	0.

Fourth (even)

x^4	y^4	z^4	x^2y^2	x^2z^2	y^2z^2	r^2x^2	r^2y^2	r^2z^2	r^4
-391.2072	-132.4673	-291.2954	-113.3892	-117.0531	-78.0091	-621.6495	-323.8656	-486.3575	-1431.8726
179.4136	45.5474	89.8164	70.2251	50.0700	33.5233	299.7086	149.2958	173.4098	622.4141

Fourth (odd)

x^3y	xy^3	xyz^2	x^3z	xy^2z	xz^3	x^2yz	y^3z	yz^3
0.	0.	0.	0.	0.	0.	2.3919	11.6899	12.0691
0.	0.	0.	0.	0.	0.	-1.9480	-13.4088	-20.6239

Expectation Values at Atomic Centers

Center			Expectation Value		
	r^{-1}	δ	xr^{-3}	yr^{-3}	zr^{-3}
H1	-1.0892	0.4258	-0.0386	0.0535	0.0222
H2	-1.0988	0.4232	-0.0337	-0.0550	0.0111
H5	-1.0476	0.4394	0.	0.0805	-0.0670
C1	-14.6979	116.3182	-0.0445	-0.0073	0.0358
N1	-18.3731	189.0523	0.	-0.1490	-0.1624

Center	$\dfrac{3x^2-r^2}{r^5}$	$\dfrac{3y^2-r^2}{r^5}$	$\dfrac{3z^2-r^2}{r^5}$	$\dfrac{3xy}{r^5}$	$\dfrac{3xz}{r^5}$	$\dfrac{3yz}{r^5}$
H1	-0.0433	0.1993	-0.1559	-0.2123	-0.0597	0.0798
H2	-0.0409	0.1988	-0.1579	0.2110	-0.0556	-0.0811
H5	-0.2466	0.3608	-0.1141	0.	0.	-0.2749
C1	0.0670	-0.1801	0.1131	-0.0153	-0.1671	0.0073
N1	0.1508	-0.6178	0.4671	0.	0.	-0.6859

Mulliken Charge

Center	S1	S2	S3	S4	X1	X2	Y1	Y2	Z1	Z2	Total
H1	0.5231	0.2970									0.8201
H2	0.5207	0.2787									0.7994
H5	0.5250	0.2045									0.7295
C1	0.0428	1.9563	0.7762	0.5452	0.8638	0.0553	0.9623	0.1998	0.7823	0.0950	6.2789
N1	0.0433	1.9559	0.8877	0.8195	0.9225	0.2589	1.1005	0.3660	0.8981	0.2211	7.4735

Overlap Populations

	Pair of Atomic Centers									
	H2	H3	H3	H4	H5	H5	C1	C1	C1	C1
	H1	H1	H2	H2	H1	H2	H1	H2	H3	H4
1A*	0.	0.	0.	0.	0.	0.	0.	0.	0.	0.
2A*	0.	0.	0.	0.	0.	0.	-0.0001	-0.0002	0.	0.
3A*	0.0002	0.	0.	0.	0.	0.	0.0109	0.0110	0.0012	0.0017
4A*	0.0030	0.0006	0.0002	0.0008	-0.0034	-0.0010	0.0573	0.0770	0.0111	0.0140
5A*	-0.0046	0.0080	-0.0003	0.0001	0.0115	-0.0004	0.1948	0.0158	0.0172	0.0027
6A*	0.0065	0.0004	0.0006	0.0095	0.0028	0.0037	0.0030	0.2170	-0.0069	-0.0090
7A*	-0.0052	0.0089	-0.0004	0.0003	-0.0208	0.0010	0.0692	0.0045	-0.0462	0.0106
8A*	-0.0110	0.0013	-0.0010	0.0094	-0.0011	0.0007	0.0212	0.	0.0189	-0.0420
Total A*	-0.0111	0.0192	-0.0010	0.0201	-0.0110	0.0041	0.3564	0.3252	-0.0047	-0.0220
1A**	0.	0.	0.	0.	0.	0.	-0.0002	-0.0002	0.	0.
2A**	0.0107	-0.0017	-0.0007	-0.0040	0.	0.	0.1789	0.1885	-0.0265	-0.0284
3A**	-0.0467	-0.0139	0.0037	-0.0125	0.	0.	0.2237	0.2514	-0.0443	-0.0319
4A**	0.0118	-0.0148	-0.0009	-0.0007	0.	0.	-0.0575	0.0164	0.0681	0.0091
Total A**	-0.0242	-0.0304	0.0021	-0.0173	0.	0.	0.3449	0.4561	-0.0028	-0.0512
Total	-0.0353	-0.0113	0.0011	0.0029	-0.0110	0.0041	0.7013	0.7813	-0.0074	-0.0732
Distance	3.4844	4.8073	5.9373	4.8073	4.2668	5.5151	2.0466	2.0466	4.1993	4.1993

Overlap Populations   (continued)

	C1	C2	N1	N1	N1	N1
	H5	C1	H1	H2	H5	C1
1A*	0.	0.	0.	0.	-0.0001	-0.0005
2A*	0.	0.0009	0.	0.	0.	-0.0003
3A*	0.0034	0.1595	0.0031	0.0036	0.0698	0.2865
4A*	-0.0345	0.2853	-0.0137	-0.0125	0.3415	-0.2197
5A*	0.0143	0.1118	-0.0058	0.0044	0.0423	0.1007
6A*	-0.0307	0.1458	-0.0022	-0.0139	0.1052	0.0532
7A*	-0.0208	0.2970	0.0138	-0.0105	0.0913	-0.0850
8A*	-0.0009	0.1978	-0.0285	0.0010	0.0111	-0.1415
Total A*	-0.0694	1.1980	-0.0333	-0.0279	0.6610	-0.0065
1A**	0.	-0.0024	0.	0.	0.	-0.0001
2A**	0.	-0.3283	0.0042	0.0060	0.	0.0579
3A**	0.	-0.2418	0.0022	-0.0022	0.	-0.0001
4A**	0.	-0.3483	-0.0583	-0.0156	0.	0.2349
Total A**	0.	-0.9207	-0.0518	-0.0118	0.	0.2926
Total	-0.0694	0.2773	-0.0851	-0.0397	0.6610	0.2860
Distance	3.8642	2.7968	4.0928	4.0928	1.8898	2.8119

Pair of Atomic Centers

Molecular Orbitals and Eigenvalues

Center		S1	S2	S3	S4	X1	X2	Y1	Y2	Z1	Z2
						Basis Function Type					

Center		S1	S2	S3	S4	X1	X2	Y1	Y2	Z1	Z2
* 1A*	H1	-0.00008	-0.00009								
-15.5539	H2	-0.00003	-0.00035								
	H3	-0.00008	-0.00009								
	H4	-0.00003	-0.00035								
	H5	-0.00005	0.00038								
	C1	-0.00000	-0.00003	-0.00020	0.00142	0.00008	0.00013	0.00005	-0.00024	-0.00022	-0.00051
	C2	-0.00000	-0.00003	-0.00020	0.00142	-0.00008	-0.00013	0.00005	-0.00024	-0.00022	-0.00051
	N1	-0.05195	-0.97905	-0.00288	-0.00425	0.	0.	-0.00159	0.00054	-0.00172	-0.00027
* 2A*	H1	0.00053	-0.00071								
-11.2660	H2	0.00057	-0.00073								
	H3	0.00053	-0.00071								
	H4	0.00057	-0.00073								
	H5	0.00009	-0.00014								
	C1	0.03670	0.69213	0.00330	0.00196	0.00079	-0.00041	0.00008	-0.00002	-0.00049	-0.00002
	C2	0.03670	0.69213	0.00330	0.00196	-0.00079	0.00041	0.00008	-0.00002	-0.00049	-0.00002
	N1	-0.00001	-0.00022	0.00036	-0.00116	0.	0.	-0.00002	0.00004	0.00021	-0.00050
* 3A*	H1	-0.03810	-0.00033								
-1.2737	H2	-0.03517	-0.00427								
	H3	-0.03810	-0.00033								
	H4	-0.03517	-0.00427								
	H5	-0.09080	0.00596								
	C1	0.00492	0.10624	-0.22618	-0.07916	-0.10257	0.01225	-0.00891	-0.00328	0.08290	-0.01009
	C2	0.00492	0.10624	-0.22618	-0.07916	0.10257	-0.01225	-0.00891	-0.00328	0.08290	-0.01009
	N1	0.00758	0.16408	-0.35254	-0.26947	0.	0.	-0.06573	-0.02153	-0.13675	-0.01318
* 4A*	H1	0.07449	0.02202								
-0.9041	H2	0.08721	0.02630								
	H3	0.07449	0.02202								
	H4	0.08721	0.02630								
	H5	-0.18126	-0.07724								
	C1	-0.00505	-0.10820	0.25128	0.18139	0.04521	0.00306	-0.02704	-0.00213	0.10179	0.00026
	C2	-0.00505	-0.10820	0.25128	0.18139	-0.04521	-0.00306	-0.02704	-0.00213	0.10179	0.00026
	N1	0.00600	0.13057	-0.28363	-0.33175	0.	0.	-0.15804	-0.01964	0.12303	-0.00791
* 5A*	H1	-0.16220	-0.09623								
-0.7040	H2	0.04921	0.00877								
	H3	-0.16220	-0.09623								
	H4	0.04921	0.00877								
	H5	-0.12244	-0.08150								
	C1	0.00080	0.01853	-0.03095	-0.03854	0.12349	-0.00266	-0.29090	-0.05962	-0.11314	-0.00464
	C2	0.00080	0.01853	-0.03095	-0.03854	-0.12349	0.00266	-0.29090	-0.05962	-0.11314	-0.00464
	N1	-0.00162	-0.03446	0.08369	0.09837	0.	0.	-0.26326	-0.05713	0.13263	0.00224
* 6A*	H1	0.01284	0.02537								
-0.6390	H2	0.17009	0.10565								
	H3	0.01284	0.02537								
	H4	0.17009	0.10565								
	H5	0.11694	0.09370								
	C1	-0.00035	-0.00889	0.00775	0.05287	-0.30165	-0.03271	-0.20216	-0.04857	0.11425	0.00585
	C2	-0.00035	-0.00889	0.00775	0.05287	0.30165	0.03271	-0.20216	-0.04857	0.11425	0.00585
	N1	-0.00137	-0.02915	0.07087	0.07321	0.	0.	0.03339	-0.00028	-0.29041	-0.06888

Center		S1	S2	S3	S4	X1	X2	Y1	Y2	Z1	Z2
* 7A*	H1	0.11177	0.11107								
-0.4926	H2	-0.02194	-0.01870								
	H3	0.11177	0.11107								
	H4	-0.02194	-0.01870								
	H5	-0.19020	-0.14627								
	C1	0.00060	0.01346	-0.02899	-0.04011	-0.29763	-0.08056	0.12442	0.02441	-0.22227	-0.05621
	C2	0.00060	0.01346	-0.02899	-0.04011	0.29763	0.08056	0.12442	0.02441	-0.22227	-0.05621
	N1	-0.00189	-0.04174	0.08969	0.20113	0.	0.	-0.34732	-0.14096	0.12321	0.04192
* 8A*	H1	-0.03805	-0.04268								
-0.3861	H2	0.08317	0.11959								
	H3	-0.03805	-0.04268								
	H4	0.08317	0.11959								
	H5	0.05195	0.01419								
	C1	-0.00009	-0.00027	0.01365	-0.11176	-0.15358	-0.06267	-0.08761	-0.01013	-0.22702	-0.05133
	C2	-0.00009	-0.00027	0.01365	-0.11176	0.15358	0.06267	-0.08761	-0.01013	-0.22702	-0.05133
	N1	0.00314	0.06476	-0.17788	-0.13885	0.	0.	0.40568	0.25729	0.37983	0.26098
9A*	H1	-0.01916	-0.34029								
0.2242	H2	-0.00705	-0.28412								
	H3	-0.01916	-0.34029								
	H4	-0.00705	-0.28412								
	H5	-0.04607	-0.86128								
	C1	0.00045	0.01233	-0.00881	0.07003	-0.11684	-0.49723	0.07106	-0.08848	0.08127	0.66806
	C2	0.00045	0.01233	-0.00881	0.07003	0.11684	0.49723	0.07106	-0.08848	0.08127	0.66806
	N1	-0.00397	-0.10304	0.09007	1.78890	0.	0.	0.16817	0.42395	0.11345	0.30554
10A*	H1	-0.00881	-0.85545								
0.2685	H2	0.00181	-0.76051								
	H3	-0.00881	-0.85545								
	H4	0.00181	-0.76051								
	H5	-0.00999	-0.89539								
	C1	-0.00353	-0.08165	0.16370	1.65424	-0.05181	-0.36925	0.09789	-0.11280	0.03644	-0.46173
	C2	-0.00353	-0.08165	0.16370	1.65424	0.05181	0.36925	0.09789	-0.11280	0.03644	-0.46173
	N1	0.00127	0.03636	-0.00533	-0.64656	0.	0.	0.08596	0.52258	-0.19247	-0.79238
11A*	H1	-0.02945	-0.38902								
0.3289	H2	-0.06090	-1.04793								
	H3	-0.02945	-0.38902								
	H4	-0.06090	-1.04793								
	H5	0.03335	1.68802								
	C1	-0.00076	-0.01683	0.04189	0.89161	0.02390	-0.96183	-0.26317	-0.00759	-0.00138	0.46838
	C2	-0.00076	-0.01683	0.04189	0.89161	-0.02390	0.96183	-0.26317	-0.00759	-0.00138	0.46838
	N1	0.00120	0.03021	-0.03391	-0.37028	0.	0.	-0.16132	-0.77814	0.16834	0.49177
12A*	H1	0.10293	1.60344								
0.3942	H2	-0.05197	-1.35364								
	H3	0.10293	1.60344								
	H4	-0.05197	-1.35364								
	H5	0.06857	-0.63260								
	C1	-0.00035	-0.00495	0.03688	-0.19301	-0.05057	0.28445	-0.23512	-1.54464	0.01160	-0.14495
	C2	-0.00035	-0.00495	0.03688	-0.19301	0.05057	-0.28445	-0.23512	-1.54464	0.01160	-0.14495
	N1	-0.00018	-0.00499	0.00189	0.12992	0.	0.	0.16776	0.89221	-0.03080	-0.13485

Center   Basis Function Type

Center		S1	S2	S3	S4	X1	X2	Y1	Y2	Z1	Z2
13A*	H1	-0.02419	0.49145								
0.4186	H2	-0.07426	0.07515								
	H3	-0.02419	0.49145								
	H4	-0.07426	0.07515								
	H5	-0.04149	-0.09388								
	C1	0.00160	0.03593	-0.07958	-0.02430	0.51466	-0.72363	0.19179	-0.49337	-0.05180	-0.01750
	C2	0.00160	0.03593	-0.07958	-0.02430	-0.51466	0.72363	0.19179	-0.49337	-0.05180	-0.01750
	N1	0.00095	0.01777	-0.06866	0.24397	0.	0.	0.04722	0.27563	-0.04567	-0.17884
14A*	H1	-0.06200	0.08691								
0.4655	H2	0.08004	0.41293								
	H3	-0.06200	0.08691								
	H4	0.08004	0.41293								
	H5	0.02542	-1.17486								
	C1	0.00016	0.00602	0.00790	-0.18605	0.19467	-0.05947	-0.49930	0.78209	0.05604	-0.25536
	C2	0.00016	0.00602	0.00790	-0.18605	-0.19467	0.05947	-0.49930	0.78209	0.05604	-0.25536
	N1	-0.00119	-0.02516	0.06749	0.43472	0.	0.	0.20672	0.09377	-0.19706	-0.19612
15A*	H1	-0.05381	0.06567								
0.6271	H2	-0.00880	-0.07740								
	H3	-0.05381	0.06567								
	H4	-0.00880	-0.07740								
	H5	-0.16033	-0.42295								
	C1	-0.00097	-0.01835	0.07287	0.50315	0.02863	0.27396	-0.08886	-0.01711	-0.71624	0.30100
	C2	-0.00097	-0.01835	0.07287	0.50315	-0.02863	-0.27396	-0.08886	-0.01711	-0.71624	0.30100
	N1	0.00415	0.07815	-0.30410	-0.14541	0.	0.	0.08025	0.33056	-0.09121	-1.27213
* 1A**	H1	0.00042	-0.00076								
-11.2653	H2	0.00046	-0.00077								
	H3	-0.00042	0.00076								
	H4	-0.00046	0.00077								
	H5	0.	0.								
	C1	0.03671	0.69230	0.00271	0.00460	0.00057	0.00062	0.00009	-0.00002	-0.00058	-0.00007
	C2	-0.03671	-0.69231	-0.00271	-0.00460	0.00057	0.00062	-0.00009	0.00002	0.00058	0.00007
	N1	0.	0.	0.	0.	-0.00028	0.00083	0.	0.	0.	0.
* 2A**	H1	0.13316	0.03598								
-0.8477	H2	0.12436	0.06706								
	H3	-0.13316	-0.03598								
	H4	-0.12436	-0.06706								
	H5	0.	0.								
	C1	-0.00616	-0.13386	0.29661	0.28985	-0.14334	0.01224	0.00907	0.03884	-0.02486	-0.00581
	C2	0.00616	0.13386	-0.29661	-0.28985	-0.14334	0.01224	-0.00907	-0.03884	0.02486	0.00581
	N1	0.	0.	0.	0.	-0.16753	-0.02345	0.	0.	0.	0.
* 3A**	H1	-0.17868	-0.13258								
-0.5385	H2	0.19003	0.12249								
	H3	0.17868	0.13258								
	H4	-0.19003	-0.12249								
	H5	0.	0.								
	C1	-0.00018	-0.00352	0.01146	-0.00189	-0.01144	-0.01405	-0.38715	-0.13335	0.01135	0.01006
	C2	0.00018	0.00352	-0.01146	0.00189	-0.01144	-0.01405	0.38715	0.13335	-0.01135	-0.01006
	N1	0.	0.	0.	0.	0.01262	0.00867	0.	0.	0.	0.

Basis Function Type

Center		S1	S2	S3	S4	X1	X2	Y1	Y2	Z1	Z2
˙ 4A**	H1	-0.08731	-0.15323								
-0.4667	H2	-0.07297	-0.02424								
	H3	0.08731	0.15323								
	H4	0.07297	0.02424								
	H5	0.	0.								
	C1	0.00004	-0.00067	-0.00955	0.15658	0.11465	0.06420	-0.01251	0.10500	-0.31932	-0.10742
	C2	-0.00004	0.00067	0.00955	-0.15658	0.11465	0.06420	0.01251	-0.10500	0.31932	0.10742
	N1	0.	0.	0.	0.	-0.53279	-0.20708	0.	0.	0.	0.
5A**	H1	-0.00503	-0.08677								
0.2363	H2	-0.00797	-0.19081								
	H3	0.00503	0.08677								
	H4	0.00797	0.19081								
	H5	0.	0.								
	C1	-0.00336	-0.09166	0.05956	2.34890	0.07896	1.53606	0.01399	-0.12242	0.13484	0.31653
	C2	0.00336	0.09166	-0.05956	-2.34890	0.07896	1.53606	-0.01399	0.12242	-0.13484	-0.31653
	N1	0.	0.	0.	0.	-0.11070	-0.02082	0.	0.	0.	0.
6A**	H1	0.02175	0.10094								
0.2769	H2	0.02279	-0.03623								
	H3	-0.02175	-0.10094								
	H4	-0.02279	0.03623								
	H5	0.	0.								
	C1	-0.00157	-0.04519	0.01328	1.41904	-0.14373	0.65007	0.00788	-0.14273	-0.24661	-1.29779
	C2	0.00157	0.04519	-0.01328	-1.41904	-0.14373	0.65007	-0.00788	0.14273	0.24661	1.29779
	N1	0.	0.	0.	0.	0.36529	1.05146	0.	0.	0.	0.
7A**	H1	0.04336	1.25762								
0.3944	H2	0.06291	1.76268								
	H3	-0.04336	-1.25762								
	H4	-0.06291	-1.76268								
	H5	0.	0.								
	C1	0.00222	0.05389	-0.08937	-1.26207	0.21255	1.43964	-0.06613	0.59679	-0.01531	-0.35677
	C2	-0.00222	-0.05389	0.08937	1.26207	0.21255	1.43964	0.06613	-0.59679	0.01531	0.35677
	N1	0.	0.	0.	0.	-0.05240	-0.33729	0.	0.	0.	0.
8A**	H1	-0.06578	-1.77607								
0.4499	H2	0.04803	1.30619								
	H3	0.06578	1.77607								
	H4	-0.04803	-1.30619								
	H5	0.	0.								
	C1	-0.00029	-0.01014	-0.00886	0.32795	-0.04841	-0.11915	-0.41592	3.72678	0.12194	-0.10497
	C2	0.00029	0.01014	0.00886	-0.32795	-0.04841	-0.11915	0.41592	-3.72678	-0.12194	0.10497
	N1	0.	0.	0.	0.	0.04686	-0.00533	0.	0.	0.	0.
9A**	H1	0.03611	-0.84955								
0.5330	H2	-0.00230	0.81734								
	H3	-0.03611	0.84955								
	H4	0.00230	-0.81734								
	H5	0.	0.								
	C1	0.00072	0.01027	-0.07932	0.03503	-0.03399	0.06558	0.31451	0.97449	0.66427	-0.88111
	C2	-0.00072	-0.01027	0.07932	-0.03503	-0.03399	0.06558	-0.31451	-0.97449	-0.66427	0.88111
	N1	0.	0.	0.	0.	-0.02847	-0.20485	0.	0.	0.	0.

Center		S1	S2	S3	S4	X1	X2	Y1	Y2	Z1	Z2
10A**	H1	-0.00166	-2.26472								
0.5529	H2	-0.01799	2.27634								
	H3	0.00166	2.26472								
	H4	0.01799	-2.27634								
	H5	0.	0.								
	C1	-0.00039	-0.00609	0.03817	-0.01035	0.01505	-0.01669	0.49617	3.38235	-0.32579	0.45350
	C2	0.00039	0.00609	-0.03817	0.01035	0.01505	-0.01669	-0.49617	-3.38236	0.32579	-0.45350
	N1	0.	0.	0.	0.	0.06332	0.04790	0.	0.	0.	0.

Overlap Matrix

First Function   Second Function and Overlap

H1S1 with
       H1S2 .68301   H2S1 .01305   H2S2 .12133   H3S1 .00035   H3S2 .02566   H5S1 .00176   H5S2 .05136
       C1S1 .00230   C1S2 .03534   C1S3 .28834   C1S4 .37448   C1X1 -.21334   C1X2 -.21262   C1Y1 .36974
       C1Y2 .36850   C1Z1 .08019   C1Z2 .07992   N1S2 .00002   N1S3 .00327   N1S4 .04935   N1X1 -.00769
       N1X2 -.11035   N1Y1 .00557   N1Y2 .07998   N1Z1 .00902   N1Z2 .12934

H1S2 with
       H2S1 .12133   H2S2 .34028   H3S1 .02566   H3S2 .12849   H5S1 .05136   H5S2 .19860   C1S1 .00774
       C1S2 .09870   C1S3 .49975   C1S4 .70870   C1X1 -.17714   C1X2 -.29815   C1Y1 .30700   C1Y2 .51673
       C1Z1 .06658   C1Z2 .11207   N1S1 .00066   N1S2 .00899   N1S3 .07140   N1S4 .20010   N1X1 -.05595
       N1X2 -.24097   N1Y1 .04055   N1Y2 .17466   N1Z1 .06558   N1Z2 .28245

H2S1 with
       H2S2 .68301   H3S1 .00001   H3S2 .00463   H4S1 .00035   H4S2 .02566   H5S1 .00003   H5S2 .00915
       C1S1 .00230   C1S2 .03534   C1S3 .28834   C1S4 .37448   C1X1 -.21334   C1X2 -.21262   C1Y1 -.36974
       C1Y2 -.36850   C1Z1 .08019   C1Z2 .07992   N1S2 .00002   N1S3 .00327   N1S4 .04935   N1X1 -.00769
       N1X2 -.11035   N1Y1 -.00557   N1Y2 -.07998   N1Z1 .00902   N1Z2 .12934

H2S2 with
       H3S1 .00463   H3S2 .04372   H4S1 .02566   H4S2 .12849   H5S1 .00915   H5S2 .06717   C1S1 .00774
       C1S2 .09870   C1S3 .49975   C1S4 .70870   C1X1 -.17714   C1X2 -.29815   C1Y1 -.30700   C1Y2 -.51673
       C1Z1 .06658   C1Z2 .11207   N1S1 .00066   N1S2 .00899   N1S3 .07140   N1S4 .20010   N1X1 -.05595
       N1X2 -.24097   N1Y1 -.04055   N1Y2 -.17466   N1Z1 .06558   N1Z2 .28245

H3S1 with
       C1S2 .00002   C1S3 .00677   C1S4 .07216   C1X1 .02580   C1X2 .21300   C1Y1 .01182   C1Y2 .09760
       C1Z1 .00256   C1Z2 .02117

H3S2 with
       C1S1 .00071   C1S2 .01003   C1S3 .08924   C1S4 .23939   C1X1 .12509   C1X2 .44206   C1Y1 .05732
       C1Y2 .20256   C1Z1 .01243   C1Z2 .04393

H4S1 with
       C1S2 .00002   C1S3 .00677   C1S4 .07216   C1X1 .02580   C1X2 .21300   C1Y1 -.01182   C1Y2 -.09760
       C1Z1 .00256   C1Z2 .02117

H4S2 with
       C1S1 .00071   C1S2 .01003   C1S3 .08924   C1S4 .23939   C1X1 .12509   C1X2 .44206   C1Y1 -.05732
       C1Y2 -.20256   C1Z1 .01243   C1Z2 .04393

H5S1 with
       H5S2 .68301   C1S2 .00009   C1S3 .01432   C1S4 .10041   C1X1 .01822   C1X2 .10229   C1Y1 .02283
       C1Y2 .12816   C1Z1 -.04100   C1Z2 -.23023   N1S1 .00272   N1S2 .03845   N1S3 .28064   N1S4 .41073
       N1Y1 .36983   N1Y2 .47359   N1Z1 -.14942   N1Z2 -.19134

H5S2 with
       C1S1 .00115   C1S2 .01588   C1S3 .12623   C1S4 .29773   C1X1 .06430   C1X2 .19626   C1Y1 .08056
       C1Y2 .24590   C1Z1 -.14472   C1Z2 -.44172   N1S1 .00682   N1S2 .08680   N1S3 .45171   N1S4 .70704
       N1Y1 .24989   N1Y2 .54296   N1Z1 -.10096   N1Z2 -.21937

C1S1 with
       C1S2 .35436   C1S3 .00805   C1S4 .01419   C2S3 .00089   C2S4 .00446   C2X1 -.00359   C2X2 -.00905
       N1S3 .00022   N1S4 .00354   N1X1 -.00072   N1X2 -.00474   N1Z1 .00126   N1Z2 .00828

C1S2 with
       C1S3 .26226   C1S4 .17745   C2S3 .01453   C2S4 .05808   C2X1 -.04877   C2X2 -.11415   N1S3 .00503
       N1S4 .04692   N1X1 -.01139   N1X2 -.05961   N1Z1 .01988   N1Z2 .10399

C1S3 with
       C1S4 .79970   C2S1 .00089   C2S2 .01453   C2S3 .16967   C2S4 .33357   C2X1 -.29625   C2X2 -.53189
       N1S1 .00067   N1S2 .01039   N1S3 .11706   N1S4 .28956   N1X1 -.10082   N1X2 -.26664   N1Z1 .17589
       N1Z2 .46516

C1S4 with
       C2S1 .00446   C2S2 .05808   C2S3 .33357   C2S4 .56054   C2X1 -.29081   C2X2 -.63610   N1S1 .00348
       N1S2 .04512   N1S3 .26356   N1S4 .49403   N1X1 -.10153   N1X2 -.28900   N1Z1 .17712   N1Z2 .50416

Overlap Matrix   (continued)

First Function   Second Function and Overlap

C1X1 with
      C1X2 .53875   C2S1 .00359   C2S2 .04877   C2S3 .29625   C2S4 .29081   C2X1 -.33536   C2X2 -.10846
      N1S1 .00137   N1S2 .01842   N1S3 .11961   N1S4 .16039   N1X1 .00283   N1X2 .14167   N1Z1 .16438
      N1Z2 .21532

C1X2 with
      C2S1 .00905   C2S2 .11415   C2S3 .53189   C2S4 .63610   C2X1 -.10846   C2X2 .06615   N1S1 .00354
      N1S2 .04485   N1S3 .22469   N1S4 .31873   N1X1 .11437   N1X2 .41289   N1Z1 .12079   N1Z2 .25936

C1Y1 with
      C1Y2 .53875   C2Y1 .14395   C2Y2 .26615   N1Y1 .09705   N1Y2 .26510

C1Y2 with
      C2Y1 .26615   C2Y2 .63877   N1Y1 .18361   N1Y2 .56156

C1Z1 with
      C1Z2 .53875   C2Z1 .14395   C2Z2 .26615   N1S1 -.00240   N1S2 -.03214   N1S3 -.20867   N1S4 -.27980
      N1X1 .16438   N1X2 .21532   N1Z1 -.18971   N1Z2 -.11053

C1Z2 with
      C2Z1 .26615   C2Z2 .63877   N1S1 -.00618   N1S2 -.07824   N1S3 -.39197   N1S4 -.55603   N1X1 .12079
      N1X2 .25936   N1Z1 -.02711   N1Z2 .10910

N1S1 with
      N1S2 .35584   N1S3 .00894   N1S4 .01446

N1S2 with
      N1S3 .27246   N1S4 .17925

N1S3 with
      N1S4 .79352

N1X1 with
      N1X2 .52739

N1Y1 with
      N1Y2 .52739

N1Z1 with
      N1Z2 .52739

C2H4O   Ethylene Oxide

Molecular Geometry   Symmetry C_{2v}

Center	Coordinates		
	X	Y	Z
H1	-2.38976580	1.74059890	0.37749744
H2	-2.38976580	-1.74059890	0.37749744
H3	2.38976580	1.74059890	0.37749744
H4	2.38976580	-1.74059890	0.37749744
C1	-1.38545001	0.	0.
C2	1.38545001	0.	0.
O1	0.	0.	-2.33336380
Mass	0.	0.	-0.81315754

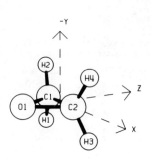

MO Symmetry Types and Occupancy

Representation	A1	A2	B1	B2
Number of MOs	16	4	12	6
Number Occupied	6	1	3	2

Energy Expectation Values

Total	-152.8012	Kinetic	153.0288
Electronic	-227.9658	Potential	-305.8301
Nuclear Repulsion	75.1645	Virial	1.9985

Electronic Potential	-380.9946
One-Electron Potential	-509.4354
Two-Electron Potential	128.4408

Moments of the Charge Distribution

First

x	y	z	Dipole Moment
0.	0.	-1.2476	1.1112
0.	0.	2.3589	

Second

x^2	y^2	z^2	xy	xz	yz	r^2
-57.8799	-24.8654	-48.1765	0.	0.	0.	-130.9218
45.8776	12.1187	32.0936	0.	0.	0.	90.0899

Third

x^3	xy^2	xz^2	x^2y	y^3	yz^2	x^2z	y^2z	z^3	xyz
0.	0.	0.	0.	0.	0.	-46.4573	-14.7403	10.0914	0.
0.	0.	0.	0.	0.	0.	45.9292	14.4292	-14.9020	0.

Fourth (even)

x^4	y^4	z^4	x^2y^2	x^2z^2	y^2z^2	r^2x^2	r^2y^2	r^2z^2	r^4
-371.1558	-112.0777	-217.7956	-107.4957	-103.7948	-56.2379	-582.4464	-275.8114	-377.8284	-1236.0861
174.6737	36.7160	56.0124	69.2099	47.6153	17.1802	291.4989	123.1061	120.8080	535.4129

Expectation Values at Atomic Centers

Center	r^{-1}	δ	Expectation Value xr^{-3}	yr^{-3}	zr^{-3}
H1	-1.0819	0.4258	-0.0370	0.0583	0.0109
C1	-14.6689	116.3574	-0.0515	0.	0.0446
O1	-22.3440	287.4027	0.	0.	-0.2747

Center	$\dfrac{3x^2-r^2}{r^5}$	$\dfrac{3y^2-r^2}{r^5}$	$\dfrac{3z^2-r^2}{r^5}$	$\dfrac{3xy}{r^5}$	$\dfrac{3xz}{r^5}$	$\dfrac{3yz}{r^5}$
H1	-0.0432	0.1977	-0.1545	-0.2129	-0.0559	0.0796
C1	0.0092	-0.2599	0.2507	0.	-0.2015	0.
O1	1.0465	-2.6139	1.5674	0.	0.	0.

Mulliken Charge

Center				Basis Function Type							Total
	S1	S2	S3	S4	X1	X2	Y1	Y2	Z1	Z2	
H1	0.5230	0.2924									0.8154
C1	0.0428	1.9563	0.7958	0.5133	0.8779	0.0474	0.9804	0.1595	0.7098	0.0624	6.1456
O1	0.0427	1.9567	0.9534	0.9484	1.0014	0.3267	1.4033	0.5334	0.9638	0.3172	8.4469

Overlap Populations

	Pair of Atomic Centers							
	H2	H3	H3	C1	C1	C2	O1	O1
	H1	H1	H2	H1	H3	C1	H1	C1
1A1	0.	0.	0.	0.	0.	0.	0.	-0.0003
2A1	0.	0.	0.	-0.0001	0.	0.0008	0.	-0.0002
3A1	0.	0.	0.	0.0043	0.0003	0.0939	0.0014	0.2658
4A1	0.0036	0.0008	0.0002	0.0806	0.0138	0.3296	-0.0101	-0.1922
5A1	0.0159	0.0047	0.0013	0.0884	-0.0169	0.1463	-0.0067	0.0648
6A1	0.0071	0.0023	0.0007	-0.0042	-0.0337	0.4041	0.0192	-0.1259
Total A1	0.0267	0.0078	0.0022	0.1688	-0.0366	0.9747	0.0038	0.0120
1A2	-0.0544	-0.0161	0.0046	0.2320	-0.0354	-0.2016	0.	0.
Total A2	-0.0544	-0.0161	0.0046	0.2320	-0.0354	-0.2016	0.	0.
1B1	0.	0.	0.	-0.0002	0.	-0.0027	0.	-0.0002
2B1	0.0093	-0.0024	-0.0006	0.1705	-0.0257	-0.3160	0.0063	0.0765
3B1	0.0160	-0.0049	-0.0014	0.0174	0.0235	-0.1864	-0.0315	0.1801
Total B1	0.0253	-0.0073	-0.0020	0.1878	-0.0021	-0.5051	-0.0252	0.2565
1B2	-0.0135	0.0037	-0.0010	0.1114	0.0136	0.1334	0.0105	0.1026
2B2	-0.0308	0.0097	-0.0029	0.0587	0.0087	0.0262	-0.0485	-0.1331
Total B2	-0.0443	0.0134	-0.0039	0.1701	0.0223	0.1596	-0.0381	-0.0305
Total	-0.0467	-0.0022	0.0009	0.7587	-0.0518	0.4277	-0.0595	0.2380
Distance	3.4812	4.7795	5.9129	2.0447	4.1743	2.7709	4.0112	2.7137

C2H4O

T-257

Molecular Orbitals and Eigenvalues

Center		S1	S2	S3	S4	X1	X2	Y1	Y2	Z1	Z2
* 1A1	H1	0.00004	0.00020								
-20.5603	H2	0.00004	0.00020								
	H3	0.00004	0.00020								
	H4	0.00004	0.00020								
	C1	-0.00000	0.00003	0.00015	-0.00092	-0.00009	-0.00006	0.	0.	0.00023	0.00030
	C2	-0.00000	0.00003	0.00015	-0.00092	0.00009	0.00006	0.	0.	0.00023	0.00030
	O1	0.05113	0.97939	0.00326	0.00300	0.	0.	0.	0.	0.00185	-0.00007
* 2A1	H1	-0.00059	0.00074								
-11.2962	H2	-0.00059	0.00074								
	H3	-0.00059	0.00074								
	H4	-0.00059	0.00074								
	C1	-0.03670	-0.69213	-0.00334	-0.00175	-0.00091	0.00051	0.	0.	0.00060	-0.00013
	C2	-0.03670	-0.69213	-0.00334	-0.00175	0.00091	-0.00051	0.	0.	0.00060	-0.00013
	O1	0.00001	0.00017	-0.00035	0.00092	0.	0.	0.	0.	-0.00030	0.00038
* 3A1	H1	-0.02437	0.00043								
-1.4284	H2	-0.02437	0.00043								
	H3	-0.02437	0.00043								
	H4	-0.02437	0.00043								
	C1	0.00404	0.08705	-0.18429	-0.04751	-0.09429	0.01792	0.	0.	0.08788	-0.01489
	C2	0.00404	0.08705	-0.18429	-0.04751	0.09429	-0.01792	0.	0.	0.08788	-0.01489
	O1	0.00887	0.19571	-0.41614	-0.36993	0.	0.	0.	0.	-0.15313	-0.02763
* 4A1	H1	-0.09239	-0.02520								
-0.9389	H2	-0.09239	-0.02520								
	H3	-0.09239	-0.02520								
	H4	-0.09239	-0.02520								
	C1	0.00595	0.12677	-0.29986	-0.17556	-0.05724	0.00403	0.	0.	-0.11016	-0.01402
	C2	0.00595	0.12677	-0.29986	-0.17556	0.05724	-0.00403	0.	0.	-0.11016	-0.01402
	O1	-0.00595	-0.13197	0.28415	0.37416	0.	0.	0.	0.	-0.11188	-0.02552
* 5A1	H1	0.10388	0.07539								
-0.6551	H2	0.10388	0.07539								
	H3	0.10388	0.07539								
	H4	0.10388	0.07539								
	C1	-0.00022	-0.00708	-0.00672	0.07517	-0.37687	-0.02663	0.	0.	0.10264	-0.01584
	C2	-0.00022	-0.00708	-0.00672	0.07517	0.37687	0.02663	0.	0.	0.10264	-0.01584
	O1	-0.00264	-0.05747	0.13577	0.17553	0.	0.	0.	0.	-0.35149	-0.12714
* 6A1	H1	-0.04296	-0.05815								
-0.4524	H2	-0.04296	-0.05815								
	H3	-0.04296	-0.05815								
	H4	-0.04296	-0.05815								
	C1	-0.00021	-0.00714	-0.00403	0.11397	0.28733	0.10022	0.	0.	0.32711	0.06017
	C2	-0.00021	-0.00714	-0.00403	0.11397	-0.28733	-0.10022	0.	0.	0.32711	0.06017
	O1	-0.00195	-0.04056	0.11329	0.06827	0.	0.	0.	0.	-0.42034	-0.25870
7A1	H1	-0.01494	-0.15549								
0.2293	H2	-0.01494	-0.15549								
	H3	-0.01494	-0.15549								
	H4	-0.01494	-0.15549								
	C1	0.00173	0.03831	-0.09332	-0.06308	-0.12722	-0.40954	0.	0.	0.09402	0.72047
	C2	0.00173	0.03831	-0.09332	-0.06308	0.12722	0.40954	0.	0.	0.09402	0.72047
	O1	-0.00337	-0.08702	0.08474	1.16077	0.	0.	0.	0.	0.28198	0.39393

		S1	S2	S3	S4	X1	X2	Y1	Y2	Z1	Z2
8A1	H1	-0.02622	-1.13023								
0.2778	H2	-0.02622	-1.13023								
	H3	-0.02622	-1.13023								
	H4	-0.02622	-1.13023								
	C1	-0.00349	-0.07719	0.18675	1.73837	-0.08437	-0.90458	0.	0.	0.04978	0.04534
	C2	-0.00349	-0.07719	0.18675	1.73837	0.08437	0.90458	0.	0.	0.04978	0.04534
	O1	0.00072	0.01913	-0.01530	-0.21133	0.	0.	0.	0.	-0.07404	-0.17271
9A1	H1	-0.05854	0.21349								
0.4077	H2	-0.05854	0.21349								
	H3	-0.05854	0.21349								
	H4	-0.05854	0.21349								
	C1	0.00173	0.03668	-0.10293	0.09549	0.53913	-0.89846	0.	0.	-0.01781	0.00358
	C2	0.00173	0.03668	-0.10293	0.09549	-0.53913	0.89846	0.	0.	-0.01781	0.00358
	O1	0.00059	0.00886	-0.05842	0.24466	0.	0.	0.	0.	-0.06476	-0.06888
10A1	H1	-0.06869	-0.16111								
0.6324	H2	-0.06869	-0.16111								
	H3	-0.06869	-0.16111								
	H4	-0.06869	-0.16111								
	C1	-0.00104	-0.02998	0.00594	0.56484	0.05454	0.02645	0.	0.	-0.79300	0.47697
	C2	-0.00104	-0.02998	0.00594	0.56484	-0.05454	-0.02645	0.	0.	-0.79300	0.47697
	O1	0.00409	0.07794	-0.30173	-0.12956	0.	0.	0.	0.	-0.37525	-0.57950
* 1A2	H1	-0.18463	-0.14125								
-0.5531	H2	0.18463	0.14125								
	H3	0.18463	0.14125								
	H4	-0.18463	-0.14125								
	C1	0.	0.	0.	0.	0.	0.	-0.39463	-0.09871	0.	0.
	C2	0.	0.	0.	0.	0.	0.	0.39463	0.09871	0.	0.
	O1	0.	0.	0.	0.	0.	0.	0.	0.	0.	0.
2A2	H1	-0.05964	-1.13399								
0.4551	H2	0.05964	1.13399								
	H3	0.05964	1.13399								
	H4	-0.05964	-1.13399								
	C1	0.	0.	0.	0.	0.	0.	-0.51751	3.17848	0.	0.
	C2	0.	0.	0.	0.	0.	0.	0.51751	-3.17848	0.	0.
	O1	0.	0.	0.	0.	0.	0.	0.	0.	0.	0.
3A2	H1	0.00030	2.65472								
0.5508	H2	-0.00030	-2.65472								
	H3	-0.00030	-2.65472								
	H4	0.00030	2.65472								
	C1	0.	0.	0.	0.	0.	0.	-0.50547	-4.12744	0.	0.
	C2	0.	0.	0.	0.	0.	0.	0.50547	4.12744	0.	0.
	O1	0.	0.	0.	0.	0.	0.	0.	0.	0.	0.
* 1B1	H1	0.00045	-0.00083								
-11.2954	H2	0.00045	-0.00083								
	H3	-0.00045	0.00083								
	H4	-0.00045	0.00083								
	C1	0.03671	0.69232	0.00265	0.00498	0.00067	0.00073	0.	0.	-0.00076	0.00003
	C2	-0.03671	-0.69232	-0.00265	-0.00498	0.00067	0.00073	0.	0.	0.00076	-0.00003
	O1	0.	0.	0.	0.	-0.00040	0.00093	0.	0.	0.	0.

Center		S1	S2	S3	S4	X1	X2	Y1	Y2	Z1	Z2
* 2B1	H1	0.12472	0.04604								
-0.8718	H2	0.12472	0.04604								
	H3	-0.12472	-0.04604								
	H4	-0.12472	-0.04604								
	C1	-0.00618	-0.13348	0.30327	0.27733	-0.13840	0.01188	0.	0.	-0.03754	0.00361
	C2	0.00618	0.13348	-0.30327	-0.27733	-0.13840	0.01188	0.	0.	0.03754	-0.00361
	O1	0.	0.	0.	0.	-0.21195	-0.05082	0.	0.	0.	0.
* 3B1	H1	-0.08528	-0.08084								
-0.5402	H2	-0.08528	-0.08084								
	H3	0.08528	0.08084								
	H4	0.08528	0.08084								
	C1	0.00046	0.00912	-0.02442	0.09198	0.13293	0.05023	0.	0.	-0.29369	-0.06575
	C2	-0.00046	-0.00912	0.02442	-0.09198	0.13293	0.05023	0.	0.	0.29369	0.06575
	O1	0.	0.	0.	0.	-0.55971	-0.23595	0.	0.	0.	0.
4B1	H1	-0.00834	-0.12321								
0.2242	H2	-0.00834	-0.12321								
	H3	0.00834	0.12321								
	H4	0.00834	0.12321								
	C1	-0.00282	-0.07698	0.05034	2.00074	0.10713	1.34705	0.	0.	0.21007	0.54187
	C2	0.00282	0.07698	-0.05034	-2.00074	0.10713	1.34705	0.	0.	-0.21007	-0.54187
	O1	0.	0.	0.	0.	-0.16783	-0.17922	0.	0.	0.	0.
5B1	H1	0.01855	0.01096								
0.2660	H2	0.01855	0.01096								
	H3	-0.01855	-0.01096								
	H4	-0.01855	-0.01096								
	C1	-0.00231	-0.06366	0.03776	1.88044	-0.13233	1.11072	0.	0.	-0.19973	-1.10626
	C2	0.00231	0.06366	-0.03776	-1.88044	-0.13233	1.11072	0.	0.	0.19973	1.10626
	O1	0.	0.	0.	0.	0.32002	0.75248	0.	0.	0.	0.
6B1	H1	-0.05780	-1.52837								
0.3894	H2	-0.05780	-1.52837								
	H3	0.05780	1.52838								
	H4	0.05780	1.52838								
	C1	-0.00224	-0.05487	0.08595	1.31200	-0.22493	-1.37298	0.	0.	0.04993	0.36913
	C2	0.00224	0.05487	-0.08595	-1.31200	-0.22493	-1.37298	0.	0.	-0.04993	-0.36913
	O1	0.	0.	0.	0.	0.06478	0.22965	0.	0.	0.	0.
7B1	H1	0.00974	-0.00137								
0.5404	H2	0.00974	-0.00137								
	H3	-0.00974	0.00137								
	H4	-0.00974	0.00137								
	C1	0.00106	0.01434	-0.11939	0.12041	0.07464	-0.01538	0.	0.	0.77916	-1.06162
	C2	-0.00106	-0.01434	0.11939	-0.12041	0.07464	-0.01538	0.	0.	-0.77916	1.06162
	O1	0.	0.	0.	0.	-0.20282	-0.01389	0.	0.	0.	0.
* 1B2	H1	0.12332	0.06189								
-0.7182	H2	-0.12332	-0.06189								
	H3	0.12332	0.06189								
	H4	-0.12332	-0.06189								
	C1	0.	0.	0.	0.	0.	0.	0.35518	0.06299	0.	0.
	C2	0.	0.	0.	0.	0.	0.	0.35518	0.06299	0.	0.
	O1	0.	0.	0.	0.	0.	0.	0.31251	0.10943	0.	0.

		S1	S2	S3	S4	X1	X2	Y1	Y2	Z1	Z2
* 2B2	H1	-0.10540	-0.11594								
-0.4525	H2	0.10540	0.11594								
	H3	-0.10540	-0.11594								
	H4	0.10540	0.11594								
	C1	0.	0.	0.	0.	0.	0.	-0.16437	-0.02438	0.	0.
	C2	0.	0.	0.	0.	0.	0.	-0.16437	-0.02438	0.	0.
	O1	0.	0.	0.	0.	0.	0.	0.67846	0.38808	0.	0.
3B2	H1	0.04760	1.29795								
0.3874	H2	-0.04760	-1.29795								
	H3	0.04760	1.29795								
	H4	-0.04760	-1.29795								
	C1	0.	0.	0.	0.	0.	0.	-0.51515	-0.94339	0.	0.
	C2	0.	0.	0.	0.	0.	0.	-0.51515	-0.94339	0.	0.
	O1	0.	0.	0.	0.	0.	0.	0.17309	0.31107	0.	0.
4B2	H1	-0.08644	-0.77012								
0.4172	H2	0.08644	0.77012								
	H3	-0.08644	-0.77012								
	H4	0.08644	0.77012								
	C1	0.	0.	0.	0.	0.	0.	-0.41910	1.40453	0.	0.
	C2	0.	0.	0.	0.	0.	0.	-0.41910	1.40453	0.	0.
	O1	0.	0.	0.	0.	0.	0.	-0.07111	-0.46536	0.	0.

Overlap Matrix

First Function   Second Function and Overlap

H1S1 with
    H1S2 .68301   H2S1 .01315   H2S2 .12172   H3S1 .00038   H3S2 .02664   C1S1 .00231   C1S2 .03550
    C1S3 .28897   C1S4 .37483   C1X1 -.21362   C1X2 -.21259   C1Y1 .37023   C1Y2 .36844   C1Z1 .08029
    C1Z2 .07991   O1S2 .00002   O1S3 .00174   O1S4 .03172   O1X1 -.00418   O1X2 -.08877   O1Y1 .00304
    O1Y2 .06466   O1Z1 .00474   O1Z2 .10070

H1S2 with
    H2S1 .12172   H2S2 .34096   H3S1 .02664   H3S2 .13156   C1S1 .00775   C1S2 .09883   C1S3 .50025
    C1S4 .70914   C1X1 -.17714   C1X2 -.29804   C1Y1 .30701   C1Y2 .51654   C1Z1 .06658   C1Z2 .11203
    O1S1 .00060   O1S2 .00807   O1S3 .05880   O1S4 .16623   O1X1 -.04207   O1X2 -.20925   O1Y1 .03064
    O1Y2 .15241   O1Z1 .04772   O1Z2 .23736

H2S1 with
    H3S1 .00001   H3S2 .00482

H2S2 with
    H3S1 .00482   H3S2 .04486

H3S1 with
    C1S2 .00002   C1S3 .00718   C1S4 .07403   C1X1 .02694   C1X2 .21591   C1Y1 .01242   C1Y2 .09955
    C1Z1 .00269   C1Z2 .02159

H3S2 with
    C1S1 .00074   C1S2 .01040   C1S3 .09168   C1S4 .24347   C1X1 .12747   C1X2 .44539   C1Y1 .05877
    C1Y2 .20535   C1Z1 .01275   C1Z2 .04454

C1S1 with
    C1S2 .35436   C1S3 .00805   C1S4 .01419   C2S3 .00095   C2S4 .00456   C2X1 -.00375   C2X2 -.00912
    O1S3 .00006   O1S4 .00288   O1X1 -.00035   O1X2 -.00496   O1Z1 .00059   O1Z2 .00836

C1S2 with
    C1S3 .26226   C1S4 .17745   C2S3 .01544   C2S4 .05929   C2X1 -.05077   C2X2 -.11492   O1S3 .00230
    O1S4 .03929   O1X1 -.00676   O1X2 -.06228   O1Z1 .01139   O1Z2 .10490

C1S3 with
    C1S4 .79970   C2S1 .00095   C2S2 .01544   C2S3 .17544   C2S4 .33901   C2X1 -.30309   C2X2 -.53397
    O1S1 .00070   O1S2 .01030   O1S3 .09917   O1S4 .26619   O1X1 -.08358   O1X2 -.27000   O1Z1 .14076
    O1Z2 .45473

C1S4 with
    C2S1 .00456   C2S2 .05929   C2S3 .33901   C2S4 .56656   C2X1 -.29271   C2X2 -.63611   O1S1 .00305
    O1S2 .03960   O1S3 .22856   O1S4 .45284   O1X1 -.07939   O1X2 -.26843   O1Z1 .13372   O1Z2 .45208

C1X1 with
    C1X2 .53875   C2S1 .00375   C2S2 .05077   C2S3 .30309   C2S4 .29271   C2X1 -.33796   C2X2 -.10290
    O1S1 .00134   O1S2 .01778   O1S3 .11145   O1S4 .17703   O1X1 -.00487   O1X2 .12162   O1Z1 .14251
    O1Z2 .24873

C1X2 with
    C2S1 .00912   C2S2 .11492   C2S3 .53397   C2S4 .63611   C2X1 -.10290   C2X2 .07736   O1S1 .00301
    O1S2 .03853   O1S3 .19895   O1S4 .31582   O1X1 .08715   O1X2 .36559   O1Z1 .09217   O1Z2 .24710

C1Y1 with
    C1Y2 .53875   C2Y1 .14844   C2Y2 .26963   O1Y1 .07975   O1Y2 .26931

C1Y2 with
    C2Y1 .26963   C2Y2 .64407   O1Y1 .14188   O1Y2 .51230

C1Z1 with
    C1Z2 .53875   C2Z1 .14844   C2Z2 .26963   O1S1 -.00226   O1S2 -.02994   O1S3 -.18771   O1S4 -.29815
    O1X1 .14251   O1X2 .24873   O1Z1 -.16027   O1Z2 -.14960

C1Z2 with
    C2Z1 .26963   C2Z2 .64407   O1S1 -.00508   O1S2 -.06490   O1S3 -.33507   O1S4 -.53190   O1X1 .09217
    O1X2 .24710   O1Z1 -.01336   O1Z2 .09614

O1S1 with
    O1S2 .35461   O1S3 .00974   O1S4 .01469

O1S2 with
    O1S3 .28286   O1S4 .18230

Overlap Matrix (continued)

First Function   Second Function and Overlap

01S3 with
     01S4   .79029

01X1 with
     01X2   .50607

01Y1 with
     01Y2   .50607

01Z1 with
     01Z2   .50607

CH4N2   Diaziridine

Molecular Geometry   Symmetry C$_2$

Center		Coordinates	
	X	Y	Z
H1	3.03049999	-0.88833855	0.35891206
H2	-3.03049999	0.88833855	0.35891206
H3	0.	1.74523120	-3.53910920
H4	0.	-1.74523120	-3.53910920
C1	0.	0.	-2.44856769
N1	-1.37101500	0.	0.
N2	1.37101500	0.	0.
Mass	0.	0.	-0.81278417

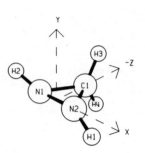

MO Symmetry Types and Occupancy

Representation	A	B
Number of MOs	20	18
Number Occupied	7	5

Energy Expectation Values

Total	-148.8430	Kinetic	148.7337
Electronic	-223.6349	Potential	-297.5767
Nuclear Repulsion	74.7919	Virial	2.0007

Electronic Potential	-372.3687
One-Electron Potential	-498.4032
Two-Electron Potential	126.0346

Moments of the Charge Distribution

First

x	y	z	Dipole Moment
0.	0.	0.9912	0.5538
0.	0.	-1.5450	

Second

x^2	y^2	z^2	xy	xz	yz	r^2
-55.7867	-22.4195	-56.3710	2.1993	0.	0.	-134.5772
44.6834	7.6700	42.9148	-5.3842	0.	0.	95.2682

Third

x^3	xy^2	xz^2	x^2y	y^3	yz^2	x^2z	y^2z	z^3	xyz
0.	0.	0.	0.	0.	0.	-37.0095	13.1506	57.5132	2.6954
0.	0.	0.	0.	0.	0.	42.9104	-14.7586	-56.0564	-6.3087

Fourth (even)

x^4	y^4	z^4	x^2y^2	x^2z^2	y^2z^2	r^2x^2	r^2y^2	r^2z^2	r^4
-342.6485	-94.1183	-354.9150	-58.6898	-102.2041	-89.0570	-503.5424	-241.8651	-546.1761	-1291.5836
218.1540	19.7997	163.3329	14.4949	42.6013	47.4452	275.2502	81.7398	253.3793	610.3693

Fourth (odd)

x^3y	xy^3	xyz^2	x^3z	xy^2z	xz^3	x^2yz	y^3z	yz^3
27.1246	4.5236	3.8517	0.	0.	0.	0.	0.	0.0001
-49.4483	-4.2489	-7.3918	0.	0.	0.	0.	0.	0.

Expectation Values at Atomic Centers

Center	r^{-1}	δ	xr^{-3}	yr^{-3}	zr^{-3}
			Expectation Value		
H1	-0.9808	0.3752	0.0389	-0.0410	0.0375
H3	-1.0948	0.4246	0.0063	0.0592	-0.0327
C1	-14.6646	116.3017	0.	0.	-0.0492
N1	-18.3343	188.4476	-0.0835	-0.1087	0.0469

	$\dfrac{3x^2-r^2}{r^5}$	$\dfrac{3y^2-r^2}{r^5}$	$\dfrac{3z^2-r^2}{r^5}$	$\dfrac{3xy}{r^5}$	$\dfrac{3xz}{r^5}$	$\dfrac{3yz}{r^5}$
H1	0.2619	-0.1007	-0.1612	-0.2796	0.1147	-0.0762
H3	-0.1710	0.1924	-0.0214	-0.0004	-0.0032	-0.2228
C1	0.0583	-0.2656	0.2073	0.0015	0.	0.
N1	1.1355	-1.7858	0.6504	-0.5533	0.1414	0.2709

Mulliken Charge

Center	S1	S2	S3	S4	X1	X2	Y1	Y2	Z1	Z2	Total
					Basis Function Type						
H1	0.4632	0.1699									0.6331
H3	0.5224	0.3015									0.8239
C1	0.0428	1.9564	0.7902	0.5191	0.7649	0.0946	0.9637	0.1600	0.7906	0.0177	6.0999
N1	0.0431	1.9560	0.8280	0.7727	0.8231	0.1348	1.3085	0.5173	0.8656	0.2439	7.4930

Overlap Populations

	H2 / H1	H3 / H1	H3 / H2	H4 / H3	C1 / H1	C1 / H3	N1 / H1	N1 / H2	N1 / H3	N1 / H4
					Pair of Atomic Centers					
1A	0.	0.	0.	0.	0.	0.	0.	-0.0002	0.	0.
2A	0.	0.	0.	0.	0.	-0.0004	0.	0.	0.	0.
3A	0.	0.	0.	0.	-0.0013	0.0055	-0.0015	0.0160	0.0005	0.0003
4A	0.0002	-0.0007	-0.0015	0.0105	-0.0227	0.2261	0.0056	0.0860	-0.0063	-0.0045
5A	0.0011	0.0017	0.0033	0.0093	0.0076	0.0800	-0.0228	0.2124	-0.0057	-0.0110
6A	0.	-0.0001	-0.0003	0.0281	0.0046	0.0440	-0.0033	0.0135	-0.0232	-0.0123
7A	0.0006	0.0005	0.0010	0.0015	-0.0149	-0.0093	-0.0001	-0.0058	0.0112	-0.0091
Total A	0.0019	0.0014	0.0025	0.0493	-0.0267	0.3458	-0.0222	0.3218	-0.0234	-0.0366
1B	0.	0.	0.	0.	0.	0.	0.	0.0001	0.	0.
2B	-0.0006	0.0001	-0.0001	0.	0.0051	0.	-0.0200	0.2639	-0.0012	0.0011
3B	-0.0001	-0.0009	0.0018	-0.0326	0.0021	0.2122	0.0008	-0.0016	0.0089	0.0192
4B	-0.0013	0.0030	-0.0058	-0.0243	-0.0448	0.0900	0.0253	0.0440	0.0392	-0.0659
5B	-0.0001	-0.0013	0.0024	-0.0648	-0.0064	0.0911	0.0031	0.0037	-0.1052	0.0340
Total B	-0.0021	0.0009	-0.0017	-0.1217	-0.0439	0.3933	0.0091	0.3100	-0.0583	-0.0117
Total	-0.0001	0.0023	0.0008	-0.0724	-0.0706	0.7391	-0.0130	0.6319	-0.0817	-0.0483
Distance	6.3160	5.5959	5.0113	3.4905	4.2255	2.0579	4.5046	1.9162	4.1774	4.1774

CH4N2

Overlap Populations (continued)

| | N1 | N2 |
	C1	N1
1A	-0.0004	0.0004
2A	-0.0004	0.
3A	0.2060	0.3356
4A	-0.1117	0.0907
5A	0.0357	0.1185
6A	-0.0279	0.3080
7A	0.0084	-0.6513
Total A	0.1097	0.2020
1B	0.	-0.0016
2B	0.0489	-0.3417
3B	0.1245	0.0695
4B	0.0966	-0.1796
5B	-0.0177	0.0659
Total B	0.2523	-0.3874
Total	0.3619	-0.1854
Distance	2.8063	2.7420

Molecular Orbitals and Eigenvalues

Center		Basis Function Type									
		S1	S2	S3	S4	X1	X2	Y1	Y2	Z1	Z2
* 1A	H1	-0.00019	0.00076								
-15.5773	H2	-0.00019	0.00076								
	H3	-0.00009	-0.00025								
	H4	-0.00009	-0.00025								
	C1	0.00000	-0.00004	-0.00041	0.00225	0.	0.	0.	0.	0.00014	0.00064
	N1	-0.03673	-0.69224	-0.00232	-0.00263	-0.00074	0.00057	-0.00077	0.00014	0.00030	0.00032
	N2	-0.03673	-0.69224	-0.00232	-0.00263	0.00074	-0.00057	0.00077	-0.00014	0.00030	0.00032
* 2A	H1	0.00010	-0.00011								
-11.2958	H2	0.00010	-0.00011								
	H3	0.00076	-0.00120								
	H4	0.00076	-0.00120								
	C1	0.05191	0.97903	0.00403	0.00418	0.	0.	0.	0.	0.00124	-0.00014
	N1	-0.00001	-0.00015	0.00035	-0.00105	0.00005	-0.00011	-0.00000	0.00000	-0.00027	0.00054
	N2	-0.00001	-0.00015	0.00035	-0.00105	-0.00005	0.00011	0.00000	-0.00000	-0.00027	0.00054
* 3A	H1	0.05854	-0.01514								
-1.3161	H2	0.05854	-0.01514								
	H3	0.03206	-0.00393								
	H4	0.03206	-0.00393								
	C1	-0.00464	-0.09906	0.22081	0.06136	0.	0.	0.	0.	0.13354	-0.02897
	N1	-0.00661	-0.14251	0.31043	0.22148	0.11376	-0.00729	0.02415	0.01313	-0.06213	-0.00442
	N2	-0.00661	-0.14251	0.31043	0.22148	-0.11376	0.00729	-0.02415	-0.01313	-0.06213	-0.00442
* 4A	H1	0.09357	0.03750								
-0.8744	H2	0.09357	0.03750								
	H3	-0.15009	-0.04533								
	H4	-0.15009	-0.04533								
	C1	0.00803	0.17201	-0.39963	-0.31480	0.	0.	0.	0.	0.09673	0.00613
	N1	-0.00304	-0.06619	0.14342	0.17228	-0.04708	-0.00397	0.03914	0.00692	0.15775	0.02171
	N2	-0.00304	-0.06619	0.14342	0.17228	0.04708	0.00397	-0.03914	-0.00692	0.15775	0.02171
* 5A	H1	0.17452	0.08458								
-0.7326	H2	0.17452	0.08458								
	H3	0.08938	0.05502								
	H4	0.08938	0.05502								
	C1	-0.00158	-0.03596	0.06794	0.12242	0.	0.	0.	0.	-0.25948	0.00367
	N1	-0.00066	-0.01438	0.03131	0.03901	-0.36273	-0.07986	0.08081	0.03029	0.08838	0.02061
	N2	-0.00066	-0.01438	0.03131	0.03901	0.36273	0.07986	-0.08081	-0.03029	0.08838	0.02061
* 6A	H1	0.05216	-0.00272								
-0.5285	H2	0.05216	-0.00272								
	H3	-0.10781	-0.10919								
	H4	-0.10781	-0.10919								
	C1	-0.00033	-0.00651	0.02308	0.08307	0.	0.	0.	0.	0.38116	0.08773
	N1	0.00071	0.01576	-0.03210	-0.06995	-0.21034	-0.10952	0.05239	0.04372	-0.35003	-0.11801
	N2	0.00071	0.01576	-0.03210	-0.06995	0.21034	0.10952	-0.05239	-0.04372	-0.35003	-0.11801
* 7A	H1	-0.06062	-0.06534								
-0.3280	H2	-0.06062	-0.06534								
	H3	-0.01284	-0.02849								
	H4	-0.01284	-0.02849								
	C1	0.00019	0.00232	-0.01985	0.13103	0.	0.	0.	0.	0.01287	0.03903
	N1	-0.00115	-0.02478	0.05897	0.10434	-0.15879	-0.10918	-0.50749	-0.35144	-0.00180	0.03246
	N2	-0.00115	-0.02478	0.05897	0.10434	0.15879	0.10918	0.50749	0.35144	-0.00180	0.03246

Center		S1	S2	S3	S4	X1	X2	Y1	Y2	Z1	Z2
8A	H1	0.00260	-0.17297								
0.2678	H2	0.00260	-0.17297								
	H3	-0.02006	-0.03981								
	H4	-0.02006	-0.03981								
	C1	0.00397	0.09708	-0.14976	-1.70603	0.	0.	0.	0.	-0.07692	-1.84771
	N1	-0.00230	-0.06298	0.03001	1.32194	-0.01666	0.10091	0.03312	0.01282	-0.08727	-0.54922
	N2	-0.00230	-0.06298	0.03001	1.32194	0.01666	-0.10091	-0.03312	-0.01282	-0.08727	-0.54922
9A	H1	0.01039	1.60538								
0.2975	H2	0.01039	1.60538								
	H3	0.00312	0.63389								
	H4	0.00312	0.63389								
	C1	0.00283	0.06262	-0.15085	-1.58624	0.	0.	0.	0.	0.19033	-0.31851
	N1	0.00164	0.03813	-0.06886	-0.72263	0.19435	0.72246	-0.13589	-0.52192	-0.05600	-0.37565
	N2	0.00164	0.03813	-0.06886	-0.72263	-0.19435	-0.72246	0.13589	0.52192	-0.05600	-0.37565
10A	H1	-0.00921	-0.77836								
0.3297	H2	-0.00921	-0.77836								
	H3	0.05181	1.71675								
	H4	0.05181	1.71675								
	C1	0.00418	0.09665	-0.19571	-2.15919	0.	0.	0.	0.	0.15421	1.43226
	N1	-0.00079	-0.02149	0.01205	0.35217	-0.10988	-0.31434	0.05519	0.25082	0.00003	-0.07957
	N2	-0.00079	-0.02149	0.01205	0.35217	0.10988	0.31434	-0.05519	-0.25082	0.00003	-0.07957
11A	H1	-0.03315	-0.34729								
0.5581	H2	-0.03315	-0.34729								
	H3	-0.09022	-0.05705								
	H4	-0.09022	-0.05705								
	C1	-0.00195	-0.04016	0.12936	1.19377	0.	0.	0.	0.	0.97836	-0.55254
	N1	0.00211	0.04054	-0.14913	0.07066	-0.16530	-0.23191	0.06213	0.08348	-0.05140	0.74892
	N2	0.00211	0.04054	-0.14913	0.07066	0.16530	0.23191	-0.06213	-0.08348	-0.05140	0.74892
* 1B	H1	0.00026	0.00015								
-15.5769	H2	-0.00026	-0.00015								
	H3	-0.00004	0.00042								
	H4	0.00004	-0.00042								
	C1	0.	0.	0.	0.	-0.00011	-0.00031	0.00001	-0.00074	0.	0.
	N1	-0.03674	-0.69229	-0.00217	-0.00331	-0.00073	-0.00120	-0.00077	0.00055	0.00036	0.00019
	N2	0.03673	0.69229	0.00217	0.00331	-0.00073	-0.00120	-0.00077	0.00055	-0.00036	-0.00019
* 2B	H1	-0.15675	-0.05974								
-0.9863	H2	0.15675	0.05974								
	H3	0.00434	-0.00406								
	H4	-0.00434	0.00406								
	C1	0.	0.	0.	0.	-0.12693	0.00088	0.01920	0.01079	0.	0.
	N1	-0.00662	-0.14340	0.31512	0.33705	-0.14389	0.01719	0.07777	0.00726	-0.02907	-0.01650
	N2	0.00662	0.14340	-0.31512	-0.33706	-0.14389	0.01719	0.07777	0.00726	0.02907	0.01650
* 3B	H1	-0.02774	-0.02752								
-0.6543	H2	0.02774	0.02752								
	H3	0.17155	0.10228								
	H4	-0.17155	-0.10228								
	C1	0.	0.	0.	0.	0.05762	0.00571	0.45381	0.08982	0.	0.
	N1	0.00121	0.02546	-0.06507	-0.05194	0.01247	0.03164	0.26156	0.07908	0.02148	0.00048
	N2	-0.00121	-0.02546	0.06507	0.05194	0.01247	0.03164	0.26156	0.07908	-0.02148	-0.00048

Center		S1	S2	S3	S4	X1	X2	Y1	Y2	Z1	Z2
* 4B	H1	0.12708	0.09506								
-0.4982	H2	-0.12708	-0.09506								
	H3	0.11704	0.09660								
	H4	-0.11704	-0.09660								
	C1	0.	0.	0.	0.	-0.37936	-0.09069	0 22294	0.06871	0.	0.
	N1	-0.00105	-0.02393	0.04615	0.14528	0.09818	0.05026	-0.21772	-0.10033	-0.26695	-0.12059
	N2	0.00105	0.02393	-0.04615	-0.14528	0.09818	0.05026	-0.21772	-0.10033	0.26695	0.12059
* 5B	H1	0.03079	0.02368								
-0.4316	H2	-0.03079	-0.02368								
	H3	-0.13603	-0.17393								
	H4	0.13602	0.17393								
	C1	0.	0.	0.	0.	-0.27817	-0.07465	-0.21346	-0.00590	0.	0.
	N1	0.00099	0.01910	-0.06362	0.02901	0.10317	0.10349	0.33405	0.16999	-0.23593	-0.12679
	N2	-0.00099	-0.01910	0.06362	-0.02901	0.10317	0.10349	0.33405	0.16999	0.23593	0.12679
6B	H1	0.11978	0.85453								
0.1686	H2	-0.11978	-0.85453								
	H3	-0.01363	0.03575								
	H4	0.01363	-0.03575								
	C1	0.	0.	0.	0.	0.06820	0.54877	0.04576	-0.19640	0.	0.
	N1	-0.00469	-0.10936	0.18991	1.09806	0.14099	-0.08970	0.08796	0.21438	0.04540	0.02739
	N2	0.00469	0.10936	-0.18991	-1.09806	0.14099	-0.08970	0.08796	0.21438	-0.04540	-0.02739
7B	H1	0.06218	0.25039								
0.2444	H2	-0.06218	-0.25039								
	H3	-0.00990	-0.00376								
	H4	0.00990	0.00376								
	C1	0.	0.	0.	0.	-0.30134	-1.16243	0.02446	-0.06215	0.	0.
	N1	-0.00036	-0.00491	0.03896	-0.13381	0.12696	0.18764	0.02942	0.08236	0.35613	0.75000
	N2	0.00036	0.00491	-0.03896	0.13381	0.12696	0.18764	0.02942	0.08236	-0.35613	-0.75000
8B	H1	-0.05224	1.50291								
0.3883	H2	0.05224	-1.50291								
	H3	0.04335	0.70531								
	H4	-0.04335	-0.70531								
	C1	0.	0.	0.	0.	0.12904	-0.23192	0.42782	-1.76520	0.	0.
	N1	0.00129	0.03679	-0.00973	-0.79656	-0.23387	-1.84890	0.10682	0.53679	-0.01626	0.13846
	N2	-0.00129	-0.03679	0.00973	0.79656	-0.23387	-1.84890	0.10682	0.53679	0.01626	-0.13846
9B	H1	0.03170	0.41676								
0.4580	H2	-0.03170	-0.41676								
	H3	-0.06088	-2.44046								
	H4	0.06088	2.44045								
	C1	0.	0.	0.	0.	0.20530	-0.17021	0.43546	3.05671	0.	0.
	N1	-0.00013	-0.00002	0.02539	-0.18143	-0.09796	-0.48241	-0.12530	-0.43428	0.00982	-0.03154
	N2	0.00013	0.00002	-0.02539	0.18143	-0.09796	-0.48241	-0.12530	-0.43428	-0.00982	0.03154
10B	H1	-0.04972	1.08815								
0.4737	H2	0.04972	-1.08815								
	H3	-0.06531	0.23414								
	H4	0.06531	-0.23414								
	C1	0.	0.	0.	0.	0.32937	-0.43248	-0.73876	0.72667	0.	0.
	N1	0.00003	0.00533	0.03014	-0.24801	-0.23017	-0.90326	0.08720	0.02878	0.03000	0.00893
	N2	-0.00003	-0.00533	-0.03014	0.24801	-0.23017	-0.90326	0.08720	0.02878	-0.03000	-0.00893

CH4N2

Center		S1	S2	S3	S4	X1	X2	Y1	Y2	Z1	Z2
						Basis Function Type					
11B	H1	-0.11569	0.77214								
0.5935	H2	0.11569	-0.77214								
	H3	-0.03151	-0.32430								
	H4	0.03151	0.32430								
	C1	0.	0.	0.	0.	-1.00852	0.95982	-0.08937	0.64189	0.	0.
	N1	0.00329	0.06532	-0.21787	-0.34917	-0.26222	-1.18078	0.01782	-0.02002	0.15687	0.12945
	N2	-0.00329	-0.06532	0.21787	0.34917	-0.26222	-1.18078	0.01782	-0.02002	-0.15687	-0.12945

Overlap Matrix

First Function   Second Function and Overlap

H1S1 with
    H1S2 .68301   H2S2 .00241   H3S1 .00002   H3S2 .00806   C1S2 .00002   C1S3 .00637   C1S4 .07024
    C1X1 .01950   C1X2 .16611   C1Y1 -.00572   C1Y2 -.04869   C1Z1 .01807   C1Z2 .15388   N1S3 .00101
    N1S4 .02801   N1X1 .00490   N1X2 .12570   N1Y1 -.00099   N1Y2 -.02537   N1Z1 .00040   N1Z2 .01025

H1S2 with
    H2S1 .00241   H2S2 .02896   H3S1 .00806   H3S2 .06202   C1S1 .00068   C1S2 .00966   C1S3 .08675
    C1S4 .23515   C1X1 .09702   C1X2 .34697   C1Y1 -.02844   C1Y2 -.10171   C1Z1 .08988   C1Z2 .32143
    N1S1 .00035   N1S2 .00489   N1S3 .04347   N1S4 .14256   N1X1 .06314   N1X2 .32585   N1Y1 -.01274
    N1Y2 -.06577   N1Z1 .00515   N1Z2 .02657

H2S1 with
    H3S1 .00018   H3S2 .01934   N1S1 .00254   N1S2 .03623   N1S3 .27102   N1S4 .40410   N1X1 -.33793
    N1X2 -.44249   N1Y1 .18090   N1Y2 .23687   N1Z1 .07309   N1Z2 .09570

H2S2 with
    H3S1 .01934   H3S2 .10756   N1S1 .00670   N1S2 .08531   N1S3 .44539   N1S4 .70025   N1X1 -.23340
    N1X2 -.50983   N1Y1 .12494   N1Y2 .27292   N1Z1 .05048   N1Z2 .11027

H3S1 with
    H3S2 .68301   H4S1 .01287   H4S2 .12061   C1S1 .00223   C1S2 .03442   C1S3 .28459   C1S4 .37234
    C1Y1 .36539   C1Y2 .36744   C1Z1 -.22832   C1Z2 -.22960   N1S2 .00001   N1S3 .00259   N1S4 .04412
    N1X1 .00356   N1X2 .05730   N1Y1 .00453   N1Y2 .07294   N1Z1 -.00918   N1Z2 -.14791

H3S2 with
    H4S1 .12061   H4S2 .33901   C1S1 .00767   C1S2 .09792   C1S3 .49679   C1S4 .70604   C1Y1 .30573
    C1Y2 .51596   C1Z1 -.19104   C1Z2 -.32241   N1S1 .00058   N1S2 .00797   N1S3 .06473   N1S4 .18713
    N1X1 .02900   N1X2 .12945   N1Y1 .03691   N1Y2 .16478   N1Z1 -.07485   N1Z2 -.33415

H4S1 with
    N1S2 .00001   N1S3 .00259   N1S4 .04412   N1X1 .00356   N1X2 .05730   N1Y1 -.00453   N1Y2 -.07294
    N1Z1 -.00918   N1Z2 -.14791

H4S2 with
    N1S1 .00058   N1S2 .00797   N1S3 .06473   N1S4 .18713   N1X1 .02900   N1X2 .12945   N1Y1 -.03691
    N1Y2 -.16478   N1Z1 -.07485   N1Z2 -.33415

C1S1 with
    C1S2 .35436   C1S3 .00805   C1S4 .01419   N1S3 .00023   N1S4 .00357   N1X1 .00072   N1X2 .00468
    N1Z1 -.00129   N1Z2 -.00835

C1S2 with
    C1S3 .26226   C1S4 .17745   N1S3 .00512   N1S4 .04722   N1X1 .01134   N1X2 .05874   N1Z1 -.02025
    N1Z2 -.10490

C1S3 with
    C1S4 .79970   N1S1 .00068   N1S2 .01054   N1S3 .11809   N1S4 .29089   N1X1 .09968   N1X2 .26243
    N1Z1 -.17803   N1Z2 -.46869

C1S4 with
    N1S1 .00350   N1S2 .04533   N1S3 .26458   N1S4 .49540   N1X1 .09992   N1X2 .28404   N1Z1 -.17845
    N1Z2 -.50729

C1X1 with
    C1X2 .53875   N1S1 -.00136   N1S2 -.01826   N1S3 -.11824   N1S4 -.15795   N1X1 .00648   N1X2 .14701
    N1Z1 .16313   N1Z2 .21269

C1X2 with
    N1S1 -.00349   N1S2 -.04413   N1S3 -.22097   N1S4 -.31322   N1X1 .11742   N1X2 .41956   N1Z1 .11922
    N1Z2 .25577

C1Y1 with
    C1Y2 .53875   N1Y1 .09782   N1Y2 .26611

C1Y2 with
    N1Y1 .18417   N1Y2 .56278

C1Z1 with
    C1Z2 .53875   N1S1 .00243   N1S2 .03261   N1S3 .21117   N1S4 .28209   N1X1 .16313   N1X2 .21269
    N1Z1 -.19352   N1Z2 -.11375

CH4N2   T-271

Overlap Matrix (continued)

First Function   Second Function and Overlap

C1Z2 with
     N1S1 .00622   N1S2 .07881   N1S3 .39464   N1S4 .55940   N1X1 .11922   N1X2 .25577   N1Z1 -.02876
     N1Z2 .10598

N1S1 with
     N1S2 .35584   N1S3 .00894   N1S4 .01446   N2S3 .00023   N2S4 .00303   N2X1 -.00138   N2X2 -.00784

N1S2 with
     N1S3 .27246   N1S4 .17925   N2S3 .00436   N2S4 .03971   N2X1 -.02028   N2X2 -.09881

N1S3 with
     N1S4 .79352   N2S1 .00023   N2S2 .00436   N2S3 .08524   N2S4 .24214   N2X1 -.17466   N2X2 -.47611

N1S4 with
     N2S1 .00303   N2S2 .03971   N2S3 .24214   N2S4 .45769   N2X1 -.23814   N2X2 -.61539

N1X1 with
     N1X2 .52739   N2S1 .00138   N2S2 .02028   N2S3 .17466   N2S4 .23814   N2X1 -.25381   N2X2 -.18945

N1X2 with
     N2S1 .00784   N2S2 .09881   N2S3 .47611   N2S4 .61539   N2X1 -.18945   N2X2 -.13070

N1Y1 with
     N1Y2 .52739   N2Y1 .07198   N2Y2 .19748

N1Y2 with
     N2Y1 .19748   N2Y2 .53704

N1Z1 with
     N1Z2 .52739   N2Z1 .07198   N2Z2 .19748

N1Z2 with
     N2Z1 .19748   N2Z2 .53704

CH2F2   Difluoromethane

Molecular Geometry   Symmetry C₂ᵥ

Center	Coordinates		
	X	Y	Z
H1	0.	1.70980300	-1.15545601
H2	0.	-1.70980300	-1.15545601
C1	0.	0.	0.
F1	2.08010101	0.	1.50299500
F2	-2.08010101	0.	1.50299500
Mass	0.	0.	1.05321233

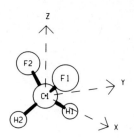

MO Symmetry Types and Occupancy

Representation	A1	A2	B1	B2
Number of MOs	16	2	10	6
Number Occupied	6	1	4	2

Energy Expectation Values

Total	-237.8687	Kinetic	237.9639
Electronic	-315.8451	Potential	-475.8327
Nuclear Repulsion	77.1760	Virial	1.9996
Electronic Potential	-553.0086		
One-Electron Potential	-721.8451		
Two-Electron Potential	168.8365		

Moments of the Charge Distribution

First

x	y	z	Dipole Moment
0.	0.	1.5339	1.1066
0.	0.	-2.6405	

Second

x^2	y^2	z^2	xy	xz	yz	r^2
-93.2415	-16.9664	-31.0821	0.	0.	0.	-141.2900
77.8828	5.8469	20.0534	0.	0.	0.	103.7831

Third

x^3	xy^2	xz^2	x^2y	y^3	yz^2	x^2z	y^2z	z^3	xyz
0.	0.	0.	0.	0.	0.	-34.2721	13.6425	32.4652	0.
0.	0.	0.	0.	0.	0.	35.0303	-12.9138	-26.9205	0.

Fourth (even)

x^4	y^4	z^4	x^2y^2	x^2z^2	y^2z^2	r^2x^2	r^2y^2	r^2z^2	r^4
-575.5640	-60.6992	-143.4113	-45.6052	-67.6277	-47.3997	-688.7970	-153.7041	-258.4388	-1100.9398
336.9847	17.0928	55.7134	0.	15.7560	28.5222	352.7407	45.6151	99.9916	498.3474

Expectation Values at Atomic Centers

Center	r^{-1}	δ	Expectation Value xr^{-3}	yr^{-3}	zr^{-3}
H1	-1.0399	0.4310	0.	0.0617	-0.0400
C1	-14.5279	116.3566	0.	0.	-0.0650
F1	-26.5692	413.5402	0.1944	0.	0.1444

Center	$\dfrac{3x^2-r^2}{r^5}$	$\dfrac{3y^2-r^2}{r^5}$	$\dfrac{3z^2-r^2}{r^5}$	$\dfrac{3xy}{r^5}$	$\dfrac{3xz}{r^5}$	$\dfrac{3yz}{r^5}$
H1	-0.1499	0.1597	-0.0099	0.	0.	-0.2243
C1	0.5294	-0.6283	0.0989	0.	0.	0.
F1	1.5963	-1.7729	0.1765	0.	2.2263	0.

Mulliken Charge

Center	S1	S2	S3	S4	X1	X2	Y1	Y2	Z1	Z2	Total
H1	0.5272	0.2984									0.8256
C1	0.0428	1.9565	0.8337	0.3625	0.5616	-0.0046	1.0476	0.0850	0.7502	0.0373	5.6725
F1	0.0435	1.9558	1.0199	0.9256	1.2124	0.4296	1.4409	0.5229	1.3079	0.4797	9.3381

Overlap Populations

Pair of Atomic Centers

	H2 H1	C1 H1	F1 H1	F1 C1	F2 F1
1A1	0.	0.	0.	-0.0002	0.
2A1	0.	-0.0004	0.	-0.0002	0.
3A1	0.0001	0.0037	0.0018	0.1488	0.0362
4A1	0.0084	0.1840	0.0022	0.0021	-0.0017
5A1	0.0077	0.0693	-0.0022	0.1143	0.0609
6A1	0.0241	0.1485	-0.0399	-0.0969	0.0336
Total A1	0.0403	0.4051	-0.0381	0.1679	0.1290
1A2	0.	0.	0.	0.	-0.0408
Total A2	0.	0.	0.	0.	-0.0408
1B1	0.	0.	0.	-0.0001	0.
2B1	0.	0.	0.	0.0801	-0.0391
3B1	0.	0.	0.	0.1369	-0.0214
4B1	0.	0.	0.	0.0247	-0.1551
Total B1	0.	0.	0.	0.2417	-0.2157
1B2	-0.0142	0.1246	0.0132	0.1451	0.0138
2B2	-0.1316	0.2418	-0.0503	-0.1422	0.0237
Total B2	-0.1458	0.3664	-0.0371	0.0030	0.0375
Total	-0.1055	0.7715	-0.0752	0.4125	-0.0899
Distance	3.4196	2.0636	3.7839	2.5663	4.1602

Molecular Orbitals and Eigenvalues

Center		S1	S2	S3	S4	X1	X2	Y1	Y2	Z1	Z2
* 1A1	H1	-0.00004	-0.00030								
-26.3057	H2	-0.00004	-0.00030								
	C1	0.00000	-0.00004	-0.00020	0.00130	-0.00000	-0.00000	0.00000	-0.00000	0.00014	0.00015
	F1	-0.03594	-0.67998	-0.00297	-0.00125	0.00061	-0.00001	-0.00000	0.00000	0.00046	-0.00007
	F2	-0.03722	-0.70420	-0.00308	-0.00128	-0.00064	0.00001	-0.00000	0.00000	0.00048	-0.00007
* 2A1	H1	0.00091	-0.00136								
-11.4324	H2	0.00091	-0.00136								
	C1	0.05192	0.97911	0.00401	0.00334	-0.00000	0.00000	-0.00000	-0.00000	0.00150	-0.00093
	F1	-0.00001	-0.00014	0.00038	-0.00071	-0.00040	0.00026	-0.00000	0.00000	-0.00032	0.00019
	F2	-0.00001	-0.00014	0.00038	-0.00071	0.00040	-0.00026	-0.00000	0.00000	-0.00032	0.00019
* 3A1	H1	0.01811	0.00274								
-1.6770	H2	0.01811	0.00274								
	C1	-0.00389	-0.08327	0.17931	0.02174	0.00000	-0.00000	0.00000	-0.00000	0.08806	-0.01386
	F1	-0.00756	-0.16264	0.36186	0.31291	-0.06943	-0.01547	-0.00000	0.00000	-0.04258	-0.01168
	F2	-0.00756	-0.16264	0.36186	0.31291	0.06943	0.01547	-0.00000	0.00000	-0.04258	-0.01168
* 4A1	H1	0.14272	0.03625								
-0.9838	H2	0.14272	0.03625								
	C1	-0.00861	-0.18019	0.46095	0.19412	0.00000	0.00000	-0.00000	-0.00000	-0.07886	-0.04505
	F1	0.00314	0.06757	-0.15539	-0.19211	-0.17378	-0.06413	-0.00000	0.00000	-0.15750	-0.05205
	F2	0.00314	0.06757	-0.15539	-0.19211	0.17378	0.06413	-0.00000	0.00000	-0.15750	-0.05205
* 5A1	H1	0.08777	0.04643								
-0.7765	H2	0.08777	0.04643								
	C1	-0.00025	-0.00750	-0.00380	0.08991	-0.00000	0.00000	-0.00000	-0.00000	-0.36286	-0.01727
	F1	-0.00114	-0.02412	0.05831	0.07381	0.39524	0.16887	0.00000	0.00000	-0.15076	-0.07416
	F2	-0.00114	-0.02412	0.05831	0.07381	-0.39524	-0.16887	0.00000	0.00000	-0.15076	-0.07416
* 6A1	H1	-0.13154	-0.08820								
-0.6228	H2	-0.13154	-0.08820								
	C1	0.00126	0.02886	-0.05293	-0.16625	0.00000	-0.00000	0.00000	0.00000	0.29557	0.00285
	F1	0.00025	0.00433	-0.01942	0.01637	0.00646	0.01153	-0.00000	-0.00000	-0.46687	-0.24678
	F2	0.00025	0.00433	-0.01942	0.01637	-0.00647	-0.01153	-0.00000	-0.00000	-0.46687	-0.24678
7A1	H1	0.03166	1.07943								
0.2596	H2	0.03166	1.07942								
	C1	-0.00107	-0.01757	0.09643	-0.81756	-0.00000	-0.00000	-0.00000	-0.00000	0.31733	1.69305
	F1	0.00204	0.04777	-0.07721	-0.47830	0.16646	0.26778	0.00000	0.00000	-0.00834	-0.11579
	F2	0.00204	0.04777	-0.07721	-0.47830	-0.16646	-0.26778	0.00000	0.00000	-0.00834	-0.11579
8A1	H1	0.00417	-1.43601								
0.3094	H2	0.00417	-1.43601								
	C1	-0.00713	-0.14760	0.45791	2.28524	-0.00000	-0.00000	0.00000	0.00000	-0.32003	-0.04531
	F1	0.00115	0.03014	-0.02215	-0.43144	0.09728	0.20765	-0.00000	-0.00000	0.07117	0.12101
	F2	0.00115	0.03014	-0.02215	-0.43144	-0.09728	-0.20765	-0.00000	-0.00000	0.07117	0.12101
9A1	H1	-0.07756	-0.74992								
0.5290	H2	-0.07756	-0.74992								
	C1	-0.00446	-0.10207	0.21439	1.78319	0.00000	-0.00000	-0.00000	-0.00000	1.04306	-1.17591
	F1	0.00227	0.04517	-0.14570	-0.19146	0.26239	0.25796	0.00000	0.00000	0.15341	0.30478
	F2	0.00227	0.04517	-0.14570	-0.19146	-0.26239	-0.25796	0.00000	0.00000	0.15341	0.30478
* 1A2	H1	0.00000	0.00000								
-0.6697	H2	-0.00000	-0.00000								
	C1	-0.00000	-0.00000	0.00000	-0.00000	0.00000	-0.00000	0.00000	0.00000	0.00000	-0.00000
	F1	0.00000	0.00000	0.00000	0.00000	-0.00000	-0.00000	-0.54152	-0.26950	-0.00000	-0.00000
	F2	-0.00000	-0.00000	0.00000	0.00000	-0.00000	-0.00000	0.54152	0.26950	-0.00000	-0.00000

Center		Basis Function Type									
		S1	S2	S3	S4	X1	X2	Y1	Y2	Z1	Z2
* 1B1	H1	-0.00000	-0.00001								
-26.3057	H2	-0.00000	-0.00001								
	C1	0.00000	-0.00000	-0.00000	0.00002	0.00008	0.00026	0.00000	-0.00000	0.00000	0.00000
	F1	-0.03722	-0.70420	-0.00321	-0.00090	0.00065	-0.00021	-0.00000	0.00000	0.00046	-0.00007
	F2	0.03594	0.67998	0.00311	0.00085	0.00063	-0.00021	-0.00000	0.00000	-0.00045	0.00006
* 2B1	H1	-0.00000	-0.00000								
-1.6093	H2	-0.00000	-0.00000								
	C1	0.00000	0.00000	-0.00000	-0.00000	0.13462	-0.01790	-0.00000	0.00000	-0.00000	0.00000
	F1	-0.00812	-0.17483	0.38845	0.34963	-0.04468	-0.01165	-0.00000	-0.00000	-0.04242	-0.00906
	F2	0.00812	0.17483	-0.38845	-0.34963	-0.04468	-0.01165	-0.00000	0.00000	0.04242	0.00906
* 3B1	H1	0.00000	0.00000								
-0.7489	H2	0.00000	0.00000								
	C1	-0.00000	-0.00000	0.00000	0.00000	-0.37582	0.01826	0.00000	-0.00000	-0.00000	-0.00000
	F1	-0.00179	-0.03767	0.09369	0.11904	0.20882	0.08177	-0.00000	-0.00000	0.40761	0.17102
	F2	0.00179	0.03767	-0.09369	-0.11904	0.20882	0.08177	0.00000	0.00000	-0.40761	-0.17102
* 4B1	H1	0.00000	0.00000								
-0.6333	H2	0.00000	0.00000								
	C1	-0.00000	-0.00000	0.00000	0.00000	0.10381	-0.05941	-0.00000	-0.00000	-0.00000	0.00000
	F1	0.00023	0.00400	-0.01775	0.01311	-0.47624	-0.23109	0.00000	0.00000	0.27859	0.13581
	F2	-0.00023	-0.00400	0.01775	-0.01311	-0.47624	-0.23109	-0.00000	-0.00000	-0.27859	-0.13580
5B1	H1	-0.00000	-0.00000								
0.2382	H2	-0.00000	-0.00000								
	C1	0.00000	0.00000	-0.00000	0.00000	-0.35026	-1.28440	-0.00000	-0.00000	-0.00000	-0.00000
	F1	-0.00212	-0.05214	0.06160	0.63173	-0.04516	-0.06846	0.00000	0.00000	-0.19377	-0.25111
	F2	0.00212	0.05214	-0.06160	-0.63172	-0.04516	-0.06846	-0.00000	0.00000	0.19377	0.25111
6B1	H1	0.00000	0.00000								
0.5851	H2	0.00000	0.00000								
	C1	0.00000	0.00000	0.00000	-0.00000	1.21529	-0.91922	0.00000	-0.00000	-0.00000	0.00000
	F1	0.00132	0.02276	-0.10848	0.10785	0.24706	0.21842	0.00000	0.00000	0.19725	0.10334
	F2	-0.00132	-0.02276	0.10848	-0.10785	0.24706	0.21842	0.00000	-0.00000	-0.19725	-0.10334
* 1B2	H1	0.12678	0.06071								
-0.7916	H2	-0.12678	-0.06071								
	C1	-0.00000	-0.00000	-0.00000	0.00000	-0.00000	0.00000	0.41194	0.06725	0.00000	-0.00000
	F1	0.00000	0.00000	0.00000	-0.00000	-0.00000	0.00000	0.37529	0.15111	-0.00000	0.00000
	F2	0.00000	0.00000	0.00000	-0.00000	0.00000	-0.00000	0.37528	0.15111	-0.00000	0.00000
* 2B2	H1	0.21315	0.23373								
-0.5473	H2	-0.21315	-0.23373								
	C1	-0.00000	-0.00000	-0.00000	0.00000	0.00000	-0.00000	0.42542	-0.00766	-0.00000	-0.00000
	F1	-0.00000	-0.00000	0.00000	-0.00000	-0.00000	0.00000	-0.38477	-0.20818	0.00000	0.00000
	F2	-0.00000	-0.00000	0.00000	-0.00000	0.00000	-0.00000	-0.38477	-0.20818	0.00000	0.00000
3B2	H1	0.06678	0.95877								
0.3704	H2	-0.06678	-0.95877								
	C1	-0.00000	-0.00000	0.00000	0.00000	0.00000	-0.00000	0.58078	-2.28716	-0.00000	-0.00000
	F1	-0.00000	-0.00000	0.00000	-0.00000	-0.00000	-0.00000	0.09570	0.31694	0.00000	0.00000
	F2	-0.00000	-0.00000	0.00000	-0.00000	0.00000	0.00000	0.09570	0.31694	0.00000	0.00000
4B2	H1	0.03649	1.80587								
0.3799	H2	-0.03649	-1.80587								
	C1	-0.00000	-0.00000	0.00000	0.00000	0.00000	-0.00000	-0.76756	-1.56063	-0.00000	-0.00000
	F1	-0.00000	-0.00000	0.00000	0.00000	-0.00000	-0.00000	0.16062	0.19679	0.00000	0.00000
	F2	-0.00000	-0.00000	0.00000	0.00000	0.00000	0.00000	0.16062	0.19679	0.00000	-0.00000

Overlap Matrix

First Function   Second Function and Overlap

H1S1 with
      H1S2 .68301   H2S1 .01515   H2S2 .12929   C1S1 .00220   C1S2 .03397   C1S3 .28273   C1S4 .37128
      C1Y1 .35554   C1Y2 .35915   C1Z1 -.24027   C1Z2 -.24271   F1S2 .00004   F1S3 .00189   F1S4 .02919
      F1X1 -.00344   F1X2 -.07473   F1Y1 .00283   F1Y2 .06143   F1Z1 -.00440   F1Z2 -.09551

H1S2 with
      H2S1 .12929   H2S2 .35407   C1S1 .00764   C1S2 .09753   C1S3 .49531   C1S4 .70471   C1Y1 .29864
      C1Y2 .50466   C1Z1 -.20182   C1Z2 -.34104   F1S1 .00069   F1S2 .00908   F1S3 .06183   F1S4 .16722
      F1X1 -.03420   F1X2 -.17719   F1Y1 .02811   F1Y2 .14565   F1Z1 -.04371   F1Z2 -.22646

C1S1 with
      C1S2 .35436   C1S3 .00805   C1S4 .01419   F1S3 .00003   F1S4 .00262   F1X1 -.00030   F1X2 -.00808
      F1Z1 -.00022   F1Z2 -.00584

C1S2 with
      C1S3 .26226   C1S4 .17745   F1S3 .00163   F1S4 .03671   F1X1 -.00743   F1X2 -.10142   F1Z1 -.00537
      F1Z2 -.07328

C1S3 with
      C1S4 .79970   F1S1 .00085   F1S2 .01188   F1S3 .09961   F1S4 .26562   F1X1 -.11786   F1X2 -.42244
      F1Z1 -.08516   F1Z2 -.30524

C1S4 with
      F1S1 .00287   F1S2 .03696   F1S3 .21336   F1S4 .43582   F1X1 -.09917   F1X2 -.37874   F1Z1 -.07166
      F1Z2 -.27366

C1X1 with
      C1X2 .53875   F1S1 .00225   F1S2 .02929   F1S3 .17763   F1S4 .30188   F1X1 -.11786   F1X2 -.12718
      F1Z1 -.13790   F1Z2 -.29322

C1X2 with
      F1S1 .00416   F1S2 .05284   F1S3 .28040   F1S4 .47820   F1X1 .01204   F1X2 .14080   F1Z1 -.07435
      F1Z2 -.23594

C1Y1 with
      C1Y2 .53875   F1Y1 .07299   F1Y2 .27862

C1Y2 with
      F1Y1 .11494   F1Y2 .46733

C1Z1 with
      C1Z2 .53875   F1S1 .00163   F1S2 .02116   F1S3 .12835   F1S4 .21812   F1X1 -.13790   F1X2 -.29322
      F1Z1 -.02665   F1Z2 .06675

C1Z2 with
      F1S1 .00301   F1S2 .03818   F1S3 .20261   F1S4 .34553   F1X1 -.07435   F1X2 -.23594   F1Z1 .06122
      F1Z2 .29685

F1S1 with
      F1S2 .35647   F1S3 .01003   F1S4 .01474   F2S4 .00003   F2X2 .00046

F1S2 with
      F1S3 .28653   F1S4 .18096   F2S4 .00047   F2X2 .00640

F1S3 with
      F1S4 .78531   F2S3 .00004   F2S4 .00687   F2X1 .00033   F2X2 .05345

F1S4 with
      F2S1 .00003   F2S2 .00047   F2S3 .00687   F2S4 .04537   F2X1 .01446   F2X2 .16668

F1X1 with
      F1X2 .49444   F2S3 -.00033   F2S4 -.01446   F2X1 -.00178   F2X2 -.07404

F1X2 with
      F2S1 -.00046   F2S2 -.00640   F2S3 -.05345   F2S4 -.16668   F2X1 -.07404   F2X2 -.35044

F1Y1 with
      F1Y2 .49444   F2Y1 .00012   F2Y2 .01143

F1Y2 with
      F2Y1 .01143   F2Y2 .09395

Overlap Matrix (continued)

First Function   Second Function and Overlap

F1Z1 with
     F1Z2 .49444   F2Z1 .00012   F2Z2 .01143

F1Z2 with
     F2Z1 .01143   F2Z2 .09395

C4H6   Bicyclobutane

Molecular Geometry   Symmetry C_{2v}

Center	Coordinates		
	X	Y	Z
H1	0.	-2.67021900	2.18852201
H2	0.	2.67021900	2.18852201
H3	3.96469900	0.	0.37511570
H4	2.22688401	0.	-2.65699100
H5	-3.96469900	0.	0.37511570
H6	-2.22688401	0.	-2.65699100
C1	0.	-1.41410100	0.60131900
C2	0.	1.41410100	0.60131900
C3	2.14052200	0.	-0.59338210
C4	-2.14052200	0.	-0.59338210
Mass	0.	0.	0.00004297

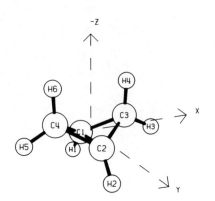

MO Symmetry Types and Occupancy

Representation	A1	A2	B1	B2
Number of MOs	22	4	14	12
Number Occupied	7	1	4	3

Energy Expectation Values

Total	-154.7889	Kinetic	155.0225
Electronic	-271.2367	Potential	-309.8114
Nuclear Repulsion	116.4478	Virial	1.9985

Electronic Potential	-426.2592
One-Electron Potential	-592.7590
Two-Electron Potential	166.4998

Moments of the Charge Distribution

First

x	y	z	Dipole Moment
0.	0.	0.4064	0.3136
0.	0.	-0.0928	

Second

x^2	y^2	z^2	xy	xz	yz	r^2
-114.2926	-58.9696	-50.5788	0.	0.	0.	-223.8410
96.3377	38.2563	32.5441	0.	0.	0.	167.1382

Third

x^3	xy^2	xz^2	x^2y	y^3	yz^2	x^2z	y^2z	z^3	xyz
0.	0.	0.	0.	0.	0.	49.3357	-40.2155	15.9318	0.
0.	0.	0.	0.	0.	0.	-47.1888	45.6363	-16.3469	0.

Fourth (even)

x^4	y^4	z^4	x^2y^2	x^2z^2	y^2z^2	r^2x^2	r^2y^2	r^2z^2	r^4
-1181.5811	-373.3436	-320.9487	-114.8525	-192.8366	-134.6946	-1489.2703	-622.8907	-648.4799	-2760.6408
795.2658	149.6605	148.6561	0.	93.8043	76.9735	889.0701	226.6340	319.4339	1435.1380

Expectation Values at Atomic Centers

Center	r^{-1}	8	xr^{-3}	yr^{-3}	zr^{-3}
H1	-1.0898	0.4159	0.	-0.0462	0.0480
H3	-1.0993	0.4178	0.0535	0.	0.0250
H4	-1.1164	0.4222	0.0084	0.	-0.0640
C1	-14.7217	116.3208	0.	-0.0887	-0.0006
C3	-14.7160	116.3145	0.0409	0.	-0.0273

Center	$\dfrac{3x^2-r^2}{r^5}$	$\dfrac{3y^2-r^2}{r^5}$	$\dfrac{3z^2-r^2}{r^5}$	$\dfrac{3xy}{r^5}$	$\dfrac{3xz}{r^5}$	$\dfrac{3yz}{r^5}$
H1	-0.1729	0.0268	0.1460	0.	0.	-0.2743
H3	0.2267	-0.1739	-0.0528	0.	0.1974	0.
H4	-0.1504	-0.1753	0.3257	0.	-0.0236	0.
C1	0.0759	0.1641	-0.2400	0.	0.	-0.1400
C3	0.0786	-0.0399	-0.0388	0.	-0.1376	0.

Mulliken Charge

Center	S1	S2	S3	S4	X1	X2	Y1	Y2	Z1	Z2	Total
H1	0.5136	0.2611									0.7747
H3	0.5166	0.2846									0.8012
H4	0.5199	0.3185									0.8384
C1	0.0428	1.9562	0.7666	0.5021	0.8308	0.0353	0.8347	0.0756	0.9520	0.2269	6.2230
C3	0.0427	1.9563	0.7649	0.5645	0.8705	0.1070	0.8390	0.1452	0.9100	0.1626	6.3627

Overlap Populations

	Pair of Atomic Centers									
	H2 H1	H3 H1	H4 H1	H4 H3	H5 H3	H5 H4	H6 H4	C1 H1	C1 H2	C1 H3
1A1	0.	0.	0.	0.	0.	0.	0.	0.	0.	0.
2A1	0.	0.	0.	0.	0.	0.	0.	0.0001	0.	0.
3A1	0.	0.	0.	0.	0.	0.	0.	0.0121	0.0018	0.0005
4A1	0.0002	-0.0005	-0.0004	0.0086	0.	0.0002	0.0075	0.0903	0.0091	-0.0097
5A1	0.0023	0.0039	0.0002	0.0027	0.0001	0.0001	0.0001	0.1177	-0.0027	-0.0038
6A1	0.0038	-0.0043	0.0015	-0.0182	0.0001	-0.0004	0.0068	0.1378	-0.0116	0.0017
7A1	0.0030	0.	-0.0030	-0.0015	0.	0.	0.0266	-0.0430	-0.0649	-0.0015
Total A1	0.0093	-0.0008	-0.0017	-0.0082	0.0003	-0.0001	0.0410	0.3149	-0.0684	-0.0128
1A2	0.	0.	0.	0.	0.	0.	0.	0.	0.	0.
Total A2	0.	0.	0.	0.	0.	0.	0.	0.	0.	0.
1B1	0.	0.	0.	0.	0.	0.	0.	0.	0.	0.
2B1	0.	0.	0.	0.0040	0.	-0.0001	-0.0025	0.	0.	-0.0002
3B1	0.	0.	0.	0.0021	0.	-0.0001	-0.0192	0.	0.	-0.0011
4B1	0.	0.	0.	-0.0248	-0.0006	0.0006	-0.0028	0.	0.	-0.0282
Total B1	0.	0.	0.	-0.0187	-0.0006	0.0005	-0.0245	0.	0.	-0.0295
1B2	0.	0.	0.	0.	0.	0.	0.	-0.0002	0.	0.
2B2	-0.0023	0.	0.	0.	0.	0.	0.	0.2703	-0.0316	0.
3B2	-0.0098	0.	0.	0.	0.	0.	0.	0.0989	0.0375	0.
Total B2	-0.0122	0.	0.	0.	0.	0.	0.	0.3690	0.0060	0.
Total	-0.0028	-0.0008	-0.0017	-0.0269	-0.0003	0.0003	0.0166	0.6839	-0.0624	-0.0422
Distance	5.3404	5.1125	5.9639	3.4948	7.9294	6.8942	4.4538	2.0241	4.3819	4.2154

Overlap Populations (continued)

	Pair of Atomic Centers								
	C1 H4	C2 C1	C3 H1	C3 H3	C3 H4	C3 H5	C3 H6	C3 C1	C4 C3
1A1	0.	-0.0001	0.	-0.0001	-0.0002	0.	0.	-0.0002	0.0003
2A1	0.	0.0018	0.	0.	0.	0.	0.	-0.0009	0.0001
3A1	0.0020	0.2692	0.0011	0.0057	0.0102	-0.0001	0.0005	0.1734	0.0306
4A1	-0.0057	0.1825	-0.0069	0.0716	0.1821	0.0013	0.0130	-0.0769	0.0436
5A1	-0.0024	0.0811	0.0059	0.1795	0.0051	-0.0009	-0.0006	0.0414	0.0095
6A1	-0.0104	0.0248	-0.0122	0.1232	0.1171	-0.0042	0.0236	-0.0207	0.0574
7A1	-0.0379	0.5795	0.0418	0.0085	0.0272	0.0001	-0.0163	-0.2441	0.1451
Total A1	-0.0544	1.1388	0.0298	0.3884	0.3415	-0.0037	0.0203	-0.1279	0.2867
1A2	0.	-0.2126	0.	0.	0.	0.	0.	0.1628	-0.0607
Total A2	0.	-0.2126	0.	0.	0.	0.	0.	0.1628	-0.0607
1B1	0.	0.	0.	-0.0002	-0.0001	0.	0.	-0.0003	-0.0003
2B1	0.0007	0.0103	0.	0.1065	0.1006	-0.0010	-0.0079	0.0704	-0.1085
3B1	-0.0129	0.0305	0.	0.0004	0.2802	0.0002	-0.0154	0.0602	-0.0538
4B1	0.0045	0.0264	0.	0.2216	0.0568	0.0216	-0.0219	0.0411	-0.2332
Total B1	-0.0077	0.0672	0.	0.3283	0.4376	0.0208	-0.0451	0.1714	-0.3957
1B2	0.	-0.0022	0.	0.	0.	0.	0.	-0.0001	0.
2B2	0.	-0.3304	0.0057	0.	0.	0.	0.	0.0577	0.0033
3B2	0.	-0.4574	-0.0340	0.	0.	0.	0.	0.1042	0.0326
Total B2	0.	-0.7900	-0.0282	0.	0.	0.	0.	0.1618	0.0360
Total	-0.0621	0.2034	0.0015	0.7167	0.7791	0.0171	-0.0249	0.3681	-0.1338
Distance	4.1923	2.8282	4.4103	2.0653	2.0654	6.1816	4.8304	2.8300	4.2810

Molecular Orbitals and Eigenvalues

Center		S1	S2	S3	S4	X1	X2	Y1	Y2	Z1	Z2
* 1A1	H1	0.00020	-0.00030								
-11.2470	H2	0.00020	-0.00030								
	H3	0.00047	-0.00065								
	H4	0.00051	-0.00082								
	H5	0.00047	-0.00065								
	H6	0.00051	-0.00082								
	C1	0.00605	0.11408	0.00092	-0.00107	0.	0.	0.00022	-0.00040	-0.00015	0.00058
	C2	0.00605	0.11408	0.00092	-0.00107	0.	0.	-0.00022	0.00040	-0.00015	0.00058
	C3	0.03620	0.68271	0.00303	0.00288	-0.00056	-0.00010	0.	0.	0.00044	-0.00027
	C4	0.03620	0.68271	0.00303	0.00288	0.00056	0.00010	0.	0.	0.00044	-0.00027
* 2A1	H1	-0.00053	-0.00019								
-11.2428	H2	-0.00053	-0.00019								
	H3	-0.00003	0.00030								
	H4	0.00001	-0.00048								
	H5	-0.00003	0.00030								
	H6	0.00001	-0.00048								
	C1	-0.03620	-0.68262	-0.00287	-0.00291	0.	0.	-0.00117	-0.00048	0.00006	0.00121
	C2	-0.03620	-0.68262	-0.00287	-0.00291	0.	0.	0.00117	0.00048	0.00006	0.00121
	C3	0.00607	0.11445	0.00021	0.00287	-0.00007	-0.00110	0.	0.	0.00005	-0.00036
	C4	0.00607	0.11445	0.00021	0.00287	0.00007	0.00110	0.	0.	0.00005	-0.00036
* 3A1	H1	0.04176	0.00189								
-1.2371	H2	0.04176	0.00189								
	H3	0.03280	-0.00384								
	H4	0.03545	0.00209								
	H5	0.03280	-0.00384								
	H6	0.03545	0.00209								
	C1	-0.00589	-0.12762	0.26874	0.10879	0.	0.	0.14492	-0.01071	-0.04338	0.00369
	C2	-0.00589	-0.12762	0.26874	0.10879	0.	0.	-0.14492	0.01071	-0.04338	0.00369
	C3	-0.00449	-0.09694	0.20509	0.08007	-0.10439	0.01364	0.	0.	0.04870	-0.00568
	C4	-0.00449	-0.09694	0.20509	0.08007	0.10439	-0.01364	0.	0.	0.04870	-0.00568
* 4A1	H1	-0.10327	-0.01584								
-0.7893	H2	-0.10327	-0.01584								
	H3	0.08260	0.02982								
	H4	0.13496	0.07650								
	H5	0.08260	0.02982								
	H6	0.13496	0.07650								
	C1	0.00326	0.07069	-0.16022	-0.16933	0.	0.	-0.03173	0.01570	-0.16883	-0.04529
	C2	0.00326	0.07069	-0.16022	-0.16933	0.	0.	0.03173	-0.01570	-0.16883	-0.04529
	C3	-0.00425	-0.09301	0.20005	0.19100	0.09036	0.00312	0.	0.	-0.14825	0.01327
	C4	-0.00425	-0.09301	0.20005	0.19100	-0.09036	-0.00312	0.	0.	-0.14825	0.01327
* 5A1	H1	0.12329	0.06849								
-0.6987	H2	0.12329	0.06849								
	H3	0.14319	0.09316								
	H4	0.01517	0.01005								
	H5	0.14319	0.09316								
	H6	0.01517	0.01005								
	C1	-0.00027	-0.00633	0.00961	0.02656	0.	0.	-0.19724	-0.01677	0.20877	0.03574
	C2	-0.00027	-0.00633	0.00961	0.02656	0.	0.	0.19724	0.01677	0.20877	0.03574
	C3	-0.00147	-0.03229	0.06940	0.10015	0.26521	0.01921	0.	0.	0.05391	0.01402
	C4	-0.00147	-0.03229	0.06940	0.10015	-0.26521	-0.01921	0.	0.	0.05391	0.01402

C4H6

Center		Basis Function Type									
		S1	S2	S3	S4	X1	X2	Y1	Y2	Z1	Z2
* 6A1	H1	0.13359	0.09087								
-0.5823	H2	0.13359	0.09087								
	H3	-0.13503	-0.07807								
	H4	0.12510	0.07396								
	H5	-0.13503	-0.07807								
	H6	0.12510	0.07396								
	C1	-0.00127	-0.02749	0.06147	0.05278	0.	0.	-0.23382	-0.04543	0.05515	0.03438
	C2	-0.00127	-0.02749	0.06147	0.05278	0.	0.	0.23382	0.04543	0.05515	0.03438
	C3	0.00081	0.01702	-0.04374	0.01944	-0.11304	-0.05920	0.	0.	-0.28261	-0.08345
	C4	0.00081	0.01702	-0.04374	0.01944	0.11304	0.05920	0.	0.	-0.28261	-0.08345
* 7A1	H1	0.01496	0.09469								
-0.3497	H2	0.01496	0.09469								
	H3	0.03090	-0.00423								
	H4	-0.13420	-0.16707								
	H5	0.03090	-0.00423								
	H6	-0.13420	-0.16707								
	C1	-0.00082	-0.01615	0.05052	-0.11147	0.	0.	-0.34377	-0.14188	-0.32123	-0.18238
	C2	-0.00082	-0.01615	0.05052	-0.11147	0.	0.	0.34377	0.14188	-0.32123	-0.18238
	C3	-0.00028	-0.00846	0.00164	0.20176	-0.07353	-0.03843	0.	0.	0.15257	0.13788
	C4	-0.00028	-0.00846	0.00164	0.20176	0.07353	0.03843	0.	0.	0.15257	0.13788
8A1	H1	-0.03761	-0.05737								
0.2504	H2	-0.03761	-0.05737								
	H3	-0.00607	-0.39457								
	H4	-0.00868	0.22269								
	H5	-0.00607	-0.39457								
	H6	-0.00868	0.22269								
	C1	-0.00279	-0.07888	0.03122	2.16705	0.	0.	-0.01417	0.53513	0.08611	-0.94938
	C2	-0.00279	-0.07888	0.03122	2.16705	0.	0.	0.01417	-0.53513	0.08611	-0.94938
	C3	0.00205	0.05853	-0.01960	-1.83679	0.09242	1.47932	0.	0.	0.01821	-0.18306
	C4	0.00205	0.05853	-0.01960	-1.83679	-0.09242	-1.47932	0.	0.	0.01821	-0.18306
9A1	H1	-0.01383	-0.12920								
0.3152	H2	-0.01383	-0.12920								
	H3	0.04452	1.49609								
	H4	-0.02399	0.77745								
	H5	0.04452	1.49609								
	H6	-0.02399	0.77745								
	C1	0.00024	-0.00096	-0.05370	0.30479	0.	0.	0.06480	-0.17687	-0.10177	0.15074
	C2	0.00024	-0.00096	-0.05370	0.30479	0.	0.	-0.06480	0.17687	-0.10177	0.15074
	C3	0.00388	0.08962	-0.18066	-1.63226	-0.19244	-0.67785	0.	0.	0.10423	-0.14063
	C4	0.00388	0.08962	-0.18066	-1.63226	0.19244	0.67785	0.	0.	0.10423	-0.14063
10A1	H1	0.07703	1.44349								
0.3560	H2	0.07703	1.44349								
	H3	0.02476	0.23254								
	H4	-0.02467	-0.96746								
	H5	0.02476	0.23254								
	H6	-0.02467	-0.96746								
	C1	0.00043	0.01201	-0.00552	-0.10127	0.	0.	0.15398	0.93150	-0.18583	-0.65210
	C2	0.00043	0.01201	-0.00552	-0.10127	0.	0.	-0.15398	-0.93150	-0.18583	-0.65210
	C3	0.00016	0.00667	0.01298	-0.44896	-0.18053	0.76624	0.	0.	-0.26664	-0.67025
	C4	0.00016	0.00667	0.01298	-0.44896	0.18053	-0.76624	0.	0.	-0.26664	-0.67025

Center		S1	S2	S3	S4	X1	X2	Y1	Y2	Z1	Z2
11A1	H1	0.07685	1.39866								
0.3800	H2	0.07685	1.39866								
	H3	0.06371	-0.29669								
	H4	0.04737	0.84888								
	H5	0.06371	-0.29669								
	H6	0.04737	0.84888								
	C1	-0.00109	-0.03882	-0.04575	-0.47764	0.	0.	-0.37017	1.79292	0.14942	-1.99147
	C2	-0.00109	-0.03882	-0.04575	-0.47764	0.	0.	0.37017	-1.79292	0.14942	-1.99147
	C3	-0.00084	-0.01238	0.08261	-1.95092	0.02488	0.61499	0.	0.	-0.11856	1.49918
	C4	-0.00084	-0.01238	0.08261	-1.95092	-0.02488	-0.61499	0.	0.	-0.11856	1.49918
12A1	H1	0.04275	2.15566								
0.4851	H2	0.04275	2.15566								
	H3	-0.02838	-1.57396								
	H4	0.03411	2.13433								
	H5	-0.02838	-1.57396								
	H6	0.03411	2.13433								
	C1	0.00132	0.03935	-0.00075	-1.17912	0.	0.	0.17561	0.85920	0.00004	-2.44292
	C2	0.00132	0.03935	-0.00075	-1.17912	0.	0.	-0.17561	-0.85920	0.00004	-2.44292
	C3	-0.00120	-0.01500	0.14350	-1.26366	-0.12710	1.49898	0.	0.	0.28415	2.71720
	C4	-0.00120	-0.01500	0.14350	-1.26366	0.12710	-1.49898	0.	0.	0.28415	2.71720
13A1	H1	0.04999	0.31506								
0.4986	H2	0.04999	0.31506								
	H3	0.05600	-0.16137								
	H4	-0.09267	-0.40232								
	H5	0.05600	-0.16137								
	H6	-0.09267	-0.40232								
	C1	0.00059	0.01103	-0.04570	0.10383	0.	0.	-0.20317	0.62688	-0.22083	0.43811
	C2	0.00059	0.01103	-0.04570	0.10383	0.	0.	0.20317	-0.62688	-0.22083	0.43811
	C3	-0.00100	-0.01628	0.09506	0.05818	0.10168	0.08786	0.	0.	0.53116	-1.22181
	C4	-0.00100	-0.01628	0.09506	0.05818	-0.10168	-0.08786	0.	0.	0.53116	-1.22181
14A1	H1	-0.10515	-0.52957								
0.5286	H2	-0.10515	-0.52957								
	H3	-0.14622	-0.54523								
	H4	0.02186	0.62463								
	H5	-0.14622	-0.54523								
	H6	0.02186	0.62463								
	C1	0.00032	-0.00499	-0.09948	0.55034	0.	0.	-0.29381	0.12524	-0.33343	0.48306
	C2	0.00032	-0.00499	-0.09948	0.55034	0.	0.	0.29381	-0.12524	-0.33343	0.48306
	C3	0.00084	0.00624	-0.12778	0.21343	-0.27757	0.72814	0.	0.	-0.07425	0.51976
	C4	0.00084	0.00624	-0.12778	0.21343	0.27757	-0.72814	0.	0.	-0.07425	0.51976
* 1A2	H1	0.	0.								
-0.4394	H2	0.	0.								
	H3	0.	0.								
	H4	0.	0.								
	H5	0.	0.								
	H6	0.	0.								
	C1	0.	0.	0.	0.	-0.36197	-0.12900	0.	0.	0.	0.
	C2	0.	0.	0.	0.	0.36197	0.12900	0.	0.	0.	0.
	C3	0.	0.	0.	0.	0.	0.	0.38948	0.10624	0.	0.
	C4	0.	0.	0.	0.	0.	0.	-0.38948	-0.10624	0.	0.

Center   Basis Function Type

C4H6

		S1	S2	S3	S4	X1	X2	Y1	Y2	Z1	Z2
2A2	H1	0.	0.								
0.3175	H2	0.	0.								
	H3	0.	0.								
	H4	0.	0.								
	H5	0.	0.								
	H6	0.	0.								
	C1	0.	0.	0.	0.	0.33232	1.83828	0.	0.	0.	0.
	C2	0.	0.	0.	0.	-0.33232	-1.83828	0.	0.	0.	0.
	C3	0.	0.	0.	0.	0.	0.	0.16253	1.59705	0.	0.
	C4	0.	0.	0.	0.	0.	0.	-0.16253	-1.59705	0.	0.
3A2	H1	0.	0.								
0.4721	H2	0.	0.								
	H3	0.	0.								
	H4	0.	0.								
	H5	0.	0.								
	H6	0.	0.								
	C1	0.	0.	0.	0.	-0.32662	1.23909	0.	0.	0.	0.
	C2	0.	0.	0.	0.	0.32662	-1.23909	0.	0.	0.	0.
	C3	0.	0.	0.	0.	0.	0.	0.44587	-0.13905	0.	0.
	C4	0.	0.	0.	0.	0.	0.	-0.44587	0.13905	0.	0.
* 1B1	H1	0.	0.								
-11.2469	H2	0.	0.								
	H3	-0.00047	0.00100								
	H4	-0.00041	0.00034								
	H5	0.00047	-0.00100								
	H6	0.00041	-0.00034								
	C1	0.	0.	0.	0.	-0.00018	0.00121	0.	0.	0.	0.
	C2	0.	0.	0.	0.	-0.00018	0.00121	0.	0.	0.	0.
	C3	-0.03670	-0.69220	-0.00279	-0.00381	0.00066	-0.00019	0.	0.	-0.00034	-0.00072
	C4	0.03670	0.69220	0.00279	0.00381	0.00066	-0.00019	0.	0.	0.00034	0.00072
* 2B1	H1	0.	0.								
-0.9347	H2	0.	0.								
	H3	-0.10757	-0.01324								
	H4	-0.08628	-0.04244								
	H5	0.10757	0.01324								
	H6	0.08628	0.04244								
	C1	0.	0.	0.	0.	-0.19145	0.03094	0.	0.	0.	0.
	C2	0.	0.	0.	0.	-0.19145	0.03094	0.	0.	0.	0.
	C3	0.00679	0.14730	-0.32465	-0.25729	0.00495	-0.02071	0.	0.	-0.03996	-0.03800
	C4	-0.00679	-0.14730	0.32465	0.25729	0.00495	-0.02071	0.	0.	0.03996	0.03800
* 3B1	H1	0.	0.								
-0.5875	H2	0.	0.								
	H3	-0.00923	0.01040								
	H4	0.19488	0.12644								
	H5	0.00923	-0.01040								
	H6	-0.19488	-0.12644								
	C1	0.	0.	0.	0.	-0.21509	-0.01009	0.	0.	0.	0.
	C2	0.	0.	0.	0.	-0.21509	-0.01009	0.	0.	0.	0.
	C3	-0.00076	-0.01617	0.03832	0.02110	0.12247	0.00847	0.	0.	-0.35902	-0.09562
	C4	0.00076	0.01617	-0.03832	-0.02110	0.12247	0.00847	0.	0.	0.35902	0.09562

Center		Basis Function Type									
		S1	S2	S3	S4	X1	X2	Y1	Y2	Z1	Z2
* 4B1	H1	0.	0.								
-0.5180	H2	0.	0.								
	H3	-0.20909	-0.18794								
	H4	0.07375	0.04881								
	H5	0.20909	0.18794								
	H6	-0.07375	-0.04881								
	C1	0.	0.	0.	0.	0.20294	0.00799	0.	0.	0.	0.
	C2	0.	0.	0.	0.	0.20294	0.00799	0.	0.	0.	0.
	C3	-0.00076	-0.01680	0.03829	0.09165	-0.37709	-0.08604	0.	0.	-0.12328	-0.03063
	C4	0.00076	0.01680	-0.03829	-0.09165	-0.37709	-0.08604	0.	0.	0.12328	0.03063
5B1	H1	0.	0.								
0.2488	H2	0.	0.								
	H3	0.04338	0.04710								
	H4	-0.03074	-0.36677								
	H5	-0.04338	-0.04710								
	H6	0.03074	0.36677								
	C1	0.	0.	0.	0.	-0.11069	-1.20998	0.	0.	0.	0.
	C2	0.	0.	0.	0.	-0.11069	-1.20998	0.	0.	0.	0.
	C3	-0.00330	-0.09332	0.03404	2.31790	0.05664	-0.94388	0.	0.	0.01519	0.36728
	C4	0.00330	0.09332	-0.03404	-2.31790	0.05664	-0.94388	0.	0.	-0.01519	-0.36728
6B1	H1	0.	0.								
0.3607	H2	0.	0.								
	H3	-0.03419	-0.40500								
	H4	-0.10313	-1.63458								
	H5	0.03419	0.40500								
	H6	0.10313	1.63458								
	C1	0.	0.	0.	0.	0.10471	0.43444	0.	0.	0.	0.
	C2	0.	0.	0.	0.	0.10471	0.43444	0.	0.	0.	0.
	C3	-0.00090	-0.01897	0.05599	0.44573	0.00631	0.86256	0.	0.	0.07639	-1.70115
	C4	0.00090	0.01897	-0.05599	-0.44573	0.00631	0.86256	0.	0.	-0.07639	1.70115
7B1	H1	0.	0.								
0.4437	H2	0.	0.								
	H3	0.01067	1.05924								
	H4	-0.00596	0.14105								
	H5	-0.01067	-1.05924								
	H6	0.00596	-0.14105								
	C1	0.	0.	0.	0.	0.10519	-0.54766	0.	0.	0.	0.
	C2	0.	0.	0.	0.	0.10519	-0.54766	0.	0.	0.	0.
	C3	0.00034	0.02001	0.06870	-0.43519	-0.58615	0.08500	0.	0.	-0.03054	-0.09659
	C4	-0.00034	-0.02001	-0.06870	0.43519	-0.58615	0.08500	0.	0.	0.03054	0.09659
8B1	H1	0.	0.								
0.4902	H2	0.	0.								
	H3	0.04065	-0.62597								
	H4	-0.00810	1.29407								
	H5	-0.04065	0.62597								
	H6	0.00810	-1.29407								
	C1	0.	0.	0.	0.	0.14572	-0.70256	0.	0.	0.	0.
	C2	0.	0.	0.	0.	0.14572	-0.70256	0.	0.	0.	0.
	C3	0.00126	0.02578	-0.08074	0.11314	0.11417	0.44828	0.	0.	0.62200	0.65027
	C4	-0.00126	-0.02578	0.08074	-0.11314	0.11417	0.44828	0.	0.	-0.62200	-0.65027

C4H6

Center		S1	S2	S3	S4	X1	X2	Y1	Y2	Z1	Z2
9B1	H1	0.	0.								
0.6052	H2	0.	0.								
	H3	-0.09403	-3.41089								
	H4	0.02297	1.40594								
	H5	0.09403	3.41089								
	H6	-0.02297	-1.40594								
	C1	0.	0.	0.	0.	0.09274	-1.93937	0.	0.	0.	0.
	C2	0.	0.	0.	0.	0.09274	-1.93937	0.	0.	0.	0.
	C3	-0.00116	-0.02581	0.06670	2.13788	-0.02585	3.41219	0.	0.	-0.11889	3.66272
	C4	0.00116	0.02581	-0.06670	-2.13788	-0.02585	3.41219	0.	0.	0.11889	-3.66272
* 1B2	H1	-0.00058	0.00094								
-11.2422	H2	0.00058	-0.00094								
	H3	0.	0.								
	H4	0.	0.								
	H5	0.	0.								
	H6	0.	0.								
	C1	-0.03670	-0.69226	-0.00293	-0.00428	0.	0.	-0.00127	-0.00034	0.00014	-0.00014
	C2	0.03670	0.69226	0.00293	0.00428	0.	0.	-0.00127	-0.00034	-0.00014	0.00014
	C3	0.	0.	0.	0.	0.	0.	0.00014	-0.00063	0.	0.
	C4	0.	0.	0.	0.	0.	0.	0.00014	-0.00063	0.	0.
* 2B2	H1	-0.16506	-0.06401								
-0.7865	H2	0.16506	0.06401								
	H3	0.	0.								
	H4	0.	0.								
	H5	0.	0.								
	H6	0.	0.								
	C1	0.00627	0.13635	-0.30360	-0.29893	0.	0.	0.12498	-0.00042	-0.04433	-0.01807
	C2	-0.00627	-0.13635	0.30360	0.29893	0.	0.	0.12498	-0.00042	0.04433	0.01807
	C3	0.	0.	0.	0.	0.	0.	0.15255	0.01193	0.	0.
	C4	0.	0.	0.	0.	0.	0.	0.15255	0.01193	0.	0.
* 3B2	H1	0.16949	0.15222								
-0.4651	H2	-0.16949	-0.15222								
	H3	0.	0.								
	H4	0.	0.								
	H5	0.	0.								
	H6	0.	0.								
	C1	0.00137	0.03034	-0.07006	-0.18911	0.	0.	-0.10521	-0.05778	0.35717	0.11539
	C2	-0.00137	-0.03034	0.07006	0.18911	0.	0.	-0.10521	-0.05778	-0.35717	-0.11539
	C3	0.	0.	0.	0.	0.	0.	0.29561	0.07557	0.	0.
	C4	0.	0.	0.	0.	0.	0.	0.29561	0.07557	0.	0.
4B2	H1	-0.01609	0.06414								
0.2322	H2	0.01609	-0.06414								
	H3	0.	0.								
	H4	0.	0.								
	H5	0.	0.								
	H6	0.	0.								
	C1	-0.00046	-0.00540	0.05341	-0.36068	0.	0.	-0.24341	-0.94222	-0.23998	-0.80980
	C2	0.00046	0.00540	-0.05341	0.36068	0.	0.	-0.24341	-0.94222	0.23998	0.80980
	C3	0.	0.	0.	0.	0.	0.	0.28890	0.65274	0.	0.
	C4	0.	0.	0.	0.	0.	0.	0.28890	0.65274	0.	0.

Center

Basis Function Type

		S1	S2	S3	S4	X1	X2	Y1	Y2	Z1	Z2
5B2	H1	-0.01057	0.03130								
0.2350	H2	0.01057	-0.03130								
	H3	0.	0.								
	H4	0.	0.								
	H5	0.	0.								
	H6	0.	0.								
	C1	-0.00370	-0.10183	0.05881	2.52967	0.	0.	0.01701	1.18355	-0.00724	-0.48276
	C2	0.00370	0.10183	-0.05881	-2.52967	0.	0.	0.01701	1.18355	0.00724	0.48276
	C3	0.	0.	0.	0.	0.	0.	0.05074	0.72542	0.	0.
	C4	0.	0.	0.	0.	0.	0.	0.05074	0.72542	0.	0.
6B2	H1	-0.08136	-2.26151								
0.4602	H2	0.08136	2.26151								
	H3	0.	0.								
	H4	0.	0.								
	H5	0.	0.								
	H6	0.	0.								
	C1	-0.00129	-0.02895	0.06965	0.48174	0.	0.	0.11078	-1.81165	-0.25231	2.51840
	C2	0.00129	0.02895	-0.06965	-0.48174	0.	0.	0.11078	-1.81165	0.25231	-2.51840
	C3	0.	0.	0.	0.	0.	0.	-0.12446	0.34720	0.	0.
	C4	0.	0.	0.	0.	0.	0.	-0.12446	0.34720	0.	0.
7B2	H1	0.00804	-2.07147								
0.4854	H2	-0.00804	2.07147								
	H3	0.	0.								
	H4	0.	0.								
	H5	0.	0.								
	H6	0.	0.								
	C1	-0.00181	-0.03979	0.09784	0.65052	0.	0.	-0.21684	-0.76639	0.47159	1.10964
	C2	0.00181	0.03979	-0.09784	-0.65052	0.	0.	-0.21684	-0.76639	-0.47159	-1.10964
	C3	0.	0.	0.	0.	0.	0.	0.23178	-0.38374	0.	0.
	C4	0.	0.	0.	0.	0.	0.	0.23178	-0.38374	0.	0.
8B2	H1	0.01792	-0.03968								
0.6575	H2	-0.01792	0.03968								
	H3	0.	0.								
	H4	0.	0.								
	H5	0.	0.								
	H6	0.	0.								
	C1	0.00022	0.01209	0.03740	-0.38303	0.	0.	0.42191	-1.27081	0.50132	-1.02522
	C2	-0.00022	-0.01209	-0.03740	0.38303	0.	0.	0.42191	-1.27081	-0.50132	1.02522
	C3	0.	0.	0.	0.	0.	0.	-0.49671	1.05974	0.	0.
	C4	0.	0.	0.	0.	0.	0.	-0.49671	1.05974	0.	0.

C4H6

Overlap Matrix

First Function   Second Function and Overlap

```
H1S1  with
   H1S2  .68301   H2S1  .00006   H2S2  .01195   H3S1  .00013   H3S2  .01673   H4S1  .00001   H4S2  .00442
   C1S1  .00244   C1S2  .03723   C1S3  .29585   C1S4  .37870   C1Y1 -.27381   C1Y2 -.26810   C1Z1  .34598
   C1Z2  .33876   C3S2  .00001   C3S3  .00410   C3S4  .05779   C3X1 -.00941   C3X2 -.10023   C3Y1 -.01174
   C3Y2 -.12503   C3Z1  .01223   C3Z2  .13026

H1S2  with
   H2S1  .01195   H2S2  .07948   H3S1  .01673   H3S2  .09821   H4S1  .00442   H4S2  .04251   C1S1  .00786
   C1S2  .10025   C1S3  .50563   C1S4  .71395   C1Y1 -.22391   C1Y2 -.37494   C1Z1  .28293   C1Z2  .47377
   C3S1  .00052   C3S2  .00737   C3S3  .07067   C3S4  .20673   C3X1 -.05627   C3X2 -.21929   C3Y1 -.07019
   C3Y2 -.27356   C3Z1  .07313   C3Z2  .28500

H2S1  with
   C1S2  .00001   C1S3  .00439   C1S4  .05958   C1Y1  .01906   C1Y2  .19602   C1Z1  .00741   C1Z2  .07617

H2S2  with
   C1S1  .00054   C1S2  .00769   C1S3  .07298   C1S4  .21095   C1Y1  .11073   C1Y2  .42578   C1Z1  .04303
   C1Z2  .16546

H3S1  with
   H3S2  .68301   H4S1  .01274   H4S2  .12009   H5S2  .00010   C1S2  .00002   C1S3  .00652   C1S4  .07097
   C1X1  .02604   C1X2  .21915   C1Y1  .00929   C1Y2  .07816   C1Z1 -.00149   C1Z2 -.01250   C3S1  .00219
   C3S2  .03383   C3S3  .28216   C3S4  .37095   C3X1  .37854   C3X2  .38290   C3Z1  .20098   C3Z2  .20329

H3S2  with
   H4S1  .12009   H4S2  .33810   H5S1  .00010   H5S2  .00376   C1S1  .00070   C1S2  .00981   C1S3  .08770
   C1S4  .23678   C1X1  .12827   C1X2  .45663   C1Y1  .04575   C1Y2  .16287   C1Z1 -.00732   C1Z2 -.02605
   C3S1  .00763   C3S2  .09742   C3S3  .49486   C3S4  .70430   C3X1  .31833   C3X2  .53815   C3Z1  .16901
   C3Z2  .28572

H4S1  with
   H4S2  .68301   H5S2  .00082   H6S1  .00103   H6S2  .04077   C1S2  .00002   C1S3  .00689   C1S4  .07268
   C1X1  .01532   C1X2  .12548   C1Y1  .00973   C1Y2  .07968   C1Z1 -.02242   C1Z2 -.18359   C3S1  .00219
   C3S2  .03383   C3S3  .28214   C3S4  .37094   C3X1  .01792   C3X2  .01813   C3Z1 -.42819   C3Z2 -.43315

H4S2  with
   H5S1  .00082   H5S2  .01470   H6S1  .04077   H6S2  .17184   C1S1  .00072   C1S2  .01014   C1S3  .08992
   C1S4  .24052   C1X1  .07380   C1X2  .25998   C1Y1  .04686   C1Y2  .16509   C1Z1 -.10798   C1Z2 -.38039
   C3S1  .00763   C3S2  .09741   C3S3  .49484   C3S4  .70429   C3X1  .01507   C3X2  .02548   C3Z1 -.36010
   C3Z2 -.60877

H5S1  with
   C3S3  .00002   C3S4  .00586   C3X1 -.00033   C3X2 -.04499   C3Z1  .00005   C3Z2  .00714

H5S2  with
   C3S1  .00002   C3S2  .00031   C3S3  .00632   C3S4  .04547   C3X1 -.01619   C3X2 -.16932   C3Z1  .00257
   C3Z2  .02686

H6S1  with
   C3S3  .00141   C3S4  .03597   C3X1 -.00766   C3X2 -.13943   C3Z1 -.00362   C3Z2 -.06588

H6S2  with
   C3S1  .00026   C3S2  .00381   C3S3  .04291   C3S4  .15112   C3X1 -.07118   C3X2 -.34143   C3Z1 -.03363
   C3Z2 -.16133

C1S1  with
   C1S2  .35436   C1S3  .00805   C1S4  .01419   C2S3  .00082   C2S4  .00435   C2Y1 -.00340   C2Y2 -.00897
   C3S3  .00081   C3S4  .00434   C3X1 -.00257   C3X2 -.00678   C3Y1 -.00170   C3Y2 -.00448   C3Z1  .00143
   C3Z2  .00378

C1S2  with
   C1S3  .26226   C1S4  .17745   C2S3  .01348   C2S4  .05663   C2Y1 -.04644   C2Y2 -.11318   C3S3  .01343
   C3S4  .05655   C3X1 -.03503   C3X2 -.08556   C3Y1 -.02314   C3Y2 -.05653   C3Z1  .01955   C3Z2  .04776

C1S3  with
   C1S4  .79970   C2S1  .00082   C2S2  .01348   C2S3  .16288   C2S4  .32705   C2Y1 -.28805   C2Y2 -.52923
   C3S1  .00081   C3S2  .01343   C3S3  .16249   C3S4  .32668   C3X1 -.21752   C3X2 -.40018   C3Y1 -.14370
   C3Y2 -.26437   C3Z1  .12141   C3Z2  .22335
```

First Function   Second Function and Overlap

C1S4 with
    C2S1 .00435   C2S2 .05663   C2S3 .32705   C2S4 .55327   C2Y1 -.28843   C2Y2 -.63595   C3S1 .00434
    C3S2 .05655   C3S3 .32668   C3S4 .55286   C3X1 -.21806   C3X2 -.48100   C3Y1 -.14406   C3Y2 -.31777
    C3Z1 .12171   C3Z2 .26846

C1X1 with
    C1X2 .53875   C2X1 .13865   C2X2 .26195   C3S1 .00257   C3S2 .03503   C3S3 .21752   C3S4 .21806
    C3X1 -.13062   C3X2 .04597   C3Y1 -.17769   C3Y2 -.14253   C3Z1 .15012   C3Z2 .12042

C1X2 with
    C2X1 .26195   C2X2 .63234   C3S1 .00678   C3S2 .08556   C3S3 .40018   C3S4 .48100   C3X1 .04597
    C3X2 .30014   C3Y1 -.14253   C3Y2 -.21922   C3Z1 .12042   C3Z2 .18521

C1Y1 with
    C1Y2 .53875   C2S1 .00340   C2S2 .04644   C2S3 .28805   C2S4 .28843   C2Y1 -.33200   C2Y2 -.11503
    C3S1 .00170   C3S2 .02314   C3S3 .14370   C3S4 .14406   C3X1 -.17769   C3X2 -.14253   C3Y1 .02096
    C3Y2 .16756   C3Z1 .09918   C3Z2 .07955

C1Y2 with
    C2S1 .00897   C2S2 .11318   C2S3 .52923   C2S4 .63595   C2Y1 -.11503   C2Y2 .05270   C3S1 .00448
    C3S2 .05653   C3S3 .26437   C3S4 .31777   C3X1 -.14253   C3X2 -.21922   C3Y1 .16756   C3Y2 .48715
    C3Z1 .07955   C3Z2 .12236

C1Z1 with
    C1Z2 .53875   C2Z1 .13865   C2Z2 .26195   C3S1 -.00143   C3S2 -.01955   C3S3 -.12141   C3S4 -.12171
    C3X1 .15012   C3X2 .12042   C3Y1 .09918   C3Y2 .07955   C3Z1 .05456   C3Z2 .19451

C1Z2 with
    C2Z1 .26195   C2Z2 .63234   C3S1 -.00378   C3S2 -.04776   C3S3 -.22335   C3S4 -.26846   C3X1 .12042
    C3X2 .18521   C3Y1 .07955   C3Y2 .12236   C3Z1 .19451   C3Z2 .52860

C3S1 with
    C3S2 .35436   C3S3 .00805   C3S4 .01419   C4S3 .00001   C4S4 .00094   C4X1 .00013   C4X2 .00416

C3S2 with
    C3S3 .26226   C3S4 .17745   C4S3 .00018   C4S4 .01297   C4X1 .00230   C4X2 .05419

C3S3 with
    C3S4 .79970   C4S1 .00001   C4S2 .00018   C4S3 .01443   C4S4 .10283   C4X1 .04760   C4X2 .31084

C3S4 with
    C4S1 .00094   C4S2 .01297   C4S3 .10283   C4S4 .25763   C4X1 .14148   C4X2 .49402

C3X1 with
    C3X2 .53875   C4S1 -.00013   C4S2 -.00230   C4S3 -.04760   C4S4 -.14148   C4X1 -.10591   C4X2 -.23653

C3X2 with
    C4S1 -.00416   C4S2 -.05419   C4S3 -.31084   C4S4 -.49402   C4X1 -.23653   C4X2 -.38498

C3Y1 with
    C3Y2 .53875   C4Y1 .01750   C4Y2 .10368

C3Y2 with
    C4Y1 .10368   C4Y2 .34988

C3Z1 with
    C3Z2 .53875   C4Z1 .01750   C4Z2 .10368

C3Z2 with
    C4Z1 .10368   C4Z2 .34988

C4H6*   Trans-butadiene

Molecular Geometry   Symmetry C_{2h}

Center	Coordinates		
	X	Y	Z
H1	-4.29159999	-2.95370001	0.
H2	-4.29159999	0.58770000	0.
H3	0.26460000	-2.95370001	0.
H4	-0.26460000	2.95370001	0.
H5	4.29159999	-0.58770000	0.
H6	4.29159999	2.95370001	0.
C1	-3.27680001	-1.18300000	0.
C2	-0.75020000	-1.18300000	0.
C3	0.75020000	1.18300000	0.
C4	3.27680001	1.18300000	0.
Mass	0.	0.	0.

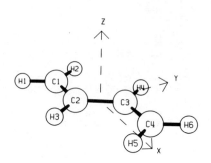

MO Symmetry Types and Occupancy

Representation	AG	AU	BG	BU
Number of MOs	22	4	4	22
Number Occupied	7	1	1	6

Energy Expectation Values

Total	-154.8648	Kinetic	154.8787
Electronic	-258.5899	Potential	-309.7435
Nuclear Repulsion	103.7251	Virial	1.9999
Electronic Potential	-413.4685		
One-Electron Potential	-567.4201		
Two-Electron Potential	153.9516		

Moments of the Charge Distribution

Second

x^2	y^2	z^2	xy	xz	yz	r^2
-227.0241	-86.1079	-22.3460	-76.2345	0.	0.	-335.4780
209.4140	69.1759	0.	75.9120	0.	0.	278.5899

Fourth (even)

x^4	y^4	z^4	x^2y^2	x^2z^2	y^2z^2	r^2x^2	r^2y^2	r^2z^2	r^4
-3551.4548	-634.7270	-98.5793	-690.7369	-180.3587	-75.9094	-4422.5505	-1401.3732	-354.8473	-6178.7711
2744.1824	351.7009	0.	525.0857	0.	0.	3269.2681	876.7867	0.	4146.0547

Fourth (odd)

x^3y	xy^3	xyz^2	x^3z	xy^2z	xz^3	x^2yz	y^3z	yz^3
-1013.4281	-414.2369	-58.3406	0.	0.	0.	0.	0.	0.
879.3882	285.8070	0.	0.	0.	0.	0.	0.	0.

Expectation Values at Atomic Centers

Center	r^{-1}	δ	xr^{-3}	yr^{-3}	zr^{-3}
H1	-1.0943	0.4205	-0.0384	-0.0575	0.
H2	-1.0924	0.4215	-0.0403	0.0575	0.
H3	-1.0920	0.4253	0.0348	-0.0626	0.
C1	-14.7328	116.4667	-0.0122	-0.0016	0.
C2	-14.7183	116.4735	0.0063	-0.0029	0.

	$\dfrac{3x^2-r^2}{r^5}$	$\dfrac{3y^2-r^2}{r^5}$	$\dfrac{3z^2-r^2}{r^5}$	$\dfrac{3xy}{r^5}$	$\dfrac{3xz}{r^5}$	$\dfrac{3yz}{r^5}$
H1	-0.0355	0.2212	-0.1858	0.2279	0.	0.
H2	-0.0351	0.2214	-0.1863	-0.2275	0.	0.
H3	-0.0356	0.2221	-0.1865	-0.2235	0.	0.
C1	-0.1258	-0.1208	0.2467	-0.0118	0.	0.
C2	-0.1790	-0.0852	0.2642	0.0283	0.	0.

Mulliken Charge

Center	S1	S2	S3	S4	X1	X2	Y1	Y2	Z1	Z2	Total
H1	0.5188	0.2941									0.8129
H2	0.5199	0.3017									0.8216
H3	0.5234	0.2937									0.8172
C1	0.0428	1.9564	0.7868	0.5378	0.9347	0.0670	0.9371	0.2080	0.6931	0.3267	6.4903
C2	0.0428	1.9564	0.8010	0.3869	0.9449	-0.0161	0.9041	0.0578	0.6896	0.2907	6.0580

Overlap Populations

	Pair of Atomic Centers									
	H2	H3	H3	H4	H4	H4	H5	H5	H6	C1
	H1	H1	H2	H1	H2	H3	H1	H2	H1	H1
1AG	0.	0.	0.	0.	0.	0.	0.	0.	0.	0.
2AG	0.	0.	0.	0.	0.	0.	0.	0.	0.	-0.0002
3AG	0.0001	0.	0.	0.	0.0001	0.	0.	0.	0.	0.0111
4AG	0.0089	-0.0016	-0.0005	-0.0001	-0.0018	0.0002	0.	0.	0.	0.1422
5AG	-0.0099	0.0032	-0.0017	0.0001	-0.0058	0.0013	0.	0.	0.	0.0476
6AG	0.0091	0.0118	0.0007	0.0006	0.0025	0.0021	0.	0.	0.	0.0241
7AG	-0.0277	-0.0127	0.0033	-0.0006	0.0109	0.0031	0.	0.0001	0.	0.1103
Total AG	-0.0195	0.0006	0.0018	0.0001	0.0059	0.0068	0.	0.0001	0.	0.3350
1AU	0.	0.	0.	0.	0.	0.	0.	0.	0.	0.
Total AU	0.	0.	0.	0.	0.	0.	0.	0.	0.	0.
1BG	0.	0.	0.	0.	0.	0.	0.	0.	0.	0.
Total BG	0.	0.	0.	0.	0.	0.	0.	0.	0.	0.
1BU	0.	0.	0.	0.	0.	0.	0.	0.	0.	0.
2BU	0.	0.	0.	0.	0.	0.	0.	0.	0.	0.
3BU	0.0024	0.0002	0.	0.	-0.0001	0.	0.	0.	0.	0.0662
4BU	0.0043	-0.0023	-0.0018	0.0001	0.0061	-0.0021	0.	0.	0.	0.0167
5BU	0.0081	0.0060	0.0005	-0.0003	-0.0017	-0.0008	0.	0.	0.	0.2180
6BU	-0.0352	-0.0083	0.0033	0.0004	-0.0108	-0.0016	0.	-0.0001	0.	0.1036
Total BU	-0.0204	-0.0044	0.0020	0.0002	-0.0065	-0.0046	0.	-0.0001	0.	0.4044
Total	-0.0399	-0.0039	0.0039	0.0003	-0.0006	0.0022	0.0000	-0.0000	0.0000	0.7393
Distance	3.5414	4.5562	5.7707	7.1494	4.6706	5.9311	8.9033	8.6633	10.4196	2.0409

Overlap Populations (continued)

Pair of Atomic Centers

	C1 H2	C1 H3	C1 H4	C1 H5	C1 H6	C2 H1	C2 H2	C2 H3	C2 H4	C2 H5
1AG	0.	0.	0.	0.	0.	0.	0.	-0.0002	0.	0.
2AG	0.	0.	0.	0.	0.	0.	0.	0.	0.	0.
3AG	0.0239	0.0048	0.0005	0.0001	0.	-0.0002	0.0033	0.0319	0.0030	0.0036
4AG	0.1177	-0.0147	-0.0022	0.0004	0.	-0.0133	-0.0166	0.0524	0.0096	-0.0075
5AG	0.1613	0.0252	-0.0128	-0.0002	-0.0001	0.0075	0.0071	0.0656	-0.0171	0.0009
6AG	0.0147	-0.0609	-0.0078	-0.0005	-0.0003	-0.0061	-0.0065	0.1250	-0.0037	0.0032
7AG	0.1059	-0.0071	0.0268	0.0010	-0.0006	-0.0148	-0.0497	0.1484	-0.0808	0.0001
Total AG	0.4234	-0.0528	0.0044	0.0007	-0.0010	-0.0268	-0.0624	0.4233	-0.0890	0.0002
1AU	0.	0.	0.	0.	0.	0.	0.	0.	0.	0.
Total AU	0.	0.	0.	0.	0.	0.	0.	0.	0.	0.
1BG	0.	0.	0.	0.	0.	0.	0.	0.	0.	0.
Total BG	0.	0.	0.	0.	0.	0.	0.	0.	0.	0.
1BU	0.	0.	0.	0.	0.	0.	0.	0.	0.	0.
2BU	-0.0001	0.	0.	0.	0.	0.	0.	0.	0.	0.
3BU	0.0508	0.0036	-0.0013	-0.0002	-0.0001	0.0125	0.0037	0.0140	0.0003	-0.0014
4BU	0.1064	0.0026	0.0066	0.0001	0.	-0.0069	-0.0052	0.2494	-0.0084	0.0118
5BU	0.0082	-0.0066	0.0121	0.0005	0.0006	-0.0271	-0.0087	0.0166	0.0024	-0.0020
6BU	0.2186	-0.0383	-0.0099	0.0011	0.0001	-0.0036	-0.0172	0.1129	0.0002	-0.0186
Total BU	0.3840	-0.0388	0.0074	0.0014	0.0006	-0.0250	-0.0273	0.3929	-0.0055	-0.0103
Total	0.8074	-0.0915	0.0119	0.0021	-0.0005	-0.0518	-0.0898	0.8162	-0.0945	-0.0100
Distance	2.0409	3.9594	5.1172	7.5918	8.6251	3.9594	3.9594	2.0409	4.1651	5.0768

Pair of Atomic Centers

	C2 H6	C2 C1	C3 C1	C3 C2	C4 C1
1AG	0.	-0.0003	-0.0001	0.0005	0.
2AG	0.	-0.0010	0.0002	0.	0.
3AG	0.	0.2082	0.0465	0.2322	0.0013
4AG	-0.0005	-0.1080	-0.0285	0.2530	0.0013
5AG	-0.0004	0.1051	-0.0001	0.0794	-0.0026
6AG	-0.0001	0.1637	-0.0313	-0.0237	0.0089
7AG	0.0047	-0.1666	-0.0212	0.2083	0.0052
Total AG	0.0038	0.2011	-0.0345	0.7496	0.0141
1AU	0.	0.2078	0.0424	0.2205	0.0045
Total AU	0.	0.2078	0.0424	0.2205	0.0045
1BG	0.	0.3296	-0.0799	-0.2230	-0.0146
Total BG	0.	0.3296	-0.0799	-0.2230	-0.0146
1BU	0.	-0.0001	0.0001	-0.0024	0.
2BU	0.	-0.0007	-0.0001	0.	0.
3BU	-0.0002	0.2746	-0.0148	-0.0017	-0.0095
4BU	0.0006	0.0106	0.0157	-0.0283	0.0002
5BU	-0.0016	0.1172	0.0157	0.0408	-0.0106
6BU	0.0013	-0.0440	-0.0169	0.0158	0.0028
Total BU	0.0001	0.3575	-0.0002	0.0242	-0.0171
Total	0.0039	1.0960	-0.0723	0.7713	-0.0130
Distance	6.5217	2.5266	4.6706	2.8016	6.9676

Molecular Orbitals and Eigenvalues

Center		S1	S2	S3	S4	X1	X2	Y1	Y2	Z1	Z2
* 1AG	H1	-0.00005	-0.00019								
-11.2510	H2	-0.00011	0.00045								
	H3	-0.00052	0.00081								
	H4	-0.00052	0.00081								
	H5	-0.00011	0.00045								
	H6	-0.00005	-0.00019								
	C1	-0.00197	-0.03717	-0.00078	0.00124	-0.00055	0.00076	0.00004	-0.00071	0.	0.
	C2	-0.03664	-0.69112	-0.00324	-0.00240	-0.00013	0.00056	-0.00040	0.00093	0.	0.
	C3	-0.03664	-0.69112	-0.00324	-0.00240	0.00013	-0.00056	0.00040	-0.00093	0.	0.
	C4	-0.00197	-0.03717	-0.00078	0.00124	0.00055	-0.00076	-0.00004	0.00071	0.	0.
* 2AG	H1	0.00048	-0.00092								
-11.2361	H2	0.00045	-0.00024								
	H3	0.00007	-0.00052								
	H4	0.00007	-0.00052								
	H5	0.00045	-0.00024								
	H6	0.00048	-0.00092								
	C1	0.03665	0.69125	0.00297	0.00336	0.00015	0.00034	0.00003	-0.00068	0.	0.
	C2	-0.00199	-0.03756	0.00048	-0.00180	-0.00059	0.00208	0.00000	-0.00017	0.	0.
	C3	-0.00199	-0.03756	0.00048	-0.00180	0.00059	-0.00208	-0.00000	0.00017	0.	0.
	C4	0.03665	0.69125	0.00297	0.00336	-0.00015	-0.00034	-0.00003	0.00068	0.	0.
* 3AG	H1	0.04205	-0.00539								
-1.0977	H2	0.04625	0.01259								
	H3	0.06795	-0.00001								
	H4	0.06795	-0.00001								
	H5	0.04625	0.01259								
	H6	0.04205	-0.00539								
	C1	-0.00440	-0.09533	0.20437	0.12966	0.09151	-0.00660	0.00816	-0.01421	0.	0.
	C2	-0.00641	-0.13822	0.30313	0.12062	-0.02586	0.04578	0.05181	-0.01230	0.	0.
	C3	-0.00641	-0.13822	0.30313	0.12062	0.02586	-0.04578	-0.05181	0.01230	0.	0.
	C4	-0.00440	-0.09533	0.20437	0.12966	-0.09151	0.00660	-0.00816	0.01421	0.	0.
* 4AG	H1	-0.12151	-0.03968								
-0.8227	H2	-0.10035	-0.06028								
	H3	0.06903	0.02784								
	H4	0.06903	0.02784								
	H5	-0.10035	-0.06028								
	H6	-0.12151	-0.03968								
	C1	0.00521	0.11318	-0.24868	-0.20510	0.09998	0.00436	0.03081	0.03493	0.	0.
	C2	-0.00358	-0.07764	0.17414	0.14864	0.18583	-0.02204	0.08701	-0.01572	0.	0.
	C3	-0.00358	-0.07764	0.17414	0.14864	-0.18583	0.02204	-0.08701	0.01572	0.	0.
	C4	0.00521	0.11318	-0.24868	-0.20510	-0.09998	-0.00436	-0.03081	-0.03493	0.	0.
* 5AG	H1	0.08333	0.03700								
-0.6448	H2	-0.15015	-0.07842								
	H3	0.09915	0.07516								
	H4	0.09915	0.07516								
	H5	-0.15015	-0.07842								
	H6	0.08333	0.03700								
	C1	0.00042	0.00879	-0.02327	-0.01543	0.09688	0.01549	-0.28666	-0.08277	0.	0.
	C2	-0.00044	-0.00938	0.02278	0.00008	-0.07917	0.00855	-0.27019	-0.01159	0.	0.
	C3	-0.00044	-0.00938	0.02278	0.00008	0.07917	-0.00855	0.27019	0.01159	0.	0.
	C4	0.00042	0.00879	-0.02327	-0.01543	-0.09688	-0.01549	0.28666	0.08277	0.	0.

Center		S1	S2	S3	S4	X1	X2	Y1	Y2	Z1	Z2
* 6AG	H1	0.10327	0.12908								
-0.5553	H2	0.05582	0.02554								
	H3	0.12013	0.09837								
	H4	0.12013	0.09837								
	H5	0.05582	0.02554								
	H6	0.10327	0.12908								
	C1	0.00087	0.01893	-0.04302	-0.06002	-0.34730	-0.05468	-0.05710	0.05710	0.	0.
	C2	0.00014	0.00251	-0.01113	0.08187	0.35858	-0.00928	-0.04301	-0.04456	0.	0.
	C3	0.00014	0.00251	-0.01113	0.08187	-0.35858	0.00928	0.04301	0.04456	0.	0.
	C4	0.00087	0.01893	-0.04302	-0.06002	0.34730	0.05468	0.05710	-0.05710	0.	0.
* 7AG	H1	0.13597	0.10689								
-0.4937	H2	-0.11713	-0.10477								
	H3	-0.15508	-0.11805								
	H4	-0.15508	-0.11805								
	H5	-0.11713	-0.10477								
	H6	0.13597	0.10689								
	C1	-0.00005	-0.00036	0.00681	-0.03562	-0.03659	-0.02337	-0.23349	-0.07849	0.	0.
	C2	-0.00076	-0.01625	0.04071	0.04330	0.06631	-0.04563	0.35681	0.11296	0.	0.
	C3	-0.00076	-0.01625	0.04071	0.04330	-0.06631	0.04563	-0.35681	-0.11296	0.	0.
	C4	-0.00005	-0.00036	0.00681	-0.03562	0.03659	0.02337	0.23349	0.07849	0.	0.
8AG	H1	-0.04880	-0.52909								
0.2800	H2	0.01609	-0.14566								
	H3	0.00167	0.34610								
	H4	0.00167	0.34610								
	H5	0.01609	-0.14566								
	H6	-0.04880	-0.52909								
	C1	0.00068	0.02737	0.04751	-1.76690	-0.02003	-1.55815	-0.00831	-0.18749	0.	0.
	C2	-0.00240	-0.06545	0.04150	2.04905	-0.00950	-1.90775	-0.17671	0.57469	0.	0.
	C3	-0.00240	-0.06545	0.04150	2.04905	0.00950	1.90775	0.17671	-0.57469	0.	0.
	C4	0.00068	0.02737	0.04751	-1.76690	0.02003	1.55815	0.00831	0.18749	0.	0.
9AG	H1	0.01477	1.31981								
0.3362	H2	0.01099	0.75517								
	H3	-0.05319	0.14806								
	H4	-0.05319	0.14806								
	H5	0.01099	0.75517								
	H6	0.01477	1.31981								
	C1	0.00342	0.08457	-0.12022	-2.25354	0.29604	-0.07538	0.01049	0.54961	0.	0.
	C2	0.00010	-0.00482	-0.05109	0.86968	-0.15127	-0.73241	0.12707	-0.28237	0.	0.
	C3	0.00010	-0.00482	-0.05109	0.86968	0.15127	0.73241	-0.12707	0.28237	0.	0.
	C4	0.00342	0.08457	-0.12022	-2.25354	-0.29604	0.07538	-0.01049	-0.54961	0.	0.
10AG	H1	0.02606	0.84819								
0.3957	H2	-0.10894	-1.79427								
	H3	0.02559	-0.44674								
	H4	0.02559	-0.44674								
	H5	-0.10894	-1.79427								
	H6	0.02606	0.84819								
	C1	-0.00060	-0.01383	0.02968	0.59341	0.11089	-0.82934	-0.18699	2.46485	0.	0.
	C2	0.00084	0.01505	-0.06822	0.83489	-0.14459	0.48594	0.04798	-1.37238	0.	0.
	C3	0.00084	0.01505	-0.06822	0.83489	0.14459	-0.48594	-0.04798	1.37238	0.	0.
	C4	-0.00060	-0.01383	0.02968	0.59341	-0.11089	0.82934	0.18699	-2.46485	0.	0.

C4H6*

Center		Basis Function Type									
		S1	S2	S3	S4	X1	X2	Y1	Y2	Z1	Z2
11AG	H1	0.01546	-0.02610								
0.4224	H2	0.02137	0.25902								
	H3	-0.06869	-2.25940								
	H4	-0.06869	-2.25940								
	H5	0.02137	0.25902								
	H6	0.01546	-0.02610								
	C1	0.00076	0.01090	-0.08030	0.51223	-0.07486	0.25667	-0.48027	1.02483	0.	0.
	C2	-0.00092	-0.02625	0.01185	1.35052	0.08909	1.42428	-0.05999	-2.37143	0.	0.
	C3	-0.00092	-0.02625	0.01185	1.35052	-0.08909	-1.42428	0.05999	2.37143	0.	0.
	C4	0.00076	0.01090	-0.08030	0.51223	0.07486	-0.25667	0.48027	-1.02483	0.	0.
12AG	H1	0.07528	1.09531								
0.4955	H2	0.04526	0.03234								
	H3	0.06994	0.25822								
	H4	0.06994	0.25822								
	H5	0.04526	0.03234								
	H6	0.07528	1.09531								
	C1	0.00046	0.01859	0.02940	-1.29466	-0.00549	0.55598	-0.00651	0.50891	0.	0.
	C2	-0.00087	-0.01764	0.05608	-0.32930	0.10586	-1.01952	-0.52049	1.24003	0.	0.
	C3	-0.00087	-0.01764	0.05608	-0.32930	-0.10586	1.01952	0.52049	-1.24003	0.	0.
	C4	0.00046	0.01859	0.02940	-1.29466	0.00549	-0.55598	0.00651	-0.50891	0.	0.
13AG	H1	0.10036	-0.44594								
0.6114	H2	0.02480	1.10324								
	H3	0.09574	3.31288								
	H4	0.09574	3.31288								
	H5	0.02480	1.10324								
	H6	0.10036	-0.44594								
	C1	-0.00259	-0.01960	0.39291	-2.38445	0.07165	0.21170	-0.39706	-1.05244	0.	0.
	C2	-0.00150	0.01272	0.39258	-1.89337	0.14916	-3.50301	0.20644	3.21540	0.	0.
	C3	-0.00150	0.01272	0.39258	-1.89337	-0.14916	3.50301	-0.20644	-3.21540	0.	0.
	C4	-0.00259	-0.01960	0.39291	-2.38445	-0.07165	-0.21170	0.39706	1.05244	0.	0.
* 1AU	H1	0.	0.								
-0.4437	H2	0.	0.								
	H3	0.	0.								
	H4	0.	0.								
	H5	0.	0.								
	H6	0.	0.								
	C1	0.	0.	0.	0.	0.	0.	0.	0.	-0.27446	-0.10693
	C2	0.	0.	0.	0.	0.	0.	0.	0.	-0.36301	-0.13116
	C3	0.	0.	0.	0.	0.	0.	0.	0.	-0.36301	-0.13116
	C4	0.	0.	0.	0.	0.	0.	0.	0.	-0.27446	-0.10693
2AU	H1	0.	0.								
0.1101	H2	0.	0.								
	H3	0.	0.								
	H4	0.	0.								
	H5	0.	0.								
	H6	0.	0.								
	C1	0.	0.	0.	0.	0.	0.	0.	0.	-0.34942	-0.56560
	C2	0.	0.	0.	0.	0.	0.	0.	0.	0.28042	0.39976
	C3	0.	0.	0.	0.	0.	0.	0.	0.	0.28042	0.39976
	C4	0.	0.	0.	0.	0.	0.	0.	0.	-0.34942	-0.56560

Center		S1	S2	S3	S4	X1	X2	Y1	Y2	Z1	Z2
3AU	H1	0.	0.								
0.4724	H2	0.	0.								
	H3	0.	0.								
	H4	0.	0.								
	H5	0.	0.								
	H6	0.	0.								
	C1	0.	0.	0.	0.	0.	0.	0.	0.	-0.47619	0.43437
	C2	0.	0.	0.	0.	0.	0.	0.	0.	-0.43970	0.36213
	C3	0.	0.	0.	0.	0.	0.	0.	0.	-0.43970	0.36213
	C4	0.	0.	0.	0.	0.	0.	0.	0.	-0.47619	0.43437
4AU	H1	0.	0.								
0.6414	H2	0.	0.								
	H3	0.	0.								
	H4	0.	0.								
	H5	0.	0.								
	H6	0.	0.								
	C1	0.	0.	0.	0.	0.	0.	0.	0.	0.55047	-0.95791
	C2	0.	0.	0.	0.	0.	0.	0.	0.	-0.54367	0.82617
	C3	0.	0.	0.	0.	0.	0.	0.	0.	-0.54367	0.82617
	C4	0.	0.	0.	0.	0.	0.	0.	0.	0.55047	-0.95791
* 1BG	H1	0.	0.								
-0.3238	H2	0.	0.								
	H3	0.	0.								
	H4	0.	0.								
	H5	0.	0.								
	H6	0.	0.								
	C1	0.	0.	0.	0.	0.	0.	0.	0.	0.38061	0.20267
	C2	0.	0.	0.	0.	0.	0.	0.	0.	0.29213	0.16657
	C3	0.	0.	0.	0.	0.	0.	0.	0.	-0.29213	-0.16657
	C4	0.	0.	0.	0.	0.	0.	0.	0.	-0.38061	-0.20267
2BG	H1	0.	0.								
0.2060	H2	0.	0.								
	H3	0.	0.								
	H4	0.	0.								
	H5	0.	0.								
	H6	0.	0.								
	C1	0.	0.	0.	0.	0.	0.	0.	0.	-0.15772	-0.71136
	C2	0.	0.	0.	0.	0.	0.	0.	0.	0.33938	1.05483
	C3	0.	0.	0.	0.	0.	0.	0.	0.	-0.33938	-1.05483
	C4	0.	0.	0.	0.	0.	0.	0.	0.	0.15772	0.71136
3BG	H1	0.	0.								
0.5410	H2	0.	0.								
	H3	0.	0.								
	H4	0.	0.								
	H5	0.	0.								
	H6	0.	0.								
	C1	0.	0.	0.	0.	0.	0.	0.	0.	-0.64314	0.57641
	C2	0.	0.	0.	0.	0.	0.	0.	0.	-0.22626	0.52904
	C3	0.	0.	0.	0.	0.	0.	0.	0.	0.22626	-0.52904
	C4	0.	0.	0.	0.	0.	0.	0.	0.	0.64314	-0.57641

Center		S1	S2	S3	S4	X1	X2	Y1	Y2	Z1	Z2
4BG	H1	0.	0.								
0.6174	H2	0.	0.								
	H3	0.	0.								
	H4	0.	0.								
	H5	0.	0.								
	H6	0.	0.								
	C1	0.	0.	0.	0.	0.	0.	0.	0.	0.37247	-0.66036
	C2	0.	0.	0.	0.	0.	0.	0.	0.	-0.73457	1.40704
	C3	0.	0.	0.	0.	0.	0.	0.	0.	0.73457	-1.40704
	C4	0.	0.	0.	0.	0.	0.	0.	0.	-0.37247	0.66036
* 1BU	H1	-0.00007	0.00028								
-11.2503	H2	-0.00002	-0.00035								
	H3	-0.00039	0.00002								
	H4	0.00039	-0.00002								
	H5	0.00002	0.00035								
	H6	0.00007	-0.00028								
	C1	-0.00211	-0.03982	-0.00060	0.00075	-0.00052	0.00046	-0.00002	0.00072	0.	0.
	C2	-0.03665	-0.69118	-0.00260	-0.00377	0.00021	-0.00041	0.00014	-0.00146	0.	0.
	C3	0.03665	0.69118	0.00260	0.00377	0.00021	-0.00041	0.00014	-0.00146	0.	0.
	C4	0.00211	0.03982	0.00060	-0.00075	-0.00052	0.00046	-0.00002	0.00072	0.	0.
* 2BU	H1	0.00041	-0.00016								
-11.2361	H2	0.00040	-0.00035								
	H3	0.00004	-0.00033								
	H4	-0.00004	0.00033								
	H5	-0.00040	0.00035								
	H6	-0.00041	0.00016								
	C1	0.03664	0.69110	0.00290	0.00293	0.00016	0.00066	0.00002	0.00029	0.	0.
	C2	-0.00213	-0.04025	0.00034	-0.00154	-0.00053	0.00122	-0.00004	-0.00016	0.	0.
	C3	0.00213	0.04025	-0.00034	0.00154	-0.00053	0.00122	-0.00004	-0.00016	0.	0.
	C4	-0.03664	-0.69110	-0.00290	-0.00293	0.00016	0.00066	0.00002	0.00029	0.	0.
* 3BU	H1	-0.07387	-0.03464								
-1.0102	H2	-0.07117	-0.01276								
	H3	-0.04809	0.00139								
	H4	0.04809	-0.00139								
	H5	0.07117	0.01276								
	H6	0.07387	0.03464								
	C1	0.00612	0.13323	-0.28383	-0.17917	-0.08559	-0.00464	0.00306	-0.02458	0.	0.
	C2	0.00422	0.09104	-0.20057	-0.06394	0.15442	-0.02053	0.04536	0.02568	0.	0.
	C3	-0.00422	-0.09104	0.20057	0.06394	0.15442	-0.02053	0.04536	0.02568	0.	0.
	C4	-0.00612	-0.13323	0.28383	0.17917	-0.08559	-0.00464	0.00306	-0.02458	0.	0.
* 4BU	H1	0.03923	0.02155								
-0.7601	H2	0.11289	0.06427								
	H3	-0.16798	-0.09318								
	H4	0.16798	0.09318								
	H5	-0.11289	-0.06427								
	H6	-0.03923	-0.02155								
	C1	-0.00241	-0.05216	0.11742	0.07892	-0.13326	-0.01324	0.10032	0.01169	0.	0.
	C2	0.00401	0.08691	-0.19545	-0.14230	-0.08724	-0.02163	0.20064	0.01784	0.	0.
	C3	-0.00401	-0.08691	0.19545	0.14230	-0.08724	-0.02163	0.20064	0.01784	0.	0.
	C4	0.00241	0.05216	-0.11742	-0.07892	-0.13326	-0.01324	0.10032	0.01169	0.	0.

Center

Basis Function Type

		S1	S2	S3	S4	X1	X2	Y1	Y2	Z1	Z2
* 5BU	H1	0.17010	0.09186								
-0.6436	H2	0.02900	0.03353								
	H3	0.05742	0.06275								
	H4	-0.05742	-0.06275								
	H5	-0.02900	-0.03353								
	H6	-0.17010	-0.09186								
	C1	-0.00109	-0.02271	0.05951	0.04926	-0.30152	-0.05023	-0.17141	-0.06704	0.	0.
	C2	0.00176	0.03787	-0.08898	-0.01135	0.23093	0.00885	-0.10210	0.02030	0.	0.
	C3	-0.00176	-0.03787	0.08898	0.01135	0.23093	0.00885	-0.10210	0.02030	0.	0.
	C4	0.00109	0.02271	-0.05951	-0.04926	-0.30152	-0.05023	-0.17141	-0.06704	0.	0.
* 6BU	H1	0.12155	0.09715								
-0.5495	H2	-0.18265	-0.14197								
	H3	-0.11399	-0.08429								
	H4	0.11399	0.08429								
	H5	0.18265	0.14197								
	H6	-0.12155	-0.09715								
	C1	-0.00019	-0.00451	0.00655	0.01033	0.12494	0.02973	-0.32256	-0.06935	0.	0.
	C2	0.00087	0.01900	-0.04331	-0.08216	-0.17387	-0.01810	0.09568	0.02631	0.	0.
	C3	-0.00087	-0.01900	0.04331	0.08216	-0.17387	-0.01810	0.09568	0.02631	0.	0.
	C4	0.00019	0.00451	-0.00655	-0.01033	0.12494	0.02973	-0.32256	-0.06935	0.	0.
7BU	H1	0.00054	-0.31764								
0.2562	H2	-0.01909	-0.40641								
	H3	-0.01480	0.08253								
	H4	0.01480	-0.08253								
	H5	0.01909	0.40641								
	H6	-0.00054	0.31764								
	C1	-0.00352	-0.09364	·0.07589	2.48592	-0.12271	0.82992	0.02126	-0.12635	0.	0.
	C2	-0.00036	-0.00956	0.00677	0.06819	0.07669	1.70905	0.02922	1.19967	0.	0.
	C3	0.00036	0.00956	-0.00677	-0.06819	0.07669	1.70905	0.02922	1.19967	0.	0.
	C4	0.00352	0.09364	-0.07589	-2.48592	-0.12271	0.82992	0.02126	-0.12635	0.	0.
8BU	H1	-0.00411	0.34573								
0.2879	H2	0.03797	0.03787								
	H3	0.03313	0.32869								
	H4	-0.03313	-0.32869								
	H5	-0.03797	-0.03787								
	H6	0.00411	-0.34573								
	C1	-0.00083	-0.02640	-0.00933	0.91759	-0.04282	1.28308	0.05647	0.46877	0.	0.
	C2	0.00325	0.09300	-0.03044	-3.26070	-0.07364	-0.04466	0.07388	-1.13557	0.	0.
	C3	-0.00325	-0.09300	0.03044	3.26070	-0.07364	-0.04466	0.07388	-1.13557	0.	0.
	C4	0.00083	0.02640	0.00933	-0.91759	-0.04282	1.28308	0.05647	0.46877	0.	0.
9BU	H1	-0.01622	-0.17777								
0.3469	H2	-0.05047	-1.14846								
	H3	0.03766	1.18332								
	H4	-0.03766	-1.18332								
	H5	0.05047	1.14846								
	H6	0.01622	0.17777								
	C1	0.00005	0.00169	0.00151	0.05061	0.00704	-1.14702	0.25404	0.21212	0.	0.
	C2	0.00085	0.01931	-0.04232	-0.14392	-0.17103	-0.69033	0.28594	0.18320	0.	0.
	C3	-0.00085	-0.01931	0.04232	0.14392	-0.17103	-0.69033	0.28594	0.18320	0.	0.
	C4	-0.00005	-0.00169	-0.00151	-0.05061	0.00704	-1.14702	0.25404	0.21212	0.	0.

Center		S1	S2	S3	S4	X1	X2	Y1	Y2	Z1	Z2
10BU	H1	-0.08551	-1.04703								
0.3785	H2	-0.03108	-0.92776								
	H3	-0.05497	-1.02265								
	H4	0.05497	1.02265								
	H5	0.03108	0.92776								
	H6	0.08551	1.04703								
	C1	-0.00120	-0.03082	0.03724	0.68619	-0.03307	-1.03603	-0.26976	0.90196	0.	0.
	C2	0.00160	0.04140	-0.04621	-1.13055	-0.08372	-0.82040	-0.08648	-2.40923	0.	0.
	C3	-0.00160	-0.04140	0.04621	1.13055	-0.08372	-0.82040	-0.08648	-2.40923	0.	0.
	C4	0.00120	0.03082	-0.03724	-0.68619	-0.03307	-1.03603	-0.26976	0.90196	0.	0.
11BU	H1	-0.08136	-1.05147								
0.4913	H2	0.09444	1.60541								
	H3	-0.11212	0.08153								
	H4	0.11212	-0.08153								
	H5	-0.09444	-1.60541								
	H6	0.08136	1.05147								
	C1	-0.00076	-0.00741	0.10374	-0.01214	0.21708	-0.02545	0.21544	-2.50014	0.	0.
	C2	0.00076	0.01468	-0.05439	0.74079	-0.18413	0.70259	0.05656	1.27708	0.	0.
	C3	-0.00076	-0.01468	0.05439	-0.74079	-0.18413	0.70259	0.05656	1.27708	0.	0.
	C4	0.00076	0.00741	-0.10374	0.01214	0.21708	-0.02545	0.21544	-2.50014	0.	0.
12BU	H1	0.08486	0.26350								
0.5220	H2	0.00768	0.73216								
	H3	0.04879	0.99039								
	H4	-0.04879	-0.99039								
	H5	-0.00768	-0.73216								
	H6	-0.08486	-0.26350								
	C1	-0.00060	0.00139	0.13179	-0.01969	0.22992	0.28752	-0.43477	0.14248	0.	0.
	C2	0.00131	0.03746	-0.01362	-0.38777	-0.34348	0.80825	-0.01485	0.94675	0.	0.
	C3	-0.00131	-0.03746	0.01362	0.38777	-0.34348	0.80825	-0.01485	0.94675	0.	0.
	C4	0.00060	-0.00139	-0.13179	0.01969	0.22992	0.28752	-0.43477	0.14248	0.	0.
13BU	H1	-0.06366	-2.25128								
0.5707	H2	0.01862	1.44553								
	H3	0.09693	2.24988								
	H4	-0.09693	-2.24988								
	H5	-0.01862	-1.44553								
	H6	0.06366	2.25128								
	C1	-0.00098	-0.01872	0.07450	-0.07497	-0.08636	-0.34549	-0.07731	-3.78132	0.	0.
	C2	0.00206	0.03838	-0.15899	0.84312	-0.00295	-0.60549	-0.42081	3.96045	0.	0.
	C3	-0.00206	-0.03838	0.15899	-0.84311	-0.00295	-0.60549	-0.42081	3.96045	0.	0.
	C4	0.00098	0.01872	-0.07450	0.07497	-0.08636	-0.34549	-0.07731	-3.78132	0.	0.

C4H6*

Overlap Matrix

First Function   Second Function and Overlap

H1S1 with
  H1S2 .68301   H2S1 .01142   H2S2 .11463   H3S1 .00076   H3S2 .03578   H4S2 .00050   H5S2 .00001
  C1S1 .00234   C1S2 .03582   C1S3 .29024   C1S4 .37555   C1X1 -.21684   C1X2 -.21514   C1Y1 -.37836
  C1Y2 -.37539   C2S2 .00006   C2S3 .01165   C2S4 .09167   C2X1 -.03851   C2X2 -.24068   C2Y1 -.01925
  C2Y2 -.12034

H1S2 with
  H2S1 .11463   H2S2 .32840   H3S1 .03578   H3S2 .15832   H4S1 .00050   H4S2 .01069   H5S1 .00001
  H5S2 .00088   H6S2 .00007   C1S1 .00777   C1S2 .09909   C1S3 .50125   C1S4 .71004   C1X1 -.17935
  C1X2 -.30148   C1Y1 -.31294   C1Y2 -.52604   C2S1 .00101   C2S2 .01399   C2S3 .11473   C2S4 .28036
  C2X1 -.14847   C2X2 -.47188   C2Y1 -.07423   C2Y2 -.23594

H2S1 with
  H2S2 .68301   H3S1 .00001   H3S2 .00609   H4S1 .00054   H4S2 .03082   H5S2 .00002   C1S1 .00234
  C1S2 .03582   C1S3 .29024   C1S4 .37555   C1X1 -.21684   C1X2 -.21514   C1Y1 .37836   C1Y2 .37539
  C2S2 .00006   C2S3 .01165   C2S4 .09167   C2X1 -.03851   C2X2 -.24068   C2Y1 .01925   C2Y2 .12034

H2S2 with
  H3S1 .00609   H3S2 .05199   H4S1 .03082   H4S2 .14415   H5S1 .00002   H5S2 .00128   C1S1 .00777
  C1S2 .09909   C1S3 .50125   C1S4 .71004   C1X1 -.17935   C1X2 -.30148   C1Y1 .31294   C1Y2 .52604
  C2S1 .00101   C2S2 .01399   C2S3 .11473   C2S4 .28036   C2X1 -.14847   C2X2 -.47188   C2Y1 .07423
  C2Y2 .23594

H3S1 with
  H3S2 .68301   H4S1 .00001   H4S2 .00467   C1S2 .00006   C1S3 .01165   C1S4 .09167   C1X1 .03851
  C1X2 .24068   C1Y1 -.01925   C1Y2 -.12034   C2S1 .00234   C2S2 .03582   C2S3 .29024   C2S4 .37555
  C2X1 .21684   C2X2 .21514   C2Y1 -.37836   C2Y2 -.37539

H3S2 with
  H4S1 .00467   H4S2 .04401   C1S1 .00101   C1S2 .01399   C1S3 .11473   C1S4 .28036   C1X1 .14847
  C1X2 .47188   C1Y1 -.07423   C1Y2 -.23594   C2S1 .00777   C2S2 .09909   C2S3 .50125   C2S4 .71004
  C2X1 .17935   C2X2 .30148   C2Y1 -.31294   C2Y2 -.52604

H4S1 with
  C1S3 .00064   C1S4 .02539   C1X1 .00270   C1X2 .07258   C1Y1 .00371   C1Y2 .09968   C2S2 .00003
  C2S3 .00733   C2S4 .07473   C2X1 .00353   C2X2 .02798   C2Y1 .03006   C2Y2 .23837

H4S2 with
  C1S1 .00016   C1S2 .00235   C1S3 .02973   C1S4 .12003   C1X1 .03460   C1X2 .19305   C1Y1 .04751
  C1Y2 .26512   C2S1 .00075   C2S2 .01053   C2S3 .09258   C2S4 .24498   C2X1 .01655   C2X2 .05760
  C2Y1 .14100   C2Y2 .49064

H5S1 with
  C1S4 .00055   C1X2 .00826   C1Y2 .00065   C2S3 .00072   C2S4 .02670   C2X1 .00498   C2X2 .12652
  C2Y1 .00059   C2Y2 .01494

H5S2 with
  C1S2 .00001   C1S3 .00052   C1S4 .00948   C1X1 .00193   C1X2 .05426   C1Y1 .00015   C1Y2 .00427
  C2S1 .00017   C2S2 .00252   C2S3 .03135   C2S4 .12409   C2X1 .06090   C2X2 .33252   C2Y1 .00719
  C2Y2 .03926

H6S1 with
  C1S4 .00007   C1X2 .00159   C1Y2 .00087   C2S3 .00001   C2S4 .00347   C2X1 .00010   C2X2 .02428
  C2Y1 .00008   C2Y2 .01992

H6S2 with
  C1S3 .00006   C1S4 .00245   C1X1 .00026   C1X2 .01689   C1Y1 .00014   C1Y2 .00923   C2S1 .00001
  C2S2 .00015   C2S3 .00362   C2S4 .03208   C2X1 .00792   C2X2 .10349   C2Y1 .00650   C2Y2 .08491

C1S1 with
  C1S2 .35436   C1S3 .00805   C1S4 .01419   C2S3 .00175   C2S4 .00552   C2X1 -.00550   C2X2 -.00964
  C3S4 .00056   C3X1 -.00003   C3X2 -.00262   C3Y1 -.00002   C3Y2 -.00154   C4S4 .00001   C4X2 -.00020
  C4Y2 -.00007

C1S2 with
  C1S3 .26226   C1S4 .17745   C2S2 .00002   C2S3 .02666   C2S4 .07132   C2X1 -.07265   C2X2 -.12105
  C3S3 .00004   C3S4 .00789   C3X1 -.00069   C3X2 -.03456   C3Y1 -.00041   C3Y2 -.02031   C4S4 .00017
  C4X2 -.00287   C4Y2 -.00103

First Function   Second Function and Overlap

C1S3 with
  C1S4 .79970   C2S1 .00175   C2S2 .02666   C2S3 .23648   C2S4 .39182   C2X1 -.36934   C2X2 -.54816
  C3S2 .00004   C3S3 .00633   C3S4 .06954   C3X1 -.02166   C3X2 -.21234   C3Y1 -.01273   C3Y2 -.12476
  C4S3 .00001   C4S4 .00345   C4X1 -.00015   C4X2 -.02964   C4Y1 -.00005   C4Y2 -.01070

C1S4 with
  C2S1 .00552   C2S2 .07132   C2S3 .39182   C2S4 .62351   C2X1 -.30763   C2X2 -.63059   C3S1 .00056
  C3S2 .00789   C3S3 .06954   C3S4 .19903   C3X1 -.09117   C3X2 -.37098   C3Y1 -.05356   C3Y2 -.21796
  C4S1 .00001   C4S2 .00017   C4S3 .00345   C4S4 .02753   C4X1 -.00840   C4X2 -.10740   C4Y1 -.00303
  C4Y2 -.03877

C1X1 with
  C1X2 .53875   C2S1 .00550   C2S2 .07265   C2S3 .36934   C2S4 .30763   C2X1 -.35283   C2X2 -.04529
  C3S1 .00003   C3S2 .00069   C3S3 .02166   C3S4 .09117   C3X1 -.04620   C3X2 -.14413   C3Y1 -.03243
  C3Y2 -.12927   C4S3 .00015   C4S4 .00840   C4X1 -.00097   C4X2 -.04652   C4Y1 -.00038   C4Y2 -.01934

C1X2 with
  C2S1 .00964   C2S2 .12105   C2S3 .54816   C2S4 .63059   C2X1 -.04529   C2X2 .18619   C3S1 .00262
  C3S2 .03456   C3S3 .21234   C3S4 .37098   C3X1 -.14413   C3X2 -.24595   C3Y1 -.12927   C3Y2 -.31284
  C4S1 .00020   C4S2 .00287   C4S3 .02964   C4S4 .10740   C4X1 -.04652   C4X2 -.24288   C4Y1 -.01934
  C4Y2 -.11004

C1Y1 with
  C1Y2 .53875   C2Y1 .19634   C2Y2 .30293   C3S1 .00002   C3S2 .00041   C3S3 .01273   C3S4 .05356
  C3X1 -.03243   C3X2 -.12927   C3Y1 -.01005   C3Y2 -.00005   C4S3 .00005   C4S4 .00303   C4X1 -.00038
  C4X2 -.01934   C4Y1 -.00007   C4Y2 .00006

C1Y2 with
  C2Y1 .30293   C2Y2 .69365   C3S1 .00154   C3S2 .02031   C3S3 .12476   C3S4 .21796   C3X1 -.12927
  C3X2 -.31284   C3Y1 -.00005   C3Y2 .10271   C4S1 .00007   C4S2 .00103   C4S3 .01070   C4S4 .03877
  C4X1 -.01934   C4X2 -.11004   C4Y1 .00006   C4Y2 .02220

C1Z1 with
  C1Z2 .53875   C2Z1 .19634   C2Z2 .30293   C3Z1 .00900   C3Z2 .07590   C4Z1 .00007   C4Z2 .00704

C1Z2 with
  C2Z1 .30293   C2Z2 .69365   C3Z1 .07590   C3Z2 .28651   C4Z1 .00704   C4Z2 .06193

C2S1 with
  C2S2 .35436   C2S3 .00805   C2S4 .01419   C3S3 .00088   C3S4 .00444   C3X1 -.00191   C3X2 -.00484
  C3Y1 -.00301   C3Y2 -.00763

C2S2 with
  C2S3 .26226   C2S4 .17745   C3S3 .01436   C3S4 .05785   C3X1 -.02593   C3X2 -.06105   C3Y1 -.04088
  C3Y2 -.09627

C2S3 with
  C2S4 .79970   C3S1 .00088   C3S2 .01436   C3S3 .16862   C3S4 .33257   C3X1 -.15798   C3X2 -.28464
  C3Y1 -.24912   C3Y2 -.44885

C2S4 with
  C3S1 .00444   C3S2 .05785   C3S3 .33257   C3S4 .55943   C3X1 -.15555   C3X2 -.34065   C3Y1 -.24529
  C3Y2 -.53718

C2X1 with
  C2X2 .53875   C3S1 .00191   C3S2 .02593   C3S3 .15798   C3S4 .15555   C3X1 .00603   C3X2 .15796
  C3Y1 -.21618   C3Y2 -.16959

C2X2 with
  C3S1 .00484   C3S2 .06105   C3S3 .28464   C3S4 .34065   C3X1 .15796   C3X2 .47324   C3Y1 -.16959
  C3Y2 -.25947

C2Y1 with
  C2Y2 .53875   C3S1 .00301   C3S2 .04088   C3S3 .24912   C3S4 .24529   C3X1 -.21618   C3X2 -.16959
  C3Y1 -.19777   C3Y2 -.00193

C2Y2 with
  C3S1 .00763   C3S2 .09627   C3S3 .44885   C3S4 .53718   C3X1 -.16959   C3X2 -.25947   C3Y1 -.00193
  C3Y2 .22863

Overlap Matrix (continued)

First Function   Second Function and Overlap

C2Z1 with
    C2Z2 .53875   C3Z1 .14312   C3Z2 .26550

C2Z2 with
    C3Z1 .26550   C3Z2 .63778

C4H6*   T-305

C3H4O   Trans-acrolein

Molecular Geometry   Symmetry C_{1h}

Center	X	Coordinates Y	Z
H1	1.78274199	0.	-1.09818000
H2	-4.10738099	0.	-0.51820500
H3	-3.86921799	0.	-5.21254599
H4	-0.34653700	0.	-5.00171602
C1	0.	0.	0.
C2	-2.32271501	0.	-1.52380900
C3	-2.17049199	0.	-4.06097603
O1	0.	0.	2.30360699
Mass	-1.08003035	0.	-0.75133511

MO Symmetry Types and Occupancy

Representation	A*	A**
Number of MOs	40	8
Number Occupied	13	2

Energy Expectation Values

Total	-190.6899	Kinetic	190.7019
Electronic	-293.6259	Potential	-381.3918
Nuclear Repulsion	102.9360	Virial	1.9999

Electronic Potential	-484.3278
One-Electron Potential	-652.7978
Two-Electron Potential	168.4699

Moments of the Charge Distribution

First

x	y	z	Dipole Moment
0.8878	0.	2.7753	1.6090
-1.0987	0.	-4.3704	

Second

x^2	y^2	z^2	xy	xz	yz	r^2
-74.5661	-19.1493	-206.4352	0.	-67.6916	0.	-300.1507
58.4086	0.	185.4938	0.	66.3049	0.	243.9025

Third

x^3	xy^2	xz^2	x^2y	y^3	yz^2	x^2z	y^2z	z^3	xyz
28.7135	3.9253	27.1935	0.	0.	0.	34.1839	8.7621	142.0209	0.
-27.2443	0.	-33.9028	0.	0.	0.	-34.7032	0.	-155.2562	0.

Fourth (even)

x^4	y^4	z^4	x^2y^2	x^2z^2	y^2z^2	r^2x^2	r^2y^2	r^2z^2	r^4
-490.5571	-73.3163	-2856.3171	-58.9953	-491.7831	-137.7538	-1041.3354	-270.0654	-3485.8540	-4797.2548
253.8127	0.	2143.2367	0.	340.7563	0.	594.5689	0.	2483.9930	3078.5619

Fourth (odd)

x^3y	xy^3	xyz^2	x^3z	xy^2z	xz^3	x^2yz	y^3z	yz^3
0.	0.	0.	-265.7700	-42.8188	-803.2768	0.	0.	0.
0.	0.	0.	151.6315	0.	680.8883	0.	0.	0.

Expectation Values at Atomic Centers

Center	r^{-1}	δ	xr^{-3}	yr^{-3}	zr^{-3}
H1	-1.0587	0.4205	0.0634	0.	-0.0275
H2	-1.0659	0.4146	-0.0566	0.	0.0321
H3	-1.0571	0.4156	-0.0531	0.	-0.0382
H4	-1.0521	0.4173	0.0579	0.	-0.0357
C1	-14.6143	116.5708	0.0062	0.	-0.0456
C2	-14.6987	116.4883	-0.0165	0.	0.0071
C3	-14.6858	116.4966	-0.0039	0.	-0.0107
O1	-22.3373	287.3191	0.0026	0.	0.2557

Center	$\dfrac{3x^2-r^2}{r^5}$	$\dfrac{3y^2-r^2}{r^5}$	$\dfrac{3z^2-r^2}{r^5}$	$\dfrac{3xy}{r^5}$	$\dfrac{3xz}{r^5}$	$\dfrac{3yz}{r^5}$
H1	0.1758	-0.1520	-0.0238	0.	-0.1983	0.
H2	0.2207	-0.1782	-0.0424	0.	-0.2115	0.
H3	0.1776	-0.1740	-0.0036	0.	0.2353	0.
H4	0.2312	-0.1761	-0.0550	0.	-0.2061	0.
C1	-0.3000	0.4886	-0.1886	0.	0.0099	0.
C2	-0.0979	0.2291	-0.1312	0.	0.0464	0.
C3	-0.1884	0.3761	-0.1877	0.	-0.0152	0.
O1	-2.5011	1.6629	0.8381	0.	-0.0413	0.

Mulliken Charge

Center	S1	S2	S3	S4	X1	X2	Y1	Y2	Z1	Z2	Total
H1	0.5190	0.3436									0.8626
H2	0.5128	0.2552									0.7680
H3	0.5142	0.2812									0.7954
H4	0.5158	0.2992									0.8150
C1	0.0429	1.9565	0.8467	0.3665	0.9470	0.0770	0.5696	0.1485	0.8878	0.0315	5.8740
C2	0.0428	1.9564	0.8009	0.4074	0.9164	0.0755	0.7146	0.3060	0.9354	-0.0279	6.1275
C3	0.0428	1.9564	0.7989	0.5300	0.9505	0.1978	0.6307	0.2926	0.9482	0.0574	6.4054
O1	0.0426	1.9567	0.9653	0.8665	1.3919	0.4921	0.9010	0.4369	1.0817	0.2175	8.3521

Overlap Populations

Pair of Atomic Centers

	H2 H1	H3 H1	H3 H2	H4 H1	H4 H2	H4 H3	C1 H1	C1 H2	C1 H3	C1 H4
1A*	0.	0.	0.	0.	0.	0.	0.	0.	0.	0.
2A*	0.	0.	0.	0.	0.	0.	-0.0002	0.	0.	0.
3A*	0.	0.	0.	0.	0.	0.	0.	0.	0.	0.
4A*	0.	0.	0.	0.	0.	0.	0.	0.	0.	0.
5A*	0.	0.	-0.0002	-0.0001	0.	-0.0003	-0.0008	^.0018	-0.0003	0.0009
6A*	0.	0.	0.	0.0001	0.	0.0013	0.0100	0.0011	0.0006	0.0029
7A*	0.	-0.0001	-0.0002	-0.0035	0.	0.0098	0.1171	0.0030	-0.0031	-0.0173
8A*	-0.0018	0.0001	-0.0019	0.0075	-0.0014	0.0048	0.2777	-0.0004	0.0008	0.0160
9A*	0.0007	0.	0.	-0.0081	-0.0011	0.0001	0.1203	-0.0243	0.	-0.0013
10A*	-0.0004	-0.0002	0.0044	0.0004	-0.0001	-0.0032	-0.0139	0.0040	0.0031	-0.0016
11A*	0.0024	0.0011	0.0104	0.0001	0.	0.0006	0.0857	0.0116	-0.0011	-0.0003
12A*	0.	0.	-0.0135	0.0001	0.0056	-0.0494	0.0003	-0.0539	0.0042	-0.0145
13A*	0.0026	-0.0005	-0.0015	0.0071	0.0003	-0.0026	0.1609	-0.0008	-0.0011	0.0024
Total A*	0.0035	0.0004	-0.0026	0.0037	0.0032	-0.0390	0.7571	-0.0578	0.0031	-0.0128
1A**	0.	0.	0.	0.	0.	0.	0.	0.	0.	0.
2A**	0.	0.	0.	0.	0.	0.	0.	0.	0.	0.
Total A**	0.	0.	0.	0.	0.	0.	0.	0.	0.	0.
Total	0.0035	0.0004	-0.0026	0.0037	0.0032	-0.0390	0.7571	-0.0578	0.0031	-0.0128
Distance	5.9186	6.9909	4.7004	4.4465	5.8520	3.5290	2.0938	4.1399	6.4916	5.0137

Pair of Atomic Centers

	C2 H1	C2 H2	C2 H3	C2 H4	C2 C1	C3 H1	C3 H2	C3 H3	C3 H4	C3 C1
1A*	0.	0.	0.	0.	0.	0.	0.	0.	0.	0.
2A*	0.	0.	0.	0.	-0.0008	0.	0.	0.	0.	0.
3A*	0.	0.	0.	0.	0.	0.	0.	-0.0002	-0.0001	0.
4A*	0.	-0.0001	0.	0.	-0.0006	0.	0.	0.	0.	0.
5A*	-0.0004	0.0012	0.0009	0.0004	0.0409	0.	0.0010	-0.0024	-0.0005	-0.0042
6A*	0.0015	0.0335	0.0092	0.0057	0.1860	0.0013	0.0028	0.0470	0.0513	0.0639
7A*	0.0066	0.0110	-0.0070	-0.0055	0.2768	-0.0064	-0.0035	0.1638	0.1265	-0.0912
8A*	-0.0151	0.2406	-0.0092	-0.0124	-0.0273	0.0085	-0.0001	0.0218	0.1136	0.0228
9A*	-0.0091	0.0508	-0.0001	-0.0128	0.0721	-0.0078	0.0167	0.0003	0.2068	-0.0220
10A*	0.0076	0.0565	-0.0224	0.0056	0.0067	0.0080	-0.0014	0.2441	0.0033	-0.0131
11A*	-0.0092	0.0722	-0.0076	-0.0001	-0.0047	-0.0140	-0.0378	0.0751	0.0007	-0.0328
12A*	-0.0005	0.3300	-0.0182	-0.0510	-0.0057	0.0002	-0.0660	0.1824	0.2883	-0.0057
13A*	-0.2039	-0.0144	0.0027	-0.0166	0.0695	0.0534	0.0123	0.0110	0.0115	0.0098
Total A*	-0.2224	0.7812	-0.0518	-0.0866	0.6128	0.0433	-0.0761	0.7428	0.8014	-0.0725
1A**	0.	0.	0.	0.	0.1122	0.	0.	0.	0.	0.0169
2A**	0.	0.	0.	0.	-0.0806	0.	0.	0.	0.	-0.0236
Total A**	0.	0.	0.	0.	0.0316	0.	0.	0.	0.	-0.0067
Total	-0.2224	0.7812	-0.0518	-0.0866	0.6444	0.0433	-0.0761	0.7428	0.8014	-0.0792
Distance	4.1275	2.0485	3.9998	4.0001	2.7779	4.9403	4.0377	2.0523	2.0523	4.6046

C3H4O

Overlap Populations (continued)

	C3 C2	O1 H1	O1 H2	O1 H3	O1 H4	O1 C1	O1 C2	O1 C3
1A*	0.	-0.0001	0.	0.	0.	-0.0013	0.0003	0.
2A*	0.	0.	0.	0.	0.	-0.0002	0.	0.
3A*	-0.0010	0.	0.	0.	0.	0.	0.	0.
4A*	-0.0008	0.	0.	0.	0.	0.	0.	0.
5A*	-0.0045	-0.0044	0.0023	0.	0.0001	0.5206	0.0406	-^.0022
6A*	0.4411	-0.0004	0.0001	0.	0.	0.0037	-0.0098	-0.0004
7A*	-0.0337	-0.0061	0.0001	0.	0.	-0.0601	0.0007	0.
8A*	-0.0082	0.0049	0.0045	0.	0.0001	0.0343	0.0144	0.0001
9A*	0.1376	0.0113	-0.0054	0.	0.0001	0.1532	0.0076	-0.0014
10A*	0.1273	-0.0003	0.0010	0.0001	-0.0001	0.0411	-0.0068	-0.0035
11A*	0.1668	-0.0256	-0.0034	-0.0002	0.	0.1225	-0.0153	0.0047
12A*	-0.1757	0.0002	-0.0119	0.	0.0001	0.0570	-0.0211	0.0013
13A*	-0.1515	-0.1104	0.0151	-0.0001	0.	-0.1672	-0.0928	0.0078
Total A*	0.4974	-0.1309	0.0023	-0.0002	0.0003	0.7035	-0.0822	0.0064
1A**	0.0402	0.	0.	0.	0.	0.3613	0.0234	0.0015
2A**	0.4706	0.	0.	0.	0.	0.0569	-0.0525	-0.0053
Total A**	0.5108	0.	0.	0.	0.	0.4182	-0.0291	-0.0038
Total	1.0082	-0.1309	0.0023	-0.0002	0.0003	1.1217	-0.1112	0.0026
Distance	2.5417	3.8406	4.9833	8.4536	7.3135	2.3036	4.4771	6.7245

Molecular Orbitals and Eigenvalues

Center		S1	S2	S3	S4	X1	X2	Y1	Y2	Z1	Z2
* 1A*	H1	0.00013	-0.00148								
-20.5648	H2	0.00001	-0.00076								
	H3	0.00003	-0.00087								
	H4	-0.00003	0.00056								
	C1	-0.00000	-0.00001	-0.00005	-0.00156	0.00004	0.00174	0.	0.	-0.00030	-0.00252
	C2	0.00000	-0.00000	-0.00006	0.00040	-0.00001	0.00004	0.	0.	0.00003	0.00249
	C3	0.00000	-0.00001	-0.00006	0.00166	0.00002	-0.00136	0.	0.	0.00007	0.00000
	O1	0.05113	0.97937	0.00309	0.00410	-0.00001	-0.00016	0.	0.	-0.00187	0.00014
* 2A*	H1	-0.00071	0.00080								
-11.3574	H2	-0.00010	0.00046								
	H3	-0.00002	0.00012								
	H4	-0.00008	0.00037								
	C1	-0.05191	-0.97904	-0.00420	-0.00336	0.00016	0.00032	0.	0.	-0.00084	0.00102
	C2	-0.00022	-0.00422	-0.00057	0.00108	-0.00050	0.00139	0.	0.	-0.00028	0.00057
	C3	0.00000	-0.00001	-0.00018	0.00034	0.00006	-0.00037	0.	0.	-0.00001	0.00039
	O1	0.00002	0.00043	-0.00065	0.00032	0.00000	-0.00004	0.	0.	0.00108	-0.00049
* 3A*	H1	-0.00006	0.00043								
-11.2838	H2	-0.00017	0.00069								
	H3	-0.00066	0.00080								
	H4	-0.00061	0.00041								
	C1	0.00002	0.00028	-0.00017	0.00042	0.00008	-0.00063	0.	0.	-0.00003	0.00058
	C2	-0.00339	-0.06388	-0.00104	0.00133	-0.00005	0.00025	0.	0.	0.00078	-0.00209
	C3	-0.05179	-0.97689	-0.00418	-0.00387	-0.00006	0.00036	0.	0.	-0.00027	-0.00045
	O1	-0.00000	-0.00000	0.00003	-0.00020	0.00001	-0.00004	0.	0.	-0.00001	-0.00003
* 4A*	H1	0.00004	-0.00004								
-11.2718	H2	0.00067	-0.00061								
	H3	0.00001	-0.00002								
	H4	0.00003	-0.00004								
	C1	-0.00024	-0.00453	0.00043	-0.00131	-0.00034	0.00067	0.	0.	-0.00024	0.00037
	C2	0.05179	0.97686	0.00429	0.00380	0.00035	0.00014	0.	0.	-0.00008	-0.00020
	C3	-0.00341	-0.06445	0.00041	-0.00163	-0.00008	0.00008	0.	0.	0.00067	-0.00089
	O1	0.00000	0.00001	0.00000	0.00011	0.00006	-0.00008	0.	0.	0.00002	-0.00010
* 5A*	H1	0.03118	-0.01483								
-1.4142	H2	0.00622	0.01758								
	H3	0.00145	-0.02022								
	H4	0.00149	0.01241								
	C1	-0.00562	-0.11938	0.26654	0.04751	-0.00508	0.02114	0.	0.	0.19788	-0.04425
	C2	-0.00095	-0.02023	0.04408	0.01742	0.02639	0.03298	0.	0.	0.01918	0.01957
	C3	-0.00019	-0.00411	0.00767	0.01182	-0.00070	-0.03790	0.	0.	0.00479	-0.00467
	O1	-0.00951	-0.20860	0.45465	0.34705	-0.00250	-0.00326	0.	0.	-0.17968	0.00578
* 6A*	H1	0.03550	0.00138								
-1.0992	H2	0.07821	-0.00899								
	H3	0.06849	0.01510								
	H4	0.07462	0.00968								
	C1	-0.00262	-0.05591	0.13062	0.06526	-0.08028	0.00689	0.	0.	-0.13368	-0.02269
	C2	-0.00769	-0.16596	0.36283	0.15186	0.04596	-0.03434	0.	0.	-0.09242	0.03178
	C3	-0.00676	-0.14614	0.31745	0.18187	0.00558	0.01354	0.	0.	0.12392	-0.00660
	O1	0.00195	0.04274	-0.09605	-0.07742	-0.01607	0.00244	0.	0.	0.00254	-0.00962

		S1	S2	S3	S4	X1	X2	Y1	Y2	Z1	Z2
* 7A*	H1	0.10555	0.04239								
-0.8935	H2	0.03970	-0.00152								
	H3	-0.12707	-0.04478								
	H4	-0.10123	-0.05953								
	C1	-0.00585	-0.12445	0.30095	0.18757	-0.06707	-0.00776	0.	0.	-0.17410	-0.02571
	C2	-0.00242	-0.05218	0.11973	0.08471	0.12464	-0.03843	0.	0.	0.22338	-0.02541
	C3	0.00625	0.13570	-0.29862	-0.22582	0.03662	0.03251	0.	0.	0.06393	-0.00387
	O1	0.00335	0.07358	-0.16580	-0.17964	-0.02711	-0.00918	0.	0.	-0.08520	-0.03549
* 8A*	H1	0.17743	0.08636								
-0.7963	H2	-0.16273	-0.08044								
	H3	0.04441	0.02325								
	H4	0.11541	0.06543								
	C1	-0.00398	-0.08449	0.20657	0.13696	0.26370	0.02967	0.	0.	-0.08737	-0.04303
	C2	0.00454	0.09863	-0.22020	-0.17275	0.17946	0.01596	0.	0.	-0.08597	-0.00955
	C3	-0.00244	-0.05292	0.11890	0.09241	0.10544	0.00997	0.	0.	-0.14280	-0.00946
	O1	0.00180	0.03932	-0.08977	-0.09937	0.13968	0.03606	0.	0.	-0.09451	-0.03054
* 9A*	H1	-0.12546	-0.06394								
-0.6783	H2	-0.07567	-0.04464								
	H3	-0.00590	0.00196								
	H4	0.16673	0.09617								
	C1	-0.00033	-0.00708	0.01684	-0.00435	-0.36988	-0.07152	0.	0.	0.01520	0.01175
	C2	0.00137	0.02974	-0.06757	-0.06296	0.22712	0.01778	0.	0.	0.19399	-0.00734
	C3	-0.00085	-0.01785	0.04559	0.04815	0.23182	0.05231	0.	0.	-0.23449	-0.02965
	O1	0.00025	0.00553	-0.01290	-0.01794	-0.30248	-0.10548	0.	0.	-0.03055	-0.01449
* 10A*	H1	0.00446	0.03896								
-0.6711	H2	-0.09125	-0.04836								
	H3	-0.18100	-0.08747								
	H4	0.02619	0.00967								
	C1	0.00243	0.04869	-0.14602	-0.05268	-0.00135	-0.01744	0.	0.	-0.24993	0.06441
	C2	-0.00110	-0.02430	0.05105	0.01725	0.18289	0.03273	0.	0.	-0.19100	-0.03809
	C3	0.00067	0.01362	-0.03865	-0.05395	0.24108	0.07841	0.	0.	0.26265	0.05300
	O1	-0.00335	-0.07349	0.16668	0.24792	0.02806	0.01591	0.	0.	0.37548	0.10799
* 11A*	H1	0.10787	0.12757								
-0.6020	H2	0.09778	0.08757								
	H3.	0.13397	0.14717								
	H4	0.01319	-0.00141								
	C1	0.00021	0.00054	-0.03479	0.08600	0.04293	-0.01918	0.	0.	-0.35516	0.07212
	C2	0.00036	0.00752	-0.02076	0.03303	-0.06444	-0.02642	0.	0.	0.31062	-0.03097
	C3	0.00069	0.01503	-0.03474	-0.04936	-0.12247	0.03872	0.	0.	-0.32801	-0.03638
	O1	-0.00211	-0.04630	0.10481	0.16621	0.01065	-0.00251	0.	0.	0.42401	0.12874
* 12A*	H1	-0.00710	0.00064								
-0.5653	H2	-0.20123	-0.13328								
	H3	0.16108	0.11069								
	H4	-0.21072	-0.16709								
	C1	-0.00120	-0.02610	0.06202	0.05874	-0.15264	-0.04844	0.	0.	-0.11495	0.00210
	C2	0.00087	0.01909	-0.04276	-0.10569	0.28107	0.08869	0.	0.	-0.16065	-0.03439
	C3	-0.00050	-0.01086	0.02504	0.01699	-0.38794	-0.11013	0.	0.	0.09669	0.01086
	O1	-0.00032	-0.00702	0.01697	0.02858	-0.22207	-0.09002	0.	0.	0.06302	0.01744

		S1	S2	S3	S4	X1	X2	Y1	Y2	Z1	Z2
• 13A•	H1	-0.18041	-0.25113								
-0.4349	H2	-0.05977	-0.04735								
	H3	0.05489	0.03437								
	H4	-0.02463	-0.02978								
	C1	-0.00021	-0.00317	0.02058	-0.05191	-0.23242	0.02462	0.	0.	-0.00724	0.03084
	C2	-0.00157	-0.03415	0.08188	0.22917	0.25418	0.11901	0.	0.	0.15313	-0.01360
	C3	0.00025	0.00627	-0.00802	-0.07526	-0.04529	-0.04766	0.	0.	-0.08478	-0.06017
	O1	-0.00005	-0.00096	0.00311	0.00016	0.62653	0.33932	0.	0.	0.00423	-0.00756
14A•	H1	-0.02939	0.34994								
0.2474	H2	-0.00007	0.28309								
	H3	-0.00971	0.50786								
	H4	0.02673	0.45975								
	C1	-0.00107	-0.03533	-0.02305	1.43317	0.01114	-1.43558	0.	0.	-0.09248	-0.36967
	C2	-0.00032	-0.00963	0.00066	0.42773	-0.02337	-0.79553	0.	0.	-0.11740	-2.51295
	C3	0.00431	0.11249	-0.11117	-3.24817	-0.02829	0.35695	0.	0.	0.15648	-1.11692
	O1	-0.00151	-0.03510	0.06488	0.27345	-0.00465	0.12610	0.	0.	-0.02796	-0.03993
15A•	H1	-0.04220	-0.53095								
0.2650	H2	0.04377	0.25269								
	H3	0.01672	0.53280								
	H4	0.02283	-0.05160								
	C1	-0.00172	-0.04972	0.01399	1.98121	0.04176	-0.83260	0.	0.	-0.13299	-0.88111
	C2	0.00391	0.11057	-0.04511	-3.76402	0.15063	-1.29689	0.	0.	-0.04309	0.36231
	C3	-0.00103	-0.03351	-0.01732	1.34053	0.08484	0.45566	0.	0.	-0.05177	1.68135
	O1	-0.00031	-0.00708	0.01415	0.06228	0.05484	0.19762	0.	0.	-0.00445	0.15242
16A•	H1	0.00838	-0.74985								
0.2950	H2	0.02519	0.29275								
	H3	-0.08047	-1.38664								
	H4	-0.00375	-1.11060								
	C1	-0.00369	-0.08818	0.15345	1.90104	0.35297	-0.56613	0.	0.	-0.17436	0.09469
	C2	0.00038	0.00916	-0.01576	0.05579	-0.10219	-0.25199	0.	0.	-0.10370	-1.09787
	C3	-0.00197	-0.04189	0.11474	0.53525	0.06502	-0.04509	0.	0.	-0.13779	-1.52502
	O1	0.00072	0.02109	-0.00028	-0.44425	-0.07690	0.01479	0.	0.	0.08337	0.20199
17A•	H1	-0.02197	-0.29405								
0.3475	H2	-0.10292	-1.49112								
	H3	-0.04718	-0.73114								
	H4	0.11285	1.61557								
	C1	-0.00223	-0.05387	0.09145	1.27763	-0.13711	0.17878	0.	0.	-0.02563	0.19408
	C2	-0.00009	-0.00303	-0.00223	0.09427	-0.25778	-1.00629	0.	0.	0.10111	0.22076
	C3	0.00024	0.00462	-0.01795	-0.19303	-0.34121	-0.89101	0.	0.	-0.08655	0.75943
	O1	0.00151	0.03771	-0.04773	-0.52451	-0.01192	-0.00969	0.	0.	0.10761	-0.24273
18A•	H1	-0.07304	-0.80792								
0.3660	H2	-0.01042	1.29473								
	H3	0.00635	0.74896								
	H4	0.06529	1.18340								
	C1	-0.00098	-0.02710	0.01826	0.14539	0.13481	0.99259	0.	0.	0.01118	1.12766
	C2	0.00070	0.00944	-0.07837	0.38429	0.13289	1.71067	0.	0.	-0.18705	-0.15440
	C3	0.00187	0.04833	-0.05370	-1.41306	0.37090	-1.18979	0.	0.	0.26283	0.32671
	O1	0.00228	0.05785	-0.06677	-0.87067	-0.09052	-0.23494	0.	0.	0.04275	0.20758

		S1	S2	S3	S4	X1	X2	Y1	Y2	Z1	Z2
19A*	H1	0.04141	-0.46387								
0.4139	H2	0.07519	0.99168								
	H3	-0.00106	-0.99686								
	H4	0.07960	0.94882								
	C1	0.00004	0.01566	0.09477	-1.83937	0.08374	0.86037	0.	0.	-0.04192	-0.25374
	C2	-0.00187	-0.04406	0.08221	0.44722	-0.23119	2.77771	0.	0.	0.32344	-0.36430
	C3	-0.00062	-0.01037	0.05551	-0.35559	0.45097	-2.84261	0.	0.	-0.30379	0.38007
	O1	-0.00166	-0.04197	0.05010	0.63335	-0.11854	-0.18703	0.	0.	-0.00477	-0.29997
20A*	H1	0.03649	0.68143								
0.4498	H2	0.03999	-0.31025								
	H3	0.08965	1.32864								
	H4	0.02824	0.04904								
	C1	-0.00004	-0.00189	-0.00637	-0.37981	0.53978	-1.73076	0.	0.	0.04161	0.83019
	C2	-0.00157	-0.03341	0.09178	-0.34809	-0.44695	0.70293	0.	0.	0.09175	-0.41092
	C3	0.00081	0.02396	-0.00433	-1.04326	-0.05184	0.83052	0.	0.	0.03791	0.73141
	O1	0.00066	0.01927	-0.00243	-0.41164	0.04612	0.30073	0.	0.	-0.00164	0.03452
21A*	H1	0.14237	2.58592								
0.4704	H2	-0.11009	-0.10013								
	H3	-0.11036	-0.85884								
	H4	0.04082	0.51902								
	C1	0.00079	0.02442	0.00365	-1.14888	-0.01035	-2.45873	0.	0.	0.10124	0.78036
	C2	-0.00018	0.00079	0.04226	-0.63341	0.16462	0.34522	0.	0.	0.03135	-0.48656
	C3	-0.00066	-0.01403	0.04093	0.03712	0.43514	-1.64581	0.	0.	0.12774	-0.38232
	O1	0.00039	0.00967	-0.01292	-0.14970	0.15487	0.48512	0.	0.	0.00225	-0.06063
22A*	H1	0.03637	2.17711								
0.5643	H2	0.14167	4.32514								
	H3	0.02225	-1.64619								
	H4	0.01110	1.78773								
	C1	-0.00258	-0.02197	0.37863	-2.43810	-0.40976	-1.97041	0.	0.	-0.12361	2.32601
	C2	0.00160	0.06379	0.10320	-2.05589	-0.18907	5.26911	0.	0.	0.00224	-2.93104
	C3	-0.00274	-0.03157	0.34478	-1.76754	-0.35362	-3.16604	0.	0.	0.13726	-0.22694
	O1	0.00069	0.02112	0.00395	-0.56320	0.09950	0.25313	0.	0.	-0.06243	-0.15720
23A*	H1	0.08865	-1.22247								
0.5758	H2	0.03838	-0.01384								
	H3	0.09309	-1.25674								
	H4	0.13942	2.16749								
	C1	-0.00194	-0.00272	0.37362	-2.77126	0.28256	1.73503	0.	0.	-0.37348	-1.62726
	C2	-0.00302	-0.02693	0.43740	-0.15518	0.12768	1.62687	0.	0.	-0.25146	3.30980
	C3	-0.00237	-0.01158	0.40557	1.00649	-0.19136	-2.96016	0.	0.	0.14175	0.34644
	O1	-0.00240	-0.06515	0.04146	1.39143	0.00842	-0.13566	0.	0.	0.06816	-0.57900
24A*	H1	-0.13368	-1.30601								
0.6670	H2	-0.03944	-0.14253								
	H3	-0.16169	-1.35368								
	H4	-0.01992	0.70782								
	C1	0.00036	-0.04638	-0.39203	0.74019	-0.36975	1.12690	0.	0.	0.16478	-2.17436
	C2	0.00519	0.07672	-0.54701	-0.79159	-0.65725	1.11231	0.	0.	-0.29776	2.49171
	C3	0.00080	-0.01206	-0.24119	2.42100	0.08360	-2.37513	0.	0.	0.28194	-0.29504
	O1	0.00040	-0.00179	-0.09487	0.57788	0.17979	-0.04367	0.	0.	0.22625	0.11412

Center		S1	S2	S3	S4	X1	X2	Y1	Y2	Z1	Z2
* 1A**	H1	0.	0.								
-0.5464	H2	0.	0.								
	H3	0.	0.								
	H4	0.	0.								
	C1	0.	0.	0.	0.	0.	0.	-0.41392	-0.11804	0.	0.
	C2	0.	0.	0.	0.	0.	0.	-0.194∪5	-0.05383	0.	0.
	C3	0.	0.	0.	0.	0.	0.	-0.11906	-0.03820	0.	0.
	O1	0.	0.	0.	0.	0.	0.	-0.49281	-0.23161	0.	0.
* 2A**	H1	0.	0.								
-0.3988	H2	0.	0.								
	H3	0.	0.								
	H4	0.	0.								
	C1	0.	0.	0.	0.	0.	0.	0.06728	0.04831	0.	0.
	C2	0.	0.	0.	0.	0.	0.	-0.43565	-0.20743	0.	0.
	C3	0.	0.	0.	0.	0.	0.	-0.42867	-0.19993	0.	0.
	O1	0.	0.	0.	0.	0.	0.	0.28671	0.17028	0.	0.
3A**	H1	0.	0.								
0.0695	H2	0.	0.								
	H3	0.	0.								
	H4	0.	0.								
	C1	0.	0.	0.	0.	0.	0.	-0.33788	-0.29220	0.	0.
	C2	0.	0.	0.	0.	0.	0.	-0.26793	-0.40564	0.	0.
	C3	0.	0.	0.	0.	0.	0.	0.39805	0.58306	0.	0.
	O1	0.	0.	0.	0.	0.	0.	0.33192	0.33587	0.	0.
4A**	H1	0.	0.								
0.1944	H2	0.	0.								
	H3	0.	0.								
	H4	0.	0.								
	C1	0.	0.	0.	0.	0.	0.	-0.38734	-0.95162	0.	0.
	C2	0.	0.	0.	0.	0.	0.	0.36632	1.02366	0.	0.
	C3	0.	0.	0.	0.	0.	0.	-0.11999	-0.69791	0.	0.
	O1	0.	0.	0.	0.	0.	0.	0.26723	0.44015	0.	0.
5A**	H1	0.	0.								
0.4519	H2	0.	0.								
	H3	0.	0.								
	H4	0.	0.								
	C1	0.	0.	0.	0.	0.	0.	-0.68714	0.61468	0.	0.
	C2	0.	0.	0.	0.	0.	0.	-0.38390	0.38476	0.	0.
	C3	0.	0.	0.	0.	0.	0.	-0.53812	0.41637	0.	0.
	O1	0.	0.	0.	0.	0.	0.	0.03416	-0.02307	0.	0.
6A**	H1	0.	0.								
0.5204	H2	0.	0.								
	H3	0.	0.								
	H4	0.	0.								
	C1	0.	0.	0.	0.	0.	0.	-0.70072	1.21667	0.	0.
	C2	0.	0.	0.	0.	0.	0.	0.27721	-0.61919	0.	0.
	C3	0.	0.	0.	0.	0.	0.	0.60881	-0.63218	0.	0.
	O1	0.	0.	0.	0.	0.	0.	-0.01858	-0.12457	0.	0.

Center		Basis Function Type									
		S1	S2	S3	S4	X1	X2	Y1	Y2	Z1	Z2
7A**	H1	0.	0.								
0.5950	H2	0.	0.								
	H3	0.	0.								
	H4	0.	0.								
	C1	0.	0.	0.	0.	0.	0.	-0.18911	0.56868	0.	0.
	C2	0.	0.	0.	0.	0.	0.	0.91017	-1.60345	0.	0.
	C3	0.	0.	0.	0.	0.	0.	-0.64205	1.11634	0.	0.
	O1	0.	0.	0.	0.	0.	0.	0.088,3	-0.08319	0.	0.

Overlap Matrix

First Function   Second Function and Overlap

H1S1 with

H1S2	.68301	H2S1	.00001	H2S2	.00477	H3S2	.00068	H4S1	.00105	H4S2	.04115	C1S1	.00203
C1S2	.03164	C1S3	.27292	C1S4	.36559	C1X1	.35743	C1X2	.36984	C1Z1	-.22018	C1Z2	-.22783
C2S2	.00003	C2S3	.00799	C2S4	.07763	C2X1	.03216	C2X2	.24398	C2Z1	.00333	C2Z2	.02529
C3S3	.00105	C3S4	.03155	C3X1	.00538	C3X2	.11353	C3Z1	.00403	C3Z2	.08509	O1S2	.00004
O1S3	.00291	O1S4	.04149	O1X1	.00497	O1X2	.08251	O1Z1	-.00949	O1Z2	-.15745		

H1S2 with

H2S1	.00477	H2S2	.04459	H3S1	.00068	H3S2	.01305	H4S1	.04115	H4S2	.17283	C1S1	.00747
C1S2	.09547	C1S3	.48742	C1S4	.69760	C1X1	.30649	C1X2	.52160	C1Z1	-.18880	C1Z2	-.32131
C2S1	.00079	C2S2	.01111	C2S3	.09636	C2S4	.25123	C2X1	.14545	C2X2	.49764	C2Z1	.01508
C2Z2	.05159	C3S1	.00021	C3S2	.00318	C3S3	.03738	C3S4	.13858	C3X1	.05647	C3X2	.28679
C3Z1	.04233	C3Z2	.21494	O1S1	.00076	O1S2	.01018	O1S3	.07175	O1S4	.19236	O1X1	.03814
O1X2	.17775	O1Z1	-.07277	O1Z2	-.33917								

H2S1 with

H2S2	.68301	H3S1	.00049	H3S2	.02963	H4S1	.00001	H4S2	.00533	C1S2	.00003	C1S3	.00777
C1S4	.07666	C1X1	-.03138	C1X2	-.24162	C1Z1	-.00396	C1Z2	-.03048	C2S1	.00229	C2S2	.03519
C2S3	.28772	C2S4	.37412	C2X1	-.37790	C2X2	-.37719	C2Z1	.21293	C2Z2	.21253	C3S2	.00004
C3S3	.00979	C3S4	.08492	C3X1	-.01810	C3X2	-.12374	C3Z1	.03311	C3Z2	.22634	O1S3	.00006
O1S4	.00554	O1X1	-.00035	O1X2	-.03646	O1Z1	-.00024	O1Z2	-.02505				

H2S2 with

H3S1	.02963	H3S2	.14063	H4S1	.00533	H4S2	.04781	C1S1	.00078	C1S2	.01091	C1S3	.09510
C1S4	.24914	C1X1	-.14367	C1X2	-.49430	C1Z1	-.01813	C1Z2	-.06236	C2S1	.00772	C2S2	.09857
C2S3	.49926	C2S4	.70826	C2X1	-.31417	C2X2	-.52904	C2Z1	.17703	C2Z2	.29810	C3S1	.00090
C3S2	.01258	C3S3	.10587	C3S4	.26655	C3X1	-.07514	C3X2	-.24708	C3Z1	.13744	C3Z2	.45193
O1S1	.00013	O1S2	.00177	O1S3	.01602	O1S4	.06407	O1X1	-.02031	O1X2	-.15401	O1Z1	-.01396
O1Z2	-.10581												

H3S1 with

H3S2	.68301	H4S1	.01176	H4S2	.11606	C1S3	.00001	C1S4	.00363	C1X1	-.00008	C1X2	-.01937
C1Z1	-.00011	C1Z2	-.02609	C2S2	.00005	C2S3	.01065	C2S4	.08814	C2X1	-.01556	C2X2	-.10182
C2Z1	-.03711	C2Z2	-.24285	C3S1	.00227	C3S2	.03488	C3S3	.28647	C3S4	.37341	C3X1	-.35808
C3X2	-.35847	C3Z1	-.24274	C3Z2	-.24301	O1X2	-.00002	O1Z2	-.00003				

H3S2 with

H4S1	.11606	H4S2	.33097	C1S1	.00001	C1S2	.00016	C1S3	.00381	C1S4	.03311	C1X1	-.00637
C1X2	-.08161	C1Z1	-.00859	C1Z2	-.10995	C2S1	.00095	C2S2	.01325	C2S3	.11009	C2S4	.27317
C2X1	-.06230	C2X2	-.20150	C2Z1	-.14860	C2Z2	-.48063	C3S1	.00770	C3S2	.09831	C3S3	.49827
C3S4	.70737	C3X1	-.29846	C3X2	-.50302	C3Z1	-.20233	C3Z2	-.34100	O1S3	.00002	O1S4	.00040
O1X1	-.00002	O1X2	-.00158	O1Z1	-.00005	O1Z2	-.00306						

H4S1 with

H4S2	.68301	C1S3	.00085	C1S4	.02886	C1X1	-.00040	C1X2	-.00926	C1Z1	-.00573	C1Z2	-.13365
C2S2	.00005	C2S3	.01065	C2S4	.08811	C2X1	.01987	C2X2	.13007	C2Z1	-.03496	C2Z2	-.22891
C3S1	.00227	C3S2	.03488	C3S3	.28646	C3S4	.37341	C3X1	.38447	C3X2	.38490	C3Z1	-.19830
C3Z2	-.19852	O1S4	.00002	O1X2	-.00003	O1Z2	-.00062						

H4S2 with

C1S1	.00019	C1S2	.00281	C1S3	.03403	C1S4	.13063	C1X1	-.00453	C1X2	-.02389	C1Z1	-.06532
C1Z2	-.34484	C2S1	.00095	C2S2	.01324	C2S3	.11005	C2S4	.27312	C2X1	.07958	C2X2	.25744
C2Z1	-.14006	C2Z2	-.45307	C3S1	.00770	C3S2	.09831	C3S3	.49827	C3S4	.70737	C3X1	.32046
C3X2	.54011	C3Z1	-.16528	C3Z2	-.27857	O1S2	.00001	O1S3	.00023	O1S4	.00282	O1X1	-.00003
O1X2	-.00081	O1Z1	-.00058	O1Z2	-.01701								

C1S1 with

C1S2	.35436	C1S3	.00805	C1S4	.01419	C2S3	.00093	C2S4	.00453	C2X1	.00310	C2X2	.00761
C2Z1	.00203	C2Z2	.00499	C3S4	.00062	C3X1	.00002	C3X2	.00152	C3Z1	.00004	C3Z2	.00284
O1S3	.00043	O1S4	.00515	O1Z1	-.00254	O1Z2	-.01280						

C1S2 with

C1S3	.26226	C1S4	.17745	C2S3	.01518	C2S4	.05896	C2X1	.04199	C2X2	.09591	C2Z1	.02755
C2Z2	.06292	C3S3	.00005	C3S4	.00861	C3X1	.00046	C3X2	.01994	C3Z1	.00085	C3Z2	.03732
O1S2	.00001	O1S3	.01042	O1S4	.06764	O1Z1	-.03962	O1Z2	-.15730				

First Function   Second Function and Overlap

C1S3 with
```
   C1S4  .79970   C2S1  .00093   C2S2  .01518   C2S3  .17386   C2S4  .33753   C2X1  .25186   C2X2  .44600
   C2Z1  .16523   C2Z2  .29260   C3S2  .00005   C3S3  .00732   C3S4  .07448   C3X1  .01326   C3X2  .12106
   C3Z1  .02480   C3Z2  .22650   O1S1  .00185   O1S2  .02569   O1S3  .18734   O1S4  .38125   O1Z1 -.25553
   O1Z2 -.60496
```

C1S4 with
```
   C2S1  .00453   C2S2  .05896   C2S3  .33753   C2S4  .56493   C2X1  .24431   C2X2  .53187   C2Z1  .16028
   C2Z2  .34893   C3S1  .00062   C3S2  .00861   C3S3  .07448   C3S4  .20826   C3X1  .05251   C3X2  .20802
   C3Z1  .09824   C3Z2  .38920   O1S1  .00413   O1S2  .05337   O1S3  .29696   O1S4  .55299   O1Z1 -.17087
   O1Z2 -.53429
```

C1X1 with
```
   C1X2  .53875   C2S1 -.00310   C2S2 -.04199   C2S3 -.25186   C2S4 -.24431   C2X1 -.19149   C2X2  .00784
   C2Z1 -.22220   C2Z2 -.17112   C3S1 -.00002   C3S2 -.00046   C3S3 -.01326   C3S4 -.05251   C3X1 -.00795
   C3X2  .01264   C3Z1 -.03378   C3Z2 -.12635   O1X1  .14303   O1X2  .36159
```

C1X2 with
```
   C2S1 -.00761   C2S2 -.09591   C2S3 -.44600   C2S4 -.53187   C2X1  .00784   C2X2  .24531   C2Z1 -.17112
   C2Z2 -.26066   C3S1 -.00152   C3S2 -.01994   C3S3 -.12106   C3S4 -.20802   C3X1  .01264   C3X2  .13653
   C3Z1 -.12635   C3Z2 -.29974   O1X1  .17447   O1X2  .59729
```

C1Y1 with
```
   C1Y2  .53875   C2Y1  .14721   C2Y2  .26868   C3Y1  .01011   C3Y2  .08017   O1Y1  .14303   O1Y2  .36159
```

C1Y2 with
```
   C2Y1  .26868   C2Y2  .64263   C3Y1  .08017   C3Y2  .29674   O1Y1  .17447   O1Y2  .59729
```

C1Z1 with
```
   C1Z2  .53875   C2S1 -.00203   C2S2 -.02755   C2S3 -.16523   C2S4 -.16028   C2X1 -.22220   C2X2 -.17112
   C2Z1  .00143   C2Z2  .15642   C3S1 -.00004   C3S2 -.00085   C3S3 -.02480   C3S4 -.09824   C3X1 -.03378
   C3X2 -.12635   C3Z1 -.05310   C3Z2 -.15623   O1S1  .00481   O1S2  .06264   O1S3  .33970   O1S4  .41936
   O1Z1 -.30153   O1Z2 -.18945
```

C1Z2 with
```
   C2S1 -.00499   C2S2 -.06292   C2S3 -.29260   C2S4 -.34893   C2X1 -.17112   C2X2 -.26066   C2Z1  .15642
   C2Z2  .47163   C3S1 -.00284   C3S2 -.03732   C3S3 -.22650   C3S4 -.38920   C3X1 -.12635   C3X2 -.29974
   C3Z1 -.15623   C3Z2 -.26408   O1S1  .00634   O1S2  .08081   O1S3  .40781   O1S4  .62102   O1Z1 -.01164
   O1Z2  .12439
```

C2S1 with
```
   C2S2  .35436   C2S3  .00805   C2S4  .01419   C3S3  .00169   C3S4  .00546   C3X1 -.00032   C3X2 -.00058
   C3Z1  .00536   C3Z2  .00960   O1S4  .00008   O1X2 -.00056   O1Z2 -.00092
```

C2S2 with
```
   C2S3  .26226   C2S4  .17745   C3S2  .00002   C3S3  .02581   C3S4  .07054   C3X1 -.00426   C3X2 -.00723
   C3Z1  .07100   C3Z2  .12052   O1S4  .00139   O1X1 -.00001   O1X2 -.00797   O1Z1 -.00001   O1Z2 -.01313
```

C2S3 with
```
   C2S4  .79970   C3S1  .00169   C3S2  .02581   C3S3  .23235   C3S4  .38848   C3X1 -.02187   C3X2 -.03279
   C3Z1  .36453   C3Z2  .54660   O1S2  .00004   O1S3  .00188   O1S4  .02869   O1X1 -.00342   O1X2 -.07130
   O1Z1 -.00564   O1Z2 -.11749
```

C2S4 with
```
   C3S1  .00546   C3S2  .07054   C3S3  .38848   C3S4  .61998   C3X1 -.01838   C3X2 -.03780   C3Z1  .30633
   C3Z2  .63010   O1S1  .00047   O1S2  .00629   O1S3  .04550   O1S4  .13216   O1X1 -.02723   O1X2 -.14848
   O1Z1 -.04487   O1Z2 -.24467
```

C2X1 with
```
   C2X2  .53875   C3S1  .00032   C3S2  .00426   C3S3  .02187   C3S4  .01838   C3X1  .19112   C3X2  .29960
   C3Z1  .03261   C3Z2  .02092   O1S1  .00002   O1S2  .00038   O1S3  .00632   O1S4  .03559   O1X1 -.00513
   O1X2 -.02267   O1Z1 -.01354   O1Z2 -.11168
```

C2X2 with
```
   C3S1  .00058   C3S2  .00723   C3S3  .03279   C3S4  .03780   C3X1  .29960   C3X2  .68878   C3Z1  .02092
   C3Z2  .03057   O1S1  .00118   O1S2  .01545   O1S3  .09180   O1S4  .18831   O1X1 -.00327   O1X2  .03881
   O1Z1 -.07085   O1Z2 -.26387
```

C2Y1 with
```
   C2Y2  .53875   C3Y1  .19308   C3Y2  .30085   O1Y1  .00309   O1Y2  .04510
```

Overlap Matrix (continued)

First Function   Second Function and Overlap

C2Y2 with
    C3Y1 .30085   C3Y2 .69061   O1Y1 .03973   O1Y2 .19894

C2Z1 with
    C2Z2 .53875   C3S1 -.00536   C3S2 -.07100   C3S3 -.36453   C3S4 -.30633   C3X1 .03261   C3X2 .02092
    C3Z1 -.35053   C3Z2 -.04786   O1S1 .00004   O1S2 .00063   O1S3 .01042   O1S4 .05865   O1X1 -.01354
    O1X2 -.11168   O1Z1 -.01922   O1Z2 -.13893

C2Z2 with
    C3S1 -.00960   C3S2 -.12052   C3S3 -.54660   C3S4 -.63010   C3X1 .02092   C3X2 .03057   C3Z1 -.04786
    C3Z2 .18114   O1S1 .00195   O1S2 .02545   O1S3 .15127   O1S4 .31031   O1X1 -.07085   O1X2 -.26387
    O1Z1 -.07702   O1Z2 -.23587

C3S1 with
    C3S2 .35436   C3S3 .00805   C3S4 .01419   O1Z2 -.00001

C3S2 with
    C3S3 .26226   C3S4 .17745   O1X2 -.00005   O1Z2 -.00013

C3S3 with
    C3S4 .79970   O1S4 .00034   O1X2 -.00166   O1Z1 -.00001   O1Z2 -.00486

C3S4 with
    O1S1 .00001   O1S2 .00016   O1S3 .00184   O1S4 .01146   O1X1 -.00111   O1X2 -.01535   O1Z1 -.00324
    O1Z2 -.04502

C3X1 with
    C3X2 .53875   O1S3 .00001   O1S4 .00058   O1X1 -.00001   O1X2 -.00040   O1Z1 -.00005   O1Z2 -.00535

C3X2 with
    O1S1 .00006   O1S2 .00084   O1S3 .00661   O1S4 .02260   O1X1 .00019   O1X2 .00904   O1Z1 -.00882
    O1Z2 -.06269

C3Y1 with
    C3Y2 .53875   O1Y1 .00001   O1Y2 .00143

C3Y2 with
    O1Y1 .00319   O1Y2 .03042

C3Z1 with
    C3Z2 .53875   O1S3 .00003   O1S4 .00169   O1X1 -.00005   O1X2 -.00535   O1Z1 -.00013   O1Z2 -.01426

C3Z2 with
    O1S1 .00018   O1S2 .00247   O1S3 .01939   O1S4 .06627   O1X1 -.00882   O1X2 -.06269   O1Z1 -.02268
    O1Z2 -.15341

O1S1 with
    O1S2 .35461   O1S3 .00974   O1S4 .01469

O1S2 with
    O1S3 .28286   O1S4 .18230

O1S3 with
    O1S4 .79029

O1X1 with
    O1X2 .50607

O1Y1 with
    O1Y2 .50607

O1Z1 with
    O1Z2 .50607

C3H4O   T-319

BF3   Boron Trifluoride

Molecular Geometry   Symmetry D$_{3h}$

Center	Coordinates		
	X	Y	Z
B1	0.	0.	0.
F1	1.22361311	0.	-2.11936009
F2	1.22361311	0.	2.11936009
F3	-2.44722626	0.	0.
Mass	0.	0.	0.

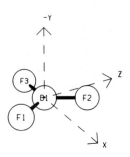

MO Symmetry Types and Occupancy

Representation	A1*	A1**	A2*	A2**	E*	E**
Number of MOs	10	4	2	0	20	4
Number Occupied	4	1	1	0	8	2

Energy Expectation Values

Total	-323.1287	Kinetic	323.4440
Electronic	-435.6218	Potential	-646.5727
Nuclear Repulsion	112.4931	Virial	1.9990
Electronic Potential	-759.0658		
One-Electron Potential	-996.9585		
Two-Electron Potential	237.8927		

Moments of the Charge Distribution

Second

x^2	y^2	z^2	xy	xz	yz	r^2
-98.4849	-13.2325	-98.4849	0.	0.	0.	-210.2024
80.8504	0.	80.8504	0.	0.	0.	161.7007

Third

x^3	xy^2	xz^2	x^2y	y^3	yz^2	x^2z	y^2z	z^3	xyz
103.5120	0.	-103.5122	0.	0.	0.	0.	0.	0.0001	0.
-98.9296	0.	98.9296	0.	0.	0.	0.	0.	0.	0.

Fourth (even)

x^4	y^4	z^4	x^2y^2	x^2z^2	y^2z^2	r^2x^2	r^2y^2	r^2z^2	r^4
-615.2008	-30.5675	-615.2011	-44.8913	-205.0670	-44.8914	-865.1591	-120.3502	-865.1595	-1850.6688
363.1546	0.	363.1546	0.	121.0515	0.	484.2061	0.	484.2061	968.4122

Fourth (odd)

x^3y	xy^3	xyz^2	x^3z	xy^2z	xz^3	x^2yz	y^3z	yz^3
0.	0.	0.	0.0001	0.	0.0001	0.	0.	0.
0.	0.	0.	0.	0.	0.	0.	0.	0.

Expectation Values at Atomic Centers

Center	r^{-1}	δ	Expectation Value xr^{-3}	yr^{-3}	zr^{-3}
B1	-11.1417	65.4487	0.	0.	0.
F1	-26.5283	413.4275	0.1342	0.	-0.2325

Center	$\dfrac{3x^2-r^2}{r^5}$	$\dfrac{3y^2-r^2}{r^5}$	$\dfrac{3z^2-r^2}{r^5}$	$\dfrac{3xy}{r^5}$	$\dfrac{3xz}{r^5}$	$\dfrac{3yz}{r^5}$
B1	-0.1109	0.2218	-0.1109	0.	C.	0.
F1	-0.5656	-0.3443	0.9099	0.	-1.2779	0.

Mulliken Charge

Center	S1	S2	S3	S4	X1	X2	Y1	Y2	Z1	Z2	Total
						Basis Function Type					
B1	0.0435	1.9562	0.6144	-0.0519	0.6283	0.0116	0.3424	-0.0037	0.6283	0.0116	4.1807
F1	0.0435	1.9557	1.0263	0.8705	1.3749	0.4644	1.3499	0.5372	1.2855	0.3652	9.2731

Overlap Populations

Pair of Atomic Centers

	F1 B1	F2 F1
1A1*	-0.0001	0.
2A1*	0.0009	0.
3A1*	0.0970	0.0203
4A1*	0.0799	0.0108
Total A1*	0.1778	0.0311
1A1**	0.1223	0.0136
Total A1**	0.1223	0.0136
1A2*	0.	-0.0486
Total A2*	0.	-0.0486
1E*	0.	0.
2E*	-0.0001	0.
3E*	0.0048	0.0020
4E*	0.2169	-0.0228
5E*	0.1282	0.0098
6E*	-0.0212	0.0012
7E*	0.0372	-0.0744
8E*	-0.0286	0.0328
Total E*	0.3372	-0.0513
1E**	0.	0.0083
2E**	0.	-0.0337
Total E**	0.	-0.0254
Total	0.6373	-0.0807
Distance	2.4472	4.2387

Molecular Orbitals and Eigenvalues

Center		S1	S2	S3	S4	X1	X2	Y1	Y2	Z1	Z2
Basis Function Type											
* 1A1*	B1	0.00000	0.00001	0.00026	0.00037	0.	0.	0.	0.	0.	0.
-26.3467	F1	-0.02987	-0.56516	-0.00258	-0.00087	0.00034	-0.00004	0.	0.	-0.00059	0.00007
	F2	-0.02987	-0.56516	-0.00258	-0.00087	0.00034	-0.00004	0.	0.	0.00059	-0.00007
	F3	-0.02987	-0.56516	-0.00258	-0.00087	-0.00068	0.00008	0.	0.	0.	0.
* 2A1*	B1	0.05294	0.97864	0.00311	-0.00072	0.	0.	0.	0.	0.	0.
-7.8835	F1	-0.00002	-0.00058	0.00079	0.00125	-0.00093	-0.00031	0.	0.	0.00160	0.00054
	F2	-0.00002	-0.00058	0.00079	0.00125	-0.00093	-0.00031	0.	0.	-0.00160	-0.00054
	F3	-0.00002	-0.00058	0.00079	0.00125	0.00185	0.00062	0.	0.	0.	0.
* 3A1*	B1	0.00494	0.10584	-0.18321	0.02327	0.	0.	0.	0.	0.	0.
-1.7315	F1	0.00629	0.13521	-0.30153	-0.25318	0.03752	0.00596	0.	0.	-0.06499	-0.01033
	F2	0.00629	0.13521	-0.30153	-0.25318	0.03752	0.00596	0.	0.	0.06499	0.01033
	F3	0.00629	0.13521	-0.30153	-0.25318	-0.07504	-0.01192	0.	0.	0.	0.
* 4A1*	B1	-0.00724	-0.14631	0.39615	-0.05946	0.	0.	0.	0.	0.	0.
-0.8782	F1	0.00232	0.04962	-0.11735	-0.15297	-0.17912	-0.06102	0.	0.	0.31025	0.10569
	F2	0.00232	0.04962	-0.11735	-0.15297	-0.17912	-0.06102	0.	0.	-0.31025	-0.10569
	F3	0.00232	0.04962	-0.11735	-0.15297	0.35824	0.12205	0.	0.	0.	0.
5A1*	B1	-0.00251	-0.08392	-0.04248	1.74740	0.	0.	0.	0.	0.	0.
0.1737	F1	0.00230	0.05406	-0.08585	-0.53421	0.05908	0.09015	0.	0.	-0.10233	-0.15614
	F2	0.00230	0.05406	-0.08585	-0.53421	0.05908	0.09015	0.	0.	0.10233	0.15614
	F3	0.00230	0.05406	-0.08585	-0.53421	-0.11816	-0.18029	0.	0.	0.	0.
6A1*	B1	-0.01509	-0.14882	2.15996	-1.09260	0.	0.	0.	0.	0.	0.
0.6538	F1	0.00183	0.03493	-0.12946	-0.09051	0.17087	0.13460	0.	0.	-0.29596	-0.23314
	F2	0.00183	0.03493	-0.12946	-0.09051	0.17087	0.13460	0.	0.	0.29596	0.23314
	F3	0.00183	0.03493	-0.12946	-0.09051	-0.34174	-0.26921	0.	0.	0.	0.
* 1A1**	B1	0.	0.	0.	0.	0.	0.	0.28084	-0.00401	0.	0.
-0.7948	F1	0.	0.	0.	0.	0.	0.	0.39101	0.15876	0.	0.
	F2	0.	0.	0.	0.	0.	0.	0.39101	0.15876	0.	0.
	F3	0.	0.	0.	0.	0.	0.	0.39101	0.15876	0.	0.
2A1**	B1	0.	0.	0.	0.	0.	0.	0.50684	0.86625	0.	0.
0.1337	F1	0.	0.	0.	0.	0.	0.	-0.22264	-0.24520	0.	0.
	F2	0.	0.	0.	0.	0.	0.	-0.22264	-0.24520	0.	0.
	F3	0.	0.	0.	0.	0.	0.	-0.22264	-0.24520	0.	0.
3A1**	B1	0.	0.	0.	0.	0.	0.	1.12326	-0.94757	0.	0.
0.2568	F1	0.	0.	0.	0.	0.	0.	-0.12212	-0.04409	0.	0.
	F2	0.	0.	0.	0.	0.	0.	-0.12212	-0.04409	0.	0.
	F3	0.	0.	0.	0.	0.	0.	-0.12212	-0.04409	0.	0.
* 1A2*	B1	0.	0.	0.	0.	0.	0.	0.	0.	0.	0.
-0.6836	F1	0.	0.	0.	0.	0.39634	0.19076	0.	0.	0.22883	0.11013
	F2	0.	0.	0.	0.	-0.39634	-0.19076	0.	0.	0.22883	0.11013
	F3	0.	0.	0.	0.	0.	0.	0.	0.	-0.45765	-0.22027
* 1E*	B1	0.	0.	0.	0.	0.00030	0.00027	0.	0.	0.00017	0.00015
-26.3467	F1	-0.00081	-0.01540	-0.00007	-0.00002	0.00004	-0.00017	0.	0.	0.00000	-0.00009
	F2	-0.03617	-0.68435	-0.00317	-0.00100	0.00045	-0.00020	0.	0.	0.00073	-0.00014
	F3	0.03698	0.69975	0.00324	0.00102	0.00088	-0.00023	0.	0.	0.00002	-0.00009
* 2E*	B1	0.	0.	0.	0.	0.00017	0.00015	0.	0.	-0.00030	-0.00027
-26.3467	F1	-0.04223	-0.79911	-0.00370	-0.00117	0.00050	-0.00013	0.	0.	-0.00087	0.00023
	F2	0.02182	0.41289	0.00191	0.00060	-0.00023	-0.00007	0.	0.	-0.00047	0.00020
	F3	0.02041	0.38622	0.00179	0.00056	0.00048	-0.00013	0.	0.	-0.00003	0.00017

Center			S1	S2	S3	S4	X1	X2	Y1	Y2	Z1	Z2
* 3E*		B1	0.	0.	0.	0.	-0.17277	0.01836	0.	0.	-0.05777	0.00614
-1.6774		F1	0.00187	0.03996	-0.09123	-0.06857	-0.00517	0.01856	0.	0.	-0.02362	0.00990
		F2	0.00701	0.14994	-0.34231	-0.25731	0.02398	0.01364	0.	0.	0.06357	-0.00481
		F3	-0.00888	-0.18990	0.43354	0.32588	0.08491	0.00336	0.	0.	-0.00527	0.00680
* 4E*		B1	0.	0.	0.	0.	-0.05777	0.00614	0.	0.	0.17276	-0.01836
-1.6774		F1	0.00917	0.19621	-0.44794	-0.33671	0.04674	-0.00197	0.	0.	-0.07432	-0.00515
		F2	-0.00621	-0.13271	0.30297	0.22775	-0.04044	0.01274	0.	0.	-0.04516	-0.01007
		F3	-0.00297	-0.06350	0.14496	0.10897	0.02839	0.00112	0.	0.	0.01576	-0.02035
* 5E*		B1	0.	0.	0.	0.	0.35394	-0.05837	0.	0.	0.22401	-0.03695
-0.8372		F1	-0.00008	-0.00190	0.00368	0.00995	0.31099	0.10729	0.	0.	0.15982	0.05704
		F2	0.00185	0.04141	-0.08013	-0.21693	-0.04565	0.00260	0.	0.	-0.40370	-0.10838
		F3	-0.00176	-0.03951	0.07645	0.20698	-0.35535	-0.08832	0.	0.	0.18693	0.06500
* 6E*		B1	0.	0.	0.	0.	0.22401	-0.03694	0.	0.	-0.35394	0.05838
-0.8372		F1	0.00209	0.04672	-0.09041	-0.24474	-0.19779	-0.04794	0.	0.	0.37099	0.09291
		F2	-0.00097	-0.02171	0.04202	0.11374	0.36573	0.11749	0.	0.	0.01434	-0.01179
		F3	-0.00112	-0.02501	0.04839	0.13100	-0.22490	-0.05590	0.	0.	-0.29536	-0.10271
* 7E*		B1	0.	0.	0.	0.	-0.00158	0.00065	0.	0.	0.14817	-0.06106
-0.7091		F1	-0.00050	-0.01056	0.02671	0.03885	-0.03701	-0.04293	0.	0.	-0.41505	-0.18039
		F2	0.00050	0.01043	-0.02638	-0.03837	0.04689	0.04790	0.	0.	-0.41594	-0.18135
		F3	0.00001	0.00013	-0.00033	-0.00048	0.00417	0.00165	0.	0.	-0.48815	-0.25954
* 8E*		B1	0.	0.	0.	0.	-0.14817	0.06106	0.	0.	-0.00158	0.00065
-0.7091		F1	-0.00028	-0.00595	0.01504	0.02189	0.46438	0.23381	0.	0.	0.04638	0.04735
		F2	-0.00029	-0.00617	0.01561	0.02271	0.46349	0.23284	0.	0.	-0.03753	-0.04350
		F3	0.00058	0.01212	-0.03065	-0.04459	0.39127	0.15464	0.	0.	0.00520	0.00276
9E*		B1	0.	0.	0.	0.	-0.02882	1.37878	0.	0.	-0.00084	0.04032
0.2255		F1	0.00088	0.02120	-0.02928	-0.23657	-0.11602	-0.18117	0.	0.	-0.04583	-0.05129
		F2	0.00098	0.02347	-0.03240	-0.26181	-0.11344	-0.17838	0.	0.	0.04209	0.04399
		F3	-0.00186	-0.04467	0.06168	0.49838	-0.03858	-0.09726	0.	0.	-0.00410	-0.00606
10E*		B1	0.	0.	0.	0.	0.00084	-0.04032	0.	0.	-0.02882	1.37877
0.2255		F1	-0.00164	-0.03934	0.05433	0.43890	-0.04061	-0.04238	0.	0.	-0.06268	-0.12336
		F2	0.00158	0.03803	-0.05252	-0.42432	0.04732	0.05290	0.	0.	-0.06525	-0.12615
		F3	0.00005	0.00131	-0.00180	-0.01457	0.00113	0.00284	0.	0.	-0.14011	-0.20727
11E*		B1	0.	0.	0.	0.	-1.74568	0.49582	0.	0.	0.07922	-0.02250
0.6641		F1	-0.00177	-0.03709	0.10194	0.30052	0.06819	-0.10207	0.	0.	0.26495	0.30006
		F2	-0.00151	-0.03168	0.08709	0.25672	0.09135	-0.07654	0.	0.	-0.24545	-0.26248
		F3	0.00328	0.06877	-0.18903	-0.55723	-0.36225	-0.57648	0.	0.	-0.01031	-0.00332
12E*		B1	0.	0.	0.	0.	-0.07922	0.02250	0.	0.	-1.74569	0.49581
0.6641		F1	0.00277	0.05799	-0.15940	-0.46996	0.25882	0.27722	0.	0.	-0.20333	-0.40132
		F2	-0.00292	-0.06111	0.16798	0.49524	-0.25158	-0.28533	0.	0.	-0.22650	-0.42685
		F3	0.00015	0.00312	-0.00858	-0.02529	-0.01644	-0.02616	0.	0.	0.22711	0.07309
* 1E**		B1	0.	0.	0.	0.	0.	0.	0.	0.	0.	0.
-0.7028		F1	0.	0.	0.	0.	0.	0.	-0.13751	-0.07114	0.	0.
		F2	0.	0.	0.	0.	0.	0.	-0.45298	-0.23434	0.	0.
		F3	0.	0.	0.	0.	0.	0.	0.59048	0.30548	0.	0.
* 2E**		B1	0.	0.	0.	0.	0.	0.	0.	0.	0.	0.
-0.7028		F1	0.	0.	0.	0.	0.	0.	-0.60244	-0.31167	0.	0.
		F2	0.	0.	0.	0.	0.	0.	0.42031	0.21744	0.	0.
		F3	0.	0.	0.	0.	0.	0.	0.18213	0.09423	0.	0.

Basis Function Type

Center

Overlap Matrix

First Function   Second Function and Overlap

B1S1 with
    B1S2 .35489   B1S3 .00732   B1S4 .01371   F1S3 .00008   F1S4 .00434   F1X1 -.00042   F1X2 -.00749
    F1Z1 .00072   F1Z2 .01297

B1S2 with
    B1S3 .24693   B1S4 .17217   F1S2 .00003   F1S3 .00545   F1S4 .06081   F1X1 -.01078   F1X2 -.09059
    F1Z1 .01867   F1Z2 .15691

B1S3 with
    B1S4 .80606   F1S1 .00226   F1S2 .02958   F1S3 .18377   F1S4 .38434   F1X1 -.08939   F1X2 -.28362
    F1Z1 .15483   F1Z2 .49125

B1S4 with
    F1S1 .00310   F1S2 .03977   F1S3 .22863   F1S4 .46932   F1X1 -.04225   F1X2 -.17812   F1Z1 .07318
    F1Z2 .30851

B1X1 with
    B1X2 .55377   F1S1 .00236   F1S2 .03024   F1S3 .16692   F1S4 .26117   F1X1 .03507   F1X2 .23913
    F1Z1 .13987   F1Z2 .29555

B1X2 with
    F1S1 .00185   F1S2 .02365   F1S3 .13044   F1S4 .24303   F1X1 .07333   F1X2 .34799   F1Z1 .03113
    F1Z2 .12091

B1Y1 with
    B1Y2 .55377   F1Y1 .11582   F1Y2 .40976

B1Y2 with
    F1Y1 .09130   F1Y2 .41780

B1Z1 with
    B1Z2 .55377   F1S1 -.00408   F1S2 -.05238   F1S3 -.28912   F1S4 -.45237   F1X1 .13987   F1X2 .29555
    F1Z1 -.12644   F1Z2 -.10214

B1Z2 with
    F1S1 -.00321   F1S2 -.04096   F1S3 -.22593   F1S4 -.42094   F1X1 .03113   F1X2 .12091   F1Z1 .03738
    F1Z2 .20837

F1S1 with
    F1S2 .35647   F1S3 .01003   F1S4 .01474   F2S4 .00002   F2Z2 -.00039

F1S2 with
    F1S3 .28653   F1S4 .18096   F2S4 .00037   F2Z2 -.00547

F1S3 with
    F1S4 .78531   F2S3 .00002   F2S4 .00573   F2Z1 -.00024   F2Z2 -.04705

F1S4 with
    F2S1 .00002   F2S2 .00037   F2S3 .00573   F2S4 .04033   F2Z1 -.01238   F2Z2 -.15334

F1X1 with
    F1X2 .49444   F2X1 .00008   F2X2 .00991

F1X2 with
    F2X1 .00991   F2X2 .08586

F1Y1 with
    F1Y2 .49444   F2Y1 .00008   F2Y2 .00991

F1Y2 with
    F2Y1 .00991   F2Y2 .08586

F1Z1 with
    F1Z2 .49444   F2S3 .00024   F2S4 .01238   F2Z1 -.00136   F2Z2 -.06700

F1Z2 with
    F2S1 .00039   F2S2 .00547   F2S3 .04705   F2S4 .15334   F2Z1 -.06700   F2Z2 -.33571

BF3   T-325

F2CO   Carbonyl Fluoride

Molecular Geometry   Symmetry C$_{2v}$

Center   Coordinates
                X   Y   Z
C1   0.   0.   0.
O1   0.   0.   2.21171999
F1   0.   2.07370001   -1.38560000
F2   0.   -2.07370001   -1.38560000

Mass   0.   0.   -0.26173325

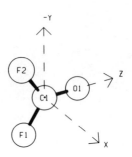

MO Symmetry Types and Occupancy

Representation   A1   A2   B1   B2
Number of MOs   20   2   6   12
Number Occupied   8   1   2   5

Energy Expectation Values

Total	-311.5244	Kinetic	311.6648
Electronic	-430.7412	Potential	-623.1892
Nuclear Repulsion	119.2167	Virial	1.9995

Electronic Potential	-742.4059
One-Electron Potential	-979.4089
Two-Electron Potential	237.0030

Moments of the Charge Distribution

First

x	y	z	Dipole Moment
0.	0.	-1.5292	0.4007
0.	0.	1.1284	

Second

x^2	y^2	z^2	xy	xz	yz	r^2
-13.9308	-94.5600	-90.5977	0.	0.	0.	-199.0885
0.	77.4042	72.0902	0.	0.	0.	149.4943

Third

x^3	xy^2	xz^2	x^2y	y^3	yz^2	x^2z	y^2z	z^3	xyz
0.	0.	0.	0.	0.	0.	-2.6598	86.5473	-111.5753	0.
0.	0.	0.	0.	0.	0.	0.	-86.9920	95.6162	0.

Fourth (even)

x^4	y^4	z^4	x^2y^2	x^2z^2	y^2z^2	r^2x^2	r^2y^2	r^2z^2	r^4
-33.2869	-565.7615	-592.9040	-43.2518	-46.0701	-179.5120	-122.6089	-788.5254	-818.4861	-1729.6203
0.	332.8559	328.1812	0.	0.	97.7674	0.	430.6233	425.9486	856.5719

Expectation Values at Atomic Centers

Center	r^{-1}	δ	xr^{-3}	yr^{-3}	zr^{-3}
			Expectation Value		
C1	-14.3764	116.3847	0.	0.	0.0080
O1	-22.2519	287.3721	0.	0.	0.2971
F1	-26.4834	413.6851	0.	0.2296	-0.1433

	$\dfrac{3x^2-r^2}{r^5}$	$\dfrac{3y^2-r^2}{r^5}$	$\dfrac{3z^2-r^2}{r^5}$	$\dfrac{3xy}{r^5}$	$\dfrac{3xz}{r^5}$	$\dfrac{3yz}{r^5}$
C1	0.0877	0.1443	-0.2319	0.	0.	0.
O1	1.1541	-2.1788	1.0247	0.	0.	0.
F1	-1.3252	1.8456	-0.5204	0.	0.	-2.4090

Mulliken Charge

Center	S1	S2	S3	S4	X1	X2	Y1	Y2	Z1	Z2	Total
					Basis Function Type						
C1	0.0428	1.9566	0.8409	0.0098	0.6945	0.0798	0.7134	0.0282	0.8609	0.0269	5.2539
O1	0.0427	1.9566	0.9769	0.8216	0.9664	0.4117	1.3595	0.4696	1.0711	0.1789	8.2551
F1	0.0436	1.9557	1.0332	0.9044	1.4058	0.5180	1.1945	0.3879	1.3546	0.4478	9.2455

Overlap Populations

	O1 C1	F1 C1	F1 O1	F2 F1
		Pair of Atomic Centers		
1A1	0.	-0.0001	0.	0.
2A1	-0.0009	0.	0.	0.
3A1	0.0010	0.0002	0.	0.
4A1	0.0318	0.1474	0.0167	0.0380
5A1	0.6109	0.0014	-0.0037	-0.0001
6A1	0.0197	0.1007	-0.0025	0.0109
7A1	-0.1496	0.0734	-0.0192	0.0359
8A1	0.1105	-0.0421	-0.0086	0.0173
Total A1	0.6233	0.2809	-0.0172	0.1021
1A2	0.	0.	0.	-0.0405
Total A2	0.	0.	0.	-0.0405
1B1	0.0751	0.1474	0.0085	0.0208
2B1	0.3287	-0.1178	-0.0343	0.0139
Total B1	0.4037	0.0296	-0.0258	0.0346
1B2	0.	-0.0001	0.	0.
2B2	0.	0.1142	-0.0012	-0.0372
3B2	0.0961	0.0937	0.0125	-0.0090
4B2	-0.0008	0.0172	0.0015	-0.1335
5B2	0.1361	-0.0395	-0.0835	-0.0280
Total B2	0.2314	0.1854	-0.0708	-0.2077
Total	1.2585	0.4959	-0.1138	-0.1115
Distance	2.2117	2.4940	4.1522	4.1474

Molecular Orbitals and Eigenvalues

Center		S1	S2	S3	S4	X1	X2	Y1	Y2	Z1	Z2
* 1A1	C1	0.00000	-0.00003	-0.00002	0.00058	0.	0.	0.	0.	-0.00011	-0.00042
-26.3961	O1	0.00000	0.00000	0.00002	0.00003	0.	0.	0.	0.	-0.00003	0.00013
	F1	-0.03658	-0.69220	-0.00306	-0.00115	0.	0.	0.00074	-0.00007	-0.00049	0.00019
	F2	-0.03658	-0.69220	-0.00306	-0.00115	0.	0.	-0.00074	0.00007	-0.00049	0.00019
* 2A1	C1	0.00000	-0.00003	-0.00020	-0.00141	0.	0.	0.	0.	-0.00042	-0.00127
-20.6477	O1	0.05113	0.97934	0.00333	0.00363	0.	0.	0.	0.	-0.00221	0.00016
	F1	0.00000	0.00001	-0.00001	0.00000	0.	0.	-0.00001	0.00005	-0.00007	0.00037
	F2	0.00000	0.00001	-0.00001	0.00000	0.	0.	0.00001	-0.00005	-0.00007	0.00037
* 3A1	C1	-0.05191	-0.97912	-0.00409	0.00052	0.	0.	0.	0.	0.00081	0.00003
-11.5871	O1	0.00002	0.00056	-0.00065	-0.00110	0.	0.	0.	0.	0.00151	0.00045
	F1	0.00001	0.00018	-0.00012	-0.00030	0.	0.	0.00051	0.00013	-0.00034	-0.00016
	F2	0.00001	0.00018	-0.00012	-0.00030	0.	0.	-0.00051	-0.00013	-0.00034	-0.00016
* 4A1	C1	-0.00484	-0.10281	0.22600	-0.01347	0.	0.	0.	0.	-0.06870	0.00440
-1.7864	O1	-0.00172	-0.03796	0.08011	0.07226	0.	0.	0.	0.	-0.04852	-0.01188
	F1	-0.00729	-0.15670	0.34907	0.29994	0.	0.	-0.07880	-0.02005	0.04487	0.01345
	F2	-0.00729	-0.15670	0.34907	0.29994	0.	0.	0.07880	0.02005	0.04487	0.01345
* 5A1	C1	-0.00453	-0.09438	0.22713	0.02843	0.	0.	0.	0.	0.29390	-0.01699
-1.5282	O1	-0.00939	-0.20517	0.45374	0.27975	0.	0.	0.	0.	-0.19197	0.02083
	F1	0.00228	0.04865	-0.11341	-0.08563	0.	0.	0.00934	-0.01424	0.02205	0.00632
	F2	0.00228	0.04865	-0.11341	-0.08563	0.	0.	-0.00934	0.01424	0.02205	0.00632
* 6A1	C1	0.00621	0.12595	-0.36593	-0.01578	0.	0.	0.	0.	0.24468	-0.01563
-0.9614	O1	-0.00257	-0.05624	0.12900	0.10114	0.	0.	0.	0.	0.07077	0.04951
	F1	-0.00323	-0.06877	0.16549	0.19261	0.	0.	0.35620	0.12786	-0.12725	-0.04373
	F2	-0.00323	-0.06877	0.16549	0.19261	0.	0.	-0.35620	-0.12786	-0.12725	-0.04373
* 7A1	C1	-0.00249	-0.04844	0.16230	0.01050	0.	0.	0.	0.	0.22783	0.00435
-0.8134	O1	0.00417	0.09162	-0.20745	-0.32258	0.	0.	0.	0.	-0.28289	-0.06465
	F1	-0.00004	-0.00115	-0.00020	0.02240	0.	0.	0.21695	0.08776	0.35814	0.16762
	F2	-0.00004	-0.00115	-0.00020	0.02240	0.	0.	-0.21695	-0.08776	0.35814	0.16762
* 8A1	C1	-0.00220	-0.04220	0.14477	0.01676	0.	0.	0.	0.	0.33123	-0.07991
-0.7070	O1	0.00321	0.07078	-0.15773	-0.26761	0.	0.	0.	0.	-0.49109	-0.11618
	F1	-0.00017	-0.00333	0.01051	0.00636	0.	0.	-0.03522	-0.03463	-0.34945	-0.15572
	F2	-0.00017	-0.00333	0.01051	0.00636	0.	0.	0.03522	0.03463	-0.34945	-0.15572
9A1	C1	-0.00455	-0.11857	0.12946	2.23289	0.	0.	0.	0.	-0.25594	-0.11285
0.2125	O1	0.00237	0.05732	-0.08808	-0.71216	0.	0.	0.	0.	0.16443	0.37230
	F1	0.00296	0.06905	-0.11567	-0.70651	0.	0.	0.24454	0.39426	-0.11324	-0.21427
	F2	0.00296	0.06905	-0.11567	-0.70651	0.	0.	-0.24454	-0.39426	-0.11324	-0.21427
10A1	C1	0.00033	0.00216	-0.05046	0.77696	0.	0.	0.	0.	0.11228	1.96466
0.3570	O1	0.00300	0.08110	-0.05326	-1.51745	0.	0.	0.	0.	-0.02657	0.28307
	F1	-0.00108	-0.02655	0.03326	0.36812	0.	0.	-0.08626	-0.15631	-0.13981	-0.21314
	F2	-0.00108	-0.02655	0.03326	0.36812	0.	0.	0.08626	0.15631	-0.13981	-0.21314
11A1	C1	-0.01020	-0.12638	1.21801	-1.86771	0.	0.	0.	0.	-0.74026	0.41125
0.5486	O1	-0.00092	-0.02502	0.01638	0.59285	0.	0.	0.	0.	0.09419	-0.52445
	F1	0.00081	0.00992	-0.09498	0.35722	0.	0.	0.30582	-0.00790	-0.17428	-0.03457
	F2	0.00081	0.00992	-0.09498	0.35722	0.	0.	-0.30582	0.00790	-0.17428	-0.03457
* 1A2	C1	0.	0.	0.	0.	0.	0.	0.	0.	0.	0.
-0.7408	O1	0.	0.	0.	0.	0.	0.	0.	0.	0.	0.
	F1	0.	0.	0.	0.	-0.54476	-0.26537	0.	0.	0.	0.
	F2	0.	0.	0.	0.	0.54476	0.26537	0.	0.	0.	0.

F2CO

Center		S1	S2	S3	S4	X1	X2	Y1	Y2	Z1	Z2
* 1B1	C1	0.	0.	0.	0.	0.33667	0.04417	0.	0.	0.	0.
-0.8370	O1	0.	0.	0.	0.	0.17883	0.04855	0.	0.	0.	0.
	F1	0.	0.	0.	0.	0.44253	0.18481	0.	0.	0.	0.
	F2	0.	0.	0.	0.	0.44253	0.18481	0.	0.	0.	0.
* 2B1	C1	0.	0.	0.	0.	-0.34486	-0.05025	0.	0.	0.	0.
-0.5896	O1	0.	0.	0.	0.	-0.56679	-0.27794	0.	0.	0.	0.
	F1	0.	0.	0.	0.	0.27015	0.16013	0.	0.	0.	0.
	F2	0.	0.	0.	0.	0.27015	0.16013	0.	0.	0.	0.
3B1	C1	0.	0.	0.	0.	0.73239	0.58460	0.	0.	0.	0.
0.1175	O1	0.	0.	0.	0.	-0.44542	-0.49731	0.	0.	0.	0.
	F1	0.	0.	0.	0.	-0.22107	-0.19279	0.	0.	0.	0.
	F2	0.	0.	0.	0.	-0.22107	-0.19279	0.	0.	0.	0.
4B1	C1	0.	0.	0.	0.	-0.80515	1.40739	0.	0.	0.	0.
0.3663	O1	0.	0.	0.	0.	-0.01108	-0.19960	0.	0.	0.	0.
	F1	0.	0.	0.	0.	-0.04712	-0.18092	0.	0.	0.	0.
	F2	0.	0.	0.	0.	-0.04712	-0.18092	0.	0.	0.	0.
* 1B2	C1	0.	0.	0.	0.	0.	0.	0.00017	0.00066	0.	0.
-26.3962	O1	0.	0.	0.	0.	0.	0.	0.00004	-0.00031	0.	0.
	F1	-0.03658	-0.69220	-0.00308	-0.00112	0.	0.	0.00079	-0.00029	-0.00046	0.00002
	F2	0.03658	0.69220	0.00308	0.00112	0.	0.	0.00079	-0.00029	0.00046	-0.00002
* 2B2	C1	0.	0.	0.	0.	0.	0.	-0.16841	0.01603	0.	0.
-1.7088	O1	0.	0.	0.	0.	0.	0.	-0.01553	0.00668	0.	0.
	F1	0.00808	0.17330	-0.38998	-0.32915	0.	0.	0.05602	0.01183	-0.04617	-0.00415
	F2	-0.00808	-0.17330	0.38998	0.32915	0.	0.	0.05602	0.01183	0.04617	0.00415
* 3B2	C1	0.	0.	0.	0.	0.	0.	0.43317	-0.03081	0.	0.
-0.8790	O1	0.	0.	0.	0.	0.	0.	0.21445	0.06938	0.	0.
	F1	0.00216	0.04612	-0.10878	-0.16109	0.	0.	-0.20064	-0.05472	0.35815	0.12651
	F2	-0.00216	-0.04612	0.10878	0.16109	0.	0.	-0.20064	-0.05472	-0.35815	-0.12651
* 4B2	C1	0.	0.	0.	0.	0.	0.	-0.11484	0.04421	0.	0.
-0.7196	O1	0.	0.	0.	0.	0.	0.	0.00668	0.00769	0.	0.
	F1	-0.00039	-0.00750	0.02557	0.00304	0.	0.	0.46653	0.21213	0.30542	0.14497
	F2	0.00039	0.00750	-0.02557	-0.00304	0.	0.	0.46653	0.21213	-0.30542	-0.14497
* 5B2	C1	0.	0.	0.	0.	0.	0.	-0.04705	-0.06989	0.	0.
-0.5731	O1	0.	0.	0.	0.	0.	0.	-0.70123	-0.32344	0.	0.
	F1	-0.00020	-0.00563	0.00033	0.07404	0.	0.	-0.11546	-0.06913	0.21647	0.11528
	F2	0.00020	0.00563	-0.00033	-0.07404	0.	0.	-0.11546	-0.06913	-0.21647	-0.11528
6B2	C1	0.	0.	0.	0.	0.	0.	-0.29514	-1.78490	0.	0.
0.3512	O1	0.	0.	0.	0.	0.	0.	0.27249	0.53952	0.	0.
	F1	-0.00250	-0.06017	0.08415	0.78764	0.	0.	0.01369	-0.06414	0.11584	0.24678
	F2	0.00250	0.06017	-0.08415	-0.78764	0.	0.	0.01369	-0.06414	-0.11584	-0.24678
7B2	C1	0.	0.	0.	0.	0.	0.	1.27336	-0.74654	0.	0.
0.5251	O1	0.	0.	0.	0.	0.	0.	-0.09991	-0.08168	0.	0.
	F1	0.00170	0.03139	-0.12598	0.00723	0.	0.	0.25426	0.27740	-0.20664	-0.14596
	F2	-0.00170	-0.03139	0.12598	-0.00723	0.	0.	0.25426	0.27740	0.20664	0.14596

Overlap Matrix

C1S1 with
```
    C1S2  .35436   C1S3  .00805   C1S4  .01419   O1S3  .00063   O1S4  .00579   O1Z1 -.00329   O1Z2 -.01343
    F1S3  .00004   F1S4  .00298   F1Y1 -.00042   F1Y2 -.00890   F1Z1  .00028   F1Z2  .00595
```

C1S2 with
```
    C1S3  .26226   C1S4  .17745   O1S2  .00003   O1S3  .01418   O1S4  .07547   O1Z1 -.04926   O1Z2 -.16431
    F1S3  .00224   F1S4  .04138   F1Y1 -.00975   F1Y2 -.11104   F1Z1  .00651   F1Z2  .07420
```

C1S3 with
```
    C1S4  .79970   O1S1  .00225   O1S2  .03089   O1S3  .21278   O1S4  .40984   O1Z1 -.27730   O1Z2 -.61680
    F1S1  .00101   F1S2  .01400   F1S3  .11263   F1S4  .28585   F1Y1 -.13228   F1Y2 -.44814   F1Z1  .08839
    F1Z2  .29944
```

C1S4 with
```
    O1S1  .00439   O1S2  .05668   O1S3  .31306   O1S4  .57573   O1Z1 -.17282   O1Z2 -.53194   F1S1  .00303
    F1S2  .03898   F1S3  .22386   F1S4  .45282   F1Y1 -.10368   F1Y2 -.39107   F1Z1  .06928   F1Z2  .26130
```

C1X1 with
```
    C1X2  .53875   O1X1  .16166   O1X2  .38380   F1X1  .08157   F1X2  .29632
```

C1X2 with
```
    O1X1  .18190   O1X2  .61607   F1X1  .11937   F1X2  .48134
```

C1Y1 with
```
    C1Y2  .53875   O1Y1  .16166   O1Y2  .38380   F1S1  .00257   F1S2  .03336   F1S3  .19847   F1S4  .32353
    F1Y1 -.13286   F1Y2 -.13330   F1Z1  .14328   F1Z2  .28706
```

C1Y2 with
```
    O1Y1  .18190   O1Y2  .61607   F1S1  .00433   F1S2  .05491   F1S3  .29041   F1S4  .49210   F1Y1  .01316
    F1Y2  .14709   F1Z1  .07097   F1Z2  .22334
```

C1Z1 with
```
    C1Z2  .53875   O1S1  .00544   O1S2  .07062   O1S3  .37028   O1S4  .43261   O1Z1 -.30864   O1Z2 -.15594
    F1S1 -.00172   F1S2 -.02229   F1S3 -.13261   F1S4 -.21618   F1Y1  .14328   F1Y2  .28706   F1Z1 -.01417
    F1Z2  .10451
```

C1Z2 with
```
    O1S1  .00639   O1S2  .08131   O1S3  .40842   O1S4  .61676   O1Z1  .00302   O1Z2  .16643   F1S1 -.00289
    F1S2 -.03669   F1S3 -.19404   F1S4 -.32881   F1Y1  .07097   F1Y2  .22334   F1Z1  .07195   F1Z2  .33211
```

O1S1 with
```
    O1S2  .35461   O1S3  .00974   O1S4  .01469   F1S4  .00004   F1Y2 -.00028   F1Z2  .00049
```

O1S2 with
```
    O1S3  .28286   O1S4  .18230   F1S4  .00060   F1Y2 -.00396   F1Z2  .00688
```

O1S3 with
```
    O1S4  .79029   F1S3  .00013   F1S4  .01029   F1Y1 -.00042   F1Y2 -.03482   F1Z1  .00072   F1Z2  .06040
```

O1S4 with
```
    F1S1  .00010   F1S2  .00136   F1S3  .01406   F1S4  .06518   F1Y1 -.01186   F1Y2 -.10046   F1Z1  .02057
    F1Z2  .17428
```

O1X1 with
```
    O1X2  .50607   F1X1  .00035   F1X2  .01748
```

O1X2 with
```
    F1X1  .01883   F1X2  .12411
```

O1Y1 with
```
    O1Y2  .50607   F1S2  .00001   F1S3  .00061   F1S4  .01193   F1Y1 -.00090   F1Y2 -.01302   F1Z1  .00216
    F1Z2  .05292
```

O1Y2 with
```
    F1S1  .00048   F1S2  .00643   F1S3  .04465   F1S4  .11386   F1Y1 -.00990   F1Y2 -.00391   F1Z1  .04983
    F1Z2  .22207
```

O1Z1 with
```
    O1Z2  .50607   F1S2 -.00001   F1S3 -.00105   F1S4 -.02069   F1Y1  .00216   F1Y2  .05292   F1Z1 -.00340
    F1Z2 -.07432
```

Overlap Matrix (continued)

First Function   Second Function and Overlap

01Z2 with
    F1S1 -.00084   F1S2 -.01115   F1S3 -.07746   F1S4 -.19752   F1Y1 .04983   F1Y2 .22207   F1Z1 -.06761
    F1Z2 -.26113

F1S1 with
    F1S2 .35647   F1S3 .01003   F1S4 .01474   F2S4 .00003   F2Y2 .00048

F1S2 with
    F1S3 .28653   F1S4 .18096   F2S4 .00049   F2Y2 .00656

F1S3 with
    F1S4 .78531   F2S3 .00004   F2S4 .00708   F2Y1 .00034   F2Y2 .05455

F1S4 with
    F2S1 .00003   F2S2 .00049   F2S3 .00708   F2S4 .04624   F2Y1 .01483   F2Y2 .16893

F1X1 with
    F1X2 .49444   F2X1 .00012   F2X2 .01170

F1X2 with
    F2X1 .01170   F2X2 .09533

F1Y1 with
    F1Y2 .49444   F2S3 -.00034   F2S4 -.01483   F2Y1 -.00186   F2Y2 -.07523

F1Y2 with
    F2S1 -.00048   F2S2 -.00656   F2S3 -.05455   F2S4 -.16893   F2Y1 -.07523   F2Y2 -.35280

F1Z1 with
    F1Z2 .49444   F2Z1 .00012   F2Z2 .01170

F1Z2 with
    F2Z1 .01170   F2Z2 .09533

N2F2   Difluorodiazine

Molecular Geometry   Symmetry C$_{2h}$

Center		Coordinates	
	X	Y	Z
N1	-1.16260000	0.	0.
N2	1.16260000	0.	0.
F1	-1.86770000	2.54249999	0.
F2	1.86770000	-2.54249999	0.
Mass	0.	0.	0.

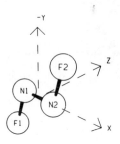

MO Symmetry Types and Occupancy

Representation	AG	AU	BG	BU
Number of MOs	16	4	4	16
Number Occupied	7	2	1	6

Energy Expectation Values

Total	-307.5104	Kinetic	307.4883
Electronic	-421.0300	Potential	-614.9988
Nuclear Repulsion	113.5196	Virial	2.0001

Electronic Potential	-728.5183
One-Electron Potential	-957.8421
Two-Electron Potential	229.3238

Moments of the Charge Distribution

Second

x^2	y^2	z^2	xy	xz	yz	r^2
-98.8316	-134.1929	-14.5336	86.3739	0.	0.	-247.5581
81.7124	116.3575	0.	-85.4753	0.	0.	198.0699

Fourth (even)

x^4	y^4	z^4	x^2y^2	x^2z^2	y^2z^2	r^2x^2	r^2y^2	r^2z^2	r^4
-505.8263	-1083.8004	-36.4138	-512.9155	-47.9243	-59.3082	-1066.6661	-1656.0241	-143.6463	-2866.3365
244.6057	752.1706	0.	405.8903	0.	0.	650.4959	1158.0609	0.	1808.5568

Fourth (odd)

x^3y	xy^3	xyz^2	x^3z	xy^2z	xz^3	x^2yz	y^3z	yz^3
404.0288	666.5886	33.5705	0.	0.	0.	0.	0.	0.
-298.1637	-552.5384	0.	0.	0.	0.	0.	0.	0.

Expectation Values at Atomic Centers

Center			Expectation Value		
	r^{-1}	δ	xr^{-3}	yr^{-3}	zr^{-3}
N1	-18.1323	189.7967	-0.1895	-0.1630	0.
F1	-26.4972	413.9385	-0.0796	0.2121	0.

Center	$\dfrac{3x^2-r^2}{r^5}$	$\dfrac{3y^2-r^2}{r^5}$	$\dfrac{3z^2-r^2}{r^5}$	$\dfrac{3xy}{r^5}$	$\dfrac{3xz}{r^5}$	$\dfrac{3yz}{r^5}$
N1	-0.6161	0.3479	0.2682	-1.0644	0.	0.
F1	-1.9221	4.0692	-2.1471	-2.0163	0.	0.

Mulliken Charge

Center				Basis Function Type							Total
	S1	S2	S3	S4	X1	X2	Y1	Y2	Z1	Z2	
N1	0.0435	1.9559	1.0070	0.7592	1.0310	0.0214	0.8027	0.0897	0.7845	0.2521	6.7470
F1	0.0436	1.9558	1.0400	0.9456	1.4336	0.4990	0.9981	0.3739	1.4491	0.5142	9.2530

Overlap Populations

	Pair of Atomic Centers			
	N2	F1	F1	F2
	N1	N1	N2	F1
1AG	0.	-0.0001	0.	0.
2AG	0.0006	-0.0001	0.	0.
3AG	0.1239	0.1324	0.0295	0.0009
4AG	0.5453	-0.1245	-0.0311	0.0001
5AG	0.0363	0.1315	-0.0081	0.
6AG	0.0762	-0.0041	-0.0552	0.0076
7AG	-0.1693	-0.1507	0.0023	0.0017
Total AG	0.6131	-0.0157	-0.0626	0.0104
1AU	0.1091	0.1050	0.0251	0.0007
2AU	0.3368	-0.1758	-0.0453	0.0007
Total AU	0.4459	-0.0708	-0.0202	0.0013
1BG	-0.0117	0.0451	-0.0113	-0.0012
Total BG	-0.0117	0.0451	-0.0113	-0.0012
1BU	0.	-0.0001	0.	0.
2BU	-0.0039	-0.0002	0.	0.
3BU	-0.0209	0.1175	-0.0227	-0.0011
4BU	-0.5076	0.1030	-0.0135	-0.0010
5BU	0.1088	0.0629	0.0485	-0.0029
6BU	-0.1175	-0.0935	-0.0533	0.0009
Total BU	-0.5410	0.1894	-0.0409	-0.0042
Total	0.5061	0.1480	-0.1350	0.0064
Distance	2.3252	2.6385	3.9556	6.3096

Molecular Orbitals and Eigenvalues

Center		Basis Function Type									
		S1	S2	S3	S4	X1	X2	Y1	Y2	Z1	Z2
* 1AG	N1	0.00000	-0.00002	-0.00014	0.00031	-0.00002	-0.00008	0.00001	0.00039	0.	0.
-26.3824	N2	0.00000	-0.00002	-0.00014	0.00031	0.00002	0.00008	-0.00001	-0.00039	0.	0.
	F1	-0.03659	-0.69223	-0.00302	-0.00101	-0.00028	0.00013	0.00072	-0.00009	0.	0.
	F2	-0.03659	-0.69223	-0.00302	-0.00101	0.00028	-0.00013	-0.00072	0.00009	0.	0.
* 2AG	N1	0.03673	0.69228	0.00229	0.00139	0.00158	-0.00074	0.00150	-0.00045	0.	0.
-15.8121	N2	0.03673	0.69228	0.00229	0.00139	-0.00158	0.00074	-0.00150	0.00045	0.	0.
	F1	-0.00000	-0.00003	0.00030	-0.00062	0.00004	-0.00014	-0.00015	0.00038	0.	0.
	F2	-0.00000	-0.00003	0.00030	-0.00062	-0.00004	0.00014	0.00015	-0.00038	0.	0.
* 3AG	N1	-0.00426	-0.09021	0.20833	0.07007	0.07114	0.00004	0.06694	-0.01085	0.	0.
-1.7405	N2	-0.00426	-0.09021	0.20833	0.07007	-0.07114	-0.00004	-0.06694	0.01085	0.	0.
	F1	-0.00672	-0.14433	0.32237	0.27965	0.02845	0.00938	-0.06561	-0.02484	0.	0.
	F2	-0.00672	-0.14433	0.32237	0.27965	-0.02845	-0.00938	0.06561	0.02484	0.	0.
* 4AG	N1	0.00635	0.13360	-0.32400	-0.15722	-0.23318	0.01031	0.01858	-0.01204	0.	0.
-1.4295	N2	0.00635	0.13360	-0.32400	-0.15722	0.23318	-0.01031	-0.01858	0.01204	0.	0.
	F1	-0.00475	-0.10217	0.22945	0.24421	-0.03145	-0.00848	0.03845	0.00113	0.	0.
	F2	-0.00475	-0.10217	0.22945	0.24421	0.03145	0.00848	-0.03845	-0.00113	0.	0.
* 5AG	N1	-0.00151	-0.03214	0.07949	0.08431	-0.32454	-0.03968	0.08679	-0.00300	0.	0.
-0.8139	N2	-0.00151	-0.03214	0.07949	0.08431	0.32454	0.03968	-0.08679	0.00300	0.	0.
	F1	0.00140	0.02937	-0.07476	-0.08057	-0.21132	-0.09144	-0.33773	-0.14936	0.	0.
	F2	0.00140	0.02937	-0.07476	-0.08057	0.21132	0.09144	0.33773	0.14936	0.	0.
* 6AG	N1	-0.00071	-0.01426	0.04372	0.05028	0.17051	-0.00688	-0.13952	0.01759	0.	0.
-0.6995	N2	-0.00071	-0.01426	0.04372	0.05028	-0.17051	0.00688	0.13952	-0.01759	0.	0.
	F1	-0.00079	-0.01571	0.04979	0.01205	-0.49690	-0.23054	0.14225	0.08499	0.	0.
	F2	-0.00079	-0.01571	0.04979	0.01205	0.49690	0.23054	-0.14225	-0.08499	0.	0.
* 7AG	N1	0.00256	0.05783	-0.11043	-0.30808	0.37095	0.10421	0.29623	0.10262	0.	0.
-0.5559	N2	0.00256	0.05783	-0.11043	-0.30808	-0.37095	-0.10421	-0.29623	-0.10262	0.	0.
	F1	-0.00019	-0.00424	0.00825	0.02440	-0.07699	-0.03974	-0.26157	-0.15111	0.	0.
	F2	-0.00019	-0.00424	0.00825	0.02440	0.07699	0.03974	0.26157	0.15111	0.	0.
8AG	N1	0.00243	0.05050	-0.14481	-0.39466	0.13811	0.18079	-0.49352	-0.87443	0.	0.
0.2973	N2	0.00243	0.05050	-0.14481	-0.39466	-0.13811	-0.18079	0.49352	0.87443	0.	0.
	F1	-0.00205	-0.04831	0.07571	0.52707	-0.01801	0.05589	-0.29560	-0.40124	0.	0.
	F2	-0.00205	-0.04831	0.07571	0.52707	0.01801	-0.05589	0.29560	0.40124	0.	0.
9AG	N1	0.00084	0.02638	0.00869	0.24001	0.45729	-1.04215	0.21615	-0.53392	0.	0.
0.6218	N2	0.00084	0.02638	0.00869	0.24001	-0.45729	1.04215	-0.21615	0.53392	0.	0.
	F1	0.00005	-0.00199	-0.02483	0.15338	0.06890	0.17759	0.02908	-0.02182	0.	0.
	F2	0.00005	-0.00199	-0.02483	0.15338	-0.06890	-0.17759	-0.02908	0.02182	0.	0.
10AG	N1	-0.00042	-0.00148	0.07126	-0.43414	-0.16874	0.48042	0.57426	-1.51514	0.	0.
0.6952	N2	-0.00042	-0.00148	0.07126	-0.43414	0.16874	-0.48042	-0.57426	1.51514	0.	0.
	F1	0.00046	-0.00063	-0.09741	0.61737	-0.24440	0.03057	0.13609	-0.20485	0.	0.
	F2	0.00046	-0.00063	-0.09741	0.61737	0.24440	-0.03057	-0.13609	0.20485	0.	0.
* 1AU	N1	0.	0.	0.	0.	0.	0.	0.	0.	-0.27710	-0.08433
-0.7911	N2	0.	0.	0.	0.	0.	0.	0.	0.	-0.27710	-0.08433
	F1	0.	0.	0.	0.	0.	0.	0.	0.	-0.43026	-0.18609
	F2	0.	0.	0.	0.	0.	0.	0.	0.	-0.43026	-0.18609
* 2AU	N1	0.	0.	0.	0.	0.	0.	0.	0.	0.44457	0.16819
-0.5694	N2	0.	0.	0.	0.	0.	0.	0.	0.	0.44457	0.16819
	F1	0.	0.	0.	0.	0.	0.	0.	0.	-0.34669	-0.18999
	F2	0.	0.	0.	0.	0.	0.	0.	0.	-0.34669	-0.18999

Center		S1	S2	S3	S4	X1	X2	Y1	Y2	Z1	Z2
3AU	N1	0.	0.	0.	0.	0.	0.	0.	0.	-0.62057	0.67868
0.5940	N2	0.	0.	0.	0.	0.	0.	0.	0.	-0.62057	0.67868
	F1	0.	0.	0.	0.	0.	0.	0.	0.	-0.11069	-0.06138
	F2	0.	0.	0.	0.	0.	0.	0.	0.	-0.11069	-0.06138
* 1BG	N1	0.	0.	0.	0.	0.	0.	0.	0.	-0.08417	-0.03087
-0.7418	N2	0.	0.	0.	0.	0.	0.	0.	0.	0.08417	0.03087
	F1	0.	0.	0.	0.	0.	0.	0.	0.	-0.53249	-0.24794
	F2	0.	0.	0.	0.	0.	0.	0.	0.	0.53249	0.24794
2BG	N1	0.	0.	0.	0.	0.	0.	0.	0.	-0.55594	-0.52720
0.0381	N2	0.	0.	0.	0.	0.	0.	0.	0.	0.55594	0.52720
	F1	0.	0.	0.	0.	0.	0.	0.	0.	0.13668	0.12064
	F2	0.	0.	0.	0.	0.	0.	0.	0.	-0.13668	-0.12064
* 1BU	N1	-0.00000	0.00002	0.00017	-0.00081	-0.00004	-0.00014	0.00001	-0.00029	0.	0.
-26.3824	N2	0.00000	-0.00002	-0.00017	0.00081	-0.00004	-0.00014	0.00001	-0.00029	0.	0.
	F1	0.03659	0.69223	0.00296	0.00115	0.00025	0.00002	-0.00071	0.00005	0.	0.
	F2	-0.03659	-0.69223	-0.00296	-0.00115	0.00025	0.00002	-0.00071	0.00005	0.	0.
* 2BU	N1	0.03674	0.69250	0.00118	0.00621	0.00140	0.00168	0.00134	-0.00001	0.	0.
-15.8105	N2	-0.03674	-0.69250	-0.00118	-0.00621	0.00140	0.00168	0.00134	-0.00001	0.	0.
	F1	-0.00000	-0.00004	0.00026	-0.00089	0.00003	-0.00024	-0.00012	0.00056	0.	0.
	F2	0.00000	0.00004	-0.00026	0.00089	0.00003	-0.00024	-0.00012	0.00056	0.	0.
* 3BU	N1	0.00172	0.03710	-0.08173	-0.05373	0.02751	-0.01684	-0.08846	0.00771	0.	0.
-1.6933	N2	-0.00172	-0.03710	0.08173	0.05373	0.02751	-0.01684	-0.08846	0.00771	0.	0.
	F1	0.00782	0.16827	-0.37465	-0.33982	-0.02253	-0.01162	0.06084	0.02451	0.	0.
	F2	-0.00782	-0.16827	0.37465	0.33982	-0.02253	-0.01162	0.06084	0.02451	0.	0.
* 4BU	N1	-0.00605	-0.12843	0.31538	0.33723	-0.14152	0.03460	0.08534	-0.02030	0.	0.
-0.9606	N2	0.00605	0.12843	-0.31538	-0.33723	-0.14152	0.03460	0.08534	-0.02030	0.	0.
	F1	0.00276	0.05806	-0.14590	-0.14444	0.01886	0.01046	-0.28564	-0.11557	0.	0.
	F2	-0.00276	-0.05806	0.14590	0.14444	0.01886	0.01046	-0.28564	-0.11557	0.	0.
* 5BU	N1	-0.00197	-0.04110	0.10616	0.08311	-0.09598	-0.04999	-0.28460	-0.05819	0.	0.
-0.8071	N2	0.00197	0.04110	-0.10616	-0.08311	-0.09598	-0.04999	-0.28460	-0.05819	0.	0.
	F1	-0.00108	-0.02279	0.05756	0.06471	-0.38289	-0.17380	0.09016	0.04218	0.	0.
	F2	0.00108	0.02279	-0.05756	-0.06471	-0.38289	-0.17380	0.09016	0.04218	0.	0.
* 6BU	N1	0.00350	0.07406	-0.18863	-0.25601	0.07854	0.01785	0.25505	0.05379	0.	0.
-0.6635	N2	-0.00350	-0.07406	0.18863	0.25601	0.07854	0.01785	0.25505	0.05379	0.	0.
	F1	0.00029	0.00587	-0.01631	-0.01151	-0.36788	-0.16949	-0.25875	-0.12942	0.	0.
	F2	-0.00029	-0.00587	0.01631	0.01151	-0.36788	-0.16949	-0.25875	-0.12942	0.	0.
7BU	N1	-0.00290	-0.06217	0.15565	0.39470	-0.10080	-0.14195	0.38952	0.37357	0.	0.
0.1264	N2	0.00290	0.06217	-0.15565	-0.39470	-0.10080	-0.14195	0.38952	0.37357	0.	0.
	F1	0.00261	0.05840	-0.11915	-0.44110	-0.14742	-0.12623	0.36693	0.38553	0.	0.
	F2	-0.00261	-0.05840	0.11915	0.44110	-0.14742	-0.12623	0.36693	0.38553	0.	0.
8BU	N1	-0.00407	-0.10068	0.13732	2.52980	0.17743	1.85846	0.02548	0.03458	0.	0.
0.3782	N2	0.00407	0.10068	-0.13732	-2.52980	0.17743	1.85846	0.02548	0.03458	0.	0.
	F1	-0.00028	-0.00562	0.01811	0.03979	-0.10054	-0.16430	0.04189	0.04268	0.	0.
	F2	0.00028	0.00562	-0.01811	-0.03979	-0.10054	-0.16430	0.04189	0.04268	0.	0.
9BU	N1	-0.00088	-0.00292	0.15261	-0.81441	0.15627	-0.25009	0.55508	-0.90454	0.	0.
0.5986	N2	0.00088	0.00292	-0.15261	0.81441	0.15627	-0.25009	0.55508	-0.90454	0.	0.
	F1	-0.00098	-0.02677	0.01243	0.56749	-0.00962	0.19912	0.20680	-0.28705	0.	0.
	F2	0.00098	0.02677	-0.01243	-0.56749	-0.00962	0.19912	0.20680	-0.28705	0.	0.

Overlap Matrix

First Function		Second Function and Overlap					

N1S1 with

N1S2	.35584	N1S3	.00894	N1S4	.01446	N2S3	.00095	N2S4	.00470	N2X1	-.00360	N2X2	-.00942
F1S3	.00001	F1S4	.00181	F1X1	.00006	F1X2	.00195	F1Y1	-.00020	F1Y2	-.00705		

N1S2 with

N1S3	.27246	N1S4	.17925	N2S2	.00001	N2S3	.01548	N2S4	.06064	N2X1	-.04871	N2X2	-.11763
F1S3	.00064	F1S4	.02498	F1X1	.00127	F1X2	.02478	F1Y1	-.00459	F1Y2	.08935		

N1S3 with

N1S4	.79352	N2S1	.00095	N2S2	.01548	N2S3	.17248	N2S4	.33807	N2X1	-.29098	N2X2	-.53358
F1S1	.00022	F1S2	.00360	F1S3	.05063	F1S4	.18750	F1X1	.02639	F1X2	.11886	F1Y1	-.09517
F1Y2	-.42860												

N1S4 with

N2S1	.00470	N2S2	.06064	N2S3	.33807	N2S4	.57006	N2X1	-.28061	N2X2	-.63391	F1S1	.00231
F1S2	.02993	F1S3	.17906	F1S4	.37931	F1X1	.03785	F1X2	.13806	F1Y1	-.13647	F1Y2	-.49784

N1X1 with

N1X2	.52739	N2S1	.00360	N2S2	.04871	N2S3	.29098	N2S4	.28061	N2X1	-.32974	N2X2	-.10690
F1S1	-.00032	F1S2	-.00445	F1S3	-.03432	F1S4	-.06896	F1X1	.02640	F1X2	.15441	F1Y1	.05677
F1Y2	.12970												

N1X2 with

N2S1	.00942	N2S2	.11763	N2S3	.53358	N2S4	.63391	N2X1	-.10690	N2X2	.06774	F1S1	-.00150
F1S2	-.01901	F1S3	-.09895	F1S4	-.16008	F1X1	.10235	F1X2	.40521	F1Y1	.06129	F1Y2	.16677

N1Y1 with

N1Y2	.52739	N2Y1	.13765	N2Y2	.26010	F1S1	.00116	F1S2	.01604	F1S3	.12375	F1S4	.24866
F1X1	.05677	F1X2	.12970	F1Y1	-.16257	F1Y2	-.27729						

N1Y2 with

N2Y1	.26010	N2Y2	.63952	F1S1	.00541	F1S2	.06855	F1S3	.35680	F1S4	.57722	F1X1	.06129
F1X2	.16677	F1Y1	-.10165	F1Y2	-.14989								

N1Z1 with

N1Z2	.52739	N2Z1	.13765	N2Z2	.26010	F1Z1	.04214	F1Z2	.19037

N1Z2 with

N2Z1	.26010	N2Z2	.63952	F1Z1	.11935	F1Z2	.45146

N2S1 with

F1S4	.00008	F1X2	.00078	F1Y2	-.00066

N2S2 with

F1S4	.00136	F1X1	.00001	F1X2	.01109	F1Y1	-.00001	F1Y2	-.00931

N2S3 with

F1S2	.00002	F1S3	.00119	F1S4	.02484	F1X1	.00343	F1X2	.09541	F1Y1	-.00288	F1Y2	-.08005

N2S4 with

F1S1	.00038	F1S2	.00512	F1S3	.03865	F1S4	.12111	F1X1	.03594	F1X2	.21277	F1Y1	-.03016
F1Y2	-.17852												

N2X1 with

F1S1	-.00001	F1S2	-.00024	F1S3	-.00573	F1S4	-.04244	F1X1	-.01015	F1X2	-.09013	F1Y1	.01010
F1Y2	.10676												

N2X2 with

F1S1	-.00153	F1S2	-.01992	F1S3	-.12088	F1S4	-.25769	F1X1	-.05597	F1X2	-.16460	F1Y1	.07586
F1Y2	.29290												

N2Y1 with

F1S1	.00001	F1S2	.00021	F1S3	.00481	F1S4	.03561	F1X1	.01010	F1X2	.10676	F1Y1	-.00659
F1Y2	-.05246												

N2Y2 with

F1S1	.00129	F1S2	.01671	F1S3	.10142	F1S4	.21621	F1X1	.07586	F1X2	.29290	F1Y1	-.02920
F1Y2	-.06125												

N2Z1 with

F1Z1	.00189	F1Z2	.03712

Overlap Matrix (continued)

First Function   Second Function and Overlap

N2Z2 with
    F1Z1 .03445   F1Z2 .18450

F1S1 with
    F1S2 .35647   F1S3 .01003   F1S4 .01474

F1S2 with
    F1S3 .28653   F1S4 .18096   F2X2 -.00001   F2Y2 .00002

F1S3 with
    F1S4 .78531   F2S4 .00001   F2X2 -.00033   F2Y2 .00044

F1S4 with
    F2S3 .00001   F2S4 .00081   F2X1 -.00004   F2X2 -.00459   F2Y1 .00005   F2Y2 .00625

F1X1 with
    F1X2 .49444   F2S4 .00004   F2X2 -.00045   F2Y2 .00074

F1X2 with
    F2S2 .00001   F2S3 .00033   F2S4 .00459   F2X1 -.00045   F2X2 -.01221   F2Y1 .00074   F2Y2 .02253

F1Y1 with
    F1Y2 .49444   F2S4 -.00005   F2X2 .00074   F2Y2 -.00092

F1Y2 with
    F2S2 -.00002   F2S3 -.00044   F2S4 -.00625   F2X1 .00074   F2X2 .02253   F2Y1 -.00092   F2Y2 -.02633

F1Z1 with
    F1Z2 .49444   F2Z2 .00009

F1Z2 with
    F2Z1 .00009   F2Z2 .00434

C3O2   Carbon Suboxide

Molecular Geometry   Symmetry D-h

Center		Coordinates	
	X	Y	Z
C1	0.	0.	0.
C2	0.	0.	-2.41888240
C3	0.	0.	2.41888240
O1	0.	0.	-4.61099458
O2	0.	0.	4.61099458
Mass	0.	0.	0.

MO Symmetry Types and Occupancy

Representation	SIGM-G	SIGM-U	PI-G-X	PI-G-Y	PI-U-X	PI-U-Y
Number of MOs	16	14	4	4	6	6
Number Occupied	6	5	1	1	2	2

Energy Expectation Values

Total	-263.1575	Kinetic	263.0413
Electronic	-385.5738	Potential	-526.1987
Nuclear Repulsion	122.4164	Virial	2.0004

Electronic Potential	-648.6151
One-Electron Potential	-862.7476
Two-Electron Potential	214.1325

Moments of the Charge Distribution

Second

x^2	y^2	z^2	xy	xz	yz	r^2
-19.7049	-19.7049	-439.4842	0.	0.	0.	-478.8941
0.	0.	410.3922	0.	0.	0.	410.3922

Fourth (even)

x^4	y^4	z^4	x^2y^2	x^2z^2	y^2z^2	r^2x^2	r^2y^2	r^2z^2	r^4
-66.9761	-66.9761	-9342.0328	-22.3254	-230.8778	-230.8778	-320.1792	-320.1792	-9803.7883	-10444.1469
0.	0.	7643.4756	0.	0.	0.	0.	0.	7643.4756	7643.4756

Expectation Values at Atomic Centers

Center	r^{-1}	δ	Expectation Value xr^{-3}	yr^{-3}	zr^{-3}
C1	-14.6713	116.0847	0.	0.	0.
C2	-14.5060	116.7376	0.	0.	0.0542
01	-22.2382	287.3707	0.	0.	-0.2897

	$\dfrac{3x^2-r^2}{r^5}$	$\dfrac{3y^2-r^2}{r^5}$	$\dfrac{3z^2-r^2}{r^5}$	$\dfrac{3xy}{r^5}$	$\dfrac{3xz}{r^5}$	$\dfrac{3yz}{r^5}$
C1	-0.2696	-0.2696	0.5393	0.	0.	0.
C2	0.2030	0.2030	-0.4060	0.	0.	0.
01	-0.4324	-0.4324	0.8648	0.	0.	0.

Mulliken Charge

Center	S1	S2	S3	S4	X1	X2	Y1	Y2	Z1	Z2	Total
C1	0.0427	1.9563	0.6694	0.2012	0.9752	0.3927	0.9752	0.3927	0.8627	0.2428	6.7109
C2	0.0429	1.9566	0.8601	-0.0202	0.6525	0.0780	0.6525	0.0780	1.0069	0.0748	5.3821
01	0.0427	1.9566	0.9772	0.8435	1.1499	0.4357	1.1499	0.4357	1.0880	0.1833	8.2624

Overlap Populations

	Pair of Atomic Centers C2 C1	C3 C2	01 C1	01 C2	01 C3	02 01
1SIGM-G	0.	0.	-0.0001	-0.0004	0.	0.
2SIGM-G	-0.0007	0.0003	0.	-0.0001	0.	0.
3SIGM-G	-0.0010	0.	0.	0.	0.	0.
4SIGM-G	-0.0002	0.0020	0.0148	0.3074	-0.0005	0.
5SIGM-G	0.3815	0.0900	0.0033	0.0218	0.0006	0.
6SIGM-G	-0.0698	0.0213	-0.0018	-0.0077	0.0007	0.
Total SIGM-G	0.3098	0.1136	0.0163	0.3210	0.0008	0.
1SIGM-U	0.	0.	0.0003	-0.0011	0.	0.
2SIGM-U	-0.0013	-0.0005	0.	-0.0003	0.	0.
3SIGM-U	0.0084	-0.0125	0.0892	0.2074	0.0005	0.
4SIGM-U	0.3803	0.0006	0.0132	0.0320	-0.0001	0.
5SIGM-U	-0.0158	-0.0305	0.0098	0.0055	-0.0005	0.
Total SIGM-U	0.3714	-0.0428	0.1125	0.2435	-0.0001	0.
1PI-G-X	0.	-0.0101	0.	0.1966	-0.0010	0.
Total PI-G-X	0.	-0.0101	0.	0.1966	-0.0010	0.
1PI-G-Y	0.	-0.0101	0.	0.1966	-0.0010	0.
Total PI-G-Y	0.	-0.0101	0.	0.1966	-0.0010	0.
1PI-U-X	0.0910	0.0150	0.0093	0.1796	0.0010	0.
2PI-U-X	0.1141	-0.0003	-0.0769	-0.0721	0.0002	0.
Total PI-U-X	0.2051	0.0148	-0.0676	0.1075	0.0012	0.
1PI-U-Y	0.0910	0.0150	0.0093	0.1796	0.0010	0.
2PI-U-Y	0.1141	-0.0003	-0.0769	-0.0721	0.0002	0.
Total PI-U-Y	0.2051	0.0148	-0.0676	0.1075	0.0012	0.
Total	1.0915	0.0802	-0.0063	1.1728	0.0010	0.0000
Distance	2.4189	4.8378	4.6110	2.1921	7.0299	9.2220

Molecular Orbitals and Eigenvalues

Center		S1	S2	S3	S4	X1	X2	Y1	Y2	Z1	Z2
* 1SIGM-G	C1	0.00000	-0.00002	-0.00014	0.00343	0.	0.	0.	0.	0.	0.
-20.6653	C2	-0.00000	0.00003	0.00022	-0.00063	0.	0.	0.	0.	-0.00036	-0.00167
	C3	-0.00000	0.00003	0.00022	-0.00063	0.	0.	0.	0.	0.00036	0.00167
	O1	-0.03615	-0.69251	-0.00229	-0.00258	0.	0.	0.	0.	-0.00157	0.00032
	O2	-0.03615	-0.69251	-0.00229	-0.00258	0.	0.	0.	0.	0.00157	-0.00032
* 2SIGM-G	C1	0.00040	0.00744	0.00194	-0.00397	0.	0.	0.	0.	0.	0.
-11.4719	C2	0.03670	0.69231	0.00298	0.00212	0.	0.	0.	0.	-0.00060	0.00109
	C3	0.03670	0.69231	0.00298	0.00212	0.	0.	0.	0.	0.00060	-0.00109
	O1	-0.00002	-0.00041	0.00065	-0.00015	0.	0.	0.	0.	0.00100	-0.00048
	O2	-0.00002	-0.00041	0.00065	-0.00015	0.	0.	0.	0.	-0.00100	0.00048
* 3SIGM-G	C1	-0.05189	-0.97874	-0.00470	-0.00513	0.	0.	0.	0.	0.	0.
-11.2877	C2	0.00031	0.00598	-0.00073	0.00263	0.	0.	0.	0.	-0.00104	0.00126
	C3	0.00031	0.00598	-0.00073	0.00263	0.	0.	0.	0.	0.00104	-0.00126
	O1	-0.00000	-0.00000	0.00002	-0.00046	0.	0.	0.	0.	-0.00006	-0.00054
	O2	-0.00000	-0.00000	0.00002	-0.00046	0.	0.	0.	0.	0.00006	0.00054
* 4SIGM-G	C1	0.00074	0.01271	-0.05963	-0.08152	0.	0.	0.	0.	0.	0.
-1.5369	C2	0.00411	0.08627	-0.19937	0.00793	0.	0.	0.	0.	0.17015	0.00464
	C3	0.00411	0.08627	-0.19937	0.00793	0.	0.	0.	0.	-0.17015	-0.00464
	O1	0.00681	0.14883	-0.32976	-0.20934	0.	0.	0.	0.	-0.13922	0.00543
	O2	0.00681	0.14883	-0.32976	-0.20934	0.	0.	0.	0.	0.13922	-0.00543
* 5SIGM-G	C1	-0.00888	-0.19576	0.38551	0.08581	0.	0.	0.	0.	0.	0.
-1.1277	C2	-0.00454	-0.09658	0.22710	0.08869	0.	0.	0.	0.	0.25517	0.03134
	C3	-0.00454	-0.09658	0.22710	0.08869	0.	0.	0.	0.	-0.25517	-0.03134
	O1	0.00164	0.03563	-0.08323	-0.02502	0.	0.	0.	0.	0.00871	0.02356
	O2	0.00164	0.03563	-0.08323	-0.02502	0.	0.	0.	0.	-0.00871	-0.02356
* 6SIGM-G	C1	0.00181	0.04482	-0.04889	-0.18828	0.	0.	0.	0.	0.	0.
-0.7573	C2	-0.00290	-0.05738	0.17887	0.08099	0.	0.	0.	0.	-0.24866	0.06878
	C3	-0.00290	-0.05738	0.17887	0.08099	0.	0.	0.	0.	0.24866	-0.06878
	O1	0.00369	0.08110	-0.18371	-0.29235	0.	0.	0.	0.	0.40566	0.08852
	O2	0.00369	0.08110	-0.18371	-0.29235	0.	0.	0.	0.	-0.40566	-0.08852
7SIGM-G	C1	-0.00588	-0.11110	0.44113	0.12536	0.	0.	0.	0.	0.	0.
0.2671	C2	0.00281	0.07360	-0.07790	-1.26160	0.	0.	0.	0.	-0.05722	0.71586
	C3	0.00281	0.07360	-0.07790	-1.26160	0.	0.	0.	0.	0.05722	-0.71586
	O1	-0.00293	-0.07025	0.11438	0.91341	0.	0.	0.	0.	0.16232	0.28309
	O2	-0.00293	-0.07025	0.11438	0.91341	0.	0.	0.	0.	-0.16232	-0.28309
8SIGM-G	C1	0.00414	0.06327	-0.41167	-3.38154	0.	0.	0.	0.	0.	0.
0.3828	C2	0.00296	0.01224	-0.51911	2.09517	0.	0.	0.	0.	-0.29154	1.67586
	C3	0.00296	0.01224	-0.51911	2.09517	0.	0.	0.	0.	0.29154	-1.67586
	O1	0.00012	-0.00246	-0.04250	0.22528	0.	0.	0.	0.	-0.11050	-0.34898
	O2	0.00012	-0.00246	-0.04250	0.22528	0.	0.	0.	0.	0.11050	0.34898
9SIGM-G	C1	0.00348	0.13099	0.16681	-5.48448	0.	0.	0.	0.	0.	0.
0.4599	C2	-0.00578	-0.10105	0.49936	1.52709	0.	0.	0.	0.	0.36326	1.89389
	C3	-0.00578	-0.10105	0.49936	1.52709	0.	0.	0.	0.	-0.36326	-1.89389
	O1	-0.00070	-0.02473	-0.02775	0.74375	0.	0.	0.	0.	-0.21462	-0.01479
	O2	-0.00070	-0.02473	-0.02775	0.74375	0.	0.	0.	0.	0.21462	0.01479
* 1SIGM-U	C1	0.	0.	0.	0.	0.	0.	0.	0.	-0.00016	0.00403
-20.6653	C2	0.00000	0.00002	-0.00003	0.00436	0.	0.	0.	0.	-0.00031	-0.00103
	C3	-0.00000	-0.00002	0.00003	-0.00436	0.	0.	0.	0.	-0.00031	-0.00103
	O1	-0.03615	-0.69251	-0.00208	-0.00348	0.	0.	0.	0.	-0.00157	-0.00009
	O2	0.03615	0.69251	0.00208	0.00348	0.	0.	0.	0.	-0.00157	-0.00009

Center		S1	S2	S3	S4	X1	X2	Y1	Y2	Z1	Z2
* 2SIGM-U	C1	0.	0.	0.	0.	0.	0.	0.	0.	0.00202	-0.00506
-11.4719	C2	-0.03671	-0.69231	-0.00208	-0.00592	0.	0.	0.	0.	0.00103	-0.00133
	C3	0.03671	0.69231	0.00208	0.00592	0.	0.	0.	0.	0.00103	-0.00133
	O1	0.00002	0.00041	-0.00070	0.00069	0.	0.	0.	0.	-0.00097	0.00066
	O2	-0.00002	-0.00041	0.00070	-0.00069	0.	0.	0.	0.	-0.00097	0.00066
* 3SIGM-U	C1	0.	0.	0.	0.	0.	0.	0.	0.	0.05493	0.14797
-1.5359	C2	0.00405	0.08504	-0.19604	0.11043	0.	0.	0.	0.	0.17498	-0.00897
	C3	-0.00405	-0.08504	0.19604	-0.11043	0.	0.	0.	0.	0.17498	-0.00897
	O1	0.00682	0.14966	-0.32635	-0.23666	0.	0.	0.	0.	-0.14079	0.00318
	O2	-0.00682	-0.14966	0.32635	0.23666	0.	0.	0.	0.	-0.14079	0.00318
* 4SIGM-U	C1	0.	0.	0.	0.	0.	0.	0.	0.	0.44237	0.07313
-0.9571	C2	0.00590	0.12710	-0.28809	0.00444	0.	0.	0.	0.	-0.24836	0.02933
	C3	-0.00590	-0.12710	0.28809	-0.00444	0.	0.	0.	0.	-0.24836	0.02933
	O1	-0.00214	-0.04607	0.11138	0.06101	0.	0.	0.	0.	-0.05683	-0.06280
	O2	0.00214	0.04607	-0.11138	-0.06101	0.	0.	0.	0.	-0.05683	-0.06280
* 5SIGM-U	C1	0.	0.	0.	0.	0.	0.	0.	0.	0.15479	0.07210
-0.7468	C2	-0.00178	-0.03420	0.11613	0.12021	0.	0.	0.	0.	-0.30136	0.10332
	C3	0.00178	0.03420	-0.11613	-0.12021	0.	0.	0.	0.	-0.30136	0.10332
	O1	0.00339	0.07457	-0.16874	-0.27865	0.	0.	0.	0.	0.40628	0.07258
	O2	-0.00339	-0.07457	0.16874	0.27865	0.	0.	0.	0.	0.40628	0.07258
6SIGM-U	C1	0.	0.	0.	0.	0.	0.	0.	0.	-0.22891	4.24952
0.2147	C2	-0.00012	-0.01843	-0.10301	3.43479	0.	0.	0.	0.	0.05745	1.70590
	C3	0.00012	0.01843	0.10301	-3.43479	0.	0.	0.	0.	0.05745	1.70590
	O1	-0.00219	-0.05079	0.09685	0.49547	0.	0.	0.	0.	0.06791	-0.13852
	O2	0.00219	0.05079	-0.09685	-0.49547	0.	0.	0.	0.	0.06791	-0.13852
7SIGM-U	C1	0.	0.	0.	0.	0.	0.	0.	0.	0.30175	-4.88356
0.5668	C2	0.00161	0.02222	-0.18226	-4.22958	0.	0.	0.	0.	-0.00011	0.86076
	C3	-0.00161	-0.02222	0.18226	4.22958	0.	0.	0.	0.	-0.00011	0.86076
	O1	-0.00192	-0.05672	0.00205	1.35406	0.	0.	0.	0.	-0.19736	0.65792
	O2	0.00192	0.05672	-0.00205	-1.35406	0.	0.	0.	0.	-0.19736	0.65792
8SIGM-U	C1	0.	0.	0.	0.	0.	0.	0.	0.	-0.73144	-5.84959
0.5800	C2	0.00354	0.11118	0.01308	-6.94659	0.	0.	0.	0.	0.31870	-1.80015
	C3	-0.00354	-0.11118	-0.01308	6.94659	0.	0.	0.	0.	0.31870	-1.80015
	O1	-0.00259	-0.05773	0.13744	0.73724	0.	0.	0.	0.	0.30305	0.92865
	O2	0.00259	0.05773	-0.13744	-0.73724	0.	0.	0.	0.	0.30305	0.92865
* 1PI-G-X	C1	0.	0.	0.	0.	0.	0.	0.	0.	0.	0.
-0.6518	C2	0.	0.	0.	0.	-0.28114	-0.04148	0.	0.	0.	0.
	C3	0.	0.	0.	0.	0.28114	0.04148	0.	0.	0.	0.
	O1	0.	0.	0.	0.	-0.43892	-0.18783	0.	0.	0.	0.
	O2	0.	0.	0.	0.	0.43892	0.18783	0.	0.	0.	0.
2PI-G-X	C1	0.	0.	0.	0.	0.	0.	0.	0.	0.	0.
0.0557	C2	0.	0.	0.	0.	-0.49699	-0.42648	0.	0.	0.	0.
	C3	0.	0.	0.	0.	0.49699	0.42648	0.	0.	0.	0.
	O1	0.	0.	0.	0.	0.31736	0.34185	0.	0.	0.	0.
	O2	0.	0.	0.	0.	-0.31736	-0.34185	0.	0.	0.	0.
3PI-G-X	C1	0.	0.	0.	0.	0.	0.	0.	0.	0.	0.
0.4683	C2	0.	0.	0.	0.	-0.62715	1.01567	0.	0.	0.	0.
	C3	0.	0.	0.	0.	0.62715	-1.01567	0.	0.	0.	0.
	O1	0.	0.	0.	0.	-0.07047	-0.07508	0.	0.	0.	0.
	O2	0.	0.	0.	0.	0.07047	0.07508	0.	0.	0.	0.

Center		S1	S2	S3	S4	X1	X2	Y1	Y2	Z1	Z2
* 1PI-G-Y	C1	0.	0.	0.	0.	0.	0.	0.	0.	0.	0.
-0.6518	C2	0.	0.	0.	0.	0.	0.	-0.28114	-0.04148	0.	0.
	C3	0.	0.	0.	0.	0.	0.	0.28114	0.04148	0.	0.
	O1	0.	0.	0.	0.	0.	0.	-0.43892	-0.18783	0.	0.
	O2	0.	0.	0.	0.	0.	0.	0.43892	0.18783	0.	0.
2PI-G-Y	C1	0.	0.	0.	0.	0.	0.	0.	0.	0.	0.
0.0557	C2	0.	0.	0.	0.	0.	0.	-0.49699	-0.42648	0.	0.
	C3	0.	0.	0.	0.	0.	0.	0.49699	0.42648	0.	0.
	O1	0.	0.	0.	0.	0.	0.	0.31736	0.34185	0.	0.
	O2	0.	0.	0.	0.	0.	0.	-0.31736	-0.34185	0.	0.
3PI-G-Y	C1	0.	0.	0.	0.	0.	0.	0.	0.	0.	0.
0.4683	C2	0.	0.	0.	0.	0.	0.	-0.62715	1.01567	0.	0.
	C3	0.	0.	0.	0.	0.	0.	0.62715	-1.01567	0.	0.
	O1	0.	0.	0.	0.	0.	0.	-0.07047	-0.07508	0.	0.
	O2	0.	0.	0.	0.	0.	0.	0.07047	0.07508	0.	0.
* 1PI-U-X	C1	0.	0.	0.	0.	-0.22017	-0.02546	0.	0.	0.	0.
-0.6878	C2	0.	0.	0.	0.	-0.31528	-0.05526	0.	0.	0.	0.
	C3	0.	0.	0.	0.	-0.31528	-0.05526	0.	0.	0.	0.
	O1	0.	0.	0.	0.	-0.36789	-0.13985	0.	0.	0.	0.
	O2	0.	0.	0.	0.	-0.36789	-0.13985	0.	0.	0.	0.
* 2PI-U-X	C1	0.	0.	0.	0.	0.53102	0.31381	0.	0.	0.	0.
-0.4086	C2	0.	0.	0.	0.	0.15558	-0.01330	0.	0.	0.	0.
	C3	0.	0.	0.	0.	0.15558	-0.01330	0.	0.	0.	0.
	O1	0.	0.	0.	0.	-0.33274	-0.19613	0.	0.	0.	0.
	O2	0.	0.	0.	0.	-0.33274	-0.19613	0.	0.	0.	0.
3PI-U-X	C1	0.	0.	0.	0.	0.60397	1.33660	0.	0.	0.	0.
0.3281	C2	0.	0.	0.	0.	-0.24541	-1.29569	0.	0.	0.	0.
	C3	0.	0.	0.	0.	-0.24541	-1.29569	0.	0.	0.	0.
	O1	0.	0.	0.	0.	0.24492	0.40289	0.	0.	0.	0.
	O2	0.	0.	0.	0.	0.24492	0.40289	0.	0.	0.	0.
4PI-U-X	C1	0.	0.	0.	0.	0.10164	-0.39177	0.	0.	0.	0.
0.3870	C2	0.	0.	0.	0.	0.68179	-0.43860	0.	0.	0.	0.
	C3	0.	0.	0.	0.	0.68179	-0.43860	0.	0.	0.	0.
	O1	0.	0.	0.	0.	-0.08801	-0.02718	0.	0.	0.	0.
	O2	0.	0.	0.	0.	-0.08801	-0.02718	0.	0.	0.	0.
5PI-U-X	C1	0.	0.	0.	0.	0.90272	-2.66933	0.	0.	0.	0.
0.6416	C2	0.	0.	0.	0.	-0.28761	1.50311	0.	0.	0.	0.
	C3	0.	0.	0.	0.	-0.28761	1.50311	0.	0.	0.	0.
	O1	0.	0.	0.	0.	-0.15374	-0.11268	0.	0.	0.	0.
	O2	0.	0.	0.	0.	-0.15374	-0.11268	0.	0.	0.	0.
* 1PI-U-Y	C1	0.	0.	0.	0.	0.	0.	-0.22017	-0.02546	0.	0.
-0.6878	C2	0.	0.	0.	0.	0.	0.	-0.31528	-0.05526	0.	0.
	C3	0.	0.	0.	0.	0.	0.	-0.31528	-0.05526	0.	0.
	O1	0.	0.	0.	0.	0.	0.	-0.36789	-0.13985	0.	0.
	O2	0.	0.	0.	0.	0.	0.	-0.36789	-0.13985	0.	0.
* 2PI-U-Y	C1	0.	0.	0.	0.	0.	0.	0.53102	0.31381	0.	0.
-0.4086	C2	0.	0.	0.	0.	0.	0.	0.15558	-0.01330	0.	0.
	C3	0.	0.	0.	0.	0.	0.	0.15558	-0.01330	0.	0.
	O1	0.	0.	0.	0.	0.	0.	-0.33274	-0.19613	0.	0.
	O2	0.	0.	0.	0.	0.	0.	-0.33274	-0.19613	0.	0.

Center		S1	S2	S3	S4	X1	X2	Y1	Y2	Z1	Z2
3PI-U-Y	C1	0.	0.	0.	0.	0.	0.	0.60397	1.33660	0.	0.
0.3281	C2	0.	0.	0.	0.	0.	0.	-0.24541	-1.29569	0.	0.
	C3	0.	0.	0.	0.	0.	0.	-0.24541	-1.29569	0.	0.
	O1	0.	0.	0.	0.	0.	0.	0.24492	0.40289	0.	0.
	O2	0.	0.	0.	0.	0.	0.	0.24492	0.40289	0.	0.
4PI-U-Y	C1	0.	0.	0.	0.	0.	0.	0.10164	-0.39177	0.	0.
0.3870	C2	0.	0.	0.	0.	0.	0.	0.68179	-0.43860	0.	0.
	C3	0.	0.	0.	0.	0.	0.	0.68179	-0.43860	0.	0.
	O1	0.	0.	0.	0.	0.	0.	-0.08801	-0.02718	0.	0.
	O2	0.	0.	0.	0.	0.	0.	-0.08801	-0.02718	0.	0.
5PI-U-Y	C1	0.	0.	0.	0.	0.	0.	0.90272	-2.66933	0.	0.
0.6416	C2	0.	0.	0.	0.	0.	0.	-0.28761	1.50311	0.	0.
	C3	0.	0.	0.	0.	0.	0.	-0.28761	1.50311	0.	0.
	O1	0.	0.	0.	0.	0.	0.	-0.15374	-0.11268	0.	0.
	O2	0.	0.	0.	0.	0.	0.	-0.15374	-0.11268	0.	0.

Overlap Matrix

First Function					Second Function and Overlap					

C1S1 with

C1S2	.35436	C1S3	.00805	C1S4	.01419	C2S3	.00225	C2S4	.00597	C2Z1	.00643	C2Z2	.00981
O1S4	.00006	O1Z2	.00085										

C1S2 with

C1S3	.26226	C1S4	.17745	C2S2	.00005	C2S3	.03339	C2S4	.07695	C2Z1	.08421	C2Z2	.12298
O1S4	.00101	O1Z1	.00001	O1Z2	.01236								

C1S3 with

C1S4	.79970	C2S1	.00225	C2S2	.03339	C2S3	.26724	C2S4	.41588	C2Z1	.39881	C2Z2	.55102
O1S2	.00002	O1S3	.00128	O1S4	.02315	O1Z1	.00478	O1Z2	.11846				

C1S4 with

C2S1	.00597	C2S2	.07695	C2S3	.41588	C2S4	.64858	C2Z1	.31226	C2Z2	.62484	O1S1	.00039
O1S2	.00527	O1S3	.03897	O1S4	.11743	O1Z1	.04644	O1Z2	.26500				

C1X1 with

C1X2	.53875	C2X1	.22081	C2X2	.31782	O1X1	.00229	O1X2	.03808

C1X2 with

C2X1	.31782	C2X2	.71515	O1X1	.03516	O1X2	.18167

C1Y1 with

C1Y2	.53875	C2Y1	.22081	C2Y2	.31782	O1Y1	.00229	O1Y2	.03808

C1Y2 with

C2Y1	.31782	C2Y2	.71515	O1Y1	.03516	O1Y2	.18167

C1Z1 with

C1Z2	.53875	C2S1	-.00643	C2S2	-.08421	C2S3	-.39881	C2S4	-.31226	C2Z1	-.35272	C2Z2	-.01712
O1S1	-.00003	O1S2	-.00050	O1S3	-.00915	O1S4	-.05788	O1Z1	-.02163	O1Z2	-.18698		

C1Z2 with

C2S1	-.00981	C2S2	-.12298	C2S3	-.55102	C2S4	-.62484	C2Z1	-.01712	C2Z2	.23563	O1S1	-.00204
O1S2	-.02673	O1S3	-.16101	O1S4	-.33852	O1Z1	-.11478	O1Z2	-.39462				

C2S1 with

C2S2	.35436	C2S3	.00805	C2S4	.01419	C3S4	.00045	C3Z1	-.00002	C3Z2	-.00263	O1S3	.00069
O1S4	.00593	O1Z1	.00346	O1Z2	.01356								

C2S2 with

C2S3	.26226	C2S4	.17745	C3S3	.00002	C3S4	.00629	C3Z1	-.00049	C3Z2	-.03478	O1S2	.00003
O1S3	.01512	O1S4	.07721	O1Z1	.05154	O1Z2	.16574						

C2S3 with

C2S4	.79970	C3S2	.00002	C3S3	.00435	C3S4	.05818	C3Z1	-.01872	C3Z2	-.22047	O1S1	.00235
O1S2	.03210	O1S3	.21848	O1S4	.41606	O1Z1	.28193	O1Z2	.61902				

C2S4 with

C3S1	.00045	C3S2	.00629	C3S3	.05818	C3S4	.17695	C3Z1	-.09221	C3Z2	-.40219	O1S1	.00445
O1S2	.05739	O1S3	.31652	O1S4	.58058	O1Z1	.17315	O1Z2	.53122				

C2X1 with

C2X2	.53875	C3X1	.00667	C3X2	.06585	O1X1	.16587	O1X2	.38859

C2X2 with

C3X1	.06585	C3X2	.26157	O1X1	.18349	O1X2	.62005

C2Y1 with

C2Y2	.53875	C3Y1	.00667	C3Y2	.06585	O1Y1	.16587	O1Y2	.38859

C2Y2 with

C3Y1	.06585	C3Y2	.26157	O1Y1	.18349	O1Y2	.62005

C2Z1 with

C2Z2	.53875	C3S1	.00002	C3S2	.00049	C3S3	.01872	C3S4	.09221	C3Z1	-.05200	C3Z2	-.20947
O1S1	-.00558	O1S2	-.07242	O1S3	-.37690	O1S4	-.43524	O1Z1	-.30977	O1Z2	-.14836		

C2Z2 with

C3S1	.00263	C3S2	.03478	C3S3	.22047	C3S4	.40219	C3Z1	-.20947	C3Z2	-.43999	O1S1	-.00639
O1S2	-.08138	O1S3	-.40837	O1S4	-.61561	O1Z1	.00623	O1Z2	.17550				

Overlap Matrix (continued)

First Function   Second Function and Overlap

C3S2 with
   O1Z2 .00006

C3S3 with
   O1S4 .00016   O1Z2 .00289

C3S4 with
   O1S1 .00001   O1S2 .00009   O1S3 .00108   O1S4 .00762   O1Z1 .00213   O1Z2 .03444

C3X1 with
   O1X2 .00081

C3X2 with
   O1X1 .00210   O1X2 .02223

C3Y1 with
   O1Y2 .00081

C3Y2 with
   O1Y1 .00210   O1Y2 .02223

C3Z1 with
   O1S3 -.00001   O1S4 -.00096   O1Z1 -.00006   O1Z2 -.00999

C3Z2 with
   O1S1 -.00012   O1S2 -.00170   O1S3 -.01396   O1S4 -.05197   O1Z1 -.01863   O1Z2 -.14170

O1S1 with
   O1S2 .35461   O1S3 .00974   O1S4 .01469

O1S2 with
   O1S3 .28286   O1S4 .18230

O1S3 with
   O1S4 .79029

O1S4 with
   O2S4 .00001   O2Z2 -.00015

O1X1 with
   O1X2 .50607

O1X2 with
   O2X2 .00011

O1Y1 with
   O1Y2 .50607

O1Y2 with
   O2Y2 .00011

O1Z1 with
   O1Z2 .50607   O2Z2 -.00001

O1Z2 with
   O2S4 .00015   O2Z1 -.00001   O2Z2 -.00194

CHF3   Trifluoromethane

Molecular Geometry   Symmetry C_{3v}

Center	Coordinates		
	X	Y	Z
H1	0.	0.	2.07494000
C1	0.	0.	0.
F1	-2.05530369	-1.18663013	-0.83886207
F2	2.05530369	-1.18663013	-0.83886207
F3	0.	2.37326026	-0.83886207
Mass	0.	0.	-0.65311500

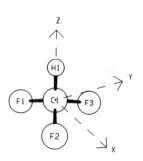

MO Symmetry Types and Occupancy

Representation	A1	A2	E
Number of MOs	16	2	24
Number Occupied	6	1	10

Energy Expectation Values

Total	-336.7157	Kinetic	336.9308
Electronic	-470.2658	Potential	-673.6465
Nuclear Repulsion	133.5501	Virial	1.9994

Electronic Potential	-807.1966
One-Electron Potential	-1070.7300
Two-Electron Potential	263.5334

Moments of the Charge Distribution

First

x	y	z	Dipole Moment
0.	0.	-0.7482	0.8834
0.	0.	1.6316	

Second

x^2	y^2	z^2	xy	xz	yz	r^2
-93.7497	-93.7497	-24.4926	0.	0.	0.	-211.9920
76.0369	76.0369	10.9332	0.	0.	0.	163.0070

Third

x^3	xy^2	xz^2	x^2y	y^3	yz^2	x^2z	y^2z	z^3	xyz
0.	0.	0.	93.9005	-93.9005	0.	13.3386	13.3386	-22.5129	0.
0.	0.	0.	-90.2277	90.2277	0.	-14.1236	-14.1236	21.8015	0.

Fourth (even)

x^4	y^4	z^4	x^2y^2	x^2z^2	y^2z^2	r^2x^2	r^2y^2	r^2z^2	r^4
-554.5806	-554.5806	-112.6504	-184.8602	-50.6636	-50.6636	-790.1043	-790.1043	-213.9775	-1794.1862
321.2007	321.2007	56.5114	107.0669	2.6234	2.6234	430.8911	430.8910	61.7583	923.5404

Fourth (odd)

x^3y	xy^3	xyz^2	x^3z	xy^2z	xz^3	x^2yz	y^3z	yz^3
0.	0.	0.	0.	0.	0.	-16.3718	16.3718	0.
0.	0.	0.	0.	0.	0.	16.7595	-16.7595	0.

T-346   CHF3

Expectation Values at Atomic Centers

Center	r^{-1}	δ	Expectation Value xr^{-3}	yr^{-3}	zr^{-3}
H1	-0.9958	0.4268	0.	0.	0.0712
C1	-14.4037	116.3252	0.	0.	0.0703
F1	-26.5262	413.6108	-0.2141	-0.1236	-0.0868

Center	$\dfrac{3x^2-r^2}{r^5}$	$\dfrac{3y^2-r^2}{r^5}$	$\dfrac{3z^2-r^2}{r^5}$	$\dfrac{3xy}{r^5}$	$\dfrac{3xz}{r^5}$	$\dfrac{3yz}{r^5}$
H1	-0.1481	-0.1481	0.2962	0.	0.	0.
C1	0.2618	0.2618	-0.5237	0.	0.	0.
F1	1.6851	-0.4916	-1.1935	1.8851	1.3436	0.7757

Mulliken Charge

Center					Basis Function Type						Total
	S1	S2	S3	S4	X1	X2	Y1	Y2	Z1	Z2	
H1	0.5221	0.2941									0.8162
C1	0.0428	1.9566	0.8548	0.1617	0.6401	0.0071	0.6401	0.0071	0.9599	0.0643	5.3346
F1	0.0435	1.9557	1.0261	0.9125	1.2089	0.4010	1.3556	0.4834	1.4024	0.4939	9.2831

Overlap Populations

	Pair of Atomic Centers			
	C1	F1	F1	F2
	H1	H1	C1	F1
1A1	0.	0.	-0.0002	0.
2A1	-0.0003	0.	0.	0.
3A1	0.0071	0.0045	0.1197	0.0251
4A1	0.1836	0.0040	0.0301	-0.0011
5A1	0.1608	0.0071	0.1066	0.0145
6A1	0.4611	-0.0998	-0.1192	0.0234
Total A1	0.8124	-0.0842	0.1371	0.0619
1A2	0.	0.	0.	-0.0560
Total A2	0.	0.	0.	-0.0560
1E	0.	0.	-0.0001	0.
2E	0.	0.	0.	0.
3E	0.	0.	0.0997	-0.0410
4E	0.	0.	0.0336	0.0138
5E	0.	0.	0.0806	-0.0095
6E	0.	0.	0.0916	0.0273
7E	0.	0.	0.0174	-0.0340
8E	0.	0.	-0.0114	0.0368
9E	0.	0.	0.0420	-0.1174
10E	0.	0.	0.0087	0.0236
Total E	0.	0.	0.3622	-0.1005
Total	0.8124	-0.0842	0.4993	-0.0946
Distance	2.0749	3.7580	2.5172	4.1106

Molecular Orbitals and Eigenvalues

Center		S1	S2	S3	S4	X1	X2	Y1	Y2	Z1	Z2
* 1A1	H1	-0.00005	-0.00032								
-26.3406	C1	0.00000	-0.00005	-0.00015	0.00156	0.	0.	0.	0.	-0.00015	-0.00019
	F1	-0.02987	-0.56517	-0.00241	-0.00126	-0.00056	-0.00003	-0.00033	-0.00002	-0.00024	0.00006
	F2	-0.02987	-0.56517	-0.00241	-0.00126	0.00056	0.00003	-0.00033	-0.00002	-0.00024	0.00006
	F3	-0.02987	-0.56517	-0.00241	-0.00126	0.	0.	0.00065	0.00003	-0.00024	0.00006
* 2A1	H1	-0.00112	0.00113								
-11.5533	C1	-0.05192	-0.97918	-0.00395	-0.00166	0.	0.	0.	0.	0.00140	-0.00084
	F1	0.00001	0.00017	-0.00027	0.00020	-0.00048	0.00003	-0.00028	0.00002	-0.00022	-0.00004
	F2	0.00001	0.00017	-0.00027	0.00020	0.00048	-0.00003	-0.00028	0.00002	-0.00022	-0.00004
	F3	0.00001	0.00017	-0.00027	0.00020	0.	0.	0.00056	-0.00003	-0.00022	-0.00004
* 3A1	H1	0.02095	0.01318								
-1.7723	C1	-0.00487	-0.10327	0.23238	0.00044	0.	0.	0.	0.	-0.06927	0.00355
	F1	-0.00594	-0.12765	0.28445	0.23939	0.06585	0.01461	0.03802	0.00844	0.02083	0.00849
	F2	-0.00594	-0.12765	0.28445	0.23939	-0.06585	-0.01461	0.03802	0.00844	0.02083	0.00849
	F3	-0.00594	-0.12765	0.28445	0.23939	0.	0.	-0.07604	-0.01687	0.02083	0.00849
* 4A1	H1	-0.14621	-0.04099								
-1.0102	C1	0.00808	0.16515	-0.46431	-0.09478	0.	0.	0.	0.	-0.09255	-0.04622
	F1	-0.00306	-0.06533	0.15361	0.17431	-0.18578	-0.06923	-0.10726	-0.03997	-0.11495	-0.03447
	F2	-0.00306	-0.06533	0.15361	0.17431	0.18578	0.06923	-0.10726	-0.03997	-0.11495	-0.03447
	F3	-0.00306	-0.06533	0.15361	0.17431	0.	0.	0.21452	0.07994	-0.11495	-0.03447
* 5A1	H1	-0.13518	-0.05531								
-0.8740	C1	-0.00021	-0.00193	0.02905	-0.06813	0.	0.	0.	0.	-0.42171	-0.03624
	F1	0.00092	0.01975	-0.04584	-0.06404	0.19611	0.07223	0.11322	0.04170	-0.22630	-0.08840
	F2	0.00092	0.01975	-0.04584	-0.06404	-0.19611	-0.07223	0.11322	0.04170	-0.22630	-0.08840
	F3	0.00092	0.01975	-0.04584	-0.06404	0.	0.	-0.22645	-0.08340	-0.22630	-0.08840
* 6A1	H1	-0.25865	-0.23407								
-0.6076	C1	0.00141	0.03116	-0.06815	-0.14848	0.	0.	0.	0.	-0.36636	0.02476
	F1	-0.00008	-0.00248	-0.00146	0.03896	0.04409	0.02073	0.02545	0.01197	0.35360	0.18643
	F2	-0.00008	-0.00248	-0.00146	0.03896	-0.04409	-0.02073	0.02545	0.01197	0.35360	0.18643
	F3	-0.00008	-0.00248	-0.00146	0.03896	0.	0.	-0.05091	-0.02394	0.35360	0.18643
7A1	H1	-0.02220	-1.31934								
0.2992	C1	-0.00600	-0.11997	0.41751	1.48240	0.	0.	0.	0.	0.56085	-0.35377
	F1	0.00117	0.02909	-0.03244	-0.37565	-0.07685	-0.17084	-0.04437	-0.09864	-0.03726	0.00845
	F2	0.00117	0.02909	-0.03244	-0.37565	0.07685	0.17084	-0.04437	-0.09864	-0.03726	0.00845
	F3	0.00117	0.02909	-0.03244	-0.37565	0.	0.	0.08874	0.19727	-0.03726	0.00845
8A1	H1	-0.01167	1.60509								
0.3388	C1	-0.00182	-0.04966	0.03636	0.26567	0.	0.	0.	0.	-0.60032	-1.66636
	F1	0.00229	0.05176	-0.10233	-0.49269	-0.15539	-0.27552	-0.08972	-0.15907	0.11375	0.15491
	F2	0.00229	0.05176	-0.10233	-0.49269	0.15539	0.27552	-0.08972	-0.15907	0.11375	0.15491
	F3	0.00229	0.05176	-0.10233	-0.49269	0.	0.	0.17943	0.31814	0.11375	0.15491
9A1	H1	-0.09325	-2.22531								
0.4478	C1	-0.00556	-0.12724	0.27593	2.58330	0.	0.	0.	0.	-0.62774	2.17982
	F1	0.00175	0.03711	-0.09652	-0.26963	-0.16108	-0.22814	-0.09300	-0.13171	-0.13710	-0.34060
	F2	0.00175	0.03711	-0.09652	-0.26963	0.16108	0.22814	-0.09300	-0.13171	-0.13710	-0.34060
	F3	0.00175	0.03711	-0.09652	-0.26963	0.	0.	0.18600	0.26343	-0.13710	-0.34060
10A1	H1	-0.22419	-1.40224								
0.6281	C1	0.01017	0.07822	-1.52951	3.58446	0.	0.	0.	0.	0.18295	0.62007
	F1	0.00064	0.01971	0.00790	-0.58911	0.22442	-0.22804	0.12957	-0.13166	-0.00599	-0.11981
	F2	0.00064	0.01971	0.00790	-0.58911	-0.22442	0.22804	0.12957	-0.13166	-0.00599	-0.11981
	F3	0.00064	0.01971	0.00790	-0.58911	0.	0.	-0.25913	0.26332	-0.00599	-0.11981

Center		S1	S2	S3	S4	X1	X2	Y1	Y2	Z1	Z2
* 1A2	H1	0.	0.								
-0.6735	C1	0.	0.	0.	0.	0.	0.	0.	0.	0.	0.
	F1	0.	0.	0.	0.	-0.23046	-0.11013	0.39917	0.19074	0.	0.
	F2	0.	0.	0.	0.	-0.23046	-0.11013	-0.39917	-0.19074	0.	0.
	F3	0.	0.	0.	0.	0.46092	0.22025	0.	0.	0.	0.
* 1E	H1	0.	0.								
-26.3506	C1	0.	0.	0.	0.	0.00011	0.00040	0.	0.	0.	0.
	F1	0.03658	0.69219	0.00316	0.00095	0.00071	-0.00025	0.00040	-0.00006	0.00027	-0.00001
	F2	-0.03658	-0.69219	-0.00316	-0.00095	0.00071	-0.00025	-0.00040	0.00006	-0.00027	0.00001
	F3	0.	0.	0.	0.	0.00003	-0.00014	0.	0.	0.	0.
* 2E	H1	0.	0.								
-26.3506	C1	0.	0.	0.	0.	0.	0.	-0.00011	-0.00040	0.	0.
	F1	-0.02112	-0.39964	-0.00183	-0.00055	-0.00040	0.00006	-0.00026	0.00018	-0.00015	0.00001
	F2	-0.02112	-0.39964	-0.00183	-0.00055	0.00040	-0.00006	-0.00026	0.00018	-0.00015	0.00001
	F3	0.04224	0.79927	0.00365	0.00109	0.	0.	-0.00094	0.00028	0.00031	-0.00002
* 3E	H1	0.	0.								
-1.6664	C1	0.	0.	0.	0.	-0.15808	0.01982	0.	0.	0.	0.
	F1	-0.00808	-0.17367	0.38759	0.33925	0.05158	0.01193	0.03876	0.00652	0.02724	0.00397
	F2	0.00808	0.17367	-0.38759	-0.33925	0.05158	0.01193	-0.03876	-0.00652	-0.02724	-0.00397
	F3	0.	0.	0.	0.	-0.01555	0.00063	0.	0.	0.	0.
* 4E	H1	0.	0.								
-1.6664	C1	0.	0.	0.	0.	0.	0.	0.15808	-0.01982	0.	0.
	F1	0.00466	0.10027	-0.22378	-0.19587	-0.03876	-0.00652	-0.00683	-0.00440	-0.01573	-0.00229
	F2	0.00466	0.10027	-0.22378	-0.19587	0.03876	0.00652	-0.00683	-0.00440	-0.01573	-0.00229
	F3	-0.00933	-0.20054	0.44756	0.39173	0.	0.	-0.07396	-0.01569	0.03145	0.00459
* 5E	H1	0.	0.								
-0.8406	C1	0.	0.	0.	0.	-0.38636	0.01825	0.	0.	0.	0.
	F1	0.00202	0.04308	-0.10278	-0.14516	0.17696	0.05553	0.29559	0.11462	0.16019	0.05494
	F2	-0.00202	-0.04308	0.10278	0.14516	0.17696	0.05553	-0.29559	-0.11462	-0.16019	-0.05494
	F3	0.	0.	0.	0.	-0.33501	-0.14301	0.	0.	0.	0.
* 6E	H1	0.	0.								
-0.8406	C1	0.	0.	0.	0.	0.	0.	0.38636	-0.01825	0.	0.
	F1	-0.00117	-0.02487	0.05934	0.08381	-0.29559	-0.11462	0.16435	0.07683	-0.09249	-0.03172
	F2	-0.00117	-0.02487	0.05934	0.08381	0.29559	0.11462	0.16435	0.07683	-0.09249	-0.03172
	F3	0.00233	0.04975	-0.11868	-0.16762	0.	0.	-0.34762	-0.12170	0.18498	0.06344
* 7E	H1	0.	0.								
-0.7243	C1	0.	0.	0.	0.	-0.10256	0.01092	0.	0.	0.	0.
	F1	0.00038	0.00777	-0.02094	-0.02156	0.08411	0.03599	-0.14274	-0.07782	0.44954	0.21540
	F2	-0.00038	-0.00777	0.02094	0.02156	0.08411	0.03599	0.14274	0.07782	-0.44954	-0.21540
	F3	0.	0.	0.	0.	0.33135	0.17078	0.	0.	0.	0.
* 8E	H1	0.	0.								
-0.7243	C1	0.	0.	0.	0.	0.	0.	0.10256	-0.01092	0.	0.
	F1	-0.00022	-0.00449	0.01209	0.01245	0.14274	0.07782	-0.24894	-0.12585	-0.25954	-0.12436
	F2	-0.00022	-0.00449	0.01209	0.01245	-0.14274	-0.07782	-0.24894	-0.12585	-0.25954	-0.12436
	F3	0.00043	0.00898	-0.02418	-0.02490	0.	0.	-0.00170	0.00894	0.51909	0.24873
* 9E	H1	0.	0.								
-0.6813	C1	0.	0.	0.	0.	-0.14213	0.07413	0.	0.	0.	0.
	F1	0.00037	0.00683	-0.02634	0.00977	0.42305	0.19613	0.02696	0.01056	-0.22675	-0.11027
	F2	-0.00037	-0.00683	0.02634	-0.00977	0.42305	0.19613	-0.02696	-0.01056	0.22675	0.11027
	F3	0.	0.	0.	0.	0.37634	0.17783	0.	0.	0.	0.

Center		S1	S2	S3	S4	X1	X2	Y1	Y2	Z1	Z2
• 10E	H1	0.	0.								
-0.6813	C1	0.	0.	0.	0.	0.	0.	0.14213	-0.07413	0.	0.
	F1	-0.00021	-0.00395	0.01521	-0.00564	-0.02696	-0.01056	-0.39191	-0.18393	0.13092	0.06366
	F2	-0.00021	-0.00395	0.01521	-0.00564	0.02696	0.01056	-0.39191	-0.18393	0.13092	0.06366
	F3	0.00043	0.00789	-0.03042	0.01128	0.	0.	-0.43861	-0.20223	-0.26183	-0.12732
11E	H1	0.	0.								
0.2721	C1	0.	0.	0.	0.	0.41361	1.43441	0.	0.	0.	0.
	F1	-0.00229	-0.05607	0.07032	0.70179	0.03982	0.07449	0.13040	0.21579	0.13161	0.14457
	F2	0.00229	0.05607	-0.07032	-0.70179	0.03982	0.07449	-0.13040	-0.21579	-0.13161	-0.14457
	F3	0.	0.	0.	0.	-0.18603	-0.29928	0.	0.	0.	0.
12E	H1	0.	0.								
0.2721	C1	0.	0.	0.	0.	0.	0.	0.41361	1.43441	0.	0.
	F1	-0.00132	-0.03237	0.04060	0.40518	0.13040	0.21579	-0.11075	-0.17469	0.07598	0.08347
	F2	-0.00132	-0.03237	0.04060	0.40518	-0.13040	-0.21579	-0.11075	-0.17469	0.07598	0.08347
	F3	0.00265	0.06474	-0.08120	-0.81036	0.	0.	0.11511	0.19907	-0.15196	-0.16694
13E	H1	0.	0.								
0.5326	C1	0.	0.	0.	0.	1.21116	-1.00361	0.	0.	0.	0.
	F1	-0.00131	-0.02236	0.10916	-0.12170	0.25628	0.24699	0.17029	0.08843	0.10899	0.06645
	F2	0.00131	0.02236	-0.10916	0.12170	0.25628	0.24699	-0.17029	-0.08843	-0.10899	-0.06645
	F3	0.	0.	0.	0.	-0.03866	0.09382	0.	0.	0.	0.
14E	H1	0.	0.								
0.5326	C1	0.	0.	0.	0.	0.	0.	1.21116	-1.00361	0.	0.
	F1	-0.00076	-0.01291	0.06303	-0.07026	0.17029	0.08843	0.05965	0.14487	0.06293	0.03836
	F2	-0.00076	-0.01291	0.06303	-0.07026	-0.17029	-0.08843	0.05965	0.14487	0.06293	0.03836
	F3	0.00151	0.02582	-0.12605	0.14053	0.	0.	0.35460	0.29804	-0.12586	-0.07673

Overlap Matrix

First Function			Second Function and Overlap							

H1S1 with

H1S2	.68301	C1S1	.00213	C1S2	.03308	C1S3	.27903	C1S4	.36915	C1Z1	.42563	C1Z2	.43383
F1S2	.00004	F1S3	.00205	F1S4	.03056	F1X1	.00367	F1X2	.07673	F1Y1	.00212	F1Y2	.04430
F1Z1	.00521	F1Z2	.10878										

H1S2 with

C1S1	.00758	C1S2	.09676	C1S3	.49235	C1S4	.70205	C1Z1	.36028	C1Z2	.61044	F1S1	.00071
F1S2	.00940	F1S3	.06372	F1S4	.17113	F1X1	.03481	F1X2	.17880	F1Y1	.02010	F1Y2	.10323
F1Z1	.04935	F1Z2	.25348										

C1S1 with

C1S2	.35436	C1S3	.00805	C1S4	.01419	F1S3	.00004	F1S4	.00286	F1X1	.00037	F1X2	.00855
F1Y1	.00022	F1Y2	.00494	F1Z1	.00015	F1Z2	.00349						

C1S2 with

C1S3	.26226	C1S4	.17745	F1S3	.00202	F1S4	.03984	F1X1	.00886	F1X2	.10683	F1Y1	.00511
F1Y2	.06168	F1Z1	.00361	F1Z2	.04360								

C1S3 with

C1S4	.79970	F1S1	.00096	F1S2	.01329	F1S3	.10833	F1S4	.27928	F1X1	.12628	F1X2	.43551
F1Y1	.07291	F1Y2	.25144	F1Z1	.05154	F1Z2	.17775						

C1S4 with

F1S1	.00298	F1S2	.03833	F1S3	.22048	F1S4	.44736	F1X1	.10123	F1X2	.38331	F1Y1	.05844
F1Y2	.22130	F1Z1	.04131	F1Z2	.15645								

C1X1 with

C1X2	.53875	F1S1	-.00244	F1S2	-.03169	F1S3	-.18970	F1S4	-.31338	F1X1	-.12382	F1X2	-.12307
F1Y1	-.11695	F1Y2	-.23882	F1Z1	-.08267	F1Z2	-.16883						

C1X2 with

F1S1	-.00423	F1S2	-.05371	F1S3	-.28437	F1S4	-.48285	F1X1	.01486	F1X2	.15157	F1Y1	-.05952
F1Y2	-.18780	F1Z1	-.04207	F1Z2	-.13276								

C1Y1 with

C1Y2	.53875	F1S1	-.00141	F1S2	-.01830	F1S3	-.10952	F1S4	-.18093	F1X1	-.11695	F1X2	-.23882
F1Y1	.01122	F1Y2	.15270	F1Z1	-.04773	F1Z2	-.09747						

C1Y2 with

F1S1	-.00244	F1S2	-.03101	F1S3	-.16418	F1S4	-.27878	F1X1	-.05952	F1X2	-.18780	F1Y1	.08358
F1Y2	.36842	F1Z1	-.02429	F1Z2	-.07665								

C1Z1 with

C1Z2	.53875	F1S1	-.00100	F1S2	-.01293	F1S3	-.07742	F1S4	-.12791	F1X1	-.08267	F1X2	-.16883
F1Y1	-.04773	F1Y2	-.09747	F1Z1	.04499	F1Z2	.22168						

C1Z2 with

F1S1	-.00173	F1S2	-.02192	F1S3	-.11607	F1S4	-.19707	F1X1	-.04207	F1X2	-.13276	F1Y1	-.02429
F1Y2	-.07665	F1Z1	.10077	F1Z2	.42267								

F1S1 with

F1S2	.35647	F1S3	.01003	F1S4	.01474	F2S4	.00004	F2X2	-.00051

F1S2 with

F1S3	.28653	F1S4	.18096	F2S4	.00054	F2X2	-.00705

F1S3 with

F1S4	.78531	F2S3	.00005	F2S4	.00769	F2X1	-.00040	F2X2	-.05784

F1S4 with

F2S1	.00004	F2S2	.00054	F2S3	.00769	F2S4	.04883	F2X1	-.01592	F2X2	-.17550

F1X1 with

F1X2	.49444	F2S3	.00040	F2S4	.01592	F2X1	-.00211	F2X2	-.07872

F1X2 with

F2S1	.00051	F2S2	.00705	F2S3	.05784	F2S4	.17550	F2X1	-.07872	F2X2	-.35950

F1Y1 with

F1Y2	.49444	F2Y1	.00014	F2Y2	.01249

CHF3

Overlap Matrix (continued)

First Function   Second Function and Overlap

F1Y2 with
    F2Y1 .01249   F2Y2 .09937

F1Z1 with
    F1Z2 .49444   F2Z1 .00014   F2Z2 .01249

F1Z2 with
    F2Z1 .01249   F2Z2 .09937

CF2N2   Difluorodiazirine

Molecular Geometry   Symmetry C₂ᵥ

Center	Coordinates		
	X	Y	Z
C1	0.	0.	0.
N1	1.22235499	0.	-2.40160701
N2	-1.22235499	0.	-2.40160701
F1	0.	2.05774900	1.39320400
F2	0.	-2.05774900	1.39320400
Mass	0.	0.	-0.18361110

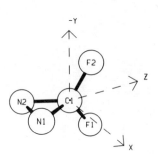

MO Symmetry Types and Occupancy

Representation	A1	A2	B1	B2
Number of MOs	22	4	12	12
Number Occupied	9	1	4	5

Energy Expectation Values

Total	-345.3999	Kinetic	345.6696
Electronic	-515.9245	Potential	-691.0695
Nuclear Repulsion	170.5246	Virial	1.9992

Electronic Potential	-861.5941
One-Electron Potential	-1159.3461
Two-Electron Potential	297.7520

Moments of the Charge Distribution

First

x	y	z	Dipole Moment
0.	0.	1.5353	0.0324
0.	0.	-1.5676	

Second

x^2	y^2	z^2	xy	xz	yz	r^2
-41.4323	-97.2931	-136.2224	0.	0.	0.	-274.9478
20.9181	76.2180	113.8296	0.	0.	0.	210.9657

Third

x^3	xy^2	xz^2	x^2y	y^3	yz^2	x^2z	y^2z	z^3	xyz
0.	0.	0.	0.	0.	0.	58.6125	-119.1018	104.1636	0.
0.	0.	0.	0.	0.	0.	-46.3963	120.1816	-82.1539	0.

Fourth (even)

x^4	y^4	z^4	x^2y^2	x^2z^2	y^2z^2	r^2x^2	r^2y^2	r^2z^2	r^4
-164.4354	-569.5015	-874.0003	-60.6815	-208.9647	-301.0306	-434.0816	-931.2137	-1383.9956	-2749.2909
31.2549	322.7321	450.1028	0.	102.9068	189.5042	134.1617	512.2363	742.5139	1388.9118

Expectation Values at Atomic Centers

Center	r^{-1}	δ	xr^{-3}	yr^{-3}	zr^{-3}
C1	-14.4087	116.1459	0.	0.	0.0168
N1	-18.1976	189.8574	0.2280	0.	-0.1417
F1	-26.5044	413.6192	0.	0.2210	0.1533

	$\dfrac{3x^2-r^2}{r^5}$	$\dfrac{3y^2-r^2}{r^5}$	$\dfrac{3z^2-r^2}{r^5}$	$\dfrac{3xy}{r^5}$	$\dfrac{3xz}{r^5}$	$\dfrac{3yz}{r^5}$
C1	-0.5946	0.3304	0.2642	0.	0.	0.
N1	-0.3928	0.8139	-0.4211	0.	0.2434	0.
F1	-1.7121	1.8215	-0.1094	0.	0.	2.4042

Mulliken Charge

Center	S1	S2	S3	S4	X1	X2	Y1	Y2	Z1	Z2	Total
C1	0.0427	1.9564	0.8231	0.1140	0.9657	0.0446	0.6940	0.0288	0.7516	0.0347	5.4556
N1	0.0435	1.9559	0.9916	0.8454	0.9979	0.0898	0.6849	0.2631	0.9797	0.1659	7.0176
F1	0.0436	1.9557	1.0269	0.9123	1.4362	0.5124	1.2001	0.3911	1.3298	0.4465	9.2545

Overlap Populations

	Pair of Atomic Centers				
	N1	N2	F1	F1	F2
	C1	N1	C1	N1	F1
1A1	0.	0.	-0.0001	0.	0.
2A1	-0.0015	0.0015	0.	0.	0.
3A1	-0.0001	0.	0.0001	0.	0.
4A1	0.0119	0.0334	0.1291	0.0158	0.0392
5A1	0.1529	0.4562	0.0117	-0.0057	0.0008
6A1	-0.0952	0.1584	0.0526	-0.0021	0.0013
7A1	-0.1189	0.1954	0.1023	-0.0041	0.0402
8A1	0.0660	0.0811	-0.0373	-0.0143	0.0298
9A1	0.0441	0.0391	-0.0710	0.0047	0.0079
Total A1	0.0592	0.9649	0.1874	-0.0056	0.1192
1A2	0.	-0.0012	0.	0.0018	-0.0418
Total A2	0.	-0.0012	0.	0.0018	-0.0418
1B1	-0.0001	-0.0026	0.	0.	0.
2B1	0.1732	-0.5810	0.0374	0.0068	0.0015
3B1	-0.0590	-0.2070	0.0864	-0.0176	0.0260
4B1	0.1124	-0.2550	-0.1266	-0.0134	0.0109
Total B1	0.2266	-1.0456	-0.0028	-0.0242	0.0383
1B2	0.	0.	-0.0001	0.	0.
2B2	0.0007	0.0001	0.1076	0.0006	-0.0379
3B2	0.0379	0.0215	0.1006	0.0061	-0.0107
4B2	0.	0.0007	0.0272	0.0015	-0.1453
5B2	0.0515	0.3537	-0.0163	-0.0371	-0.0142
Total B2	0.0901	0.3760	0.2191	-0.0288	-0.2081
Total	0.3759	0.2942	0.4036	-0.0568	-0.0924
Distance	2.6948	2.4447	2.4850	4.4865	4.1155

Molecular Orbitals and Eigenvalues

Center		S1	S2	S3	S4	X1	X2	Y1	Y2	Z1	Z2
* 1A1	C1	0.00000	-0.00003	-0.00007	0.00087	0.	0.	0.	0.	0.00012	0.00035
-26.3726	N1	-0.00000	0.00000	0.00006	-0.00012	0.00001	-0.00002	0.	0.	0.00000	-0.00011
	N2	-0.00000	0.00000	0.00006	-0.00012	-0.00001	0.00002	0.	0.	0.00000	-0.00011
	F1	-0.03658	-0.69219	-0.00306	-0.00122	0.	0.	0.00071	-0.00004	0.00052	-0.00017
	F2	-0.03658	-0.69219	-0.00306	-0.00122	0.	0.	-0.00071	0.00004	0.00052	-0.00017
* 2A1	C1	-0.00000	-0.00009	-0.00061	0.00548	0.	0.	0.	0.	-0.00035	-0.00280
-15.7494	N1	-0.03673	-0.69229	-0.00157	-0.00413	0.00179	-0.00054	0.	0.	-0.00114	-0.00061
	N2	-0.03673	-0.69229	-0.00157	-0.00413	-0.00179	0.00054	0.	0.	-0.00114	-0.00061
	F1	-0.00000	-0.00000	0.00020	-0.00067	0.	0.	0.00002	0.00012	-0.00011	0.00078
	F2	-0.00000	-0.00000	0.00020	-0.00067	0.	0.	-0.00002	-0.00012	-0.00011	0.00078
* 3A1	C1	0.05191	0.97902	0.00471	0.00086	0.	0.	0.	0.	-0.00022	-0.00055
-11.5465	N1	-0.00001	-0.00023	0.00025	-0.00045	-0.00010	-0.00004	0.	0.	0.00049	-0.00013
	N2	-0.00001	-0.00023	0.00025	-0.00045	0.00010	0.00004	0.	0.	0.00049	-0.00013
	F1	-0.00001	-0.00018	0.00018	0.00019	0.	0.	-0.00052	-0.00018	-0.00037	0.00004
	F2	-0.00001	-0.00018	0.00018	0.00019	0.	0.	0.00052	0.00018	-0.00037	0.00004
* 4A1	C1	0.00486	0.10236	-0.23434	0.04910	0.	0.	0.	0.	-0.06133	-0.01370
-1.7716	N1	0.00148	0.03197	-0.06693	-0.05911	0.03816	0.00203	0.	0.	-0.03365	-0.01331
	N2	0.00148	0.03197	-0.06693	-0.05911	-0.03816	-0.00203	0.	0.	-0.03365	-0.01331
	F1	0.00718	0.15449	-0.34249	-0.29959	0.	0.	0.07641	0.01936	0.04661	0.01922
	F2	0.00718	0.15449	-0.34249	-0.29959	0.	0.	-0.07641	-0.01936	0.04661	0.01922
* 5A1	C1	0.00295	0.06073	-0.15305	0.04068	0.	0.	0.	0.	0.21051	-0.03678
-1.5702	N1	0.00658	0.14052	-0.31629	-0.17196	0.18800	-0.00673	0.	0.	-0.07402	-0.00862
	N2	0.00658	0.14052	-0.31629	-0.17196	-0.18800	0.00673	0.	0.	-0.07402	-0.00862
	F1	-0.00249	-0.05312	0.12317	0.09225	0.	0.	-0.01402	0.00966	0.01098	0.00721
	F2	-0.00249	-0.05312	0.12317	0.09225	0.	0.	0.01402	-0.00966	0.01098	0.00721
* 6A1	C1	-0.00734	-0.15010	0.42583	0.15120	0.	0.	0.	0.	0.06608	-0.06268
-0.9763	N1	0.00215	0.04656	-0.10460	-0.12771	0.09591	0.00846	0.	0.	0.16976	0.01908
	N2	0.00215	0.04656	-0.10460	-0.12771	-0.09591	-0.00846	0.	0.	0.16976	0.01908
	F1	0.00316	0.06753	-0.15835	-0.19563	0.	0.	-0.27127	-0.09619	-0.15689	-0.04670
	F2	0.00316	0.06753	-0.15835	-0.19563	0.	0.	0.27127	0.09619	-0.15689	-0.04670
* 7A1	C1	0.00101	0.02211	-0.05151	-0.15909	0.	0.	0.	0.	0.36713	0.04669
-0.8447	N1	-0.00200	-0.04283	0.10189	0.16319	-0.06246	-0.01470	0.	0.	-0.23582	-0.02191
	N2	-0.00200	-0.04283	0.10189	0.16319	0.06246	0.01470	0.	0.	-0.23582	-0.02191
	F1	0.00086	0.01817	-0.04469	-0.05588	0.	0.	-0.32056	-0.12737	0.19426	0.08423
	F2	0.00086	0.01817	-0.04469	-0.05588	0.	0.	0.32056	0.12737	0.19426	0.08423
* 8A1	C1	0.00076	0.01398	-0.05670	-0.05132	0.	0.	0.	0.	0.25188	-0.05993
-0.7107	N1	-0.00120	-0.02636	0.05642	0.08848	0.05378	0.00460	0.	0.	-0.26736	-0.05200
	N2	-0.00120	-0.02636	0.05642	0.08848	-0.05378	-0.00460	0.	0.	-0.26736	-0.05200
	F1	0.00035	0.00678	-0.02300	-0.00021	0.	0.	0.05393	0.03813	-0.43849	-0.20697
	F2	0.00035	0.00678	-0.02300	-0.00021	0.	0.	-0.05393	-0.03813	-0.43849	-0.20697
* 9A1	C1	0.00028	-0.00081	-0.05465	0.32436	0.	0.	0.	0.	-0.02445	-0.10105
-0.6251	N1	-0.00269	-0.05832	0.13293	0.14732	0.51138	0.13852	0.	0.	0.03510	-0.03343
	N2	-0.00269	-0.05832	0.13293	0.14732	-0.51138	-0.13852	0.	0.	0.03510	-0.03343
	F1	0.00005	0.00218	0.00640	-0.05566	0.	0.	0.02828	0.03049	0.09658	0.07479
	F2	0.00005	0.00218	0.00640	-0.05566	0.	0.	-0.02828	-0.03049	0.09658	0.07479
10A1	C1	0.00645	0.15156	-0.29210	-2.32310	0.	0.	0.	0.	-0.09933	-0.42437
0.2705	N1	-0.00225	-0.04740	0.12951	0.36198	0.04289	-0.04750	0.	0.	0.18640	0.41225
	N2	-0.00225	-0.04740	0.12951	0.36198	-0.04289	0.04750	0.	0.	0.18640	0.41225
	F1	-0.00306	-0.07128	0.11991	0.75453	0.	0.	-0.23615	-0.41102	-0.09652	-0.15738
	F2	-0.00306	-0.07128	0.11991	0.75453	0.	0.	0.23615	0.41102	-0.09652	-0.15738

Center		S1	S2	S3	S4	X1	X2	Y1	Y2	Z1	Z2
11A1	C1	0.00331	0.07482	-0.17640	-2.12509	0.	0.	0.	0.	0.35786	2.29112
0.3162	N1	-0.00252	-0.06606	0.05660	1.51051	-0.03252	-0.06476	0.	0.	0.03014	0.74589
	N2	-0.00252	-0.06606	0.05660	1.51051	0.03252	0.06476	0.	0.	0.03014	0.74589
	F1	0.00083	0.01954	-0.03085	-0.23358	0.	0.	0.04794	0.09650	-0.14051	-0.30428
	F2	0.00083	0.01954	-0.03085	-0.23358	0.	0.	-0.04794	-0.09650	-0.14051	-0.30428
12A1	C1	0.00267	0.00704	-0.50006	1.55781	0.	0.	0.	0.	1.01130	-1.66288
0.5111	N1	-0.00038	0.00010	0.07512	-0.72392	-0.26379	0.24502	0.	0.	0.08511	0.10981
	N2	-0.00038	0.00010	0.07512	-0.72392	0.26379	-0.24502	0.	0.	0.08511	0.10981
	F1	0.00093	0.01799	-0.06320	-0.06782	0.	0.	0.10309	0.15385	0.12044	0.27778
	F2	0.00093	0.01799	-0.06320	-0.06782	0.	0.	-0.10309	-0.15385	0.12044	0.27778
13A1	C1	-0.00793	-0.10677	0.88810	-1.99266	0.	0.	0.	0.	0.44215	-0.04936
0.5642	N1	-0.00135	-0.03519	0.03361	0.27518	0.33966	-0.67235	0.	0.	-0.21215	0.55521
	N2	-0.00135	-0.03519	0.03361	0.27518	-0.33966	0.67235	0.	0.	-0.21215	0.55521
	F1	0.00082	0.01101	-0.09000	0.28624	0.	0.	0.20690	-0.02373	0.15126	0.02479
	F2	0.00082	0.01101	-0.09000	0.28624	0.	0.	-0.20690	0.02373	0.15126	0.02479
14A1	C1	-0.00842	-0.06711	1.26521	-1.98205	0.	0.	0.	0.	-0.22285	0.23112
0.6775	N1	-0.00015	0.01078	0.10959	0.53545	-0.31317	0.79526	0.	0.	-0.24245	0.22560
	N2	-0.00015	0.01078	0.10959	0.53545	0.31317	-0.79526	0.	0.	-0.24245	0.22560
	F1	-0.00053	-0.01366	0.01240	0.35316	0.	0.	0.15379	-0.16957	0.04776	-0.09455
	F2	-0.00053	-0.01366	0.01240	0.35316	0.	0.	-0.15379	0.16957	0.04776	-0.09455
* 1A2	C1	0.	0.	0.	0.	0.	0.	0.	0.	0.	0.
-0.7306	N1	0.	0.	0.	0.	0.	0.	0.01023	0.01790	0.	0.
	N2	0.	0.	0.	0.	0.	0.	-0.01023	-0.01790	0.	0.
	F1	0.	0.	0.	0.	0.54484	0.26355	0.	0.	0.	0.
	F2	0.	0.	0.	0.	-0.54484	-0.26355	0.	0.	0.	0.
2A2	C1	0.	0.	0.	0.	0.	0.	0.	0.	0.	0.
0.0221	N1	0.	0.	0.	0.	0.	0.	0.52562	0.51179	0.	0.
	N2	0.	0.	0.	0.	0.	0.	-0.52562	-0.51179	0.	0.
	F1	0.	0.	0.	0.	-0.04903	-0.01521	0.	0.	0.	0.
	F2	0.	0.	0.	0.	0.04903	0.01521	0.	0.	0.	0.
* 1B1	C1	0.	0.	0.	0.	-0.00020	-0.00056	0.	0.	0.	0.
-15.7483	N1	0.03674	0.69245	0.00150	0.00457	-0.00182	-0.00099	0.	0.	0.00113	-0.00017
	N2	-0.03674	-0.69245	-0.00150	-0.00457	-0.00182	-0.00099	0.	0.	-0.00113	0.00017
	F1	0.	0.	0.	0.	-0.00003	0.00014	0.	0.	0.	0.
	F2	0.	0.	0.	0.	-0.00003	0.00014	0.	0.	0.	0.
* 2B1	C1	0.	0.	0.	0.	-0.33412	-0.02002	0.	0.	0.	0.
-0.9790	N1	0.00651	0.13923	-0.33059	-0.37049	-0.09643	0.02043	0.	0.	-0.12169	-0.01836
	N2	-0.00651	-0.13923	0.33059	0.37049	-0.09643	0.02043	0.	0.	0.12169	0.01836
	F1	0.	0.	0.	0.	-0.14884	-0.04459	0.	0.	0.	0.
	F2	0.	0.	0.	0.	-0.14884	-0.04459	0.	0.	0.	0.
* 3B1	C1	0.	0.	0.	0.	-0.15504	-0.04059	0.	0.	0.	0.
-0.7741	N1	-0.00374	-0.08010	0.19094	0.26763	0.08492	0.00637	0.	0.	-0.03280	0.00390
	N2	0.00374	0.08010	-0.19094	-0.26763	0.08492	0.00637	0.	0.	0.03280	-0.00390
	F1	0.	0.	0.	0.	-0.45788	-0.20486	0.	0.	0.	0.
	F2	0.	0.	0.	0.	-0.45788	-0.20486	0.	0.	0.	0.
* 4B1	C1	0.	0.	0.	0.	0.47410	0.03102	0.	0.	0.	0.
-0.4851	N1	0.00213	0.04354	-0.12237	-0.04648	-0.14267	-0.11091	0.	0.	0.40734	0.18945
	N2	-0.00213	-0.04354	0.12237	0.04648	-0.14267	-0.11091	0.	0.	-0.40734	-0.18945
	F1	0.	0.	0.	0.	-0.22948	-0.13931	0.	0.	0.	0.
	F2	0.	0.	0.	0.	-0.22948	-0.13931	0.	0.	0.	0.

Center   Basis Function Type

CF2N2

Center		S1	S2	S3	S4	X1	X2	Y1	Y2	Z1	Z2
5B1	C1	0.	0.	0.	0.	0.48638	1.34560	0.	0.	0.	0.
0.2568	N1	0.00257	0.06193	-0.09377	-1.02604	-0.00479	0.21349	0.	0.	-0.32620	-0.66041
	N2	-0.00257	-0.06193	0.09377	1.02604	-0.00479	0.21349	0.	0.	0.32620	0.66041
	F1	0.	0.	0.	0.	-0.19746	-0.25301	0.	0.	0.	0.
	F2	0.	0.	0.	0.	-0.19746	-0.25301	0.	0.	0.	0.
6B1	C1	0.	0.	0.	0.	-0.59375	0.34782	0.	0.	0.	0.
0.2970	N1	0.00223	0.05780	-0.05543	-1.34991	0.26849	0.98776	0.	0.	0.26916	0.25438
	N2	-0.00223	-0.05780	0.05543	1.34991	0.26849	0.98776	0.	0.	-0.26916	-0.25438
	F1	0.	0.	0.	0.	0.03050	-0.03397	0.	0.	0.	0.
	F2	0.	0.	0.	0.	0.03050	-0.03397	0.	0.	0.	0.
7B1	C1	0.	0.	0.	0.	-0.76319	1.68150	0.	0.	0.	0.
0.5058	N1	-0.00325	-0.06941	0.18560	1.11712	-0.16935	-1.46700	0.	0.	0.03285	-0.16651
	N2	0.00325	0.06941	-0.18560	-1.11712	-0.16935	-1.46700	0.	0.	-0.03285	0.16651
	F1	0.	0.	0.	0.	-0.06940	-0.18904	0.	0.	0.	0.
	F2	0.	0.	0.	0.	-0.06940	-0.18904	0.	0.	0.	0.
8B1	C1	0.	0.	0.	0.	0.10599	-0.43962	0.	0.	0.	0.
0.6823	N1	0.00209	0.02718	-0.23705	0.66028	0.62009	-0.49159	0.	0.	-0.46242	1.04259
	N2	-0.00209	-0.02718	0.23705	-0.66028	0.62009	-0.49159	0.	0.	0.46242	-1.04259
	F1	0.	0.	0.	0.	-0.05181	0.03683	0.	0.	0.	0.
	F2	0.	0.	0.	0.	-0.05181	0.03683	0.	0.	0.	0.
* 1B2	C1	0.	0.	0.	0.	0.	0.	-0.00013	-0.00075	0.	0.
-26.3726	N1	0.	0.	0.	0.	0.	0.	-0.00001	0.00020	0.	0.
	N2	0.	0.	0.	0.	0.	0.	-0.00001	0.00020	0.	0.
	F1	0.03658	0.69219	0.00309	0.00116	0.	0.	-0.00076	0.00028	-0.00049	0.00003
	F2	-0.03658	-0.69219	-0.00309	-0.00116	0.	0.	-0.00076	0.00028	0.00049	-0.00003
* 2B2	C1	0.	0.	0.	0.	0.	0.	-0.16962	0.02406	0.	0.
-1.6915	N1	0.	0.	0.	0.	0.	0.	-0.00651	-0.00221	0.	0.
	N2	0.	0.	0.	0.	0.	0.	-0.00651	-0.00221	0.	0.
	F1	0.00806	0.17315	-0.38770	-0.33335	0.	0.	0.05384	0.00920	0.04874	0.00643
	F2	-0.00806	-0.17315	0.38770	0.33335	0.	0.	0.05384	0.00920	-0.04874	-0.00643
* 3B2	C1	0.	0.	0.	0.	0.	0.	0.42752	-0.03522	0.	0.
-0.8569	N1	0.	0.	0.	0.	0.	0.	0.12275	0.04268	0.	0.
	N2	0.	0.	0.	0.	0.	0.	0.12275	0.04268	0.	0.
	F1	0.00217	0.04634	-0.11051	-0.15846	0.	0.	-0.20126	-0.05788	-0.36387	-0.13434
	F2	-0.00217	-0.04634	0.11051	0.15846	0.	0.	-0.20126	-0.05788	0.36387	0.13434
* 4B2	C1	0.	0.	0.	0.	0.	0.	-0.11805	0.06506	0.	0.
-0.7017	N1	0.	0.	0.	0.	0.	0.	0.02560	0.00668	0.	0.
	N2	0.	0.	0.	0.	0.	0.	0.02560	0.00668	0.	0.
	F1	-0.00030	-0.00563	0.02136	-0.00732	0.	0.	0.47296	0.21612	-0.28891	-0.14061
	F2	0.00030	0.00563	-0.02136	0.00732	0.	0.	0.47296	0.21612	0.28891	0.14061
* 5B2	C1	0.	0.	0.	0.	0.	0.	0.00401	0.06209	0.	0.
-0.5800	N1	0.	0.	0.	0.	0.	0.	0.47027	0.18513	0.	0.
	N2	0.	0.	0.	0.	0.	0.	0.47027	0.18513	0.	0.
	F1	0.00014	0.00413	0.00081	-0.05808	0.	0.	0.06728	0.03674	0.20296	0.10601
	F2	-0.00014	-0.00413	-0.00081	0.05808	0.	0.	0.06728	0.03674	-0.20296	-0.10601
6B2	C1	0.	0.	0.	0.	0.	0.	-0.48351	-1.78873	0.	0.
0.3425	N1	0.	0.	0.	0.	0.	0.	0.18469	0.37935	0.	0.
	N2	0.	0.	0.	0.	0.	0.	0.18469	0.37935	0.	0.
	F1	-0.00277	-0.06631	0.09564	0.83283	0.	0.	-0.03180	-0.10458	-0.16824	-0.27535
	F2	0.00277	0.06631	-0.09564	-0.83283	0.	0.	-0.03180	-0.10458	0.16824	0.27535

Center		S1	S2	S3	S4	X1	X2	Y1	Y2	Z1	Z2
7B2	C1	0.	0.	0.	0.	0.	0.	1.15263	-0.93455	0.	0.
0.5273	N1	0.	0.	0.	0.	0.	0.	0.15833	-0.10376	0.	0.
	N2	0.	0.	0.	0.	0.	0.	0.15833	-0.10376	0.	0.
	F1	0.00104	0.01787	-0.08575	0.10270	0.	0.	0.25885	0.23366	0.17682	0.09927
	F2	-0.00104	-0.01787	0.08575	-0.10270	0.	0.	0.25885	0.23366	-0.17682	-0.09927

Overlap Matrix

First Function   Second Function and Overlap

C1S1 with
C1S2 .35436	C1S3 .00805	C1S4 .01419	N1S3 .00034	N1S4 .00405	N1X1 -.00089	N1X2 -.00461
N1Z1 .00175	N1Z2 .00906	F1S3 .00005	F1S4 .00303	F1Y1 -.00044	F1Y2 -.00894	F1Z1 -.00029
F1Z2 -.00605						

C1S2 with
C1S3 .26226	C1S4 .17745	N1S3 .00727	N1S4 .05331	N1X1 -.01352	N1X2 -.05773	N1Z1 .02656
N1Z2 .11343	F1S3 .00233	F1S4 .04200	F1Y1 -.01001	F1Y2 -.11146	F1Z1 -.00677	F1Z2 -.07546

C1S3 with
C1S4 .79970	N1S1 .00091	N1S2 .01376	N1S3 .13974	N1S4 .31755	N1X1 -.10447	N1X2 -.25203
N1Z1 .20525	N1Z2 .49517	F1S1 .00103	F1S2 .01429	F1S3 .11433	F1S4 .28843	F1Y1 -.13318
F1Y2 -.44809	F1Z1 -.09017	F1Z2 -.30338				

C1S4 with
N1S1 .00383	N1S2 .04952	N1S3 .28494	N1S4 .52237	N1X1 -.09584	N1X2 -.26567	N1Z1 .18831
N1Z2 .52197	F1S1 .00305	F1S2 .03923	F1S3 .22518	F1S4 .45495	F1Y1 -.10349	F1Y2 -.38973
F1Z1 -.07006	F1Z2 -.26387					

C1X1 with
C1X2 .53875	N1S1 .00152	N1S2 .02016	N1S3 .12358	N1S4 .15346	N1X1 .02840	N1X2 .18435
N1Z1 .16821	N1Z2 .20025	F1X1 .08270	F1X2 .29857			

C1X2 with
N1S1 .00333	N1S2 .04215	N1S3 .20916	N1S4 .29220	N1X1 .13909	N1X2 .46797	N1Z1 .11063
N1Z2 .23314	F1X1 .11992	F1X2 .48308				

C1Y1 with
C1Y2 .53875	N1Y1 .11402	N1Y2 .28627	F1S1 .00259	F1S2 .03365	F1S3 .19973	F1S4 .32390
F1Y1 -.13168	F1Y2 -.12776	F1Z1 -.14515	F1Z2 -.28864			

C1Y2 with
N1Y1 .19539	N1Y2 .58663	F1S1 .00432	F1S2 .05476	F1S3 .28952	F1S4 .49022	F1Y1 .01486
F1Y2 .15276	F1Z1 -.07114	F1Z2 -.22364				

C1Z1 with
C1Z2 .53875	N1S1 -.00299	N1S2 -.03962	N1S3 -.24280	N1S4 -.30151	N1X1 .16821	N1X2 .20025
N1Z1 -.21647	N1Z2 -.10717	F1S1 .00176	F1S2 .02279	F1S3 .13523	F1S4 .21930	F1Y1 -.14515
F1Y2 -.28864	F1Z1 -.01557	F1Z2 .10314				

C1Z2 with
N1S1 -.00655	N1S2 -.08281	N1S3 -.41095	N1S4 -.57410	N1X1 .11063	N1X2 .23314	N1Z1 -.02196
N1Z2 .12856	F1S1 .00292	F1S2 .03708	F1S3 .19602	F1S4 .33190	F1Y1 -.07114	F1Y2 -.22364
F1Z1 .07176	F1Z2 .33166					

N1S1 with
N1S2 .35584	N1S3 .00894	N1S4 .01446	N2S3 .00065	N2S4 .00418	N2X1 .00279	N2X2 .00902
F1S4 .00002	F1X2 .00009	F1Y2 -.00016	F1Z2 -.00029			

N1S2 with
N1S3 .27246	N1S4 .17925	N2S3 .01099	N2S4 .05409	N2X1 .03855	N2X2 .11286	F1S4 .00030
F1X2 .00139	F1Y2 -.00235	F1Z2 -.00433				

N1S3 with
N1S4 .79352	N2S1 .00065	N2S2 .01099	N2S3 .14270	N2S4 .30896	N2X1 .25454	N2X2 .52034
F1S3 .00017	F1S4 .00873	F1X1 .00023	F1X2 .01615	F1Y1 -.00039	F1Y2 -.02718	F1Z1 -.00072
F1Z2 -.05013						

N1S4 with
N2S1 .00418	N2S2 .05409	N2S3 .30896	N2S4 .53726	N2X1 .26994	N2X2 .63240	F1S1 .00015
F1S2 .00206	F1S3 .01751	F1S4 .06719	F1X1 .00667	F1X2 .05056	F1Y1 -.01122	F1Y2 -.08511
F1Z1 -.02070	F1Z2 -.15696					

N1X1 with
N1X2 .52739	N2S1 -.00279	N2S2 -.03855	N2S3 -.25454	N2S4 -.26994	N2X1 -.31152	N2X2 -.13501
F1S2 -.00001	F1S3 -.00044	F1S4 -.00639	F1X1 -.00001	F1X2 .00719	F1Y1 .00069	F1Y2 .01504
F1Z1 .00128	F1Z2 .02773					

Overlap Matrix (continued)

First Function   Second Function and Overlap

N1X2 with
   N2S1 -.00902   N2S2 -.11286   N2S3 -.52034   N2S4 -.63240   N2X1 -.13501   N2X2 .00711   F1S1 -.00030
   F1S2 -.00388   F1S3 -.02547   F1S4 -.06262   F1X1 .01041   F1X2 .08047   F1Y1 .01304   F1Y2 .06026
   F1Z1 .02405   F1Z2 .11113

N1Y1 with
   N1Y2 .52739   N2Y1 .11523   N2Y2 .24145   F1S2 .00002   F1S3 .00075   F1S4 .01077   F1X1 .00069
   F1X2 .01504   F1Y1 -.00077   F1Y2 -.00919   F1Z1 -.00215   F1Z2 -.04669

N1Y2 with
   N2Y1 .24145   N2Y2 .61007   F1S1 .00050   F1S2 .00653   F1S3 .04288   F1S4 .10542   F1X1 .01304
   F1X2 .06026   F1Y1 -.00380   F1Y2 .01482   F1Z1 -.04049   F1Z2 -.18707

N1Z1 with
   N1Z2 .52739   N2Z1 .11523   N2Z2 .24145   F1S2 .00003   F1S3 .00138   F1S4 .01985   F1X1 .00128
   F1X2 .02773   F1Y1 -.00215   F1Y2 -.04669   F1Z1 -.00357   F1Z2 -.06997

N1Z2 with
   N2Z1 .24145   N2Z2 .61007   F1S1 .00092   F1S2 .01204   F1S3 .07908   F1S4 .19442   F1X1 .02405
   F1X2 .11113   F1Y1 -.04049   F1Y2 -.18707   F1Z1 -.05651   F1Z2 -.22873

F1S1 with
   F1S2 .35647   F1S3 .01003   F1S4 .01474   F2S4 .00003   F2Y2 .00051

F1S2 with
   F1S3 .28653   F1S4 .18096   F2S4 .00053   F2Y2 .00698

F1S3 with
   F1S4 .78531   F2S3 .00004   F2S4 .00761   F2Y1 .00039   F2Y2 .05739

F1S4 with
   F2S1 .00003   F2S2 .00053   F2S3 .00761   F2S4 .04848   F2Y1 .01577   F2Y2 .17461

F1X1 with
   F1X2 .49444   F2X1 .00014   F2X2 .01238

F1X2 with
   F2X1 .01238   F2X2 .09883

F1Y1 with
   F1Y2 .49444   F2S3 -.00039   F2S4 -.01577   F2Y1 -.00207   F2Y2 -.07825

F1Y2 with
   F2S1 -.00051   F2S2 -.00698   F2S3 -.05739   F2S4 -.17461   F2Y1 -.07825   F2Y2 -.35862

F1Z1 with
   F1Z2 .49444   F2Z1 .00014   F2Z2 .01238

F1Z2 with
   F2Z1 .01238   F2Z2 .09883

CF4   Carbon Tetrafluoride

Molecular Geometry   Symmetry T_d

Center	Coordinates		
	X	Y	Z
C1	0.	0.	0.
F1	1.44345750	1.44345750	1.44345750
F2	-1.44345750	1.44345750	-1.44345750
F3	1.44345750	-1.44345750	-1.44345750
F4	-1.44345750	-1.44345750	1.44345750
Mass	0.	0.	0.

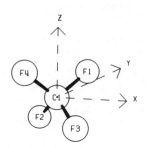

MO Symmetry Types and Occupancy

Representation	A1	A2	E	F1	F2
Number of MOs	10	0	4	6	30
Number Occupied	4	0	2	3	12

Energy Expectation Values

Total	-435.5554	Kinetic	435.8603
Electronic	-640.9890	Potential	-871.4157
Nuclear Repulsion	205.4336	Virial	1.9993

Electronic Potential	-1076.8493
One-Electron Potential	-1450.0541
Two-Electron Potential	373.2048

Moments of the Charge Distribution

Second

x^2	y^2	z^2	xy	xz	yz	r^2
-95.1250	-95.1250	-95.1250	0.	0.	0.	-285.3749
75.0085	75.0085	75.0085	0.	0.	0.	225.0255

Third

x^3	xy^2	xz^2	x^2y	y^3	yz^2	x^2z	y^2z	z^3	xyz
0.	0.	0.	0.	0.	0.	0.	0.	0.	-111.9867
0.	0.	0.	0.	0.	0.	0.	0.	0.	108.2716

Fourth (even)

x^4	y^4	z^4	x^2y^2	x^2z^2	y^2z^2	r^2x^2	r^2y^2	r^2z^2	r^4
-384.4960	-384.4960	-384.4960	-234.2907	-234.2907	-234.2907	-853.0775	-853.0774	-853.0774	-2559.2322
156.2854	156.2854	156.2854	156.2854	156.2854	156.2854	468.8563	468.8563	468.8563	1406.5689

Expectation Values at Atomic Centers

Center			Expectation Value		
	r^{-1}	δ	xr^{-3}	yr^{-3}	zr^{-3}
C1	-14.2761	116.2252	0.	0.	0.
F1	-26.4859	413.6805	0.1594	0.1594	0.1594

	$\dfrac{3x^2-r^2}{r^5}$	$\dfrac{3y^2-r^2}{r^5}$	$\dfrac{3z^2-r^2}{r^5}$	$\dfrac{3xy}{r^5}$	$\dfrac{3xz}{r^5}$	$\dfrac{3yz}{r^5}$
C1	0.	0.	0.	0.	0.	0.
F1	0.	0.	0.	1.7715	1.7715	1.7715

Mulliken Charge

Center				Basis Function Type							Total
	S1	S2	S3	S4	X1	X2	Y1	Y2	Z1	Z2	
C1	0.0427	1.9567	0.8525	-0.0171	0.7160	0.0269	0.7160	0.0269	0.7160	0.0269	5.0634
F1	0.0436	1.9557	1.0326	0.9017	1.3214	0.4455	1.3214	0.4455	1.3214	0.4455	9.2341

Overlap Populations

Pair of Atomic Centers

	F1	F2
	C1	F1
1A1	-0.0002	0.
2A1	0.0004	0.
3A1	0.1019	0.0196
4A1	0.0459	0.0001
Total A1	0.1481	0.0197
1E	0.	0.0287
2E	0.	-0.0043
Total E	0.	0.0244
1F1	0.	-0.0225
2F1	0.	-0.0311
3F1	0.	-0.0311
Total F1	0.	-0.0848
1F2	-0.0001	0.
2F2	-0.0001	0.
3F2	-0.0001	0.
4F2	0.0597	0.0202
5F2	0.0597	-0.0209
6F2	0.0597	-0.0209
7F2	0.0638	0.0141
8F2	0.0638	0.0009
9F2	0.0638	0.0009
10F2	0.0258	0.0197
11F2	0.0258	-0.0378
12F2	0.0258	-0.0378
Total F2	0.4477	-0.0616
Total	0.5958	-0.1023
Distance	2.5001	4.0827

Molecular Orbitals and Eigenvalues

Center		Basis Function Type									
		S1	S2	S3	S4	X1	X2	Y1	Y2	Z1	Z2
* 1A1	C1	-0.00000	0.00007	0.00008	-0.00189	0.	0.	0.	0.	0.	0.
-26.3910	F1	0.02587	0.48945	0.00202	0.00129	-0.00037	-0.00004	-0.00037	-0.00004	-0.00037	-0.00004
	F2	0.02587	0.48945	0.00202	0.00129	0.00037	0.00004	-0.00037	-0.00004	0.00037	0.00004
	F3	0.02587	0.48945	0.00202	0.00129	-0.00037	-0.00004	0.00037	0.00004	0.00037	0.00004
	F4	0.02587	0.48945	0.00202	0.00129	0.00037	0.00004	0.00037	0.00004	-0.00037	-0.00004
* 2A1	C1	0.05193	0.97926	0.00365	-0.00092	0.	0.	0.	0.	0.	0.
-11.6761	F1	-0.00001	-0.00018	0.00001	0.00072	-0.00038	-0.00024	-0.00038	-0.00024	-0.00038	-0.00024
	F2	-0.00001	-0.00018	0.00001	0.00072	0.00038	0.00024	-0.00038	-0.00024	0.00038	0.00024
	F3	-0.00001	-0.00018	0.00001	0.00072	-0.00038	-0.00024	0.00038	0.00024	0.00038	0.00024
	F4	-0.00001	-0.00018	0.00001	0.00072	0.00038	0.00024	0.00038	0.00024	-0.00038	-0.00024
* 3A1	C1	0.00560	0.11774	-0.27363	0.00829	0.	0.	0.	0.	0.	0.
-1.8558	F1	0.00498	0.10688	-0.23794	-0.19856	0.04343	0.01039	0.04343	0.01039	0.04343	0.01039
	F2	0.00498	0.10688	-0.23794	-0.19856	-0.04343	-0.01039	0.04343	0.01039	-0.04343	-0.01039
	F3	0.00498	0.10688	-0.23794	-0.19856	0.04343	0.01039	-0.04343	-0.01039	-0.04343	-0.01039
	F4	0.00498	0.10688	-0.23794	-0.19856	-0.04343	-0.01039	-0.04343	-0.01039	0.04343	0.01039
* 4A1	C1	-0.00762	-0.15167	0.46806	-0.00678	0.	0.	0.	0.	0.	0.
-1.0347	F1	0.00302	0.06418	-0.15392	-0.16106	-0.14399	-0.05305	-0.14399	-0.05305	-0.14399	-0.05305
	F2	0.00302	0.06418	-0.15392	-0.16106	0.14399	0.05305	-0.14399	-0.05305	0.14399	0.05305
	F3	0.00302	0.06418	-0.15392	-0.16106	-0.14399	-0.05305	0.14399	0.05305	0.14399	0.05305
	F4	0.00302	0.06418	-0.15392	-0.16106	0.14399	0.05305	0.14399	0.05305	-0.14399	-0.05305
5A1	C1	0.00664	0.15911	-0.28856	-2.52957	0.	0.	0.	0.	0.	0.
0.3090	F1	-0.00294	-0.06744	0.12311	0.69828	-0.12184	-0.25631	-0.12184	-0.25631	-0.12184	-0.25631
	F2	-0.00294	-0.06744	0.12311	0.69828	0.12184	0.25631	-0.12184	-0.25631	0.12184	0.25631
	F3	-0.00294	-0.06744	0.12311	0.69828	-0.12184	-0.25631	0.12184	0.25631	0.12184	0.25631
	F4	-0.00294	-0.06744	0.12311	0.69828	0.12184	0.25631	0.12184	0.25631	-0.12184	-0.25631
6A1	C1	-0.01228	-0.13535	1.56878	-2.36866	0.	0.	0.	0.	0.	0.
0.5369	F1	0.00011	-0.00522	-0.05947	0.49163	0.15879	-0.09138	0.15879	-0.09138	0.15879	-0.09138
	F2	0.00011	-0.00522	-0.05947	0.49163	-0.15879	0.09138	0.15879	-0.09138	-0.15879	0.09138
	F3	0.00011	-0.00522	-0.05947	0.49163	0.15879	-0.09138	-0.15879	0.09138	-0.15879	0.09138
	F4	0.00011	-0.00522	-0.05947	0.49163	-0.15879	0.09138	-0.15879	0.09138	0.15879	-0.09138
* 1E	C1	0.	0.	0.	0.	0.	0.	0.	0.	0.	0.
-0.7842	F1	0.	0.	0.	0.	0.26475	0.12883	-0.26475	-0.12883	0.	0.
	F2	0.	0.	0.	0.	-0.26475	-0.12883	-0.26475	-0.12883	0.	0.
	F3	0.	0.	0.	0.	0.26475	0.12883	0.26475	0.12883	0.	0.
	F4	0.	0.	0.	0.	-0.26475	-0.12883	0.26475	0.12883	0.	0.
* 2E	C1	0.	0.	0.	0.	0.	0.	0.	0.	0.	0.
-0.7842	F1	0.	0.	0.	0.	0.15286	0.07438	0.15286	0.07438	-0.30571	-0.14876
	F2	0.	0.	0.	0.	-0.15286	-0.07438	0.15286	0.07438	0.30571	0.14876
	F3	0.	0.	0.	0.	0.15286	0.07438	-0.15286	-0.07438	0.30571	0.14876
	F4	0.	0.	0.	0.	-0.15286	-0.07438	-0.15286	-0.07438	-0.30571	-0.14876
* 1F1	C1	0.	0.	0.	0.	0.	0.	0.	0.	0.	0.
-0.7129	F1	0.	0.	0.	0.	0.28399	0.13283	0.	0.	-0.28399	-0.13283
	F2	0.	0.	0.	0.	-0.28399	-0.13283	0.	0.	0.28399	0.13283
	F3	0.	0.	0.	0.	-0.28399	-0.13283	0.	0.	-0.28399	-0.13283
	F4	0.	0.	0.	0.	0.28399	0.13283	0.	0.	0.28399	0.13283
* 2F1	C1	0.	0.	0.	0.	0.	0.	0.	0.	0.	0.
-0.7129	F1	0.	0.	0.	0.	0.	0.	-0.28399	-0.13283	0.28399	0.13283
	F2	0.	0.	0.	0.	0.	0.	0.28399	0.13283	0.28399	0.13283
	F3	0.	0.	0.	0.	0.	0.	0.28399	0.13283	-0.28399	-0.13283
	F4	0.	0.	0.	0.	0.	0.	-0.28399	-0.13283	-0.28399	-0.13283

Center		S1	S2	S3	S4	X1	X2	Y1	Y2	Z1	Z2
						Basis Function Type					
* 3F1	C1	0.	0.	0.	0.	0.	0.	0.	0.	0.	0.
-0.7129	F1	0.	0.	0.	0.	-0.28399	-0.13283	0.28399	0.13283	0.	0.
	F2	0.	0.	0.	0.	-0.28399	-0.13283	-0.28399	-0.13283	0.	0.
	F3	0.	0.	0.	0.	0.28399	0.13283	0.28399	0.13283	0.	0.
	F4	0.	0.	0.	0.	0.28399	0.13283	-0.28399	-0.13283	0.	0.
* 1F2	C1	0.	0.	0.	0.	0.	0.	-0.00014	-0.00068	0.	0.
-26.3911	F1	0.02587	0.48945	0.00222	0.00076	-0.00036	0.00004	-0.00041	0.00027	-0.00036	0.00004
	F2	0.02587	0.48945	0.00222	0.00076	0.00036	-0.00004	-0.00041	0.00027	0.00036	-0.00004
	F3	-0.02587	-0.48945	-0.00222	-0.00076	0.00036	-0.00004	-0.00041	0.00027	-0.00036	0.00004
	F4	-0.02587	-0.48945	-0.00222	-0.00076	-0.00036	0.00004	-0.00041	0.00027	0.00036	-0.00004
* 2F2	C1	0.	0.	0.	0.	-0.00014	-0.00068	0.	0.	0.	0.
-26.3911	F1	0.02587	0.48945	0.00222	0.00076	-0.00041	0.00027	-0.00036	0.00004	-0.00036	0.00004
	F2	-0.02587	-0.48945	-0.00222	-0.00076	-0.00041	0.00027	-0.00036	0.00004	-0.00036	0.00004
	F3	0.02587	0.48945	0.00222	0.00076	-0.00041	0.00027	0.00036	-0.00004	0.00036	-0.00004
	F4	-0.02587	-0.48945	-0.00222	-0.00076	-0.00041	0.00027	-0.00036	0.00004	0.00036	-0.00004
* 3F2	C1	0.	0.	0.	0.	0.	0.	0.	0.	-0.00014	-0.00068
-26.3911	F1	0.02587	0.48945	0.00222	0.00076	-0.00036	0.00004	-0.00036	0.00004	-0.00041	0.00027
	F2	-0.02587	-0.48945	-0.00222	-0.00076	-0.00036	0.00004	0.00036	-0.00004	-0.00041	0.00027
	F3	-0.02587	-0.48945	-0.00222	-0.00076	0.00036	-0.00004	-0.00036	0.00004	-0.00041	0.00027
	F4	0.02587	0.48945	0.00222	0.00076	0.00036	-0.00004	0.00036	-0.00004	-0.00041	0.00027
* 4F2	C1	0.	0.	0.	0.	0.	0.	0.17774	-0.01600	0.	0.
-1.7143	F1	-0.00568	-0.12201	0.27363	0.23247	-0.03605	-0.00483	-0.01894	-0.00796	-0.03605	-0.00483
	F2	-0.00568	-0.12201	0.27363	0.23247	0.03605	0.00483	-0.01894	-0.00796	0.03605	0.00483
	F3	0.00568	0.12201	-0.27363	-0.23247	0.03605	0.00483	-0.01894	-0.00796	-0.03605	-0.00483
	F4	0.00568	0.12201	-0.27363	-0.23247	-0.03605	-0.00483	-0.01894	-0.00796	0.03605	0.00483
* 5F2	C1	0.	0.	0.	0.	0.17774	-0.01600	0.	0.	0.	0.
-1.7143	F1	-0.00568	-0.12201	0.27363	0.23247	-0.01894	-0.00796	-0.03605	-0.00483	-0.03605	-0.00483
	F2	0.00568	0.12201	-0.27363	-0.23247	-0.01894	-0.00796	0.03605	0.00483	-0.03605	-0.00483
	F3	-0.00568	-0.12201	0.27363	0.23247	-0.01894	-0.00796	0.03605	0.00483	0.03605	0.00483
	F4	0.00568	0.12201	-0.27363	-0.23247	-0.01894	-0.00796	-0.03605	-0.00483	0.03605	0.00483
* 6F2	C1	0.	0.	0.	0.	0.	0.	0.	0.	0.17774	-0.01600
-1.7143	F1	-0.00568	-0.12201	0.27363	0.23247	-0.03605	-0.00483	-0.03605	-0.00483	-0.01894	-0.00796
	F2	0.00568	0.12201	-0.27363	-0.23247	-0.03605	-0.00483	0.03605	0.00483	-0.01894	-0.00796
	F3	0.00568	0.12201	-0.27363	-0.23247	0.03605	0.00483	-0.03605	-0.00483	-0.01894	-0.00796
	F4	-0.00568	-0.12201	0.27363	0.23247	0.03605	0.00483	0.03605	0.00483	-0.01894	-0.00796
* 7F2	C1	0.	0.	0.	0.	0.	0.	0.41511	-0.01390	0.	0.
-0.9148	F1	0.00161	0.03449	-0.08067	-0.11986	-0.21339	-0.07276	0.05586	0.03260	-0.21339	-0.07276
	F2	0.00161	0.03449	-0.08067	-0.11986	0.21339	0.07276	0.05586	0.03260	0.21339	0.07276
	F3	-0.00161	-0.03449	0.08067	0.11986	0.21339	0.07276	0.05586	0.03260	-0.21339	-0.07276
	F4	-0.00161	-0.03449	0.08067	0.11986	-0.21339	-0.07276	0.05586	0.03260	0.21339	0.07276
* 8F2	C1	0.	0.	0.	0.	0.41511	-0.01390	0.	0.	0.	0.
-0.9148	F1	0.00161	0.03449	-0.08067	-0.11986	0.05586	0.03260	-0.21339	-0.07276	-0.21339	-0.07276
	F2	-0.00161	-0.03449	0.08067	0.11986	0.05586	0.03260	0.21339	0.07276	-0.21339	-0.07276
	F3	0.00161	0.03449	-0.08067	-0.11986	0.05586	0.03260	0.21339	0.07276	0.21339	0.07276
	F4	-0.00161	-0.03449	0.08067	0.11986	0.05586	0.03260	-0.21339	-0.07276	0.21339	0.07276
* 9F2	C1	0.	0.	0.	0.	0.	0.	0.	0.	0.41511	-0.01390
-0.9148	F1	0.00161	0.03449	-0.08067	-0.11986	-0.21339	-0.07276	-0.21339	-0.07276	0.05586	0.03260
	F2	-0.00161	-0.03449	0.08067	0.11986	-0.21339	-0.07276	0.21339	0.07276	0.05586	0.03260
	F3	-0.00161	-0.03449	0.08067	0.11986	0.21339	0.07276	-0.21339	-0.07276	0.05586	0.03260
	F4	0.00161	0.03449	-0.08067	-0.11986	0.21339	0.07276	0.21339	0.07276	0.05586	0.03260

Center		Basis Function Type									
		S1	S2	S3	S4	X1	X2	Y1	Y2	Z1	Z2
* 10F2	C1	0.	0.	0.	0.	0.	0.	0.19407	-0.08550	0.	0.
-0.7223	F1	0.00039	0.00737	-0.02575	0.00121	-0.02333	-0.00572	-0.38235	-0.17412	-0.02333	-0.00572
	F2	0.00039	0.00737	-0.02575	0.00121	0.02333	0.00572	-0.38235	-0.17412	0.02333	0.00572
	F3	-0.00039	-0.00737	0.02575	-0.00121	0.02333	0.00572	-0.38235	-0.17412	-0.02333	-0.00572
	F4	-0.00039	-0.00737	0.02575	-0.00121	-0.02333	-0.00572	-0.38235	-0.17412	0.02333	0.00572
* 11F2	C1	0.	0.	0.	0.	0.19407	-0.08550	0.	0.	0.	0.
-0.7223	F1	0.00039	0.00737	-0.02575	0.00121	-0.38235	-0.17412	-0.02333	-0.00572	-0.02333	-0.00572
	F2	-0.00039	-0.00737	0.02575	-0.00121	-0.38235	-0.17412	0.02333	0.00572	-0.02333	-0.00572
	F3	0.00039	0.00737	-0.02575	0.00121	-0.38235	-0.17412	0.02333	0.00572	0.02333	0.00572
	F4	-0.00039	-0.00737	0.02575	-0.00121	-0.38235	-0.17412	-0.02333	-0.00572	0.02333	0.00572
* 12F2	C1	0.	0.	0.	0.	0.	0.	0.	0.	0.19407	-0.08550
-0.7223	F1	0.00039	0.00737	-0.02575	0.00121	-0.02333	-0.00572	-0.02333	-0.00572	-0.38235	-0.17412
	F2	-0.00039	-0.00737	0.02575	-0.00121	-0.02333	-0.00572	0.02333	0.00572	-0.38235	-0.17412
	F3	-0.00039	-0.00737	0.02575	-0.00121	0.02333	0.00572	-0.02333	-0.00572	-0.38235	-0.17412
	F4	0.00039	0.00737	-0.02575	0.00121	0.02333	0.00572	0.02333	0.00572	-0.38235	-0.17412
13F2	C1	0.	0.	0.	0.	0.	0.	0.60782	1.44795	0.	0.
0.3177	F1	0.00186	0.04404	-0.06798	-0.51965	0.13673	0.20352	-0.08025	-0.07791	0.13673	0.20352
	F2	0.00186	0.04404	-0.06798	-0.51965	-0.13673	-0.20352	-0.08025	-0.07791	-0.13673	-0.20352
	F3	-0.00186	-0.04404	0.06798	0.51965	-0.13673	-0.20352	-0.08025	-0.07791	0.13673	0.20352
	F4	-0.00186	-0.04404	0.06798	0.51965	0.13673	0.20352	-0.08025	-0.07791	-0.13673	-0.20352
14F2	C1	0.	0.	0.	0.	0.60782	1.44795	0.	0.	0.	0.
0.3177	F1	0.00186	0.04404	-0.06798	-0.51965	-0.08025	-0.07791	0.13673	0.20352	0.13673	0.20352
	F2	-0.00186	-0.04404	0.06798	0.51965	-0.08025	-0.07791	-0.13673	-0.20352	0.13673	0.20352
	F3	0.00186	0.04404	-0.06798	-0.51965	-0.08025	-0.07791	-0.13673	-0.20352	-0.13673	-0.20352
	F4	-0.00186	-0.04404	0.06798	0.51965	-0.08025	-0.07791	0.13673	0.20352	-0.13673	-0.20352
15F2	C1	0.	0.	0.	0.	0.	0.	0.	0.	0.60782	1.44795
0.3177	F1	0.00186	0.04404	-0.06798	-0.51965	0.13673	0.20352	0.13673	0.20352	-0.08025	-0.07791
	F2	-0.00186	-0.04404	0.06798	0.51965	0.13673	0.20352	-0.13673	-0.20352	-0.08025	-0.07791
	F3	-0.00186	-0.04404	0.06798	0.51965	-0.13673	-0.20352	0.13673	0.20352	-0.08025	-0.07791
	F4	0.00186	0.04404	-0.06798	-0.51965	-0.13673	-0.20352	-0.13673	-0.20352	-0.08025	-0.07791
16F2	C1	0.	0.	0.	0.	0.	0.	-1.12486	1.28838	0.	0.
0.4687	F1	-0.00063	-0.00895	0.06497	-0.17055	-0.13152	-0.05244	-0.12957	-0.19214	-0.13152	-0.05244
	F2	-0.00063	-0.00895	0.06497	-0.17055	0.13152	0.05244	-0.12957	-0.19214	0.13152	0.05244
	F3	0.00063	0.00895	-0.06497	0.17055	0.13152	0.05244	-0.12957	-0.19214	-0.13152	-0.05244
	F4	0.00063	0.00895	-0.06497	0.17055	-0.13152	-0.05244	-0.12957	-0.19214	0.13152	0.05244
17F2	C1	0.	0.	0.	0.	-1.12486	1.28838	0.	0.	0.	0.
0.4687	F1	-0.00063	-0.00895	0.06497	-0.17055	-0.12957	-0.19214	-0.13152	-0.05244	-0.13152	-0.05244
	F2	0.00063	0.00895	-0.06497	0.17055	-0.12957	-0.19214	0.13152	0.05244	-0.13152	-0.05244
	F3	-0.00063	-0.00895	0.06497	-0.17055	-0.12957	-0.19214	0.13152	0.05244	0.13152	0.05244
	F4	0.00063	0.00895	-0.06497	0.17055	-0.12957	-0.19214	-0.13152	-0.05244	0.13152	0.05244
18F2	C1	0.	0.	0.	0.	0.	0.	0.	0.	-1.12486	1.28838
0.4687	F1	-0.00063	-0.00895	0.06497	-0.17055	-0.13152	-0.05244	-0.13152	-0.05244	-0.12957	-0.19214
	F2	0.00063	0.00895	-0.06497	0.17055	-0.13152	-0.05244	0.13152	0.05244	-0.12957	-0.19214
	F3	0.00063	0.00895	-0.06497	0.17055	0.13152	0.05244	-0.13152	-0.05244	-0.12957	-0.19214
	F4	-0.00063	-0.00895	0.06497	-0.17055	0.13152	0.05244	0.13152	0.05244	-0.12957	-0.19214

Overlap Matrix

First Function   Second Function and Overlap

```
C1S1 with
     C1S2  .35436   C1S3  .00805   C1S4  .01419   F1S3  .00004   F1S4  .00295   F1X1 -.00028   F1X2 -.00615
     F1Y1 -.00028   F1Y2 -.00615   F1Z1 -.00028   F1Z2 -.00615

C1S2 with
     C1S3  .26226   C1S4  .17745   F1S3  .00218   F1S4  .04097   F1X1 -.00663   F1X2 -.07669   F1Y1 -.00663
     F1Y2 -.07669   F1Z1 -.00663   F1Z2 -.07669

C1S3 with
     C1S4  .79970   F1S1  .00100   F1S2  .01381   F1S3  .11148   F1S4  .28410   F1X1 -.09117   F1X2 -.31033
     F1Y1 -.09117   F1Y2 -.31033   F1Z1 -.09117   F1Z2 -.31033

C1S4 with
     F1S1  .00302   F1S2  .03881   F1S3  .22296   F1S4  .45138   F1X1 -.07188   F1X2 -.27142   F1Y1 -.07188
     F1Y2 -.27142   F1Z1 -.07188   F1Z2 -.27142

C1X1 with
     C1X2  .53875   F1S1  .00177   F1S2  .02296   F1S3  .13683   F1S4  .22384   F1X1 -.02202   F1X2  .08773
     F1Y1 -.10283   F1Y2 -.20706   F1Z1 -.10283   F1Z2 -.20706

C1X2 with
     F1S1  .00300   F1S2  .03809   F1S3  .20150   F1S4  .34163   F1X1  .06769   F1X2  .31860   F1Y1 -.05130
     F1Y2 -.16155   F1Z1 -.05130   F1Z2 -.16155

C1Y1 with
     C1Y2  .53875   F1S1  .00177   F1S2  .02296   F1S3  .13683   F1S4  .22384   F1X1 -.10283   F1X2 -.20706
     F1Y1 -.02202   F1Y2  .08773   F1Z1 -.10283   F1Z2 -.20706

C1Y2 with
     F1S1  .00300   F1S2  .03809   F1S3  .20150   F1S4  .34163   F1X1 -.05130   F1X2 -.16155   F1Y1  .06769
     F1Y2  .31860   F1Z1 -.05130   F1Z2 -.16155

C1Z1 with
     C1Z2  .53875   F1S1  .00177   F1S2  .02296   F1S3  .13683   F1S4  .22384   F1X1 -.10283   F1X2 -.20706
     F1Y1 -.10283   F1Y2 -.20706   F1Z1 -.02202   F1Z2  .08773

C1Z2 with
     F1S1  .00300   F1S2  .03809   F1S3  .20150   F1S4  .34163   F1X1 -.05130   F1X2 -.16155   F1Y1 -.05130
     F1Y2 -.16155   F1Z1  .06769   F1Z2  .31860

F1S1 with
     F1S2  .35647   F1S3  .01003   F1S4  .01474   F2S4  .00004   F2X2  .00038   F2Z2  .00038

F1S2 with
     F1S3  .28653   F1S4  .18096   F2S4  .00058   F2X2  .00526   F2Z2  .00526

F1S3 with
     F1S4  .78531   F2S3  .00005   F2S4  .00819   F2X1  .00032   F2X2  .04273   F2Z1  .00032   F2Z2  .04273

F1S4 with
     F2S1  .00004   F2S2  .00058   F2S3  .00819   F2S4  .05086   F2X1  .01188   F2X2  .12769   F2Z1  .01188
     F2Z2  .12769

F1X1 with
     F1X2  .49444   F2S3 -.00032   F2S4 -.01188   F2X1 -.00108   F2X2 -.03415   F2Z1 -.00124   F2Z2 -.04727

F1X2 with
     F2S1 -.00038   F2S2 -.00526   F2S3 -.04273   F2S4 -.12769   F2X1 -.03415   F2X2 -.13099   F2Z1 -.04727
     F2Z2 -.23351

F1Y1 with
     F1Y2  .49444   F2Y1  .00016   F2Y2  .01312

F1Y2 with
     F2Y1  .01312   F2Y2  .10252

F1Z1 with
     F1Z2  .49444   F2S3 -.00032   F2S4 -.01188   F2X1 -.00124   F2X2 -.04727   F2Z1 -.00108   F2Z2 -.03415

F1Z2 with
     F2S1 -.00038   F2S2 -.00526   F2S3 -.04273   F2S4 -.12769   F2X1 -.04727   F2X2 -.23351   F2Z1 -.03415
     F2Z2 -.13099
```

N2O4   Nitrogen Tetroxide

Molecular Geometry   Symmetry D_{2h}

Center   Coordinates
                X   Y   Z
N1   0.   -1.65353291 0.
N2   0.   1.65353291 0.
O1   -2.05038080 -2.53037781 0.
O2   2.05038080 -2.53037781 0.
O3   -2.05038080   2.53037781 0.
O4   2.05038080   2.53037781 0.

Mass   0.   0.   0.

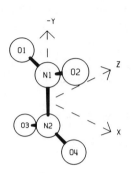

MO Symmetry Types and Occupancy

Representation	AG	AU	B1G	B1U	B2G	B2U	B3G	B3U
Number of MOs	14	2	10	4	2	14	4	10
Number Occupied	6	1	4	1	1	5	1	4

Energy Expectation Values

Total	-407.8354	Kinetic	408.4423
Electronic	-647.3336	Potential	-816.2776
Nuclear Repulsion	239.4982	Virial	1.9985

Electronic Potential   -1055.7759
One-Electron Potential   -1441.3808
Two-Electron Potential   385.6049

Moments of the Charge Distribution

Second

x^2	y^2	z^2	xy	xz	yz	r^2
-164.4960	-270.4317	-22.1207	0.	0.	0.	-457.0484
134.5300	243.1684	0.	0.	0.	0.	377.6983

Fourth (even)

x^4	y^4	z^4	x^2y^2	x^2z^2	y^2z^2	r^2x^2	r^2y^2	r^2z^2	r^4
-1058.4441	-2258.4502	-58.4314	-1110.8039	-87.9734	-137.7615	-2257.2214	-3507.0156	-284.1664	-6048.4036
565.5722	1416.5317	0.	861.3701	0.	0.	1426.9423	2277.9017	0.	3704.8440

Expectation Values at Atomic Centers

Center			Expectation Value		
	r^{-1}	δ	xr^{-3}	yr^{-3}	zr^{-3}
N1	-17.9890	189.1933	0.	0.0796	0.
O1	-22.2043	287.6481	-0.2839	-0.1198	0.

	$\dfrac{3x^2-r^2}{r^5}$	$\dfrac{3y^2-r^2}{r^5}$	$\dfrac{3z^2-r^2}{r^5}$	$\dfrac{3xy}{r^5}$	$\dfrac{3xz}{r^5}$	$\dfrac{3yz}{r^5}$
N1	-0.1508	-0.1281	0.2789	0.	0.	0.
O1	1.3719	-1.6174	0.2455	1.8620	0.	0.

Mulliken Charge

Center				Basis Function Type							Total
	S1	S2	S3	S4	X1	X2	Y1	Y2	Z1	Z2	
N1	0.0433	1.9561	0.9804	0.4144	1.0137	-0.0128	0.9488	0.1794	0.8508	0.2457	6.6200
O1	0.0427	1.9566	0.9830	0.9225	0.9907	0.1935	1.2812	0.3681	1.0512	0.4006	8.1900

Overlap Populations

	Pair of Atomic Centers					
	N2	O1	O1	O2	O3	O3
	N1	N1	N2	O1	O1	O2
1AG	0.	-0.0004	0.	0.	0.	0.
2AG	0.0002	-0.0001	0.	0.	0.	0.
3AG	0.0591	0.1579	0.0105	0.0184	0.0016	0.0002
4AG	0.4227	-0.0537	-0.0223	0.0157	0.0009	0.0001
5AG	-0.0089	-0.0603	0.0086	-0.0035	0.0186	0.0001
6AG	0.4106	-0.2488	-0.1220	0.0512	0.0905	0.0136
Total AG	0.8837	-0.2053	-0.1252	0.0819	0.1116	0.0140
1AU	0.	0.	0.	-0.0467	-0.0151	0.0021
Total AU	0.	0.	0.	-0.0467	-0.0151	0.0021
1B1G	0.	-0.0009	0.0001	-0.0001	0.	0.
2B1G	0.0028	0.1071	0.0045	-0.0344	-0.0069	0.0007
3B1G	0.0012	0.0390	-0.0034	0.0003	-0.0002	0.0010
4B1G	-0.0155	0.0199	-0.0183	-0.0969	-0.0438	-0.0010
Total B1G	-0.0116	0.1651	-0.0171	-0.1310	-0.0509	0.0007
1B1U	0.0792	0.1121	0.0083	0.0068	0.0020	0.0002
Total B1U	0.0792	0.1121	0.0083	0.0068	0.0020	0.0002
1B2G	0.	0.	0.	-0.0467	0.0151	-0.0021
Total B2G	0.	0.	0.	-0.0467	0.0151	-0.0021
1B2U	0.	-0.0004	0.	0.0001	0.	0.
2B2U	-0.0011	-0.0001	0.	0.	0.	0.
3B2U	-0.0694	0.1821	-0.0155	0.0242	-0.0026	-0.0004
4B2U	-0.1891	-0.1433	0.0213	-0.0015	-0.0071	0.0003
5B2U	-0.5458	0.1715	-0.0669	0.0147	-0.0261	-0.0012
Total B2U	-0.8054	0.2099	-0.0612	0.0375	-0.0358	-0.0013
1B3G	-0.0992	0.1329	-0.0106	0.0083	-0.0024	-0.0003
Total B3G	-0.0992	0.1329	-0.0106	0.0083	-0.0024	-0.0003
1B3U	0.	-0.0003	0.	0.	0.	0.
2B3U	0.0024	0.1325	0.0004	-0.0266	0.0073	-0.0007
3B3U	0.0162	-0.0214	0.0010	0.0008	-0.0008	-0.0004
4B3U	0.0111	0.0160	0.0134	-0.0936	0.0294	0.0012
Total B3U	0.0297	0.1267	0.0149	-0.1194	0.0359	0.0001
Total	0.0765	0.5413	-0.1909	-0.2094	0.0603	0.0134
Distance	3.3071	2.2300	4.6593	4.1008	5.0608	6.5136

Molecular Orbitals and Eigenvalues

Center		S1	S2	S3	S4	X1	X2	Y1	Y2	Z1	Z2
* 1AG	N1	-0.00000	-0.00006	-0.00026	0.00247	0.	0.	-0.00006	-0.00110	0.	0.
-20.6981	N2	-0.00000	-0.00006	-0.00026	0.00247	0.	0.	0.00006	0.00110	0.	0.
	O1	-0.02556	-0.48969	-0.00157	-0.00213	-0.00105	-0.00037	-0.00048	0.00025	0.	0.
	O2	-0.02556	-0.48969	-0.00157	-0.00213	0.00105	0.00037	-0.00048	0.00025	0.	0.
	O3	-0.02556	-0.48969	-0.00157	-0.00213	-0.00105	-0.00037	0.00048	-0.00025	0.	0.
	O4	-0.02556	-0.48969	-0.00157	-0.00213	0.00105	0.00037	0.00048	-0.00025	0.	0.
* 2AG	N1	0.03673	0.69233	0.00252	0.00147	0.	0.	-0.00050	0.00016	0.	0.
-15.9566	N2	0.03673	0.69233	0.00252	0.00147	0.	0.	0.00050	-0.00016	0.	0.
	O1	-0.00001	-0.00022	0.00037	-0.00054	0.00038	-0.00024	0.00016	-0.00020	0.	0.
	O2	-0.00001	-0.00022	0.00037	-0.00054	-0.00038	0.00024	0.00016	-0.00020	0.	0.
	O3	-0.00001	-0.00022	0.00037	-0.00054	0.00038	-0.00024	-0.00016	0.00020	0.	0.
	O4	-0.00001	-0.00022	0.00037	-0.00054	-0.00038	0.00024	-0.00016	0.00020	0.	0.
* 3AG	N1	0.00604	0.12654	-0.30792	-0.07962	0.	0.	0.05871	-0.02679	0.	0.
-1.7722	N2	0.00604	0.12654	-0.30792	-0.07962	0.	0.	-0.05871	0.02679	0.	0.
	O1	0.00376	0.08183	-0.18112	-0.11695	-0.09403	-0.01048	-0.03745	0.00454	0.	0.
	O2	0.00376	0.08183	-0.18112	-0.11695	0.09403	0.01048	-0.03745	0.00454	0.	0.
	O3	0.00376	0.08183	-0.18112	-0.11695	-0.09403	-0.01048	0.03745	-0.00454	0.	0.
	O4	0.00376	0.08183	-0.18112	-0.11695	0.09403	0.01048	0.03745	-0.00454	0.	0.
* 4AG	N1	0.00390	0.08151	-0.21161	-0.18522	0.	0.	-0.31745	-0.07646	0.	0.
-1.0717	N2	0.00390	0.08151	-0.21161	-0.18522	0.	0.	0.31745	0.07646	0.	0.
	O1	-0.00298	-0.06552	0.14639	0.16861	-0.00006	0.00124	-0.11344	-0.01201	0.	0.
	O2	-0.00298	-0.06552	0.14639	0.16861	0.00006	-0.00124	-0.11344	-0.01201	0.	0.
	O3	-0.00298	-0.06552	0.14639	0.16861	-0.00006	0.00124	0.11344	0.01201	0.	0.
	O4	-0.00298	-0.06552	0.14639	0.16861	0.00006	-0.00124	0.11344	0.01201	0.	0.
* 5AG	N1	-0.00269	-0.05298	0.17055	0.10460	0.	0.	-0.26244	-0.05981	0.	0.
-0.8851	N2	-0.00269	-0.05298	0.17055	0.10460	0.	0.	0.26244	0.05981	0.	0.
	O1	0.00297	0.06609	-0.14209	-0.23376	0.22973	0.05057	-0.07362	-0.03004	0.	0.
	O2	0.00297	0.06609	-0.14209	-0.23376	-0.22973	-0.05057	-0.07362	-0.03004	0.	0.
	O3	0.00297	0.06609	-0.14209	-0.23376	0.22973	0.05057	0.07362	0.03004	0.	0.
	O4	0.00297	0.06609	-0.14209	-0.23376	-0.22973	-0.05057	0.07362	0.03004	0.	0.
* 6AG	N1	0.00061	0.01892	0.00640	-0.32305	0.	0.	-0.27904	-0.05886	0.	0.
-0.4854	N2	0.00061	0.01892	0.00640	-0.32305	0.	0.	0.27904	0.05886	0.	0.
	O1	-0.00001	-0.00078	-0.00243	0.06101	0.02272	0.02425	0.35617	0.20276	0.	0.
	O2	-0.00001	-0.00078	-0.00243	0.06101	-0.02272	-0.02425	0.35617	0.20276	0.	0.
	O3	-0.00001	-0.00078	-0.00243	0.06101	0.02272	0.02425	-0.35617	-0.20276	0.	0.
	O4	-0.00001	-0.00078	-0.00243	0.06101	-0.02272	-0.02425	-0.35617	-0.20276	0.	0.
7AG	N1	0.00269	0.07473	-0.03783	-2.18719	0.	0.	0.28756	0.87738	0.	0.
0.3839	N2	0.00269	0.07473	-0.03783	-2.18719	0.	0.	-0.28756	-0.87738	0.	0.
	O1	-0.00282	-0.06538	0.12903	0.91336	0.22162	0.74168	0.01088	-0.00021	0.	0.
	O2	-0.00282	-0.06538	0.12903	0.91336	-0.22162	-0.74168	0.01088	-0.00021	0.	0.
	O3	-0.00282	-0.06538	0.12903	0.91336	0.22162	0.74168	-0.01088	0.00021	0.	0.
	O4	-0.00282	-0.06538	0.12903	0.91336	-0.22162	-0.74168	-0.01088	0.00021	0.	0.
* 1AU	N1	0.	0.	0.	0.	0.	0.	0.	0.	0.	0.
-0.5239	N2	0.	0.	0.	0.	0.	0.	0.	0.	0.	0.
	O1	0.	0.	0.	0.	0.	0.	0.	0.	-0.37930	-0.20544
	O2	0.	0.	0.	0.	0.	0.	0.	0.	0.37930	0.20544
	O3	0.	0.	0.	0.	0.	0.	0.	0.	0.37930	0.20544
	O4	0.	0.	0.	0.	0.	0.	0.	0.	-0.37930	-0.20544

Center		S1	S2	S3	S4	X1	X2	Y1	Y2	Z1	Z2
* 1B1G	N1	0.	0.	0.	0.	-0.00021	-0.00467	0.	0.	0.	0.
-20.6982	N2	0.	0.	0.	0.	0.00021	0.00467	0.	0.	0.	0.
	O1	-0.02556	-0.48968	-0.00123	-0.00417	-0.00113	-0.00023	-0.00052	-0.00036	0.	0.
	O2	0.02556	0.48968	0.00123	0.00417	-0.00113	-0.00023	0.00052	0.00036	0.	0.
	O3	0.02556	0.48968	0.00123	0.00417	0.00113	0.00023	-0.00052	-0.00036	0.	0.
	O4	-0.02556	-0.48968	-0.00123	-0.00417	0.00113	0.00023	0.00052	0.00036	0.	0.
* 2B1G	N1	0.	0.	0.	0.	-0.27294	0.05880	0.	0.	0.	0.
-1.5418	N2	0.	0.	0.	0.	0.27294	-0.05880	0.	0.	0.	0.
	O1	-0.00497	-0.10908	0.23644	0.20431	0.08500	0.00620	0.04502	0.00224	0.	0.
	O2	0.00497	0.10908	-0.23644	-0.20431	0.08500	0.00620	-0.04502	-0.00224	0.	0.
	O3	0.00497	0.10908	-0.23644	-0.20431	-0.08500	-0.00620	0.04502	0.00224	0.	0.
	O4	-0.00497	-0.10908	0.23644	0.20431	-0.08500	-0.00620	-0.04502	-0.00224	0.	0.
* 3B1G	N1	0.	0.	0.	0.	-0.33204	0.05030	0.	0.	0.	0.
-0.7919	N2	0.	0.	0.	0.	0.33204	-0.05030	0.	0.	0.	0.
	O1	0.00272	0.05915	-0.13946	-0.16753	0.22079	0.05859	0.16383	0.05551	0.	0.
	O2	-0.00272	-0.05915	0.13946	0.16753	0.22079	0.05859	-0.16383	-0.05551	0.	0.
	O3	-0.00272	-0.05915	0.13946	0.16753	-0.22079	-0.05859	0.16383	0.05551	0.	0.
	O4	0.00272	0.05915	-0.13946	-0.16753	-0.22079	-0.05859	-0.16383	-0.05551	0.	0.
* 4B1G	N1	0.	0.	0.	0.	0.02670	-0.10646	0.	0.	0.	0.
-0.5580	N2	0.	0.	0.	0.	-0.02670	0.10646	0.	0.	0.	0.
	O1	-0.00016	-0.00277	0.01258	-0.05458	-0.23689	-0.11116	0.33301	0.15137	0.	0.
	O2	0.00016	0.00277	-0.01258	0.05458	-0.23689	-0.11116	-0.33301	-0.15137	0.	0.
	O3	0.00016	0.00277	-0.01258	0.05458	0.23689	0.11116	0.33301	0.15137	0.	0.
	O4	-0.00016	-0.00277	0.01258	-0.05458	0.23689	0.11116	-0.33301	-0.15137	0.	0.
5B1G	N1	0.	0.	0.	0.	-0.28701	-2.72527	0.	0.	0.	0.
0.5230	N2	0.	0.	0.	0.	0.28701	2.72527	0.	0.	0.	0.
	O1	0.00238	0.06660	-0.03018	-1.75386	0.01182	-0.65482	0.01923	-0.52757	0.	0.
	O2	-0.00238	-0.06660	0.03018	1.75386	0.01182	-0.65482	-0.01923	0.52757	0.	0.
	O3	-0.00238	-0.06660	0.03018	1.75386	-0.01182	0.65482	0.01923	-0.52757	0.	0.
	O4	0.00238	0.06660	-0.03018	-1.75386	-0.01182	0.65482	-0.01923	0.52757	0.	0.
* 1B1U	N1	0.	0.	0.	0.	0.	0.	0.	0.	0.39056	0.11010
-0.8529	N2	0.	0.	0.	0.	0.	0.	0.	0.	0.39056	0.11010
	O1	0.	0.	0.	0.	0.	0.	0.	0.	0.22906	0.06741
	O2	0.	0.	0.	0.	0.	0.	0.	0.	0.22906	0.06741
	O3	0.	0.	0.	0.	0.	0.	0.	0.	0.22906	0.06741
	O4	0.	0.	0.	0.	0.	0.	0.	0.	0.22906	0.06741
2B1U	N1	0.	0.	0.	0.	0.	0.	0.	0.	0.39873	0.34718
-0.0077	N2	0.	0.	0.	0.	0.	0.	0.	0.	0.39873	0.34718
	O1	0.	0.	0.	0.	0.	0.	0.	0.	-0.30236	-0.28888
	O2	0.	0.	0.	0.	0.	0.	0.	0.	-0.30236	-0.28888
	O3	0.	0.	0.	0.	0.	0.	0.	0.	-0.30236	-0.28888
	O4	0.	0.	0.	0.	0.	0.	0.	0.	-0.30236	-0.28888
3B1U	N1	0.	0.	0.	0.	0.	0.	0.	0.	-0.59813	0.75925
0.5176	N2	0.	0.	0.	0.	0.	0.	0.	0.	-0.59813	0.75925
	O1	0.	0.	0.	0.	0.	0.	0.	0.	-0.12058	-0.05548
	O2	0.	0.	0.	0.	0.	0.	0.	0.	-0.12058	-0.05548
	O3	0.	0.	0.	0.	0.	0.	0.	0.	-0.12058	-0.05548
	O4	0.	0.	0.	0.	0.	0.	0.	0.	-0.12058	-0.05548

Center		S1	S2	S3	S4	X1	X2	Y1	Y2	Z1	Z2
* 1B2G	N1	0.	0.	0.	0.	0.	0.	0.	0.	0.	0.
-0.5350	N2	0.	0.	0.	0.	0.	0.	0.	0.	0.	0.
	O1	0.	0.	0.	0.	0.	0.	0.	0.	-0.37172	-0.20637
	O2	0.	0.	0.	0.	0.	0.	0.	0.	0.37172	0.20637
	O3	0.	0.	0.	0.	0.	0.	0.	0.	-0.37172	-0.20637
	O4	0.	0.	0.	0.	0.	0.	0.	0.	0.37172	0.20637
* 1B2U	N1	-0.00000	-0.00007	-0.00024	0.00237	0.	0.	-0.00009	-0.00108	0.	0.
-20.6981	N2	0.00000	0.00007	0.00024	-0.00237	0.	0.	-0.00009	-0.00108	0.	0.
	O1	-0.02556	-0.48969	-0.00147	-0.00233	-0.00105	-0.00042	-0.00044	0.00008	0.	0.
	O2	-0.02556	-0.48969	-0.00147	-0.00233	0.00105	0.00042	-0.00044	0.00008	0.	0.
	O3	0.02556	0.48969	0.00147	0.00233	0.00105	0.00042	-0.00044	0.00008	0.	0.
	O4	0.02556	0.48969	0.00147	0.00233	-0.00105	-0.00042	-0.00044	0.00008	0.	0.
* 2B2U	N1	-0.03673	-0.69233	-0.00200	-0.00324	0.	0.	0.00061	-0.00188	0.	0.
-15.9565	N2	0.03673	0.69233	0.00200	0.00324	0.	0.	0.00061	-0.00188	0.	0.
	O1	0.00001	0.00023	-0.00021	0.00024	-0.00036	0.00010	-0.00019	0.00045	0.	0.
	O2	0.00001	0.00023	-0.00021	0.00024	0.00036	-0.00010	-0.00019	0.00045	0.	0.
	O3	-0.00001	-0.00023	0.00021	-0.00024	0.00036	-0.00010	-0.00019	0.00045	0.	0.
	O4	-0.00001	-0.00023	0.00021	-0.00024	-0.00036	0.00010	-0.00019	0.00045	0.	0.
* 3B2U	N1	-0.00570	-0.11953	0.28925	0.09447	0.	0.	-0.10445	0.04225	0.	0.
-1.7345	N2	0.00570	0.11953	-0.28925	-0.09447	0.	0.	-0.10445	0.04225	0.	0.
	O1	-0.00399	-0.08736	0.19011	0.13578	0.10029	0.01333	0.03317	-0.00221	0.	0.
	O2	-0.00399	-0.08736	0.19011	0.13578	-0.10029	-0.01333	0.03317	-0.00221	0.	0.
	O3	0.00399	0.08736	-0.19011	-0.13578	-0.10029	-0.01333	0.03317	-0.00221	0.	0.
	O4	0.00399	0.08736	-0.19011	-0.13578	0.10029	0.01333	0.03317	-0.00221	0.	0.
* 4B2U	N1	0.00462	0.09433	-0.26881	-0.21981	0.	0.	0.06668	-0.05129	0.	0.
-0.9087	N2	-0.00462	-0.09433	0.26881	0.21981	0.	0.	0.06668	-0.05129	0.	0.
	O1	-0.00401	-0.08796	0.19926	0.26379	-0.19146	-0.04496	-0.06368	-0.00622	0.	0.
	O2	-0.00401	-0.08796	0.19926	0.26379	0.19146	0.04496	-0.06368	-0.00622	0.	0.
	O3	0.00401	0.08796	-0.19926	-0.26379	0.19146	0.04496	-0.06368	-0.00622	0.	0.
	O4	0.00401	0.08796	-0.19926	-0.26379	-0.19146	-0.04496	-0.06368	-0.00622	0.	0.
* 5B2U	N1	0.00186	0.04131	-0.08480	-0.18095	0.	0.	-0.27907	-0.20162	0.	0.
-0.7312	N2	-0.00186	-0.04131	0.08480	0.18095	0.	0.	-0.27907	-0.20162	0.	0.
	O1	0.00016	0.00401	-0.00442	-0.04604	0.12433	0.03377	-0.30115	-0.08986	0.	0.
	O2	0.00016	0.00401	-0.00442	-0.04604	-0.12433	-0.03377	-0.30115	-0.08986	0.	0.
	O3	-0.00016	-0.00401	0.00442	0.04604	-0.12433	-0.03377	-0.30115	-0.08986	0.	0.
	O4	-0.00016	-0.00401	0.00442	0.04604	0.12433	0.03377	-0.30115	-0.08986	0.	0.
6B2U	N1	-0.00404	-0.09094	0.18942	0.83406	0.	0.	0.54296	0.24127	0.	0.
0.0317	N2	0.00404	0.09094	-0.18942	-0.83406	0.	0.	0.54296	0.24127	0.	0.
	O1	0.00095	0.02086	-0.04963	-0.16425	-0.06644	-0.08998	-0.24270	-0.22729	0.	0.
	O2	0.00095	0.02086	-0.04963	-0.16425	0.06644	0.08998	-0.24270	-0.22729	0.	0.
	O3	-0.00095	-0.02086	0.04963	0.16425	0.06644	0.08998	-0.24270	-0.22729	0.	0.
	O4	-0.00095	-0.02086	0.04963	0.16425	-0.06644	-0.08998	-0.24270	-0.22729	0.	0.
7B2U	N1	-0.00187	-0.04928	0.04761	1.32867	0.	0.	-0.26392	-1.48162	0.	0.
0.3888	N2	0.00187	0.04928	-0.04761	-1.32867	0.	0.	-0.26392	-1.48162	0.	0.
	O1	0.00268	0.06399	-0.10831	-0.94386	-0.22064	-0.68437	0.06459	0.05385	0.	0.
	O2	0.00268	0.06399	-0.10831	-0.94386	0.22064	0.68437	0.06459	0.05385	0.	0.
	O3	-0.00268	-0.06399	0.10831	0.94386	0.22064	0.68437	0.06459	0.05385	0.	0.
	O4	-0.00268	-0.06399	0.10831	0.94386	-0.22064	-0.68437	0.06459	0.05385	0.	0.

		S1	S2	S3	S4	X1	X2	Y1	Y2	Z1	Z2
8B2U	N1	0.00404	0.07317	-0.32703	-1.06829	0.	0.	0.51219	-1.76805	0.	0.
0.5956	N2	-0.00404	-0.07317	0.32703	1.06829	0.	0.	0.51219	-1.76805	0.	0.
	O1	-0.00026	-0.00729	0.00262	-0.08579	0.09784	0.15300	0.16112	0.31624	0.	0.
	O2	-0.00026	-0.00729	0.00262	-0.08579	-0.09784	-0.15300	0.16112	0.31624	0.	0.
	O3	0.00026	0.00729	-0.00262	0.08579	-0.09784	-0.15300	0.16112	0.31624	0.	0.
	O4	0.00026	0.00729	-0.00262	0.08579	0.09784	0.15300	0.16112	0.31624	0.	0.
* 1B3G	N1	0.	0.	0.	0.	0.	0.	0.	0.	-0.38427	-0.13694
-0.7998	N2	0.	0.	0.	0.	0.	0.	0.	0.	0.38427	0.13694
	O1	0.	0.	0.	0.	0.	0.	0.	0.	-0.25395	-0.07387
	O2	0.	0.	0.	0.	0.	0.	0.	0.	-0.25395	-0.07387
	O3	0.	0.	0.	0.	0.	0.	0.	0.	0.25395	0.07387
	O4	0.	0.	0.	0.	0.	0.	0.	0.	0.25395	0.07387
2B3G	N1	0.	0.	0.	0.	0.	0.	0.	0.	-0.50748	-0.41124
0.0659	N2	0.	0.	0.	0.	0.	0.	0.	0.	0.50748	0.41124
	O1	0.	0.	0.	0.	0.	0.	0.	0.	0.28762	0.29276
	O2	0.	0.	0.	0.	0.	0.	0.	0.	0.28762	0.29276
	O3	0.	0.	0.	0.	0.	0.	0.	0.	-0.28762	-0.29276
	O4	0.	0.	0.	0.	0.	0.	0.	0.	-0.28762	-0.29276
3B3G	N1	0.	0.	0.	0.	0.	0.	0.	0.	-0.51308	1.21880
0.6719	N2	0.	0.	0.	0.	0.	0.	0.	0.	0.51308	-1.21880
	O1	0.	0.	0.	0.	0.	0.	0.	0.	-0.22203	-0.07486
	O2	0.	0.	0.	0.	0.	0.	0.	0.	-0.22203	-0.07486
	O3	0.	0.	0.	0.	0.	0.	0.	0.	0.22203	0.07486
	O4	0.	0.	0.	0.	0.	0.	0.	0.	0.22203	0.07486
* 1B3U	N1	0.	0.	0.	0.	-0.00020	-0.00168	0.	0.	0.	0.
-20.6982	N2	0.	0.	0.	0.	-0.00020	-0.00168	0.	0.	0.	0.
	O1	-0.02556	-0.48968	-0.00145	-0.00244	-0.00105	-0.00004	-0.00044	-0.00036	0.	0.
	O2	0.02556	0.48968	0.00145	0.00244	-0.00105	-0.00004	0.00044	0.00036	0.	0.
	O3	-0.02556	-0.48968	-0.00145	-0.00244	-0.00105	-0.00004	0.00044	0.00036	0.	0.
	O4	0.02556	0.48968	0.00145	0.00244	-0.00105	-0.00004	-0.00044	-0.00036	0.	0.
* 2B3U	N1	0.	0.	0.	0.	-0.27329	0.02262	0.	0.	0.	0.
-1.5617	N2	0.	0.	0.	0.	-0.27329	0.02262	0.	0.	0.	0.
	O1	-0.00485	-0.10662	0.23032	0.18339	0.08225	0.00235	0.04823	0.00828	0.	0.
	O2	0.00485	0.10662	-0.23032	-0.18339	0.08225	0.00235	-0.04823	-0.00828	0.	0.
	O3	-0.00485	-0.10662	0.23032	0.18339	0.08225	0.00235	0.04823	0.00828	0.	0.
	O4	0.00485	0.10662	-0.23032	-0.18339	0.08225	0.00235	0.04823	0.00828	0.	0.
* 3B3U	N1	0.	0.	0.	0.	0.32278	0.01481	0.	0.	0.	0.
-0.8078	N2	0.	0.	0.	0.	0.32278	0.01481	0.	0.	0.	0.
	O1	-0.00287	-0.06292	0.14369	0.21370	-0.20025	-0.05125	-0.17092	-0.05423	0.	0.
	O2	0.00287	0.06292	-0.14369	-0.21370	-0.20025	-0.05125	0.17092	0.05423	0.	0.
	O3	-0.00287	-0.06292	0.14369	0.21370	-0.20025	-0.05125	0.17092	0.05423	0.	0.
	O4	0.00287	0.06292	-0.14369	-0.21370	-0.20025	-0.05125	-0.17092	-0.05423	0.	0.
* 4B3U	N1	0.	0.	0.	0.	-0.06288	0.10320	0.	0.	0.	0.
-0.6112	N2	0.	0.	0.	0.	-0.06288	0.10320	0.	0.	0.	0.
	O1	-0.00001	-0.00117	-0.00585	0.06763	0.24311	0.10892	-0.30107	-0.12444	0.	0.
	O2	0.00001	0.00117	0.00585	-0.06763	0.24311	0.10892	0.30107	0.12444	0.	0.
	O3	-0.00001	-0.00117	-0.00585	0.06763	0.24311	0.10892	0.30107	0.12444	0.	0.
	O4	0.00001	0.00117	0.00585	-0.06763	0.24311	0.10892	-0.30107	-0.12444	0.	0.

Center						Basis Function Type					
		S1	S2	S3	S4	X1	X2	Y1	Y2	Z1	Z2
5B3U	N1	0.	0.	0.	0.	-0.16624	-1.55058	0.	0.	0.	0.
0.4184	N2	0.	0.	0.	0.	-0.16624	-1.55058	0.	0.	0.	0.
	O1	0.00256	0.06345	-0.08780	-1.12553	0.00501	-0.40174	-0.14158	-0.40990	0.	0.
	O2	-0.00256	-0.06345	0.08780	1.12553	0.00501	-0.40174	0.14158	0.40990	0.	0.
	O3	0.00256	0.06345	-0.08780	-1.12553	0.00501	-0.40174	0.14158	0.40990	0.	0.
	O4	-0.00256	-0.06345	0.08780	1.12553	0.00501	-0.40174	-0.14158	-0.40990	0.	0.
6B3U	N1	0.	0.	0.	0.	0.65236	-0.27934	0.	0.	0.	0.
0.6981	N2	0.	0.	0.	0.	0.65236	-0.27934	0.	0.	0.	0.
	O1	-0.00083	-0.02268	0.01446	0.14901	0.07677	0.47201	-0.13386	0.32231	0.	0.
	O2	0.00083	0.02268	-0.01446	-0.14901	0.07677	0.47201	0.13386	-0.32231	0.	0.
	O3	-0.00083	-0.02268	0.01446	0.14901	0.07677	0.47201	0.13386	-0.32231	0.	0.
	O4	0.00083	0.02268	-0.01446	-0.14901	0.07677	0.47201	-0.13386	0.32231	0.	0.

Overlap Matrix

First Function			Second Function and Overlap				

N1S1 with

N1S2	.35584	N1S3	.00894	N1S4	.01446	N2S3	.00002	N2S4	.00149	N2Y1	-.00027	N2Y2	-.00537
O1S3	.00046	O1S4	.00447	O1X1	.00227	O1X2	.00967	O1Y1	.00097	O1Y2	.00414		

N1S2 with

N1S3	.27246	N1S4	.17925	N2S3	.00057	N2S4	.02002	N2Y1	-.00467	N2Y2	-.06883	O1S3	.00925
O1S4	.05809	O1X1	.03302	O1X2	.11984	O1Y1	.01412	O1Y2	.05125				

N1S3 with

N1S4	.79352	N2S1	.00002	N2S2	.00057	N2S3	.02697	N2S4	.14103	N2Y1	-.07264	N2Y2	-.36559
O1S1	.00103	O1S2	.01561	O1S3	.15079	O1S4	.33067	O1X1	.22734	O1X2	.51486	O1Y1	.09722
O1Y2	.22018												

N1S4 with

N2S1	.00149	N2S2	.02002	N2S3	.14103	N2S4	.32082	N2Y1	-.16895	N2Y2	-.54175	O1S1	.00416
O1S2	.05376	O1S3	.29885	O1S4	.54165	O1X1	.20224	O1X2	.55342	O1Y1	.08649	O1Y2	.23667

N1X1 with

N1X2	.52739	N2X1	.02634	N2X2	.12657	O1S1	-.00323	O1S2	-.04307	O1S3	-.25543	O1S4	-.30111
O1X1	-.24268	O1X2	-.09639	O1Y1	-.15513	O1Y2	-.16243						

N1X2 with

N2X1	.12657	N2X2	.40483	O1S1	-.00721	O1S2	-.09108	O1S3	-.43518	O1S4	-.59375	O1X1	-.03298
O1X2	.13316	O1Y1	-.10300	O1Y2	-.20658								

N1Y1 with

N1Y2	.52739	N2S1	.00027	N2S2	.00467	N2S3	.07264	N2S4	.16895	N2Y1	-.13975	N2Y2	-.23334
O1S1	-.00138	O1S2	-.01842	O1S3	-.10924	O1S4	-.12877	O1X1	-.15513	O1X2	-.16243	O1Y1	.05372
O1Y2	.21397												

N1Y2 with

N2S1	.00537	N2S2	.06883	N2S3	.36559	N2S4	.54175	N2Y1	-.23334	N2Y2	-.32734	O1S1	-.00308
O1S2	-.03895	O1S3	-.18610	O1S4	-.25392	O1X1	-.10300	O1X2	-.20658	O1Y1	.16381	O1Y2	.52788

N1Z1 with

N1Z2	.52739	N2Z1	.02634	N2Z2	.12657	O1Z1	.12006	O1Z2	.28343

N1Z2 with

N2Z1	.12657	N2Z2	.40483	O1Z1	.20786	O1Z2	.61622

N2S1 with

O1S4	.00004	O1X2	.00027	O1Y2	.00055

N2S2 with

O1S4	.00065	O1X2	.00383	O1Y2	.00781

N2S3 with

O1S3	.00022	O1S4	.01196	O1X1	.00058	O1X2	.03456	O1Y1	.00119	O1Y2	.07051

N2S4 with

O1S1	.00013	O1S2	.00183	O1S3	.01758	O1S4	.07298	O1X1	.01270	O1X2	.09485	O1Y1	.02591
O1Y2	.19355												

N2X1 with

O1S2	-.00001	O1S3	-.00078	O1S4	-.01214	O1X1	-.00085	O1X2	-.00637	O1Y1	-.00282	O1Y2	-.05425

N2X2 with

O1S1	-.00045	O1S2	-.00608	O1S3	-.04189	O1S4	-.10362	O1X1	-.00316	O1X2	.02797	O1Y1	-.04902
O1Y2	-.20706												

N2Y1 with

O1S2	-.00002	O1S3	-.00159	O1S4	-.02478	O1X1	-.00282	O1X2	-.05425	O1Y1	-.00522	O1Y2	-.09048

N2Y2 with

O1S1	-.00093	O1S2	-.01241	O1S3	-.08547	O1S4	-.21144	O1X1	-.04902	O1X2	-.20706	O1Y1	-.07917
O1Y2	-.29307												

N2Z1 with

O1Z1	.00053	O1Z2	.02022

N2Z2 with

O1Z1	.02087	O1Z2	.12944

Overlap Matrix (continued)

First Function   Second Function and Overlap

01S1 with
      01S2 .35463   01S3 .00974   01S4 .01469   02S4 .00013   02X2 -.00124   03S4 .00001   03Y2 -.00023

01S2 with
      01S3 .28286   01S4 .18230   02S4 .00188   02X1 -.00002   02X2 -.01680   03S4 .00017   03Y2 -.00334

01S3 with
      01S4 .79029   02S3 .00044   02S4 .02108   02X1 -.00283   02X2 -.11753   03S3 .00001   03S4 .00315
      03Y1 -.00010   03Y2 -.03162

01S4 with
      02S1 .00013   02S2 .00188   02S3 .02108   02S4 .09299   02X1 -.03891   02X2 -.27493   03S1 .00001
      03S2 .00017   03S3 .00315   03S4 .02685   03Y1 -.00778   03Y2 -.11637

01X1 with
      01X2 .50607   02S2 .00002   02S3 .00283   02S4 .03891   02X1 -.01107   02X2 -.13628   03X1 .00004
      03X2 .00665

01X2 with
      02S1 .00124   02S2 .01680   02S3 .11753   02S4 .27493   02X1 -.13628   02X2 -.43007   03X1 .00665
      03X2 .06477

01Y1 with
      01Y2 .50607   02Y1 .00098   02Y2 .02918   03S3 .00010   03S4 .00778   03Y1 -.00071   03Y2 -.05048

01Y2 with
      02Y1 .02918   02Y2 .16578   03S1 .00023   03S2 .00334   03S3 .03162   03S4 .11637   03Y1 -.05048
      03Y2 -.28976

01Z1 with
      01Z2 .50607   02Z1 .00098   02Z2 .02918   03Z1 .00004   03Z2 .00665

01Z2 with
      02Z1 .02918   02Z2 .16578   03Z1 .00665   03Z2 .06477

02S1 with
      03X2 .00001   03Y2 -.00001

02S2 with
      03X2 .00008   03Y2 -.00010

02S3 with
      03S4 .00008   03X2 .00139   03Y2 -.00171

02S4 with
      03S3 .00008   03S4 .00250   03X1 .00020   03X2 .01219   03Y1 -.00025   03Y2 -.01504

02X1 with
      03S4 -.00020   03X2 -.00184   03Y2 .00277

02X2 with
      03S1 -.00001   03S2 -.00008   03S3 -.00139   03S4 -.01219   03X1 -.00184   03X2 -.02785   03Y1 .00277
      03Y2 .04763

02Y1 with
      03S4 .00025   03X2 .00277   03Y2 -.00302

02Y2 with
      03S1 .00001   03S2 .00010   03S3 .00171   03S4 .01504   03X1 .00277   03X2 .04763   03Y1 -.00302
      03Y2 -.04804

02Z1 with
      03Z2 .00040

02Z2 with
      03Z1 .00040   03Z2 .01074

C2F4   Tetrafluoroethylene

Molecular Geometry   Symmetry D_{2h}

Center   Coordinates

Center	X	Y	Z
C1	0.	1.24062090	0.
C2	0.	-1.24062090	0.
F1	0.	2.59200001	2.08094001
F2	0.	2.59200001	-2.08094001
F3	0.	-2.59200001	-2.08094001
F4	0.	-2.59200001	2.08094001
Mass	0.	0.	0.

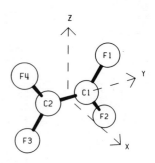

MO Symmetry Types and Occupancy

Representation	AG	AU	B1G	B1U	B2G	B2U	B3G	B3U
Number of MOs	14	2	4	10	10	4	2	14
Number Occupied	6	1	1	4	4	2	1	5

Energy Expectation Values

Total	-473.3331	Kinetic	473.6869
Electronic	-718.9671	Potential	-947.0200
Nuclear Repulsion	245.6340	Virial	1.9993

Electronic Potential	-1192.6540
One-Electron Potential	-1619.0098
Two-Electron Potential	426.3558

Moments of the Charge Distribution

Second

x^2	y^2	z^2	xy	xz	yz	r^2
-21.7148	-286.4845	-182.4463	0.	0.	0.	-490.6457
0.	260.3344	155.8912	0.	0.	0.	416.2256

Fourth (even)

x^4	y^4	z^4	x^2y^2	x^2z^2	y^2z^2	r^2x^2	r^2y^2	r^2z^2	r^4
-57.9616	-2401.3225	-1125.1072	-125.0789	-81.4991	-1269.6712	-264.5395	-3796.0726	-2476.2774	-6536.8893
0.	1653.3868	675.0575	0.	0.	1047.3495	0.	2700.7362	1722.4069	4423.1432

Expectation Values at Atomic Centers

Center			Expectation Value		
	r^{-1}	8	xr^{-3}	yr^{-3}	zr^{-3}
C1	-14.4679	116.2527	0.	-0.0379	0.
F1	-26.5035	413.5955	0.	0.1474	0.2240

	$\dfrac{3x^2-r^2}{r^5}$	$\dfrac{3y^2-r^2}{r^5}$	$\dfrac{3z^2-r^2}{r^5}$	$\dfrac{3xy}{r^5}$	$\dfrac{3xz}{r^5}$	$\dfrac{3yz}{r^5}$
C1	-0.5022	0.0444	0.4578	0.	0.	0.
F1	-1.7527	-0.1117	1.8644	0.	0.	2.4271

Mulliken Charge

Center				Basis Function Type							Total
	S1	S2	S3	S4	X1	X2	Y1	Y2	Z1	Z2	
C1	0.0427	1.9564	0.7951	0.0497	0.9355	0.1769	0.8408	0.0524	0.6374	-0.0142	5.4728
F1	0.0435	1.9557	1.0276	0.9183	1.4370	0.5068	1.3286	0.4588	1.1968	0.3905	9.2636

Overlap Populations

	Pair of Atomic Centers					
	C2	F1	F1	F2	F3	F3
	C1	C1	C2	F1	F1	F2
1AG	0.	-0.0001	0.	0.	0.	0.
2AG	0.0011	0.	0.	0.	0.	0.
3AG	0.0168	0.0777	0.0078	0.0182	0.0002	0.0027
4AG	0.5853	0.0094	0.0045	-0.0009	0.0001	0.
5AG	0.0608	0.0659	-0.0020	0.0179	-0.0001	0.0050
6AG	0.2288	-0.0490	-0.0397	0.0170	0.0015	0.0251
Total AG	0.8928	0.1039	-0.0294	0.0522	0.0017	0.0329
1AU	0.	0.	0.	-0.0196	0.0003	-0.0043
Total AU	0.	0.	0.	-0.0196	0.0003	-0.0043
1B1G	-0.0334	0.0616	-0.0090	0.0145	-0.0003	-0.0032
Total B1G	-0.0334	0.0616	-0.0090	0.0145	-0.0003	-0.0032
1B1U	0.	0.	0.	0.	0.	0.
2B1U	0.0068	0.0518	0.	-0.0173	-0.0002	0.0029
3B1U	0.0575	0.0529	0.0107	-0.0061	-0.0010	0.0059
4B1U	0.0021	0.0129	0.0041	-0.0672	0.0003	0.0114
Total B1U	0.0663	0.1175	0.0148	-0.0906	-0.0009	0.0203
1B2G	0.	-0.0002	0.	0.	0.	0.
2B2G	-0.0056	0.0460	0.0021	-0.0182	0.0001	-0.0024
3B2G	-0.0292	0.0850	-0.0081	-0.0121	0.0015	-0.0078
4B2G	-0.0015	0.0094	-0.0052	-0.0642	-0.0004	-0.0152
Total B2G	-0.0364	0.1403	-0.0111	-0.0945	0.0013	-0.0254
1B2U	0.0800	0.0800	0.0098	0.0098	0.0002	0.0021
2B2U	0.4802	-0.2106	-0.0359	0.0141	0.0003	0.0033
Total B2U	0.5603	-0.1306	-0.0261	0.0239	0.0004	0.0054
1B3G	0.	0.	0.	-0.0197	-0.0004	0.0044
Total B3G	0.	0.	0.	-0.0197	-0.0004	0.0044
1B3U	0.	0.	0.	0.	0.	0.
2B3U	-0.0017	0.	0.	0.	0.	0.
3B3U	-0.0083	0.0277	0.0375	0.0211	-0.0003	-0.0039
4B3U	-0.0601	0.0583	0.0010	0.0079	-0.0004	-0.0010
5B3U	-0.0757	0.0441	-0.0238	0.0253	-0.0007	-0.0250
Total B3U	-0.1458	0.1300	0.0147	0.0544	-0.0015	-0.0299
Total	1.3038	0.4227	-0.0462	-0.0793	0.0009	0.0000
Distance	2.4812	2.4812	4.3611	4.1619	6.6479	5.1840

Molecular Orbitals and Eigenvalues

Center		S1	S2	S3	S4	X1	X2	Y1	Y2	Z1	Z2
* 1AG	C1	-0.00000	0.00002	-0.00002	-0.00048	0.	0.	-0.00008	-0.00049	0.	0.
-26.3770	C2	-0.00000	0.00002	-0.00002	-0.00048	0.	0.	0.00008	0.00049	0.	0.
	F1	0.02587	0.48946	0.00215	0.00085	0.	0.	-0.00037	0.00019	-0.00051	0.00002
	F2	0.02587	0.48946	0.00215	0.00085	0.	0.	-0.00037	0.00019	0.00051	-0.00002
	F3	0.02587	0.48946	0.00215	0.00085	0.	0.	0.00037	-0.00019	0.00051	-0.00002
	F4	0.02587	0.48946	0.00215	0.00085	0.	0.	0.00037	-0.00019	-0.00051	0.00002
* 2AG	C1	-0.03669	-0.69204	-0.00391	-0.00016	0.	0.	0.00000	0.00049	0.	0.
-11.4901	C2	-0.03669	-0.69204	-0.00391	-0.00016	0.	0.	-0.00000	-0.00049	0.	0.
	F1	0.00001	0.00013	-0.00021	0.00004	0.	0.	0.00028	-0.00021	0.00036	0.00004
	F2	0.00001	0.00013	-0.00021	0.00004	0.	0.	0.00028	-0.00021	-0.00036	-0.00004
	F3	0.00001	0.00013	-0.00021	0.00004	0.	0.	-0.00028	0.00021	-0.00036	-0.00004
	F4	0.00001	0.00013	-0.00021	0.00004	0.	0.	-0.00028	0.00021	0.00036	0.00004
* 3AG	C1	-0.00338	-0.07169	0.15769	0.00677	0.	0.	0.04208	-0.01264	0.	0.
-1.7640	C2	-0.00338	-0.07169	0.15769	0.00677	0.	0.	-0.04208	0.01264	0.	0.
	F1	-0.00524	-0.11258	0.25091	0.21374	0.	0.	-0.03220	-0.00875	-0.05448	-0.01287
	F2	-0.00524	-0.11258	0.25091	0.21374	0.	0.	-0.03220	-0.00875	0.05448	0.01287
	F3	-0.00524	-0.11258	0.25091	0.21374	0.	0.	0.03220	0.00875	0.05448	0.01287
	F4	-0.00524	-0.11258	0.25091	0.21374	0.	0.	0.03220	0.00875	-0.05448	-0.01287
* 4AG	C1	-0.00680	-0.14414	0.34371	0.07447	0.	0.	-0.24252	-0.01582	0.	0.
-1.1456	C2	-0.00680	-0.14414	0.34371	0.07447	0.	0.	0.24252	0.01582	0.	0.
	F1	0.00223	0.04735	-0.11354	-0.10447	0.	0.	-0.10071	-0.03105	-0.05708	-0.03354
	F2	0.00223	0.04735	-0.11354	-0.10447	0.	0.	-0.10071	-0.03105	0.05708	0.03354
	F3	0.00223	0.04735	-0.11354	-0.10447	0.	0.	0.10071	0.03105	0.05708	0.03354
	F4	0.00223	0.04735	-0.11354	-0.10447	0.	0.	0.10071	0.03105	-0.05708	-0.03354
* 5AG	C1	-0.00115	-0.01988	0.09300	-0.01127	0.	0.	0.29260	-0.02497	0.	0.
-0.8847	C2	-0.00115	-0.01988	0.09300	-0.01127	0.	0.	-0.29260	0.02497	0.	0.
	F1	0.00139	0.02980	-0.06862	-0.10160	0.	0.	0.05888	0 03957	-0.29472	-0.10858
	F2	0.00139	0.02980	-0.06862	-0.10160	0.	0.	0.05888	0.03957	0.29472	0.10858
	F3	0.00139	0.02980	-0.06862	-0.10160	0.	0.	-0.05888	-0.03957	0.29472	0.10858
	F4	0.00139	0.02980	-0.06862	-0.10160	0.	0.	-0.05888	-0.03957	-0.29472	-0.10858
* 6AG	C1	-0.00143	-0.03165	0.07002	0.09044	0.	0.	-0.27434	0.04295	0.	0.
-0.6852	C2	-0.00143	-0.03165	0.07002	0.09044	0.	0.	0.27434	-0.04295	0.	0.
	F1	-0.00026	-0.00503	0.01688	0.00412	0.	0.	0.35160	0.16831	-0.01212	-0.01961
	F2	-0.00026	-0.00503	0.01688	0.00412	0.	0.	0.35160	0.16831	0.01212	0.01961
	F3	-0.00026	-0.00503	0.01688	0.00412	0.	0.	-0.35160	-0.16831	0.01212	0.01961
	F4	-0.00026	-0.00503	0.01688	0.00412	0.	0.	-0.35160	-0.16831	-0.01212	-0.01961
7AG	C1	-0.00081	-0.04632	-0.14567	1.33617	0.	0.	0.40298	0.63995	0.	0.
0.2695	C2	-0.00081	-0.04632	-0.14567	1.33617	0.	0.	-0.40298	-0.63995	0.	0.
	F1	0.00221	0.05169	-0.08527	-0.55785	0.	0.	0.02389	0.05943	0.15751	0.32220
	F2	0.00221	0.05169	-0.08527	-0.55785	0.	0.	0.02389	0.05943	-0.15751	-0.32220
	F3	0.00221	0.05169	-0.08527	-0.55785	0.	0.	-0.02389	-0.05943	-0.15751	-0.32220
	F4	0.00221	0.05169	-0.08527	-0.55785	0.	0.	-0.02389	-0.05943	0.15751	0.32220
8AG	C1	-0.00054	-0.02472	-0.06076	-0.41556	0.	0.	0.47067	-1.59946	0.	0.
0.4832	C2	-0.00054	-0.02472	-0.06076	-0.41556	0.	0.	-0.47067	1.59946	0.	0.
	F1	0.00011	-0.00131	-0.03149	0.21676	0.	0.	0.16904	0.24792	0.04933	-0.05739
	F2	0.00011	-0.00131	-0.03149	0.21676	0.	0.	0.16904	0.24792	-0.04933	0.05739
	F3	0.00011	-0.00131	-0.03149	0.21676	0.	0.	-0.16904	-0.24792	-0.04933	0.05739
	F4	0.00011	-0.00131	-0.03149	0.21676	0.	0.	-0.16904	-0.24792	0.04933	-0.05739

Center		S1	S2	S3	S4	X1	X2	Y1	Y2	Z1	Z2
9AG	C1	0.00847	0.08759	-1.13525	0.83072	0.	0.	-0.27808	0.00081	0.	0.
0.6188	C2	0.00847	0.08759	-1.13525	0.83072	0.	0.	0.27808	-0.00081	0.	0.
	F1	-0.00059	-0.01035	0.04706	-0.10164	0.	0.	-0.00729	-0.05467	-0.24942	-0.01399
	F2	-0.00059	-0.01035	0.04706	-0.10164	0.	0.	-0.00729	-0.05467	0.24942	0.01399
	F3	-0.00059	-0.01035	0.04706	-0.10164	0.	0.	0.00729	0.05467	0.24942	0.01399
	F4	-0.00059	-0.01035	0.04706	-0.10164	0.	0.	0.00729	0.05467	-0.24942	-0.01399
* 1AU	C1	0.	0.	0.	0.	0.	0.	0.	0.	0.	0.
-0.7254	C2	0.	0.	0.	0.	0.	0.	0.	0.	0.	0.
	F1	0.	0.	0.	0.	-0.38757	-0.18574	0.	0.	0.	0.
	F2	0.	0.	0.	0.	0.38757	0.18574	0.	0.	0.	0.
	F3	0.	0.	0.	0.	-0.38757	-0.18574	0.	0.	0.	0.
	F4	0.	0.	0.	0.	0.38757	0.18574	0.	0.	0.	0.
* 1B1G	C1	0.	0.	0.	0.	0.14807	0.03329	0.	0.	0.	0.
-0.7834	C2	0.	0.	0.	0.	-0.14807	-0.03329	0.	0.	0.	0.
	F1	0.	0.	0.	0.	0.36075	0.15721	0.	0.	0.	0.
	F2	0.	0.	0.	0.	0.36075	0.15721	0.	0.	0.	0.
	F3	0.	0.	0.	0.	-0.36075	-0.15721	0.	0.	0.	0.
	F4	0.	0.	0.	0.	-0.36075	-0.15721	0.	0.	0.	0.
2B1G	C1	0.	0.	0.	0.	0.62563	0.69440	0.	0.	0.	0.
0.1477	C2	0.	0.	0.	0.	-0.62563	-0.69440	0.	0.	0.	0.
	F1	0.	0.	0.	0.	-0.18365	-0.17859	0.	0.	0.	0.
	F2	0.	0.	0.	0.	-0.18365	-0.17859	0.	0.	0.	0.
	F3	0.	0.	0.	0.	0.18365	0.17859	0.	0.	0.	0.
	F4	0.	0.	0.	0.	0.18365	0.17859	0.	0.	0.	0.
3B1G	C1	0.	0.	0.	0.	-0.62795	1.71579	0.	0.	0.	0.
0.4794	C2	0.	0.	0.	0.	0.62795	-1.71579	0.	0.	0.	0.
	F1	0.	0.	0.	0.	-0.04904	-0.17730	0.	0.	0.	0.
	F2	0.	0.	0.	0.	-0.04904	-0.17730	0.	0.	0.	0.
	F3	0.	0.	0.	0.	0.04904	0.17730	0.	0.	0.	0.
	F4	0.	0.	0.	0.	0.04904	0.17730	0.	0.	0.	0.
* 1B1U	C1	0.	0.	0.	0.	0.	0.	0.	0.	0.00006	0.00026
-26.3770	C2	0.	0.	0.	0.	0.	0.	0.	0.	0.00006	0.00026
	F1	-0.02587	-0.48946	-0.00219	-0.00074	0.	0.	0.00033	0.00001	0.00054	-0.00019
	F2	0.02587	0.48946	0.00219	0.00074	0.	0.	-0.00033	-0.00001	0.00054	-0.00019
	F3	0.02587	0.48946	0.00219	0.00074	0.	0.	0.00033	0.00001	0.00054	-0.00019
	F4	-0.02587	-0.48946	-0.00219	-0.00074	0.	0.	-0.00033	-0.00001	0.00054	-0.00019
* 2B1U	C1	0.	0.	0.	0.	0.	0.	0.	0.	-0.11500	0.01752
-1.6935	C2	0.	0.	0.	0.	0.	0.	0.	0.	-0.11500	0.01752
	F1	0.00567	0.12190	-0.27176	-0.23821	0.	0.	0.03379	0.00628	0.03891	0.00536
	F2	-0.00567	-0.12190	0.27176	0.23821	0.	0.	-0.03379	-0.00628	0.03891	0.00536
	F3	-0.00567	-0.12190	0.27176	0.23821	0.	0.	0.03379	0.00628	0.03891	0.00536
	F4	0.00567	0.12190	-0.27176	-0.23821	0.	0.	-0.03379	-0.00628	0.03891	0.00536
* 3B1U	C1	0.	0.	0.	0.	0.	0.	0.	0.	-0.29359	0.02045
-0.8587	C2	0.	0.	0.	0.	0.	0.	0.	0.	-0.29359	0.02045
	F1	-0.00161	-0.03408	0.08321	0.10956	0.	0.	0.26886	0.10379	0.14634	0.04137
	F2	0.00161	0.03408	-0.08321	-0.10956	0.	0.	-0.26886	-0.10379	0.14634	0.04137
	F3	0.00161	0.03408	-0.08321	-0.10956	0.	0.	0.26886	0.10379	0.14634	0.04137
	F4	-0.00161	-0.03408	0.08321	0.10956	0.	0.	-0.26886	-0.10379	0.14634	0.04137

Center		S1	S2	S3	S4	X1	X2	Y1	Y2	Z1	Z2
* 4B1U	C1	0.	0.	0.	0.	0.	0.	0.	0.	0.08477	-0.04117
-0.7101	C2	0.	0.	0.	0.	0.	0.	0.	0.	0.08477	-0.04117
	F1	0.00022	0.00403	-0.01619	0.00687	0.	0.	0.20767	0.09836	-0.32943	-0.14925
	F2	-0.00022	-0.00403	0.01619	-0.00687	0.	0.	-0.20767	-0.09836	-0.32943	-0.14925
	F3	-0.00022	-0.00403	0.01619	-0.00687	0.	0.	0.20767	0.09836	-0.32943	-0.14925
	F4	0.00022	0.00403	-0.01619	0.00687	0.	0.	-0.20767	-0.09836	-0.32943	-0.14925
5B1U	C1	0.	0.	0.	0.	0.	0.	0.	0.	-0.40398	-0.60887
0.2187	C2	0.	0.	0.	0.	0.	0.	0.	0.	-0.40398	-0.60887
	F1	-0.00198	-0.04577	0.08064	0.44177	0.	0.	-0.19802	-0.21362	-0.07746	-0.10516
	F2	0.00198	0.04577	-0.08064	-0.44177	0.	0.	0.19802	0.21362	-0.07746	-0.10516
	F3	0.00198	0.04577	-0.08064	-0.44177	0.	0.	-0.19802	-0.21362	-0.07746	-0.10516
	F4	-0.00198	-0.04577	0.08064	0.44177	0.	0.	0.19802	0.21362	-0.07746	-0.10516
6B1U	C1	0.	0.	0.	0.	0.	0.	0.	0.	0.73913	-0.72180
0.5158	C2	0.	0.	0.	0.	0.	0.	0.	0.	0.73913	-0.72180
	F1	0.00012	-0.00021	-0.02619	0.17048	0.	0.	0.10160	-0.00516	0.20405	0.13482
	F2	-0.00012	0.00021	0.02619	-0.17048	0.	0.	-0.10160	0.00516	0.20405	0.13482
	F3	-0.00012	0.00021	0.02619	-0.17048	0.	0.	0.10160	-0.00516	0.20405	0.13482
	F4	0.00012	-0.00021	-0.02619	0.17048	0.	0.	-0.10160	0.00516	0.20405	0.13482
* 1B2G	C1	0.	0.	0.	0.	0.	0.	0.	0.	0.00013	0.00126
-26.3770	C2	0.	0.	0.	0.	0.	0.	0.	0.	-0.00013	-0.00126
	F1	-0.02587	-0.48946	-0.00210	-0.00109	0.	0.	0.00037	-0.00007	0.00056	-0.00022
	F2	0.02587	0.48946	0.00210	0.00109	0.	0.	-0.00037	0.00007	0.00056	-0.00022
	F3	-0.02587	-0.48946	-0.00210	-0.00109	0.	0.	-0.00037	0.00007	-0.00056	0.00022
	F4	0.02587	0.48946	0.00210	0.00109	0.	0.	0.00037	-0.00007	-0.00056	0.00022
* 2B2G	C1	0.	0.	0.	0.	0.	0.	0.	0.	-0.11638	0.02571
-1.6822	C2	0.	0.	0.	0.	0.	0.	0.	0.	0.11638	-0.02571
	F1	0.00576	0.12367	-0.27766	-0.23867	0.	0.	0.03027	0.00176	0.03819	0.00706
	F2	-0.00576	-0.12367	0.27766	0.23867	0.	0.	-0.03027	-0.00176	0.03819	0.00706
	F3	0.00576	0.12367	-0.27766	-0.23867	0.	0.	-0.03027	-0.00176	-0.03819	-0.00706
	F4	-0.00576	-0.12367	0.27766	0.23867	0.	0.	0.03027	0.00176	-0.03819	-0.00706
* 3B2G	C1	0.	0.	0.	0.	0.	0.	0.	0.	0.28468	-0.07741
-0.7949	C2	0.	0.	0.	0.	0.	0.	0.	0.	-0.28468	0.07741
	F1	0.00125	0.02625	-0.06753	-0.07550	0.	0.	-0.27569	-0.10457	-0.19479	-0.06732
	F2	-0.00125	-0.02625	0.06753	0.07550	0.	0.	0.27569	0.10457	-0.19479	-0.06732
	F3	0.00125	0.02625	-0.06753	-0.07550	0.	0.	0.27569	0.10457	0.19479	0.06732
	F4	-0.00125	-0.02625	0.06753	0.07550	0.	0.	-0.27569	-0.10457	0.19479	0.06732
* 4B2G	C1	0.	0.	0.	0.	0.	0.	0.	0.	0.05827	-0.03946
-0.6955	C2	0.	0.	0.	0.	0.	0.	0.	0.	-0.05827	0.03946
	F1	0.00020	0.00371	-0.01311	0.00214	0.	0.	0.24469	0.11841	-0.31397	-0.14414
	F2	-0.00020	-0.00371	0.01311	-0.00214	0.	0.	-0.24469	-0.11841	-0.31397	-0.14414
	F3	0.00020	0.00371	-0.01311	0.00214	0.	0.	-0.24469	-0.11841	0.31397	0.14414
	F4	-0.00020	-0.00371	0.01311	-0.00214	0.	0.	0.24469	0.11841	0.31397	0.14414
5B2G	C1	0.	0.	0.	0.	0.	0.	0.	0.	0.24164	2.50722
0.4798	C2	0.	0.	0.	0.	0.	0.	0.	0.	-0.24164	-2.50722
	F1	0.00205	0.05164	-0.05345	-0.84727	0.	0.	0.00454	0.17843	-0.02105	0.07181
	F2	-0.00205	-0.05164	0.05345	0.84727	0.	0.	-0.00454	-0.17843	-0.02105	0.07181
	F3	0.00205	0.05164	-0.05345	-0.84727	0.	0.	-0.00454	-0.17843	0.02105	-0.07181
	F4	-0.00205	-0.05164	0.05345	0.84727	0.	0.	0.00454	0.17843	0.02105	-0.07181

Center		S1	S2	S3	S4	X1	X2	Y1	Y2	Z1	Z2
* 1B2U	C1	0.	0.	0.	0.	0.25897	0.03358	0.	0.	0.	0.
-0.8323	C2	0.	0.	0.	0.	0.25897	0.03358	0.	0.	0.	0.
	F1	0.	0.	0.	0.	0.31205	0.12806	0.	0.	0.	0.
	F2	0.	0.	0.	0.	0.31205	0.12806	0.	0.	0.	0.
	F3	0.	0.	0.	0.	0.31205	0.12806	0.	0.	0.	0.
	F4	0.	0.	0.	0.	0.31205	0.12806	0.	0.	0.	0.
* 2B2U	C1	0.	0.	0.	0.	-0.51053	-0.15584	0.	0.	0.	0.
-0.4342	C2	0.	0.	0.	0.	-0.51053	-0.15584	0.	0.	0.	0.
	F1	0.	0.	0.	0.	0.23751	0.16692	0.	0.	0.	0.
	F2	0.	0.	0.	0.	0.23751	0.16692	0.	0.	0.	0.
	F3	0.	0.	0.	0.	0.23751	0.16692	0.	0.	0.	0.
	F4	0.	0.	0.	0.	0.23751	0.16692	0.	0.	0.	0.
3B2U	C1	0.	0.	0.	0.	0.55674	-0.72910	0.	0.	0.	0.
0.3802	C2	0.	0.	0.	0.	0.55674	-0.72910	0.	0.	0.	0.
	F1	0.	0.	0.	0.	0.05414	0.09995	0.	0.	0.	0.
	F2	0.	0.	0.	0.	0.05414	0.09995	0.	0.	0.	0.
	F3	0.	0.	0.	0.	0.05414	0.09995	0.	0.	0.	0.
	F4	0.	0.	0.	0.	0.05414	0.09995	0.	0.	0.	0.
* 1B3G	C1	0.	0.	0.	0.	0.	0.	0.	0.	0.	0.
-0.7320	C2	0.	0.	0.	0.	0.	0.	0.	0.	0.	0.
	F1	0.	0.	0.	0.	-0.38474	-0.18670	0.	0.	0.	0.
	F2	0.	0.	0.	0.	0.38474	0.18670	0.	0.	0.	0.
	F3	0.	0.	0.	0.	0.38474	0.18670	0.	0.	0.	0.
	F4	0.	0.	0.	0.	-0.38474	-0.18670	0.	0.	0.	0.
* 1B3U	C1	0.00000	-0.00002	0.00000	-0.00004	0.	0.	0.00010	0.00040	0.	0.
-26.3770	C2	-0.00000	0.00002	-0.00000	0.00004	0.	0.	0.00010	0.00040	0.	0.
	F1	-0.02587	-0.48946	-0.00216	-0.00080	0.	0.	0.00036	-0.00015	0.00052	-0.00006
	F2	-0.02587	-0.48946	-0.00216	-0.00080	0.	0.	0.00036	-0.00015	-0.00052	0.00006
	F3	0.02587	0.48946	0.00216	0.00080	0.	0.	0.00036	-0.00015	0.00052	-0.00006
	F4	0.02587	0.48946	0.00216	0.00080	0.	0.	0.00036	-0.00015	-0.00052	0.00006
* 2B3U	C1	0.03671	0.69250	0.00236	0.00265	0.	0.	0.00160	-0.00113	0.	0.
-11.4882	C2	-0.03671	-0.69250	-0.00236	-0.00265	0.	0.	0.00160	-0.00113	0.	0.
	F1	-0.00001	-0.00013	0.00016	-0.00011	0.	0.	-0.00026	-0.00001	-0.00037	0.00016
	F2	-0.00001	-0.00013	0.00016	-0.00011	0.	0.	-0.00026	-0.00001	0.00037	-0.00016
	F3	0.00001	0.00013	-0.00016	0.00011	0.	0.	-0.00026	-0.00001	-0.00037	0.00016
	F4	0.00001	0.00013	-0.00016	0.00011	0.	0.	-0.00026	-0.00001	0.00037	-0.00016
* 3B3U	C1	0.00267	0.05648	-0.12631	0.10772	0.	0.	-0.08765	-0.03427	0.	0.
-1.7457	C2	-0.00267	-0.05648	0.12631	-0.10772	0.	0.	-0.08765	-0.03427	0.	0.
	F1	0.00541	0.11631	-0.25865	-0.22827	0.	0.	0.02780	0.01613	0.05525	0.01498
	F2	0.00541	0.11631	-0.25865	-0.22827	0.	0.	0.02780	0.01613	-0.05525	-0.01498
	F3	-0.00541	-0.11631	0.25865	0.22827	0.	0.	0.02780	0.01613	0.05525	0.01498
	F4	-0.00541	-0.11631	0.25865	0.22827	0.	0.	0.02780	0.01613	-0.05525	-0.01498
* 4B3U	C1	0.00434	0.09088	-0.23529	-0.07763	0.	0.	-0.17739	0.03760	0.	0.
-0.9293	C2	-0.00434	-0.09088	0.23529	0.07763	0.	0.	-0.17739	0.03760	0.	0.
	F1	-0.00223	-0.04748	0.11478	0.13338	0.	0.	0.08370	0.02737	0.27050	0.09755
	F2	-0.00223	-0.04748	0.11478	0.13338	0.	0.	0.08370	0.02737	-0.27050	-0.09755
	F3	0.00223	0.04748	-0.11478	-0.13338	0.	0.	0.08370	0.02737	0.27050	0.09755
	F4	0.00223	0.04748	-0.11478	-0.13338	0.	0.	0.08370	0.02737	-0.27050	-0.09755

Basis Function Type

		S1	S2	S3	S4	X1	X2	Y1	Y2	Z1	Z2
• 5B3U	C1	-0.00137	-0.02765	0.08534	0.05862	0.	0.	-0.04234	-0.04229	0.	0.
-0.7524	C2	0.00137	0.02765	-0.08534	-0.05862	0.	0.	-0.04234	-0.04229	0.	0.
	F1	0.00016	0.00313	-0.01050	0.00297	0.	0.	-0.35399	-0.16397	0.12915	0.06077
	F2	0.00016	0.00313	-0.01050	0.00297	0.	0.	-0.35399	-0.16397	-0.12915	-0.06077
	F3	-0.00016	-0.00313	0.01050	-0.00297	0.	0.	-0.35399	-0.16397	0.12915	0.06077
	F4	-0.00016	-0.00313	0.01050	-0.00297	0.	0.	-0.35399	-0.16397	-0.12915	-0.06077
6B3U	C1	-0.00394	-0.07780	0.27614	-0.50249	0.	0.	0.05368	1.26748	0.	0.
0.2544	C2	0.00394	0.07780	-0.27614	0.50249	0.	0.	0.05368	1.26748	0.	0.
	F1	0.00219	0.05055	-0.08947	-0.48971	0.	0.	0.02353	-0.02039	0.16057	0.22375
	F2	0.00219	0.05055	-0.08947	-0.48971	0.	0.	0.02353	-0.02039	-0.16057	-0.22375
	F3	-0.00219	-0.05055	0.08947	0.48971	0.	0.	0.02353	-0.02039	0.16057	0.22375
	F4	-0.00219	-0.05055	0.08947	0.48971	0.	0.	0.02353	-0.02039	-0.16057	-0.22375
7B3U	C1	0.00576	0.14466	-0.20241	-4.74814	0.	0.	-0.09047	2.36380	0.	0.
0.3789	C2	-0.00576	-0.14466	0.20241	4.74814	0.	0.	-0.09047	2.36380	0.	0.
	F1	-0.00076	-0.01661	0.03882	0.14365	0.	0.	-0.17015	-0.33595	-0.15425	-0.22431
	F2	-0.00076	-0.01661	0.03882	0.14365	0.	0.	-0.17015	-0.33595	0.15425	0.22431
	F3	0.00076	0.01661	-0.03882	-0.14365	0.	0.	-0.17015	-0.33595	-0.15425	-0.22431
	F4.	0.00076	0.01661	-0.03882	-0.14365	0.	0.	-0.17015	-0.33595	0.15425	0.22431

Overlap Matrix

First Function	Second Function and Overlap

C1S1 with

C1S2	.35436	C1S3	.00805	C1S4	.01419	C2S3	.00195	C2S4	.00571	C2Y1	.00588	C2Y2	.00972
F1S3	.00005	F1S4	.00305	F1Y1	-.00029	F1Y2	-.00590	F1Z1	-.00045	F1Z2	-.00909		

C1S2 with

C1S3	.26226	C1S4	.17745	C2S2	.00003	C2S3	.02935	C2S4	.07367	C2Y1	.07737	C2Y2	.12192
F1S3	.00237	F1S4	.04226	F1Y1	-.00666	F1Y2	-.07355	F1Z1	-.01026	F1Z2	-.11326		

C1S3 with

C1S4	.79970	C2S1	.00195	C2S2	.02935	C2S3	.24915	C2S4	.40191	C2Y1	.38178	C2Y2	.54963
F1S1	.00104	F1S2	.01441	F1S3	.11506	F1S4	.28952	F1Y1	-.08800	F1Y2	-.29522	F1Z1	-.13551
F1Z2	-.45459												

C1S4 with

C2S1	.00571	C2S2	.07367	C2S3	.40191	C2S4	.63408	C2Y1	.30974	C2Y2	.62842	F1S1	.00306
F1S2	.03934	F1S3	.22574	F1S4	.45585	F1Y1	-.06813	F1Y2	-.25641	F1Z1	-.10491	F1Z2	-.39484

C1X1 with

C1X2	.53875	C2X1	.20639	C2X2	.30919	F1X1	.08317	F1X2	.29951

C1X2 with

C2X1	.30919	C2X2	.70274	F1X1	.12016	F1X2	.48382

C1Y1 with

C1Y2	.53875	C2S1	-.00588	C2S2	-.07737	C2S3	-.38178	C2S4	-.30974	C2Y1	-.35335	C2Y2	-.03362
F1S1	.00172	F1S2	.02226	F1S3	.13195	F1S4	.21351	F1Y1	-.00988	F1Y2	.11505	F1Z1	-.14329
F1Z2	-.28406												

C1Y2 with

C2S1	-.00972	C2S2	-.12192	C2S3	-.54963	C2S4	-.62842	C2Y1	-.03362	C2Y2	.20693	F1S1	.00284
F1S2	.03604	F1S3	.19051	F1S4	.32246	F1Y1	.07475	F1Y2	.34114	F1Z1	-.06991	F1Z2	-.21971

C1Z1 with

C1Z2	.53875	C2Z1	.20639	C2Z2	.30919	F1S1	.00264	F1S2	.03427	F1S3	.20318	F1S4	.32878
F1Y1	-.14329	F1Y2	-.28406	F1Z1	-.13747	F1Z2	-.13790							

C1Z2 with

C2Z1	.30919	C2Z2	.70274	F1S1	.00437	F1S2	.05550	F1S3	.29336	F1S4	.49655	F1Y1	-.06991
F1Y2	-.21971	F1Z1	.01250	F1Z2	.14549								

C2S1 with

F1S4	.00003	F1Y2	-.00050	F1Z2	-.00027

C2S2 with

F1S4	.00063	F1Y2	-.00777	F1Z2	-.00422

C2S3 with

F1S2	.00004	F1S3	.00149	F1S4	.02148	F1Y1	-.00394	F1Y2	-.09255	F1Z1	-.00214	F1Z2	-.05025

C2S4 with

F1S1	.00046	F1S2	.00604	F1S3	.04166	F1S4	.11862	F1Y1	-.03637	F1Y2	-.21151	F1Z1	-.01974
F1Z2	-.11484												

C2X1 with

F1X1	.00234	F1X2	.03601

C2X2 with

F1X1	.03178	F1X2	.17125

C2Y1 with

F1S1	.00005	F1S2	.00071	F1S3	.00952	F1S4	.05199	F1Y1	-.01595	F1Y2	-.13487	F1Z1	-.00993
F1Z2	-.09278												

C2Y2 with

F1S1	.00185	F1S2	.02383	F1S3	.14117	F1S4	.29954	F1Y1	-.06470	F1Y2	-.23496	F1Z1	-.05238
F1Z2	-.22056												

C2Z1 with

F1S1	.00003	F1S2	.00039	F1S3	.00517	F1S4	.02823	F1Y1	-.00993	F1Y2	-.09278	F1Z1	-.00305
F1Z2	-.01436												

C2F4

First Function   Second Function and Overlap

C2Z2 with
 F1S1 .00100 F1S2 .01294 F1S3 .07665 F1S4 .16264 F1Y1 -.05238 F1Y2 -.22056 F1Z1 .00334
 F1Z2 .05150

F1S1 with
 F1S2 .35647 F1S3 .01003 F1S4 .01474 F2S4 .00003 F2Z2 .00046

F1S2 with
 F1S3 .28653 F1S4 .18096 F2S4 .00047 F2Z2 .00637 F3Y2 .00001 F3Z2 .00001

F1S3 with
 F1S4 .78531 F2S3 .00004 F2S4 .00685 F2Z1 .00032 F2Z2 .05330 F3Y2 .00017 F3Z2 .00014

F1S4 with
 F2S1 .00003 F2S2 .00047 F2S3 .00685 F2S4 .04526 F2Z1 .01441 F2Z2 .16639 F3S4 .00037
 F3Y1 .00002 F3Y2 .00323 F3Z1 .00001 F3Z2 .00259

F1X1 with
 F1X2 .49444 F2X1 .00012 F2X2 .01140 F3X2 .00004

F1X2 with
 F2X1 .01140 F2X2 .09377 F3X1 .00004 F3X2 .00238

F1Y1 with
 F1Y2 .49444 F2Y1 .00012 F2Y2 .01140 F3S4 -.00002 F3Y2 -.00037 F3Z2 -.00033

F1Y2 with
 F2Y1 .01140 F2Y2 .09377 F3S2 -.00001 F3S3 -.00017 F3S4 -.00323 F3Y1 -.00037 F3Y2 -.01512
 F3Z1 -.00033 F3Z2 -.01406

F1Z1 with
 F1Z2 .49444 F2S3 -.00032 F2S4 -.01441 F2Z1 -.00177 F2Z2 -.07389 F3S4 -.00001 F3Y2 -.00033
 F3Z2 -.00023

F1Z2 with
 F2S1 -.00046 F2S2 -.00637 F2S3 -.05330 F2S4 -.16639 F2Z1 -.07389 F2Z2 -.35013 F3S2 -.00001
 F3S3 -.00014 F3S4 -.00259 F3Y1 -.00033 F3Y2 -.01406 F3Z1 -.00023 F3Z2 -.00890

F2S1 with
 F3Y2 .00004

F2S2 with
 F3S4 .00002 F3Y2 .00063

F2S3 with
 F3S4 .00050 F3Y2 .00798

F2S4 with
 F3S2 .00002 F3S3 .00050 F3S4 .00821 F3Y1 .00146 F3Y2 .04721

F2X1 with
 F3X2 .00146

F2X2 with
 F3X1 .00146 F3X2 .02542

F2Y1 with
 F3S4 -.00146 F3Y1 -.00003 F3Y2 -.01535

F2Y2 with
 F3S1 -.00004 F3S2 -.00063 F3S3 -.00798 F3S4 -.04721 F3Y1 -.01535 F3Y2 -.16128

F2Z1 with
 F3Z2 .00146

F2Z2 with
 F3Z1 .00146 F3Z2 .02542

REFERENCES

1. Enrico Clementi, "Tables of Atomic Functions," Suppl. I.B.M. J. Res. and Dev. $\underline{9}$, 2 (1965).
2. S. Fraga and G. Malli, <u>Many-Electron Systems: Properties and Interactions</u>, W. B. Saunders Co., Philadelphia, 1968; Charlotte Froese, "Hartree-Fock Parameters for the Atoms Helium to Radon," March, 1966, University of British Columbia, unpublished; and S. Fraga, J. Thorhallsson, and C. Fisk, Can. J. Phys. $\underline{47}$, 1415 (1969).
3. C. A. Coulson and A. Streitwiser, <u>Dictionary of π-Electron Calculations</u>, W. H. Freeman and Co., San Francisco, 1965.
4. A. D. McLean and M. Yoshimine, "Tables of Linear Molecule Wave Functions," Suppl. I.B.M. J. Res. and Dev., November, 1967.
5. M. Krauss, "Compendium of <u>ab-initio</u> Calculations of Molecular Energies and Properties," NBS Technical Note 438, December, 1967; also see R. G. Clark and E. T. Stewart, Quart. Rev. $\underline{24}$, 95 (1970).
6. C. C. J. Roothaan, Rev. Mod. Phys., $\underline{23}$, 69 (1951).
7. S. F. Boys, Proc. Roy. Soc. (London), $\underline{A200}$, 542 (1950).
8. I. G. Csizmadia, M. C. Harrison, J. W. Moskowitz, and B. T. Sutcliffe, Theor. Chim. Acta., $\underline{6}$, 191 (1966).
9. J. W. Moskowitz and M. C. Harrison, J. Chem. Phys., $\underline{42}$, 1726 (1965); $\underline{43}$, 3350 (1965). J. W. Moskowitz, J. Chem. Phys. 43, 60 (1965).
10. I. Shavitt, in <u>Methods of Computational Physics</u>, B. Alder et al, eds., Academic Press Inc., New York, 1965, Vol. 2, p. 1.
11. POLYATOM (Version-2) System of Programs for Quantitative Theoretical Chemistry: Part 1, Description of Programs; Part 2, Listing of Programs, by D. B. Neumann, H. Basch, R. L. Kornegay, L. C. Snyder, J. W. Moskowitz, C. Hornback, and S. P. Liebmann, Program 199, Quantum Chemistry Program Exchange, Indiana University, Bloomington, Indiana.
12. S. Huzinaga, J. Chem. Phys., $\underline{42}$, 1293 (1965).
13. J. L. Whitten, J. Chem. Phys., $\underline{44}$, 359 (1966).
14. The term, "double-zeta" seems to have its origin in J. W. Richardson, Rev. Mod. Phys., $\underline{32}$, 461 (1960). See also W. Huo, J. Chem. Phys., $\underline{43}$, 624 (1965).
15. D. B. Neumann and J. W. Moskowitz, J. Chem. Phys., $\underline{50}$, 2216 (1969).
16. R. S. Mulliken, J. Chem. Phys., $\underline{23}$, 1833 (1955).
17. J. C. Slater, <u>Quantum Theory of Molecules and Solids</u>, <u>Vol. 1: Electronic Structure of Molecules</u> McGraw-Hill, New York, 1963.
18. M. Born and J. R. Oppenheimer, Ann. Physik, $\underline{84}$, 457 (1927).
19. H. Margenau and G. M. Murphy, <u>The Mathematics of Physics and Chemistry</u>, D. Van Nostrand, Princeton, 1956.
20. P. O. Lowdin, Advan. Chem. Phys., $\underline{2}$, 207 (1959).
21. W. Pauli, Z. Physik, $\underline{31}$, 765 (1925).
22. S. M. Blinder, Amer. J. Phys., $\underline{33}$, 431 (1965).

23. See, for example, R. K. Nesbet, Rev. Mod. Phys., 35, 552 (1963);
    T. H. Dunning, Chem. Phys. Lett., 8, 169 (1971).
24. T. Koopmans, Physica, 1, 104 (1933).
25. L. Brillouin, Actualites Sci. Ind. 71, Secs. 159, 160 (1933-34).
26. C. Moller and M. S. Plesset, Phys. Rev., 46, 618 (1934).
27. S. Fraga, Theor. Chim. Acta., 2, 406 (1964).
28. J. Goodisman, Theor. Chim. Acta., 15, 165 (1969).
29. A. D. McLean, J. Chem. Phys., 40, 2774 (1964).
30. U. Kaldor and I. Shavitt, J. Chem. Phys., 44, 1823 (1966).
31. H. Margenau and N. R. Kestner, Theory of Intermolecular Forces,
    Pergamon Press, 1969.
32. W. H. Flygare, Rec. Chem. Prog., 28, 63 (1967).
33. A. D. Buckingham and H. C. Longuet-Higgins, Mol. Phys., 14, 63
    (1968).
34. D. W. Davies, The Theory of the Electric and Magnetic Properties
    of Molecules, John Wiley and Sons, New York, 1966.
35. F. N. H. Robinson, Bell System Tech. J., 46, 913 (1967).
36. T. K. Ha and C. T. O'Konski, Chem. Phys. Lett., 3, 603 (1969).
37. H. Basch, Chem. Phys. Lett., 5, 337 (1970).
38. C. A. Coulson and H. L. Strauss, Proc. Roy. Soc., 269, 443 (1962).
39. S. I. Chan and T. P. Das, J. Chem. Phys., 37, 1527 (1962).
40. H. Basch and A. P. Ginsberg, J. Phys. Chem., 73, 854 (1969).
41. T. P. Das and E. L. Hahn, Nuclear Quadrupole Resonance Spectroscopy,
    Academic Press, New York, 1958.
42. R. L. Matcha, J. Chem. Phys., 47, 4595 (1967).
43. A. B. Anderson, N. C. Handy, and R. G. Parr, J. Chem. Phys., 50,
    3634 (1969).
44. J. F. Harrison, J. Chem. Phys., 48, 2379 (1968).
45. P. O. Lowdin, J. Chem. Phys., 18, 365 (1950); E. R. Davidson, J.
    Chem. Phys., 46, 3320 (1967).
46. R. E. Christoffersen and K. A. Baker, Chem. Phys. Lett., 8, 4
    (1971).
47. E. Clementi and W. Von Niessen, J. Chem. Phys., 54, 521 (1971).
48. M. J. Fernberg and K. Ruedenberg, J. Chem. Phys., 54, 1495 (1971).
49. G. Malli and S. Fraga, Theor. Chim. Acta., 5, 275 (1966).
50. G. Malli and S. Fraga, Theor. Chim. Acta., 5, 284 (1966).
51. Factors required to convert expectation values in refs. 49 and 50
    to atomic units were taken from the book by Fraga and Malli of
    ref. 2: Avagadro's Number = 6.02486×10^{23}, and $\alpha = 137.0373$.
52. W. M. Huo, J. Chem. Phys., 43, 624 (1965).
53. H. Basch, M. B. Robin, and N. A. Kuebler, J. Chem. Phys., 47, 1201
    (1967).
54. A. D. McLean and M. Yoshimine, J. Chem. Phys., 47, 3256 (1967).
55. H. Basch, M. B. Robin, and N. A. Kuebler, J. Chem. Phys., 49, 5007
    (1968); 50, 5048 (1969).
56. M. B. Robin, H. Basch, N. A. Kuebler, B. E. Kaplan, and J. Meinwald,
    J. Chem. Phys., 48, 5037 (1968).

REFERENCES

57. J. R. Lombardi, W. Klemperer, M. B. Robin, H. Basch, and N. A. Kuebler, J. Chem. Phys., 51, 33 (1969).
58. M. B. Robin, H. Basch, N. A. Kuebler, K. B. Wiberg, and G. B. Ellison, J. Chem. Phys., 51, 45 (1969).
59. H. Basch, M. B. Robin, N. A. Kuebler, C. Baker, and D. W. Turner, J. Chem. Phys., 51, 52 (1969).
60. J. L. Whitten and M. Hackmeyer, J. Chem. Phys., 51, 5584 (1969).
61. M. Yaris, A. Moscowitz, and R. S. Berry, J. Chem. Phys., 49, 3150 (1968).
62. L. C. Snyder and H. Basch, J. Am. Chem. Soc., 91, 2189 (1969).
63. L. C. Snyder, J. Chem. Phys., 46, 3602 (1967).
64. L. C. Snyder, "Thermochemistry and the Electronic Structure of Molecules," The Robert A. Welch Foundation Research Bulletin, 29, August, 1971.
65. K. Siegbahn et al., ESCA Applied to Free Molecules, North Holland Publishing Company, Amsterdam, 1969.
66. H. Basch and L. C. Snyder, Chem. Phys. Lett., 3, 333 (1969).
67. L. C. Snyder, J. Chem. Phys., 55, 95 (1971).
68. H. Basch, Chem. Phys. Lett., 5, 337 (1970).
69. W. H. Flygare and J. Goodisman, J. Chem. Phys., 49, 3122 (1968).
70. C. R. Brundle, D. W. Turner, M. B. Robin, and H. Basch, Chem. Phys. Lett. 3, 292 (1909).
71. C. R. Brundle, M. B. Robin, and H. Basch, J. Chem. Phys., 53, 2196 (1970).
72. G. Herzberg, Molecular Spectra and Molecular Structure II. Infrared and Raman Spectra of Polyatomic Molecules, D. Van Nostrand Co., New York, 1954.
73. G. Herzberg, Molecular Spectra and Molecular Structure I. Spectra of Diatomic Molecules, D. Van Nostrand Co., New York, 1966.
74. Assumed: See W. E. Palke and W. N. Lipscomb, J. Chem. Phys., 45, 3948 (1966).
75. D. R. J. Boyd and H. W. Thompson, Trans. Faraday Soc., 49, 1281 (1953).
76. G. Herzberg, Molecular Spectra and Structure III. Electronic Spectra of Polyatomic Molecules, D. Van Nostrand Co., New York, 1966.
77. D. Neumann and J. W. Moskowitz, J. Chem. Phys., 49, 2056 (1968).
78. L. S. Bartell and B. L. Carroll, J. Chem. Phys., 42, 1135 (1965).
79. W. H. Fink and L. C. Allen, J. Chem. Phys., 46, 2261 (1967).
80. A. Veillard, Theor. Chim. Acta, 5, 413 (1966).
81. R. H. Hunt, R. A. Leacock, C. W. Peters, and K. T. Hecht, J. Chem. Phys., 42, 1931 (1965); R. L. Redington, W. B. Olson, and P. C. Cross, J. Chem. Phys., 36, 1311 (1962).
82. P. Venkateswarlu and W. Gordy, J. Chem. Phys., 23, 1200 (1955).
83. C. C. Costain, J. Chem. Phys., 29, 864 (1958).
84. A. Almenningen, O. Bastiansen, and M. Traetteberg, Acta Chemica Scandinavica, 13, 1699 (1959).

85. A. P. Cox, L. F. Thomas, and J. Sheridan, Spectr. Acta, 15, 542 (1959).
86. J. K. Tyler, L. F. Thomas, and J. Sheridan, Proc. Chem. Soc. (London), 155 (1959).
87. W. Gordy, H. Ring, and A. B. Burg, Phys. Rev., 78, 512 (1950).
88. L. F. Thomas, E. I. Sherrard, and J. Sheridan, Trans. Faraday Soc., 51, 619 (1955).
89. M. Kessler, H. Ring, R. Trambarulo, and W. Gordy, Phys. Rev., 79, 54 (1950).
90. P. H. Kasai, R. J. Myers, D. F. Eggers, Jr., K. B. Wiberg, J. Chem. Phys., 30, 519 (1959).
91. L. Pierce and V. Dobyus, J. Am. Chem. Soc., 84, 2651 (1962).
92. R. Trambarulo, S. N. Ghosh, C. A. Burrus, and W. Gordy, J. Chem. Phys., 21, 851 (1953).
93. "JANAF" Thermochemical Tables, D. R. Stull, ed., The Dow Chemical Co., Midland, Mich., 1964.
94. C. C. Costain and J. M. Dowling, J. Chem. Phys., 32, 158 (1960).
95. L. E. Sutton, "Interatomic Distances," Chem. Soc. (London) Spec. Publ. 11 (1958), and Spec. Publ. 18 (1965).
96. O. Bastiansen, F. N. Fritsch, and K. Hedberg, Acta Cryst., 17, 538 (1964).
97. T. E. Turner, V. C. Fiora, and W. M. Kendrick, J. Chem. Phys., 25, 1966 (1955).
98. G. L. Cunningham, Jr., A. W. Boyd, R. J. Meyers, W. D. Gwinn, and W. I. LeVan, J. Chem. Phys., 19, 676 (1951).
99. Assumed trans-structure with $R(N-N) = 1.451$ Å, $R(C-N) = 1.485$ Å, $R(N-H) = 1.014$ Å, $R(C-H) = 1.089$ Å, <HCH = 116°, <NCN = 58.5°, and <NNH = 150°.
100. R. P. Blickensderfer, J. H. S. Wang, and W. H. Glygare, J. Chem. Phys., 51, 3196 (1969).
101. K. W. Cox, M. D. Harmony, G. Nelson, and K. B. Wiberg, J. Chem. Phys., 50, 1976 (1969).
102. R. J. Buenker and J. L. Whitten, J. Chem. Phys., 49, 5381 (1968).
103. E. A. Cherniak and C. C. Costain, J. Chem. Phys., 45, 104 (1966).
104. A. H. Nielsen, J. Chem. Phys., 22, 659 (1954).
105. R. K. Bohn and S. H. Bauer, Inorg. Chem., 6, 309 (1967).
106. R. L. Livingston and C. N. R. Rao, J. Am. Chem. Soc., 81, 285 (1959).
107. S. N. Ghosh, R. Trambarulo, and W. Gordy, J. Chem. Phys., 20, 605 (1952).
108. J. L. Hencher and S. H. Bauer, J. Am. Chem. Soc., 89, 5527 (1967).
109. C. G. Thorton, Dissertation Abstr., 14, 604 (1954).
110. D. W. Smith and K. Hedberg, J. Chem. Phys., 25, 1282 (1956).
111. J. A. Young, Dissertation Abstr., 16, 460 (1956).